Mirror Geometry of Lie Algebras, Lie Groups and Homogeneous Spaces

Mathematics and Its Applications

Managing Editor:

M. HAZEWINKEL

Centre for Mathematics and Computer Science, Amsterdam, The Netherlands

Volume 573

Mirror Geometry of Lie Algebras, Lie Groups and Homogeneous Spaces

by

Lev V. Sabinin

Faculty of Science,
Morelos State University, Morelos, Cuernavaca, Mexico
and Friendship University,
Moscow, Russia

KLUWER ACADEMIC PUBLISHERS
DORDRECHT / BOSTON / LONDON

A C.I.P. Catalogue record for this book is available from the Library of Congress.

ISBN 978-90-481-6676-3 (PB)
ISBN 978-1-4020-2545-7 (e-book)

Published by Kluwer Academic Publishers,
P.O. Box 17, 3300 AA Dordrecht, The Netherlands.

Sold and distributed in North, Central and South America
by Kluwer Academic Publishers,
101 Philip Drive, Norwell, MA 02061, U.S.A.

In all other countries, sold and distributed
by Kluwer Academic Publishers,
P.O. Box 322, 3300 AH Dordrecht, The Netherlands.

Printed on acid-free paper

TABLE OF CONTENTS

On the artistic and poetic fragments of the book

It is evident to us that the way of writing a mathematical treatise in a monotonous logically-didactic manner subsumed in the contemporary world is very harmful and belittles the greatness of Mathematics, which is authentic basis of the Transcendental Being of Universe.

Everyone touched by Mathematical Creativity knows that Images, Words, Sounds, and Colours fly above the ocean of logic in the process of exploration, constituting the real body of concepts, theorems, and proofs. But all this abundance disappears and, alas, does so tracelessly for the reader of a modern mathematical treatise.

Therefore the attempts at attracting a reader to this majestic Irrationality appear to be natural and justified.

Thus the inclusion of poetic inscriptions into mathematical works has already (and long ago) been used by different authors. Attempts to draw (not to illustrate only) Mathematics is already habitual amongst intellectuals. In this connection let us note the remarkable pathological-topological-anatomical graphics of the Moscow topologist A.T. Fomenko.

In our treatise we also make use of graphics and drawings (mental images-faces of super-mathematical reality, created by sweet dreams of mirages of pure logic) and words (poetic inscriptions) which Eternity whispered during our aspirations to learn the beauty of Irrationality.

The reader should not search for any direct relations between our artistic poetic substance and certain parts and sections of the treatise. It is to be considered as some general super-mathematical-philosophical body of the treatise as a whole.

Lev Sabinin

INTRODUCTION

Сфероид медленно вращался
Сметая сонмы странных Числ…
Миры, Века, катились вниз
И лик Творца обозначался.

Стремясь постигнуть тайный смысл
Сметенья стонов странных Числ,
Значков причудливый альянс,
Я погружался в сонный Транс.
Гримассы Жуть и Боли Смысл
На лицах прочитал я Числ.

Квантасмагор

INTRODUCTION

As K. Nomizu has justly noted [K. Nomizu, 56], Differential Geometry ever will
be initiating newer and newer aspects of the theory of Lie groups. This monograph
is devoted to just some such aspects of Lie groups and Lie algebras.

New differential geometric problems came into being in connection with so called
subsymmetric spaces, subsymmetries, and mirrors introduced in our works dating
back to 1957 [L.V. Sabinin, 58a,59a,59b].

In addition, the exploration of mirrors and systems of mirrors is of interest in the
case of symmetric spaces. Geometrically, the most rich in content there appeared
to be the homogeneous Riemannian spaces with systems of mirrors generated by
commuting subsymmetries, in particular, so called tri-symmetric spaces introduced
in [L.V. Sabinin, 61b].

As to the concrete geometric problem which needs be solved and which is solved
in this monograph, we indicate, for example, the problem of the classification of
all tri-symmetric spaces with simple compact groups of motions.

Passing from groups and subgroups connected with mirrors and subsymmetries
to the corresponding Lie algebras and subalgebras leads to an important new
concept of the involutive sum of Lie algebras [L.V. Sabinin, 65].

This concept is directly concerned with unitary symmetry of elementary particles (see [L.V. Sabinin, 95,85] and Appendix 1).

The first examples of involutive (even iso-involutive) sums appeared in the exploration of homogeneous Riemannian spaces with $ds^2 > 0$ and axial symmetry.
The consideration of spaces with $(n-1)$-dimensional mirrors [L.V. Sabinin, 59b]
again led to iso-involutive sums.

The construction of the so called hyper-involutive decomposition (sum) can be
dated back to 1960–62, see, for example, the short presentation of our report at
the International Congress of Mathematicians (1962, Stockholm) in volume 13
of Transactions of the Seminar on Vector and Tensor Analysis (1966, Moscow
University) and [L.V. Sabinin, 67].

Furthermore, a very important heuristic role was played by the work of Shirokov
[P.A. Shirokov, 57], in which the algebraic structure of the curvature tensor of
the symmetric space $SU(n+1)/U(n)$ was given, and by the work of Rosenfeld
[B.A. Rosenfeld 57]. That allowed us to construct characteristic iso-involutive
decompositions for all classical Lie algebras [L.V. Sabinin, 65, 68].

In this way the apparatus for direct exploration of symmetric spaces of rank 1
with compact Lie groups of motions was introduced (avoiding the well known indi-
rect approach connected with the Root Method and the examination of E. Cartan's
list of all symmetric spaces with compact simple Lie groups of motions).

The most difficult, certainly, and fundamental element of the suggested theory was the understanding (1966) of the role of principal unitary and special unitary automorphisms of Lie groups and Lie algebras [L.V. Sabinin, 67,69,70]. The above work solved the problem of introducing the 'standard' mirrors into a homogeneous Riemannian space with $ds^2 > 0$.

Indeed, in this case the stationary subgroup is compact and, taking its principal unitary involutive automorphism (which is possible, except for trivial subcases), we can generate a 'standard' subsymmetry and a 'standard' mirror in a Riemannian space. Analogously, one can introduce systems of 'standard' mirrors in a Riemannian space with $ds^2 > 0$ and, furthermore, with their help, explore geometric properties of homogeneous Riemannian spaces.

The detailed consideration of involutive sums of Lie algebras has shown, however, that their role is more significant than the role of the only convenient auxiliary apparatus for solving some differential-geometric problems. We may talk about the theory of independent interest and it is natural to call it 'Mirror geometry of Lie algebras', or 'Mirror calculus'; the role and significance of which is comparable with the role and significance of the well known 'Root Method' in the theory of Lie algebras.

Part I and II of this treatise are devoted to the presentation of Mirror Geometry over the reals.

A Lie algebra $\mathfrak{g}$ has the group of automorphisms $\mathrm{Aut}(\mathfrak{g})$ and consequently generates the geometry in the sense of F. Klein. In the case of a semi-simple compact Lie algebra $\mathfrak{g}$ the group $\mathrm{Int}(\mathfrak{g}) \subset \mathrm{Aut}(\mathfrak{g})$ is a compact linear Lie group, which allows us to use some knowledge from the theory of compact Lie groups (however, we need not too much from that theory, and necessary results can be proved without the above theory). We may regard Cartan's theorem on the existence of non-trivial inner involutive automorphism of a simple compact non-one-dimensional Lie algebra as the typical theorem of the Lie algebras geometry (the proof follows immediately from the existence of non-trivial involutive elements in $\mathrm{Int}(\mathfrak{g})$).

In an ordinary Euclidean space a plane can be defined as a set of all points immobile under the action of some involutive automorphism. Thus the maximal subset of elements immobile under the action of an involutive automorphism, that is, some involutive subalgebra $\mathfrak{l} \subset \mathfrak{g}$ in a compact Lie algebra, may be regarded as an analogue of a plane in an Euclidean space.

Let us now consider the problem of a canonical base of a compact Lie algebra. From the point of view of Classical Invariants' theory the problem of the classification of Lie algebras is connected with the finding of a base in which the structure tensor has a sufficiently simple form (canonical form).

In order to clarify what has just been said, let us consider a simple problem of that kind, namely, the problem of a canonical form for a bilinear form in a centered-Euclidean space. As is easily seen, here the determination of a canonical base is reduced to the finding of commuting isometric involutive automorphisms and the subsequent choice of a base in such a way that the above involutive automorphisms have the basis vectors as proper vectors. For this it is enough to find the 'standard' involutive automorphisms (for example, connected with reflections with respect to hyperplanes) of that type; other involutive automorphisms can be obtained as products of the 'standard' involutive automorphisms.

Any two commuting 'standard' involutive automorphisms S_1, S_2 generate the third automorphism $S_3 = S_1 S_2 = S_2 S_1$ (non-standard, in general) and consequently a discrete commutative group $\{\mathrm{Id}, S_1, S_2, S_3\}$, the so called involutive group $\chi(S_1, S_2, S_3)$.

Returning to a compact Lie algebra $\mathfrak{g}$ we see that the construction described above is valid here, and in a natural way we have the notion of involutive group $\chi(S_1, S_2, S_3) \subset \mathrm{Aut}(\mathfrak{g})$.

The only problem in the consideration presented above is to introduce, reasonably, the 'standard' involutive automorphisms for any compact semi-simple Lie algebra.

With any involutive group $\chi(S_1, S_2, S_3)$ of a Lie algebra $\mathfrak{g}$ one may associate in a natural way the decomposition

$$\mathfrak{g} = \mathfrak{l}_1 + \mathfrak{l}_2 + \mathfrak{l}_3, \qquad \mathfrak{l}_1 \cap \mathfrak{l}_2 = \mathfrak{l}_2 \cap \mathfrak{l}_3 = \mathfrak{l}_3 \cap \mathfrak{l}_1 = \mathfrak{l}_0,$$

where $\mathfrak{l}_1, \mathfrak{l}_2, \mathfrak{l}_3$ are involutive algebras of the involutive automorphisms S_1, S_2, S_3, respectively, ($\mathfrak{l}_\alpha = \{\zeta \in \mathfrak{g} \mid S_\alpha \zeta = \zeta\}$). This is a so called involutive decomposition (involutive sum). As well, one can introduce the corresponding involutive base (in fact, a set of involutive bases) whose vectors are proper vectors for S_1, S_2, S_3. Thus if we are interested in a canonical base of the Lie algebra $\mathfrak{g}$ then it is an involutive base of some involutive group.

Among involutive groups $\chi(S_1, S_2, S_3)$ one may select two special classes, namely: iso-involutive groups, $\chi(S_1, S_2, S_3; \varphi)$, where S_1 and S_2 are conjugated by $\varphi \in \mathrm{Aut}\,\mathfrak{g}$, $\varphi^2 = S_3$, and hyper-involutive groups, $\chi(S_1, S_2, S_3; p)$, where S_1 and S_2, S_2 and S_3, S_3 and S_1 are conjugated by $p \in \mathrm{Aut}\,\mathfrak{g}$. They generate, respectively, iso-involutive sums, iso-involutive bases and hyper-involutive sums, hyper-involutive bases.

We show that any arbitrarily taken non-trivial simple compact Lie algebra $\mathfrak{g}$ ($\dim \mathfrak{g} \neq 1$) with an involutive automorphism S_1 has iso-involutive groups $\chi(S_1, S_2, S_3; \varphi)$.

This result turns iso-involutive sums into an instrument of exploration of Lie algebras.

Hyper-involutive sums are not universal to the same extent, but in appropriate cases serve also as an effective apparatus of investigation. One sufficient condition for the existence of hyper-involutive sums for a simple compact Lie algebra $\mathfrak{g}$ is: there exists a three-dimensional simple subalgebra $\mathfrak{b} \subset \mathfrak{g}$ such that the restriction $\mathrm{Int}_\mathfrak{g}\mathfrak{b}$ of $\mathrm{Int}\mathfrak{g}$ to $\mathfrak{b}$ is isomorphic to $SO(3)$.

Now we pass to the problem of determination of 'standard' involutive automorphisms of a simple compact Lie algebra $\mathfrak{g}$ ($\dim \mathfrak{g} \neq 1$).

An involutive algebra $\mathfrak{l} \subset \mathfrak{g}$ (and the corresponding involutive automorphism S) is called principal if it contains a simple three-dimensional ideal $\mathfrak{b}$, that is, $\mathfrak{l} = \mathfrak{b} \oplus \tilde{\mathfrak{l}}$. In this case, if $\mathrm{Int}_\mathfrak{g}(\mathfrak{b}) \cong SO(3)$ then we say that $\mathfrak{l}$ (and S) is principal orthogonal and if $\mathrm{Int}_\mathfrak{g}(\mathfrak{b}) \cong SU(2)$ then we say that $\mathfrak{l}$ (and S) is principal unitary.

By means of involutive decompositions we prove the main theorem: any simple compact Lie algebra $\mathfrak{g}$ ($\dim \mathfrak{g} \neq 1$) has a non-trivial principal involutive automorphism. If $\dim \mathfrak{g} \neq 3$ then $\mathfrak{g}$ has a non-trivial principal unitary involutive automorphism.

This solves the problem of introducing 'standard' involutive automorphisms which may be regarded as principal.

One may introduce also the broader class of special involutive algebras and involutive automorphisms, in particular, the unitary special involutive algebras and involutive automorphisms.

We give, furthermore, the simple classification of principal unitary involutive automorphisms: principal di-unitary, principal unitary central, principal unitary of index 1, exceptional principal unitary. Using the apparatus of involutive sums and involutive bases we explore all these types. As a result the type of principal unitary involutive automorphism defines, in general, the type of simple compact Lie algebra. For example, if a simple compact Lie algebra has a principal unitary non-central involutive automorphism of index 1 then $\mathfrak{g}$ is isomorphic to $sp(n)$, $n > 1$. For the Lie algebra $\mathfrak{g}_2$ the construction is presented in a hyper-involutive base up to the numerical values of structural constants, that is, the problem is completely solved in the sense of the classical theory of invariants. For other types of exceptional Lie algebras we determine the basis involutive sums and the structure of their involutive algebras.

Furthermore, we consider the problem of the classification of special unitary non-principal involutive authomorphisms. The principle of the involutive duality of principal unitary and special unitary non-principal involutive automorphisms is established. This principle allows us to define all special simple unitary subalgebras and all special unitary involutive automorphisms for simple compact Lie algebras.

For all simple compact Lie algebras $\mathfrak{g}$ ($\dim \mathfrak{g} \neq 1$), except

$$so(3) \cong su(2) \cong sp(1)\,, \qquad so(5) \cong sp(2)\,, \qquad su(3)\,,$$
$$so(6) \cong su(4)\,, \qquad so(7)\,, \qquad so(8)\,, \qquad \mathfrak{g}_2\,,$$

we construct the basis iso-involutive decomposition $\mathfrak{g} = \mathfrak{l}_1 + \mathfrak{l}_2 + \mathfrak{l}_3$, where $\mathfrak{l}_1$ and $\mathfrak{l}_2$ are principal unitary involutive Lie algebras and $\mathfrak{l}_3$ is a special unitary involutive Lie algebra. By the type of such involutive sum the type of $\mathfrak{g}$ is uniquely defined. For each of the Lie algebras which have been excluded above we construct the basis hyper-involutive decomposition which uniquely characterizes any of them.

Furthermore, for simple compact Lie algebras we consider the possibility of constructing hyper-involutive sums with principal unitary involutive automorphisms.

Using the procedure of involutive reconstruction of basis involutive sums we prove that principal unitary hyper-involutive sums exist and are unique for Lie algebras $so(n)$ ($n > 5$), $su(n)$ ($n > 2$), $\mathfrak{f}_4$, $\mathfrak{e}_6$, $\mathfrak{e}_7$, $\mathfrak{e}_8$ (all these involutive sums are found) and do not exist for $sp(n)$.

It is shown that for $sp(n)$ ($n > 2$) one can construct hyper-involutive sums with special unitary involutive algebras. An analogous construction is valid for $so(n)$ ($n > 8$), $su(n)$ ($n > 4$), $\mathfrak{f}_4$, $\mathfrak{e}_6$, $\mathfrak{e}_7$, $\mathfrak{e}_8$. All such involutive sums are found as well.

Thus the suggested theory is a theory of structures of a new type for compact real Lie algebras and is related to discrete involutive groups of automorphisms and the corresponding involutive decompositions.

Let us now turn to possible geometric applications, which, in particular, may be found in Part III and IV.

First of all we note that since we deal with involutive automorphisms, all, or almost all, proved results may be reformulated in terms of symmetric spaces [E. Cartan, 49,52], [B.A. Rosenfeld, 57], [L.V. Sabinin, 59c], [S. Helgason, 62,78], [A.P. Shirokov, 57] and in terms of mirrors in homogeneous spaces.

Such applications are concentrated at the beginning of Part III after some necessary definitions. Despite that here many results have been obtained simply as a reformulation of theorems of Part I and II from the language of Lie algebras into the language of Lie groups, homogeneous spaces, and mirrors, those are very interesting. (For example, the characterization of symmetric spaces of rank 1 by the properties of geodesic mirrors.)

In addition, the theory of Part I and II implies two new interesting types of symmetric spaces—principal and special—and allows us to explore geometric properties of their mirrors.

Furthermore, in Section III.5 we consider applications of Mirror Geometry to some problems of simple compact Lie groups. Thus it is shown that $\mathrm{Int}(\mathfrak{g}_2)$ and $\mathrm{Int}(\mathfrak{f}_4)$ are the only simple compact connected Lie groups of types G_2 and F_4, respectively.

Moreover, it is shown how, knowing involutive decompositions for simple compact Lie algebras, one may find out their inner involutive automorphisms (in the cases of $\mathfrak{g}_2$, $\mathfrak{f}_4$, $\mathfrak{e}_6$, $\mathfrak{e}_8$).

Lastly, the final sections of Part III (III.6–III.8) are devoted to the complete classification of tri-symmetric spaces with a simple compact Lie group of motions. The solution of this problem, when treated by conventional methods, had serious difficulties. Indeed, the first part of this problem, the definition of involutive groups $\{\mathrm{Id}, S_1, S_2, S_3\}$ of automorphisms, is already not trivial. Since, even if all S_α are inner automorphisms, they can not be generated by a one Cartan subgroup, it is necessary to bring into consideration the normalizers of maximal tori ([Seminar Sophus Lie, 62] Ch. 20). But the determination of normalizers of maximal tori in exceptional simple compact Lie groups is a complicated problem owing to the absence of good matrix models. However, the theory developed in Part I and II gives a natural apparatus for solving the above problem.

The classification shows, in particular, that all non-trivial non-symmetric tri-symmetric spaces have isomorphic basis mirrors (in hyper-symmetric decomposition) and have irreducible Lie groups of motions if they are maximal. Their mirrors possess remarkable geometric properties being either principal or central. In this relation we note that in [O.V. Manturov, 66] two spaces, $G_2/SU(3)$ and $E_7/F_4 \times SO(3)$, with irreducible groups of motions have not been found.

The results of Part II belong to the area in which strong methods and detailed theories existed earlier. Therefore we naturally need some comparisons.

The theory of compact Lie algebras has been established mainly by the work of Lie [S. Lie, 1888,1890,1893], Killing [W. Killing, 1888,1889a,1889b,1890], E. Cartan [E. Cartan, 49,52], H. Weyl [H. Weyl, 25, 26a,b,c, 47], Van der Warden [B.L. Van der Warden 33], Dynkin [E.B. Dynkin, 47], Gantmacher [Gantmacher 39a,b] *etc.*, and is well known.

We now intend to compare the well known 'Root Method' with our new theory, which we briefly call 'Mirror Geometry'.

First of all, Mirror Geometry deals with new types of structures (involutive sums) and is introduced independently of the Root Method. Thus these two theories seem to be different. But since Mirror Geometry leads us to the classification of simple compact Lie algebras (through the classification of principal unitary involutive automorphisms) we need some comparisons.

The Root Method has a complex nature and the classification of real simple compact Lie algebras require the supplementary theory (the theory of real forms). Mirror Geometry has a real nature.

The determination of involutive automorphisms of Lie algebras in the Root Method requires a supplementary theory. In Mirror Geometry, owing to the construction, any type of simple compact Lie algebra appears together with two (generally speaking) involutive automorphisms, principal unitary and its dual, special unitary. This is of importance for exceptional Lie algebras (for example, in the case of $\mathfrak{f}_4$ there are no other involutive automorphisms).

The Root Method gives the description of a compact simple Lie algebra by the type of root system which is a rather complicated invariant of a Lie algebra.

Mirror Geometry gives the description of a compact simple Lie algebra by the type of principal unitary involutive automorphism being a simple algebraic-geometric characteristic of a Lie algebra.

The Root Method does not give a classification of simple compact Lie algebras in the sense of the Invariant Theory, that is, does not give the method of construction of a canonical base: there the problem of classification is solved by the 'guessing' of a concrete Lie algebra with an admissible root system.

Mirror Geometry is, in essence, the method of determination of a canonical base in a Lie algebra.

The problems in applications to tri-symmetric spaces of rank 1, for example, can be solved in the 'Root Method' by the observation of all possibilities of the list of E. Cartan. Mirror Geometry gives a direct approach to symmetric spaces of rank 1, avoiding the general classification. Moreover, any theorem of Mirror Geometry is a theorem of the theory of symmetric spaces (after some trivial reformulation). This is not valid for the Root Method.

The Root Method is not effective in the theory of homogeneous Riemannian spaces with mirrors (that is, all cases of homogeneous Riemannian spaces with $ds^2 > 0$ and non-trivial isotropy group). Mirror Geometry gives in this case the system of standard mirrors.

Of course, there are some problems when the possibilities of the Root Method are obviously effective but the possibilities of Mirror Geometry are not yet evident enough. Perhaps, here we need more systematic development in the future.

One may ask whether Mirror Geometry can be obtained from the Root Method. The simple example of an iso-involutive sum of index 1 and of type 1 for a simple compact Lie algebra demonstrates that Mirror Geometry and the Root Method are in some sense opposite. Indeed, the constructions of the Root Method depend on 'regular vectors', whereas in the above example the conjugating automorphism is generated by a singular vector.

Let us note in this relation that, probably, the procedure of inclusion of a system of n roots into a system of $n + 1$ roots in the Root Method is connected with the construction of some involutive sum in Mirror Geometry.

We realize that many proofs in this treatise may be perfected as well as the whole presentation of the subject. This is a profound work for the future.

Finally, we would like to mention those who have influenced so much our mathematical research in Lie groups and Homogeneous spaces. On the one hand, this is Academician A.I. Malc'ev who shaped our algebraic interests during our collaboration in the Siberian Scientific Center. On the other hand, we should mention Professor P.K. Rashevski (and his Moscow Geometric School) who has helped us to learn the beauty and flavor of the theory of homogeneous spaces.

Well known results of the general theory of Lie groups and Lie algebras sometimes are used without references and may be found in [N. Jacobson, 64], [L.S. Pontryagin, 54,79], [P.K. Rashevski, 53], [B.A. Rosenfeld, 55], [Seminar Sophus Lie, 62], [S. Helgason, 64,78], [I.G. Chebotarev, 40], [C. Chevalley, 48,58a,58b], [L.P. Eisenhart, 48], [J.-P. Serre, 69].

This book could not appear without assistance of my permanent collaborator Dr. L. Sbitneva, who also has prepared the manuscript for publication and carried out much editorial work. I am much indebted to her.

This book has been prepared in the frames of Research Project supported by Mexican National Council of Science and Technology (CONACYT).

I am delighted that this book appeared in the celebrated innovative series edited by Professor M. Hasewinkel.

Lev V. Sabinin

September 2002
Cuernavaca, Morelos,
    Mexico

PART ONE

Это - миазмы,
Марева измы,
Сонмы страданий Духов Игры
Пляшут Фантазмы -
Тьмы катаклизмы,
Цепь Мирозданий рушит Миры.

Синие Блики, красные Тени,
Царствует властно
Ложный Экстаз.
Странные Лики полные Лени...
Смотрит бесстрастно
Вечность на нас.

Квантасмагор

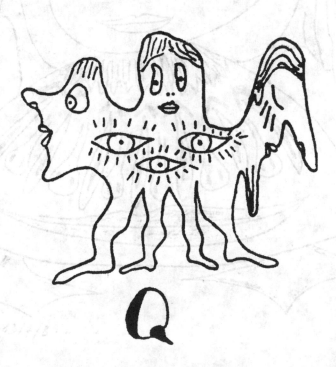

CHAPTER I.1

PRELIMINARIES

In this treatise we consider finite-dimensional real Lie algebras only. We frequently say 'algebra' instead of 'Lie algebra' since almost no other type of algebra appears in this book (if any does, we clearly indicate it).

I.1.1. We recall first the following basic definition:
A vector space $\mathfrak{g}$ over a field F equipped with a bilinear multiplication

$$(\xi, \eta) \in \mathfrak{g} \times \mathfrak{g} \mapsto [\xi \eta] \in \mathfrak{g}$$

is called a Lie algebra if for any $\xi, \eta, \zeta \in \mathfrak{g}$ the identities

$$\left.\begin{array}{c} [\xi\,\xi] = 0, \\[2mm] [\xi\,[\eta\,\zeta]] + [\eta\,[\zeta\,\xi]] + [\zeta\,[\xi\,\eta]] = 0 \quad \textit{(Jacobi identity)} \end{array}\right\} \tag{I.0}$$

are satisfied.

If $[\xi\,\xi] = 0$ then $[\xi\,\eta] = -[\eta\,\xi]$ (skew-symmetry).

Indeed,

$$0 = [(\xi + \eta)(\xi + \eta)] = [\xi\,\xi] + [\xi\,\eta] + [\eta\,\xi] + [\eta\,\eta] = [\xi\,\eta] + [\eta\,\xi].$$

And if $[\xi\,\eta] = -[\eta\,\xi]$, $\operatorname{ch} F \neq 2$, then $[\zeta\,\zeta] = 0$. Indeed, in this case $[\xi\,\xi] = -[\xi\,\xi]$, or $2[\xi\,\xi] = 0$, or $[\xi\,\xi] = 0$.

Thus we may replace the identity $[\xi\,\xi] = 0$ in (I.0) by $[\xi\,\eta] = -[\eta\,\xi]$ (skew-symmetry) since we consider $F = \mathbb{R}$ ($\operatorname{ch}\mathbb{R} = 0$).

I.1.2. Let $e_1, \ldots, e_r$ be a base in Lie algebra $\mathfrak{g}$, then its structure is completely defined by the structure tensor C_{IJ}^K,

$$[e_I\,e_J] = C_{IJ}^K e_K, \quad C_{IJ}^K = -C_{JI}^K; \quad I, J, K = 1, \ldots, r, \tag{I.1}$$

and by the Jacobi identities

$$[e_I\,[e_J\,e_K]] + [e_K\,[e_I\,e_J]] + [e_J\,[e_K\,e_I]] = 0 \tag{I.2}$$

which may be presented as $C_{<JK}^P C_{I>P}^R = 0$.

(Note that $A_{<JKI>} \stackrel{\text{def}}{=} A_{JKI} + A_{KIJ} + A_{IJK}$, a so called cyclic sum.)

3

I.1.3. Any linear bijection $A : \mathfrak{g} \to \mathfrak{g}$ of a Lie algebra $\mathfrak{g}$ such that

$$[(A\eta)(A\zeta)] = A[\eta\,\zeta], \quad \forall\,\eta,\,\zeta \in \mathfrak{g}\,,$$

is called an *automorphism* of a Lie algebra $\mathfrak{g}$.

All automorphisms of a Lie algebra $\mathfrak{g}$ constitute its *adjoint group* Aut($\mathfrak{g}$), which is linear.

The maximal connected subgroup Int($\mathfrak{g}$) $\subset$ Aut($\mathfrak{g}$) is a linear group which is called *the group of inner automorphisms* (or *connected adjoint group*) of a Lie algebra $\mathfrak{g}$.

The structure tensor C_{IK}^R is invariant under the action of Aut($\mathfrak{g}$).

I.1.4. A linear map $D : \mathfrak{g} \to \mathfrak{g}$ is said to be a *differentiation* (or *derivation*) of a Lie algebra $\mathfrak{g}$ if, for any $\eta, \zeta \in \mathfrak{g}$,

$$D[\eta\,\zeta] = [(D\eta)\zeta] + [\eta(D\zeta)]\,.$$

The set $\partial(\mathfrak{g})$ of all differentiations of a Lie algebra $\mathfrak{g}$ constitutes a Lie algebra with respect to the natural structure of a vector space and the operation of multiplication

$$[D_1 D_2] = D_1 D_2 - D_2 D_1\,.$$

It is called a *Lie algebra of differentiations* of $\mathfrak{g}$.

It is easy to see that

$$D_\eta : \zeta \mapsto [\eta\,\zeta]$$

is a differentiation of a Lie algebra $\mathfrak{g}$. It is called an *inner differentiation*.

The set of all inner differentiations ad($\mathfrak{g}$) $\subset \partial(\mathfrak{g})$ is a subalgebra (and ideal) in $\partial(\mathfrak{g})$, it is called the *algebra of inner differentiations* of a Lie algebra $\mathfrak{g}$.

The correspondence $\zeta \mapsto D_\zeta$ ($\zeta \in \mathfrak{g}$) defines a so called *adjoint representation* of an algebra Lie $\mathfrak{g}$ on itself, since by the Jacobi identity

$$D_{[\xi\,\eta]} = D_\xi D_\eta - D_\eta D_\xi = [D_\xi D_\eta]\,.$$

I.1.5. We say that $A \in$ Aut($\mathfrak{g}$) is an *involutive automorphism* (*invomorphism*) if $A^2 = \text{Id}$.

Let $\mathfrak{g}$ be a Lie algebra. The subset

$$\mathfrak{l} = \{\zeta \in \mathfrak{g} \mid A\zeta = \zeta\}$$

is a subalgebra which is called an *involutive algebra* (*invoalgebra*) (*of involutive automorphism A*). The pair $\mathfrak{g}/\mathfrak{l}$ is called an *involutive pair* (*invopair*).

As is well known, transformations from Int($\mathfrak{g}$) are of the form

$$\eta \mapsto e^{c(u)}\eta, \quad c(u) = \|C_{JK}^I u^K\|\,. \tag{I.3}$$

Henceforth we intend to consider compact Lie algebras only.

I.1.6. A Lie algebra $\mathfrak{g}$ over $\mathbb{R}$ is said to be *compact* if there exists a bilinear symmetric positive-definite form $(\xi, \eta) \in \mathfrak{g} \times \mathfrak{g} \mapsto f(\xi, \eta) \in \mathbb{R}$ such that, for any $\xi, \eta, \zeta \in \mathfrak{g}$,

$$f([\xi\,\eta], \zeta) + f(\eta, [\xi\,\zeta]) = 0.$$

A compact Lie algebra $\mathfrak{g}$ is the unique sum decomposition (of ideals) $\mathfrak{g} = \tilde{\mathfrak{g}} \oplus \mathfrak{z}$, where $\mathfrak{z}$ is the centre and $\tilde{\mathfrak{g}}$ is the maximal compact semi-simple ideal of $\mathfrak{g}$.

Any compact semi-simple Lie algebra $\mathfrak{g}$ is a unique direct sum decomposition of its simple semi-simple ideals.

Any differentiation of a compact semi-simple Lie algebra is its inner differentiation.

I.1.7. For any compact semi-simple Lie algebra $\mathfrak{g}$ one may introduce a positive-definite Cartan metric tensor

$$a_{IJ} = C_{IK}^P C_{PJ}^K, \qquad a_{IJ} = a_{JI}. \tag{I.4}$$

Then we may consider an orthonormal base in which

$$a_{IJ} = \delta_{IJ} = \begin{cases} 1, & I = J, \\ 0, & I \neq J. \end{cases} \tag{I.5}$$

The matrices of the form $C_{JK}^I u^K$ are matrices of differentiations from $\mathrm{ad}(\mathfrak{g})$. Because

$$a([\xi\,\eta], \zeta) + a(\eta, [\xi\,\zeta]) = 0,$$

where $a(\xi, \eta) = a_{IJ}\xi^I\eta^J$, we obtain in a base (I.5)

$$C_{JK}^I = -C_{IK}^J. \tag{I.6}$$

Since the Cartan metric (I.4) is invariant under the action of an involutive automorphism A, the matrix of A in the orthonormal base, see (I.5), is given by a symmetric orthogonal matrix. It implies that with respect to the Cartan metric one may choose an orthonormal base in which A has the diagonal matrix with ± 1 along the principal diagonal. We call such an orthogonal base *canonical* for A.

I.1.8. Let $\mathfrak{g}/\mathfrak{l}$ be an involutive pair of involutive automorphism A and

$$\mathfrak{m} = \{\eta \in \mathfrak{g} \mid A\eta = -\eta\},$$

then $\mathfrak{m}$ is a linear subspace of $\mathfrak{g}$, orthogonal to $\mathfrak{l}$ with respect to the Cartan metric (I.4). Taking into account the decomposition

$$\xi = \frac{1}{2}(\mathrm{Id} - A)\xi + \frac{1}{2}(\mathrm{Id} + A)\xi,$$

we obtain

$$\mathfrak{g} = \mathfrak{m}\dot{+}\mathfrak{l}, \quad \mathfrak{m} \perp \mathfrak{l}, \quad A\eta = \begin{cases} \eta, & \eta \in \mathfrak{l}, \\ -\eta, & \eta \in \mathfrak{m}, \end{cases} \quad \mathfrak{g}-\mathfrak{l} \overset{\text{def}}{=} \mathfrak{m}. \tag{I.7}$$

Let us also note the well known result that any not one-dimensional compact simple Lie algebra has a non-trivial involutive automorphism $A \neq \mathrm{Id}$.

I.1.9. We denote the restriction of $\mathrm{Int}(\mathfrak{g})$ to a subalgebra $\mathfrak{h} \subset \mathfrak{g}$ by $\mathrm{Int}_\mathfrak{g}(\mathfrak{h})$, in particular, $\mathrm{Int}_\mathfrak{g}(\mathfrak{g}) = \mathrm{Int}(\mathfrak{g})$. If $\mathfrak{g}$ is compact simple and not one-dimensional then $\mathrm{Int}(\mathfrak{g})$ is compact linear Lie group; for a semi-simple subalgebra $\mathfrak{h} \subset \mathfrak{g}$ then $\mathrm{Int}_\mathfrak{g}(\mathfrak{h})$ is compact. Thus if $\mathfrak{l} \subset \mathfrak{g}$ is involutive algebra then $\mathrm{Int}_\mathfrak{g}(\mathfrak{l})$ and $\mathrm{Int}_\mathfrak{l}(\mathfrak{l})$ are compact.

We denote the restriction of $\mathrm{Int}_\mathfrak{g}(\mathfrak{h})$ to some invariant subspace $\mathfrak{k} \subset \mathfrak{g}$ by $\mathrm{Int}_\mathfrak{g}^\mathfrak{k}(\mathfrak{h})$. If $\mathrm{Int}_\mathfrak{g}(\mathfrak{h})$ is compact then $\mathrm{Int}_\mathfrak{g}^\mathfrak{k}(\mathfrak{h})$ is compact, in particular, $\mathrm{Int}_\mathfrak{g}^\mathfrak{h}(\mathfrak{h})$ is compact.

Evidently, if $\mathfrak{g}$ is a compact simple and not one-dimensional Lie algebra with an involutive algebra $\mathfrak{l}$ then $\mathrm{Int}_\mathfrak{g}^{\mathfrak{g}-\mathfrak{l}}(\mathfrak{l})$ is isomorphic to $\mathrm{Int}_\mathfrak{g}(\mathfrak{l})$.

We denote the restriction of the adjoint representation (that is of the algebra of inner differentiations) of an algebra Lie $\mathfrak{g}$ to its subalgebra $\mathfrak{h}$ by $\mathrm{ad}_\mathfrak{g}(\mathfrak{h})$, and the restriction of $\mathrm{ad}_\mathfrak{g}(\mathfrak{h})$ to an invariant subspace $\mathfrak{k} \subset \mathfrak{g}$ by $\mathrm{ad}_\mathfrak{g}^\mathfrak{k}(\mathfrak{h})$.

Then for a compact simple not one-dimensional algebra Lie $\mathfrak{g}$ we have

$$\mathrm{Int}_\mathfrak{g}^\mathfrak{k}(\mathfrak{h}) = e^{\mathrm{ad}_\mathfrak{g}^\mathfrak{k}(\mathfrak{h})},$$

as is well known.

I.1.10. Definition 1. We say that an involutive automorphism A of a Lie algebra $\mathfrak{g}$ is principal if its involutive algebra $\mathfrak{l}$ has a simple three-dimensional ideal $\mathfrak{b}$.

Respectively, we say in this case that $\mathfrak{l}$ is a principal involutive algebra and $\mathfrak{g}/\mathfrak{l}$ is a principal involutive pair.

If A is a principal involutive automorphism of a Lie algebra $\mathfrak{g}$ then $\mathrm{Int}_\mathfrak{g}(\mathfrak{b})$ is a three-dimensional compact simple connected Lie group, and consequently is isomorphic to either $SO(3)$ or $SU(2)$.

I.1.11. Definition 2. Let A be a principal involutive automorphism of a Lie algebra $\mathfrak{g}$ with an involutive algebra $\mathfrak{l}$, and let $\mathfrak{b}$ be a simple three-dimensional ideal of $\mathfrak{l}$.

We say that A is principal orthogonal (or of type O) if $\mathrm{Int}_\mathfrak{g}(\mathfrak{b}) \cong SO(3)$, and that A is principal unitary (or of type U) if $\mathrm{Int}_\mathfrak{g}(\mathfrak{b}) \cong SU(2)$.

Respectively, we distinguish between orthogonal and unitary principal involutive algebras $\mathfrak{l}$ and involutive pairs $\mathfrak{g}/\mathfrak{l}$.

I.1.12. Definition 3. We say that an involutive automorphism A of a Lie algebra $\mathfrak{g}$ is central if its involutive algebra $\mathfrak{l}$ has a non-trivial centre.

In this case we say that $\mathfrak{l}$ is a central involutive algebra and $\mathfrak{g}/\mathfrak{l}$ is a central involutive pair.

I.1.13. Definition 4. A principal involutive automorphism A of a Lie algebra $\mathfrak{g}$ is called principal di-unitary (or of type U^2) if its involutive algebra $\mathfrak{l} = \mathfrak{p} \oplus \mathfrak{q} \oplus \tilde{\mathfrak{l}}$ (direct sum decomposition of ideals), where $\mathrm{Int}_\mathfrak{g}(\mathfrak{p}) \cong SU(2)$, $\mathrm{Int}_\mathfrak{g}(\mathfrak{q}) \cong SU(2)$.

In this case we also say that $\mathfrak{l}$ is a principal di-unitary involutive algebra and $\mathfrak{g}/\mathfrak{l}$ is a principal di-unitary involutive pair.

I.1.14. Definition 5. A unitary but not di-unitary principal involutive automorphism A of a Lie algebra $\mathfrak{g}$ is called mono-unitary (or of type U^1).

In this case we also say that $\mathfrak{l}$ is a principal mono-unitary involutive algebra and $\mathfrak{g}/\mathfrak{l}$ is a principal mono-unitary involutive pair.

I.1.15. Definition 6. An involutive automorphism A of a Lie algebra $\mathfrak{g}$ is said to be special if its involutive algebra $\mathfrak{l}$ has a principal involutive automorphism of type O.

In this case we also say that $\mathfrak{l}$ is a special involutive algebra and $\mathfrak{g}/\mathfrak{l}$ is a special involutive pair.

I.1.16. Definition 7. Let $\mathfrak{l}$ be a special involutive algebra of a Lie algebra $\mathfrak{g}$ and $\mathfrak{b}$ be a three-dimensional simple ideal of its principal involutive algebra such that $\mathrm{Int}_{\mathfrak{l}}(\mathfrak{b}) \cong SO(3)$.

We say that an involutive algebra $\mathfrak{l}$ is orthogonal special (or of type O) if $\mathrm{Int}_{\mathfrak{g}}(\mathfrak{b}) \cong SO(3)$ and is unitary special (or of type U) if $\mathrm{Int}_{\mathfrak{g}}(\mathfrak{b}) \cong SU(2)$.

Respectively, we distinguish between orthogonal and unitary special involutive automorphisms and involutive pairs.

I.1.17. Definition 8. Let $\mathfrak{g}$ be a compact simple Lie algebra and $\mathfrak{l} \subset \mathfrak{g}$ its involutive algebra of an involutive automorphism A.

An ideal $\mathfrak{k}$ of $\mathfrak{l}$ is called a special unitary (or of type U) subalgebra of an involutive automorphism A in $\mathfrak{g}$ if there exists a principal orthogonal involutive pair $\mathfrak{k}/\mathfrak{b} \oplus \tilde{\mathfrak{k}}$ such that $A \in \mathrm{Int}_{\mathfrak{g}}(\mathfrak{b}) \cong SU(2)$.

In this case $\mathfrak{l}$ is obviously a special unitary involutive algebra of $\mathfrak{g}$, see I.1.15 (Definition 6) and I.1.16 (Definition 7).

I.1.18. Definition 9. Let $\mathfrak{g}/\mathfrak{l}$ be an involutive pair of an involutive automorphism A, and $\mathfrak{t} \subset \mathfrak{m} = \mathfrak{g} \overset{\sim}{-} \mathfrak{l}$ be a maximal subalgebra in $\mathfrak{m}$. Then $\min\limits_{\mathfrak{t} \subset \mathfrak{m}}(\dim \mathfrak{t})$ and $\max\limits_{\mathfrak{t} \subset \mathfrak{m}}(\dim \mathfrak{t})$, respectively, are called the lower and the upper indices of an involutive automorphism A, an involutive algebra $\mathfrak{l}$, and involutive pair $\mathfrak{g}/\mathfrak{l}$.

If

$$\min_{\mathfrak{t} \subset \mathfrak{m}}(\dim \mathfrak{t}) = \max_{\mathfrak{t} \subset \mathfrak{m}}(\dim \mathfrak{t}) = r$$

then we say that r is the index (rank) of an involutive automorphism A, an involutive algebra $\mathfrak{l}$, and involutive pair $\mathfrak{g}/\mathfrak{l}$.

An involutive pair $\mathfrak{g}/\mathfrak{l}$ is called *irreducible* if $\mathrm{ad}\, _{\mathfrak{g}}^{\mathfrak{g}-\mathfrak{l}}(\mathfrak{l})$ is irreducible.

I.1.19. Definition 10. We say that an involutive pair $\mathfrak{g}/\mathfrak{l}$ is elementary if either $\mathfrak{g}$ is simple and semi-simple or $\mathfrak{g} = \mathfrak{p} \times \mathfrak{p}$ (direct product of ideals), where $\mathfrak{p}$ is simple and semi-simple, and $\mathfrak{l} = \triangle(\mathfrak{p} \times \mathfrak{p}) = \{(\xi, \xi) \mid \xi \in \mathfrak{p}\}$ is the diagonal algebra of the canonical involutive automorphism in $\mathfrak{p} \times \mathfrak{p}$.

Evidently an elementary involutive pair is irreducible.

Note that if a Lie algebra $\mathfrak{g}$ is compact and semi-simple and $\mathfrak{g}/\mathfrak{l}$ is irreducible (in particular, elementary) then $\mathfrak{l}$ has at most a one-dimensional centre.

I.1.20. Definition 11. We say that an involutive pair $\mathfrak{g}/\mathfrak{l}$ is a sum of involutive pairs $\mathfrak{g}_1/\mathfrak{l}_1, \ldots, \mathfrak{g}_m/\mathfrak{l}_m$ and write

$$\mathfrak{g}/\mathfrak{l} = \mathfrak{g}_1/\mathfrak{l}_1 + \cdots + \mathfrak{g}_m/\mathfrak{l}_m = \sum_{\alpha=1}^{m} (\mathfrak{g}_\alpha/\mathfrak{l}_\alpha)$$

if there are direct sum decompositions of ideals:

$$\mathfrak{g} = \mathfrak{g}_1 \oplus \cdots \oplus \mathfrak{g}_m, \qquad \mathfrak{l} = \mathfrak{l}_1 \oplus \cdots \oplus \mathfrak{l}_m \quad (\mathfrak{l}_\alpha \subset \mathfrak{g}_\alpha).$$

If $\mathfrak{g}$ is compact then an involutive pair $\mathfrak{g}/\mathfrak{l}$ has a unique decomposition:

$$\mathfrak{g}/\mathfrak{l} = \mathfrak{g}_0/\mathfrak{l}_0 + \sum_{\alpha=1}^{m} (\mathfrak{g}_\alpha/\mathfrak{l}_\alpha),$$

where $\mathfrak{g}_0$ is a centre of $\mathfrak{g}$ and $\mathfrak{g}_1/\mathfrak{l}_1, \ldots, \mathfrak{g}_m/\mathfrak{l}_m$ are elementary involutive pairs. See, for example, [S. Helgason 62,78].

I.1.21. Definition 12. Let $S_1, S_2, S_3 \in \text{Aut}(\mathfrak{g})$ be pair-wise different commuting involutive automorphisms of a Lie algebra $\mathfrak{g}$ such that the product of any two of them is equal to the third. Then Id, S_1, S_2, S_3 constitute a discrete subgroup $\chi(S_1, S_2, S_3) \subset \text{Aut}(\mathfrak{g})$ which is called an involutive group of the algebra $\mathfrak{g}$.

I.1.22. Definition 13. An involutive group $\chi(S_1, S_2, S_3)$ of a Lie algebra $\mathfrak{g}$ is called an iso-involutive group and is denoted $\chi(S_1, S_2, S_3; \varphi)$ if

$$S_3 = \varphi^2, \qquad \varphi \in \text{Int}_\mathfrak{g}(\{t\zeta\}_{t\in\mathbb{R}}), \qquad S_1\zeta = -\zeta, \qquad S_2\zeta = -\zeta, \quad \zeta \neq 0.$$

In this case obviously $\varphi^{-1}S_1\varphi = S_2$, $\varphi^{-1}S_2\varphi = S_1$.

I.1.23. Remark. The condition $S_2\zeta = -\zeta$ in I.1.22 (Definition 13) may be obtained from the others. We have put this only for symmetry of I.1.22. (Definition 13).

I.1.24. Definition 14. An involutive group $\chi(S_1, S_2, S_3)$ of a Lie algebra $\mathfrak{g}$ is called a hyper-involutive group and is denoted $\chi(S_1, S_2, S_3; p)$ if there exists $p \in \text{Aut}(\mathfrak{g})$ such that

$$p^{-1}S_1 p = S_2, \qquad p^{-1}S_2 p = S_3, \qquad p^{-1}S_3 p = S_1, \qquad p^3 = \text{Id}.$$

I.1.25. Definition 15. We say that a Lie algebra $\mathfrak{g}$ is an involutive sum (invosum) of subalgebras $\mathfrak{l}_1, \mathfrak{l}_2, \mathfrak{l}_3$ if $\mathfrak{g} = \mathfrak{l}_1 + \mathfrak{l}_2 + \mathfrak{l}_3$ and $\mathfrak{l}_1, \mathfrak{l}_2, \mathfrak{l}_3$ are involutive algebras of involutive automorphisms S_1, S_2, S_3, respectively, of an involutive group $\chi(S_1, S_2, S_3)$ of $\mathfrak{g}$.

I.1.26. Definition 16. An involutive sum $\mathfrak{g} = \mathfrak{l}_1 + \mathfrak{l}_2 + \mathfrak{l}_3$ of a Lie algebra $\mathfrak{g}$ is called an iso-involutive sum (iso-invosum), or iso-involutive decomposition, if the corresponding involutive group $\chi(S_1, S_2, S_3)$ is an iso-involutive group $\chi(S_1, S_2, S_3; \varphi)$.

I.1.27. Definition 17. An involutive sum of a Lie algebra $\mathfrak{g}$, $\mathfrak{g} = \mathfrak{l}_1 + \mathfrak{l}_2 + \mathfrak{l}_3$, is called hyper-involutive sum (hyper-invosum), or hyper-involutive decomposition, if the corresponding involutive group $\chi(S_1, S_2, S_3)$ is a hyper-involutive group $\chi(S_1, S_2, S_3; p)$.

CHAPTER I.2

CURVATURE TENSOR OF AN INVOLUTIVE PAIR.
CLASSICAL INVOLUTIVE PAIRS OF INDEX 1

I.2.1. Let $\mathfrak{g}$ be a Lie algebra and $\mathfrak{l}$ be its involutive algebra of an involutive automorphism $S \in \mathrm{Aut}(\mathfrak{g})$.

Then

$$\mathfrak{g} = \mathfrak{m} \dot{+} \mathfrak{l}, \qquad \mathfrak{m} = \{\xi \in \mathfrak{g} \mid S\xi = -\xi\}, \qquad \mathfrak{l} = \{\xi \in \mathfrak{g} \mid S\xi = \xi\},$$

see (I.7), and using the involutive automorphism S we obtain

$$[\mathfrak{m}\,\mathfrak{m}] \subset \mathfrak{l}, \qquad [\mathfrak{m}\,\mathfrak{l}] \subset \mathfrak{m}, \qquad [\mathfrak{l}\,\mathfrak{l}] \subset \mathfrak{l}.$$

For this reason we can define the multilinear map $R: \mathfrak{m} \times \mathfrak{m} \times \mathfrak{m} \to \mathfrak{m}$,

$$R(\xi; \eta, \zeta) = -[\xi\,[\eta\,\zeta]]; \quad \xi, \eta, \zeta \in \mathfrak{m}. \tag{I.8}$$

Evidently

$$\left.\begin{array}{c} R(\xi; \eta, \eta) = 0, \\[6pt] R(\xi; \eta, \zeta) + R(\zeta; \xi, \eta) + R(\eta; \zeta, \xi) = 0, \\[6pt] R_{\tau, \mu} R(\xi; \eta, \zeta) = R(R_{\tau, \mu}\xi; \eta, \zeta) + R(\xi; R_{\tau, \mu}\eta, \zeta) + R(\xi; \eta, R_{\tau, \mu}\zeta) \\[6pt] (R_{\tau, \mu}: \nu \mapsto R(\nu; \tau, \mu)) \end{array}\right\} \tag{I.9}$$

which shows that $\mathfrak{m}$ equipped with the ternary operation $R(\xi; \eta, \zeta)$ is a Lie triple system. See [K. Yamaguti 58b], [L.V. Sabinin 99].

I.2.2. Definition 18. Let $\mathfrak{g}$ be a Lie algebra, and $\mathfrak{l}$ be its involutive algebra of an involutive automorphism S. A multilinear operator

$$R: (\xi, \eta, \zeta) \mapsto R(\xi; \eta, \zeta) = -[\xi\,[\eta\,\zeta]],$$
$$R: \mathfrak{m} \times \mathfrak{m} \times \mathfrak{m} \to \mathfrak{m}, \qquad \mathfrak{m} = \mathfrak{g} \dot{-} \mathfrak{l}, \tag{I.10}$$

is called the curvature tensor of the involutive automorphism S, involutive algebra $\mathfrak{l}$, involutive pair $\mathfrak{g}/\mathfrak{l}$.

In fact, the curvature tensor of an involutive pair is the curvature tensor for some symmetric space in a non-holonomic frame, as is well known (see [S. Helgason 62,78].

I.2.3. If an involutive pair $\mathfrak{g}/\mathfrak{l}$ is exact, that is, there is no non-zero ideal of $\mathfrak{g}$ in $\mathfrak{l}$ (in particular, elementary) then $\mathrm{ad}_{\mathfrak{g}}^{\mathfrak{m}}(\mathfrak{l})$ is a maximal Lie algebra of endomorphisms Φ of $\mathfrak{m}$ such that

$$\Phi R(\xi; \eta, \zeta) = R(\Phi\xi; \eta, \zeta) + R(\xi; \Phi\eta, \zeta) + R(\xi; \eta, \Phi\zeta).$$

For this reason, knowing for an exact (in particular elementary) involutive pair $\mathfrak{g}/\mathfrak{l}$ its curvature tensor $R(\xi; \eta, \zeta)$ we may restore $\mathrm{ad}_{\mathfrak{g}}^{\mathfrak{m}}(\mathfrak{l})$, and, further, $\mathrm{ad}_{\mathfrak{g}}(\mathfrak{l})$, since $\mathrm{ad}_{\mathfrak{g}}^{\mathfrak{m}}(\mathfrak{l})$ is a faithful representation of $\mathfrak{l}$ on $\mathfrak{m}$. Because $\mathfrak{m} = \mathfrak{g} \dot{-} \mathfrak{l}$ is invariant under the action of $\mathrm{ad}(\mathfrak{g})$, we may restore $\mathrm{ad}(\mathfrak{g})$, and consequently $\mathfrak{g}/\mathfrak{l}$.

Thus we have:

I.2.4. Theorem 1. *An exact (in particular, elementary) involutive pair $\mathfrak{g}/\mathfrak{l}$ is uniquely defined by its curvature tensor.*

Taking a base $X_1, \ldots, X_n$ on $\mathfrak{m} = \mathfrak{g} \dot{-} \mathfrak{l}$, we have

$$R(X_i; X_j, X_k) = -[X_i [X_j X_k]]$$
$$= R_{i,jk}^q X_q. \tag{I.11}$$

Thus we also say that $R_{i,jk}^q$ is a curvature tensor for the involutive pair $\mathfrak{g}/\mathfrak{l}$.

The following theorem holds.

I.2.5. Theorem 2. *There exists a unique (up to isomorphism) elementary involutive pair $\mathfrak{g}/\mathfrak{l}$ such that*

$$R_{i,jk}^q = \delta_k^q g_{ij} - \delta_j^q g_{ik},$$

where $g_{ij} = g_{ji}$ is positive-definite ($i, j, k, q = 1, \ldots, n$; $n > 1$).

In this case

$$\mathfrak{g}/\mathfrak{l} \cong so(n+1)/so(n)$$

(with the natural embedding).

Proof. One may consider a sphere in $\mathbb{R}^{n+1}$ as the symmetric space $SO(n+1)/SO(n)$ and pass from groups to corresponding algebras. Another way is the direct verification that $so(n+1)/so(n)$ has the prescribed form of curvature tensor. After that one should take into account I.2.4. (Theorem 1). ■

I.2.6. Theorem 3. *There exists a unique (up to isomorphism) elementary involutive pair $\mathfrak{g}/\mathfrak{l}$ such that*

$$R_{i,jk}^q = 2b_i^q b_{jk} + b_k^q b_{ji} - b_j^q b_{ki} + \delta_k^q g_{ij} - \delta_j^q g_{ik},$$

where $g_{ij} = g_{ji}$ is positive-definite,

$$b_{ij} = -b_{ji} \qquad b_{ij} = b_i^s g_{sj}, \qquad b_q^i b_j^q = -\delta_j^i$$
$$(i, j, q, s = 1, \ldots, 2n).$$

In this case

$$\mathfrak{g}/\mathfrak{l} \cong su(n+1)/u(n)$$

(with the natural embedding).

Proof. One may consider a sphere in the unitary space $\mathbb{C}^{n+1}$ as the real symmetric space $SU(n+1)/U(n)$ and pass from groups to the corresponding Lie algebras.

Another way is the direct verification that $su(n+1)/u(n)$ has the prescribed form of curvature tensor. After that one should take into account I.2.4 (Theorem 1). ∎

I.2.7. Theorem 4. *There exists a unique (up to isomorphism) elementary involutive pair $\mathfrak{g}/\mathfrak{l}$ such that*

$$R^q_{i,jk} = 2b^q_{i\lambda}b^\lambda_{jk} + b^q_{k\lambda}b^\lambda_{ji} - b^q_{j\lambda}b^\lambda_{ki} + \delta^q_k g_{ij} - \delta^q_j g_{ik},$$

where $g_{ij} = g_{ji}$ is positive-definite,

$$b^\lambda_{ij} = -b^\lambda_{ji}, \qquad b^\lambda_{ij} = b^s_{i\lambda}g_{sj}, \qquad b^i_{q\lambda}b^q_{j\mu} + b^i_{q\mu}b^q_{j\lambda} = -2\delta^i_j\delta_{\lambda\mu}$$

$$(i,j,k,q,s = 1,\dots,4m; \quad \lambda,\mu = 1,2,3).$$

In this case

$$\mathfrak{g}/\mathfrak{l} \cong sp(n+1)/sp(n) \oplus sp(1)$$

(with the natural embedding).

Proof. One may consider a sphere in $\mathbb{H}^{n+1}$ ($\mathbb{H}$ stands for the algebra of the quaternions) as the real symmetric space $Sp(n+1)/Sp(n) \times Sp(1)$ and pass from groups to the corresponding Lie algebras.

Another way is the direct verification that $sp(n+1)/sp(n) \oplus sp(1)$ has the prescribed form of the curvature tensor. After that one should take into account I.2.4 (Theorem 1). ∎

I.2.8. Remark. Let us note the isomorphisms well known from the theory of classical Lie algebras:

$$so(3)/so(2) \cong su(2)/u(1) \cong sp(1)/u(1), \qquad so(5)/so(4) \cong sp(2)/sp(1) \oplus sp(1).$$

See, for example, [S. Helgason, 62,78].

CHAPTER I.3

ISO-INVOLUTIVE SUMS OF LIE ALGEBRAS

I.3.1. Let S_1 and S_2 be two different and non-trivial (that is, not equal to Id) commutative involutive automorphisms of a Lie algebra $\mathfrak{g}$. Then $S_1 S_2 = S_2 S_1 = S_3$ is a non-trivial involutive automorphism different from S_1, S_2. The elements $\mathrm{Id}, S_1, S_2, S_3$ constitute an involutive group $\chi(S_1, S_2, S_3) \subset \mathrm{Aut}(\mathfrak{g})$.

Thus in order to define an involutive group it is sufficient to take two different non-trivial commutative involutive automorphisms. From the pair-wise commutativity of S_1, S_2, S_3 of an arbitrary involutive group $\chi(S_1, S_2, S_3)$ and (I.7), which is satisfied for any involutive automorphism, it follows that any involutive group $\chi(S_1, S_2, S_3)$ of a Lie algebra $\mathfrak{g}$ generates an involutive sum

$$\mathfrak{g} = \mathfrak{l}_1 + \mathfrak{l}_2 + \mathfrak{l}_3 .$$

Moreover,

$$\mathfrak{l}_1 \cap \mathfrak{l}_2 = \mathfrak{l}_2 \cap \mathfrak{l}_3 = \mathfrak{l}_3 \cap \mathfrak{l}_1 = \mathfrak{l}_1 \cap \mathfrak{l}_2 \cap \mathfrak{l}_3 = \mathfrak{l}_0 ,$$

and $\mathfrak{l}_\alpha / \mathfrak{l}_0$ ($\alpha = 1, 2, 3$) are involutive pairs of involutive automorphisms σ_α ($\alpha = 1, 2, 3$), where

$$\sigma_1 = S_2|_{\mathfrak{l}_1} = S_3|_{\mathfrak{l}_1}, \qquad \sigma_2 = S_1|_{\mathfrak{l}_2} = S_3|_{\mathfrak{l}_2}, \qquad \sigma_3 = S_1|_{\mathfrak{l}_3} = S_2|_{\mathfrak{l}_3}$$

(here by $S_\alpha|_{\mathfrak{l}_\beta}$ we denote the restriction of an involutive automorphism S_α on $\mathfrak{l}_\beta$).

In addition, we have

$$\mathfrak{l}_\alpha = \mathfrak{l}_0 \dot{+} \mathfrak{m}_\alpha \quad (\alpha = 1, 2, 3),$$

where, by (I.7),

$$\mathfrak{m}_\alpha = \{\zeta \in \mathfrak{m}_\alpha \mid \sigma_\alpha \zeta = -\zeta\}.$$

Finally, because of the invariance of the Cartan metric with respect to $\chi(S_1, S_2, S_3)$, it follows that $\mathfrak{l}_0, \mathfrak{m}_1, \mathfrak{m}_2, \mathfrak{m}_3$ are pairwise orthogonal.

We have arrived to the theorems:

I.3.2. Theorem 5. *Any involutive group is defined by two non-trivial commuting involutive automorphisms.*

12

I.3.3. Theorem 6. *Any involutive group* $\chi(S_1, S_2, S_3) \subset \mathrm{Aut}(\mathfrak{g})$, $\mathfrak{g}$ *being a Lie algebra, generates an involutive sum* $\mathfrak{g} = \mathfrak{l}_1 + \mathfrak{l}_2 + \mathfrak{l}_3$, *where* $\mathfrak{l}_\alpha$ *is the involutive algebra of the involutive automorphism* S_α ($\alpha = 1, 2, 3$).

Moreover,

$$\mathfrak{l}_1 \cap \mathfrak{l}_2 = \mathfrak{l}_2 \cap \mathfrak{l}_3 = \mathfrak{l}_3 \cap \mathfrak{l}_1 = \mathfrak{l}_1 \cap \mathfrak{l}_2 \cap \mathfrak{l}_3 = \mathfrak{l}_0,$$

and $\mathfrak{l}_\alpha / \mathfrak{l}_0$ ($\alpha = 1, 2, 3$) *are involutive pairs of the involutive automorphisms* $\sigma_\alpha = S_\mu|_{\mathfrak{l}_\alpha}$ ($\alpha \neq \mu$, $\alpha = 1, 2, 3$), *that is, the restrictions of the involutive automorphisms* S_μ *on* $\mathfrak{l}_\alpha$.

In addition, $\mathfrak{l}_\alpha = \mathfrak{l}_0 \dot{+} \mathfrak{m}_\alpha$ ($\alpha = 1, 2, 3$), *where* $\mathfrak{m}_\alpha = \{\zeta \in \mathfrak{l}_\alpha \mid \sigma_\alpha \zeta = -\zeta\}$, *and* $\mathfrak{l}_0, \mathfrak{m}_1, \mathfrak{m}_2, \mathfrak{m}_3$ *are pair-wise orthogonal with respect to the Cartan metric of the Lie algebra* $\mathfrak{g}$.

I.3.4. Corollary. In the notations of I.3.3 (Theorem 6) we have

$$\mathfrak{g} = \mathfrak{l}_0 \dot{+} \mathfrak{m}_1 \dot{+} \mathfrak{m}_2 \dot{+} \mathfrak{m}_3.$$

I.3.5. Theorem 7. *Let* $\chi(S_1, S_2, S_3; \varphi)$ *be an iso-involutive group of a Lie algebra* $\mathfrak{g}$, *and* $\mathfrak{g} = \mathfrak{l}_1 + \mathfrak{l}_2 + \mathfrak{l}_3$ *be the corresponding involutive sum.*

Then $\varphi \mathfrak{l}_1 = \mathfrak{l}_2$, $\varphi \mathfrak{l}_2 = \mathfrak{l}_1$, *and the automorphisms* $\varphi|_{\mathfrak{l}_3}$, $\varphi|_{\mathfrak{l}_0}$, *the restrictions of* φ *on* $\mathfrak{l}_3$ *and on* $\mathfrak{l}_0 = \mathfrak{l}_1 \cap \mathfrak{l}_2 \cap \mathfrak{l}_3$, *respectively, are involutive automorphisms.*

Proof. By the definition of an involutive group we have $S_1 \varphi = \varphi S_2$, whence $\varphi \mathfrak{l}_2 = \mathfrak{l}_1$. But then $\varphi \mathfrak{l}_1 = \varphi^2 \mathfrak{l}_2 = S_3 \mathfrak{l}_2 = \mathfrak{l}_2$. Consequently $\varphi \mathfrak{l}_0 = \varphi(\mathfrak{l}_1 \cap \mathfrak{l}_2) = \mathfrak{l}_0$. Furthermore, since $\varphi^2 = S_3$ it follows that $\varphi \mathfrak{l}_3 = \mathfrak{l}_3$ and $\varphi^2 \zeta = S_3 \zeta = \zeta$ for $\zeta \in \mathfrak{l}_3$. Thus $\varphi|_{\mathfrak{l}_3}$ is an involutive automorphism. But $\mathfrak{l}_0 \subset \mathfrak{l}_3$, which implies that $\varphi|_{\mathfrak{l}_0}$ is an involutive automorphism as well. ∎

I.3.6. Definition 19. An iso-involutive group $\chi(S_1, S_2, S_3; \varphi)$ is said to be of type 1 if $\varphi|_{\mathfrak{l}_3}$ (the restriction of φ to the involutive algebra $\mathfrak{l}_3$ of the involutive automorphism S_3) is the identity automorphism.

In this case we also say that $\mathfrak{g} = \mathfrak{l}_1 + \mathfrak{l}_2 + \mathfrak{l}_3$ is an iso-involutive sum of type 1.

I.3.7. Definition 20. Let $\chi(S_1, S_2, S_3; \varphi)$ be an iso-involutive group of a Lie algebra $\mathfrak{g}$, where $\varphi \in \mathrm{Int}_\mathfrak{g}(\{t\zeta\})$, and $\mathfrak{l}_3$ be the involutive algebra of the involutive automorphism S_3. And let $\chi(\sigma_1, \sigma_2, \sigma_3; \psi)$ be an iso-involutive group of the Lie algebra $\mathfrak{l}_3$ such that $\sigma_\alpha = S_\alpha|_{\mathfrak{l}_3}$ ($\alpha \neq 3$), that is, the restrictions of S_1 and S_2 to $\mathfrak{l}_3$, $\sigma_3 = \varphi|_{\mathfrak{l}_3}$, where $\varphi = \theta^2$, $\theta \in \mathrm{Int}_\mathfrak{g}(\{t\zeta\}_{t\in\mathbb{R}})$, $\psi = \theta|_{\mathfrak{l}_3}$.

Then we say that $\chi(\sigma_1, \sigma_2, \sigma_3; \psi)$ is a derived iso-involutive group for $\chi(S_1, S_2, S_3; \varphi)$ and denote it $\chi^{(1)}(S_1, S_2, S_3; \varphi)$.

One may consider the derived iso-involutive group $\chi^{(2)}(S_1, S_2, S_3; \varphi)$ of $\chi^{(1)}(S_1, S_2, S_3; \varphi)$ and so on.

I.3.8. Theorem 8. *If* $\chi(S_1, S_2, S_2; \varphi)$ *is not of type 1 then* $\chi^{(1)}(S_1, S_2, S_3; \varphi)$ *exists, otherwise it does not exist.*

Proof. Let us consider the conjugate subgroup $\mathrm{Int}_\mathfrak{g}(\{t\zeta\}_{t\in\mathbb{R}})$ of the involutive group $\chi(S_1, S_2, S_3; \varphi)$, see I.1.22 (Definition 13). Then $\varphi \in \mathrm{Int}_\mathfrak{g}(\{t\zeta\})$.

We take $\theta \in \mathrm{Int}_\mathfrak{g}(\{t\zeta\}_{t\in\mathbb{R}})$ such that $\theta^2 = \varphi$ (which is obviously possible).

Furthermore, evidently $S_1\theta S_1 = \theta^{-1}$ and passing to the restrictions to $\mathfrak{l}_3$ we have $\sigma_1\psi\sigma_1 = \psi^{-1}$, $\psi^2 = \sigma_3$ ($\psi \neq \mathrm{Id}$ by construction), $\psi \in \mathrm{Int}_\mathfrak{g}^{\mathfrak{l}_3}(\{t\zeta\}_{t\in\mathbb{R}}) = \mathrm{Int}_{\mathfrak{l}_3}(\{t\zeta\}_{t\in\mathbb{R}})$.

Finally, $\psi^{-1}\sigma_1\psi = \sigma_2$ is an involutive automorphism and $\sigma_1\sigma_2 = \sigma_2\sigma_1 = \sigma_3$. Thus we have constructed an involutive group with the desired properties. ∎

I.3.9. Let $\mathfrak{g}$ be a compact semi-simple Lie algebra and $\mathfrak{g}/\mathfrak{l}$ be a non-trivial involutive pair (i.e., $\mathfrak{g} \neq \mathfrak{l}$) of an involutive automorphism S. Then, see (I.7), $\mathfrak{g} = \mathfrak{m}\dot{+}\mathfrak{l}$. Any vector $\zeta \in \mathfrak{m}$ generates a one-dimensional subgroup $\mathrm{Int}_\mathfrak{g}(\{t\zeta\}) \subset \mathrm{Int}(\mathfrak{g})$ of the form $\theta(t) = e^{c(t\zeta)}$, see (I.3), in some base, or, what is the same, $\theta(t) = e^{\mathrm{ad}_\mathfrak{g}(t\zeta)}$. The involutive automorphism S of the involutive pair $\mathfrak{g}/\mathfrak{l}$ transforms ζ into $(-\zeta)$, thus $S\theta(t)S = \theta(-t) = [\theta(t)]^{-1}$.

Since $\mathrm{Int}(\mathfrak{g})$ is compact, because $\mathfrak{g}$ is compact, the closure $\overline{\mathrm{Int}_\mathfrak{g}(\{t\zeta\})}$ in $\mathrm{Int}(\mathfrak{g})$ is a torus, that is, a compact connected abelian group. Moreover, $\overline{\mathrm{Int}_\mathfrak{g}(\{t\zeta\})}$ is closed in $\mathrm{Int}(\mathfrak{g})$, and consequently is a Lie group. For this reason $\overline{\mathrm{Int}_\mathfrak{g}(\{t\zeta\})} = \mathrm{Int}_\mathfrak{g}(\mathfrak{k})$, where $\mathfrak{k}$ is a commutative subalgebra of $\mathfrak{g}$.

For $\theta \in \mathrm{Int}_\mathfrak{g}(\mathfrak{k})$ evidently $S\theta S = \theta^{-1}$, which implies $\mathfrak{k} \subset \mathfrak{m}$. As is well known, any torus (in our case $\mathrm{Int}_\mathfrak{g}(\mathfrak{k})$) contains non-trivial involutive elements which effectively act on $\mathfrak{g}$ (otherwise $\mathrm{Int}(\mathfrak{g})$ possesses the non-trivial centre, which is impossible for a semi-simple Lie algebra).

Lastly, through any element of a torus, in particular, through an involutive element, there passes at least a one-dimensional subgroup. But any one-dimensional subgroup containing an involutive element is compact.

We obtained the following lemma:

I.3.10. Lemma 1. *Let $\mathfrak{g}$ be a compact semi-simple Lie algebra, and $\mathfrak{g}/\mathfrak{l}$ be its involutive pair of an involutive automorphism $S \neq \mathrm{Id}$. Then there exists $\zeta \in \mathfrak{g}$ such that $S\zeta = -\zeta$, and $\mathrm{Int}_\mathfrak{g}(\{t\zeta\})$ is compact.*

I.3.11. Lemma 2. *If $\mathfrak{g}$ is a compact semi-simple Lie algebra and S is its non-trivial involutive automorphism then for any $m \in \mathbb{N}$ there exists $\theta_m \in \mathrm{Int}(\mathfrak{g})$ such that $(\theta_{m+1})^2 = \theta_m$ $\theta_0 = S$, $S\theta_m S = (\theta_m)^{-1}$, $(\theta_m)^k \neq \mathrm{Id}$, for $0 < k < m$.*

Indeed, owing to the compactness of a one-dimensional subgroup from I.3.10 (Lemma 1) it contains elements θ_m with the required properties.

I.3.12. In a compact semi-simple Lie algebra $\mathfrak{g}$, starting from its non-trivial involutive automorphism, $S_1 \neq \mathrm{Id}$, we can construct the iso-involutive group $\chi(S_1, S_2, S_3; \varphi)$ and its corresponding iso-involutive decomposition $\mathfrak{g} = \mathfrak{l}_1\dot{+}\mathfrak{l}_2\dot{+}\mathfrak{l}_3$.

This turns iso-involutive sums into a new apparatus for exploring Lie algebras. Indeed, we have:

I.3.13. Theorem 9. *Let $\mathfrak{g}$ be a compact semi-simple Lie algebra. If $\mathfrak{l}_1$ is its involutive algebra of an involutive automorphism $S_1 \neq \mathrm{Id}$ then there exists an iso-involutive group $\chi(S_1, S_2, S_3; \varphi)$ of $\mathfrak{g}$ and the corresponding involutive decomposition $\mathfrak{g} = \mathfrak{l}_1 + \mathfrak{l}_2 + \mathfrak{l}_3$.*

Proof. By I.3.10. (Lemma 1) and I.3.11. (Lemma 2) there exists an automorphism φ such that

$$\varphi^4 = \mathrm{Id}, \qquad \varphi \neq \mathrm{Id}, \qquad \varphi^2 \neq \mathrm{Id}, \qquad \varphi^3 \neq \mathrm{Id}, \qquad \varphi \in \mathrm{Int}_{\mathfrak{g}}(\{t\zeta\}),$$
$$S_1\zeta = -\zeta, \qquad S_1\varphi S_1 = \varphi^{-1}.$$

Then $\varphi^2 = \varphi^{-2} = S_3 \neq \mathrm{Id}$ and S_3 is an involutive automorphism. And, since $\varphi^{-1}S_1\varphi = S_2 \neq \mathrm{Id}$, S_2 is an involutive automorphism as well.

Furthermore,

$$S_1 S_2 = S_1\left(\varphi^{-1}S_1\varphi\right) = S_1\left(\varphi^{-1}S_1\right)\varphi = S_1\left(S_1\varphi\right)\varphi = \varphi^2 = S_3,$$
$$S_2 S_1 = \left(\varphi^{-1}S_1\varphi\right)S_1 = \varphi^{-1}\left(S_1\varphi\right)S_1 = \varphi^{-1}\left(\varphi^{-1}S_1\right)S_1 = \varphi^{-2} = S_3.$$

The non-trivial commuting involutive automorphisms S_1 and S_2 generate an involutive group of Lie algebra $\mathfrak{g}$. But also $S_1\zeta = -\zeta$, $\varphi\zeta = \zeta$, which results in

$$S_2\,\zeta = \varphi^{-1}S_1(\varphi\zeta) = \varphi^{-1}(S_1\zeta) = -(\varphi^{-1}\zeta) = -\zeta.$$

Therefore S_1 and S_2 generate an iso-involutive group $\chi(S_1, S_2, S_3; \varphi)$; and consequently, see I.3.3 (Theorem 6), the corresponding iso-involutive sum $\mathfrak{g} = \mathfrak{l}_1 + \mathfrak{l}_2 + \mathfrak{l}_3$. ∎

I.3.14. Definition 21. We say that an iso-involutive group $\chi(S_1, S_2, S_3; \varphi)$ and its involutive sum $\mathfrak{g} = \mathfrak{l}_1 + \mathfrak{l}_2 + \mathfrak{l}_3$ is of lower index 1 if $\varphi \in \mathrm{Int}_{\mathfrak{g}}(\{t\zeta\})$, where $\{t\zeta\}_{t\in\mathbb{R}}$ is a maximal one-dimensional subalgebra in $\mathfrak{m} = \mathfrak{g} \dot{-} \mathfrak{l}$.

I.3.15. Remark. In this case evidently $\mathfrak{g}/\mathfrak{l}_1$ and $\mathfrak{g}/\mathfrak{l}_2$ are involutive pairs of the lower index 1.

I.3.16. Theorem 10. *Let $\mathfrak{g}$ be a compact semi-simple Lie algebra, $\mathfrak{g}/\mathfrak{l}_1$ be an involutive pair of an involutive automorphism S_1 of lower index 1, and let $\{t\zeta\}_{t\in\mathbb{R}}$ be a one-dimensional maximal subalgebra in $\mathfrak{m} = \mathfrak{g} \dot{-} \mathfrak{l}_1$.*

Then there exists an iso-involutive group $\chi(S_1, S_2, S_3; \varphi)$ and the corresponding involutive sum of lower index 1 such that $\varphi \in \mathrm{Int}_{\mathfrak{g}}(\{t\zeta\})$.

Proof. Let $\mathfrak{g}$ be a compact semi-simple Lie algebra and $\mathfrak{g}/\mathfrak{l}_1$ be an involutive pair of the involutive automorphism S_1 of lower index 1, see I.1.18 (Definition 9). Then, taking an one-dimensional maximal subalgebra $\{t\zeta\} = \mathfrak{t} \subset \mathfrak{g} \dot{-} \mathfrak{l}_1$, we have evidently that $\mathrm{Int}\{t\zeta\}_{t\in\mathbb{R}}$ is compact. Indeed, then $\overline{\mathrm{Int}_{\mathfrak{g}}(\{t\zeta\})} = \mathrm{Int}_{\mathfrak{g}}(\mathfrak{t})$, where $\mathfrak{t} \subset \mathfrak{m} = \mathfrak{g} \dot{-} \mathfrak{l}_1$, $\{t\zeta\} \subset \mathfrak{t}$. But since $\{t\zeta\}_{t\in\mathbb{R}}$ is a maximal subalgebra in $\mathfrak{m} = \mathfrak{g} \dot{-} \mathfrak{l}_1$ we have $\{t\zeta\}_{t\in\mathbb{R}} = \mathfrak{t}$ and $\overline{\mathrm{Int}_{\mathfrak{g}}(\{t\zeta\})} = \mathrm{Int}_{\mathfrak{g}}(\{t\zeta\})$ is closed in $\mathrm{Int}(\mathfrak{g})$, and consequently is compact.

Using, furthermore, $\mathrm{Int}_{\mathfrak{g}}(\{t\zeta\})$ as a conjugating subgroup, by I.3.13 (Theorem 9) we obtain an involutive group $\chi(S_1, S_2, S_3; \varphi)$ and the corresponding involutive decomposition $\mathfrak{g} = \mathfrak{l}_1 + \mathfrak{l}_2 + \mathfrak{l}_3$ of lower index 1. ∎

CHAPTER I.4

ISO-INVOLUTIVE BASE AND STRUCTURE EQUATIONS

I.4.1. Definition 22. We say that a base of a Lie algebra $\mathfrak{g}$ is invariant with respect to an iso-involutive group $\chi(S_1, S_2, S_3; \varphi)$ if in this base S_1, S_2, S_3, φ have canonical forms.

We also say that a base of a Lie algebra $\mathfrak{g}$ is an iso-involutive base (iso-invobase) if it is invariant (up to the multiplication by (± 1)) with respect to $\chi(S_1, S_2, S_3; \varphi)$ and all its derived iso-involutive groups.

I.4.2. Theorem 11. *Let* $\chi(S_1, S_2, S_3; \varphi)$ *be an iso-involutive group of a Lie algebra* $\mathfrak{g}$, *then there exists an involutive base of* $\mathfrak{g}$ *which is invariant with respect to* $\chi(S_1, S_2, S_3; \varphi)$.·

If, in addition, $\mathfrak{g}$ *is compact semi-simple then there exists an involutive base which is orthogonal with respect to the Cartan metric of* $\mathfrak{g}$.

Proof. By I.3.3 (Theorem 6) and I.3.4 (Corollary 1) we have a decomposition

$$\mathfrak{g} = \mathfrak{l}_0 \dotplus \mathfrak{m}_1 \dotplus \mathfrak{m}_2 \dotplus \mathfrak{m}_3,$$

by I.3.5 (Theorem 7) $\varphi \mathfrak{m}_1 = \mathfrak{m}_2$ also.

Taking an arbitrary base in $\mathfrak{m}_1$ we induce a base in $\mathfrak{m}_2$ by the automorphism φ. Furthermore, if $\varphi|_{\mathfrak{l}_0}$, $\varphi|_{\mathfrak{l}_3}$ are the identity involutive automorphisms then we take bases in $\mathfrak{l}_0$ and $\mathfrak{m}_3$ arbitrarily.

The union of all the above bases is evidently an involutive base in $\mathfrak{g}$ which is invariant with respect to $\chi(S_1, S_2, S_3; \varphi)$.

If $\varphi|_{\mathfrak{l}_3} \neq \mathrm{Id}$, see I.3.5 (Theorem 7), then $\varphi|_{\mathfrak{l}_0}$, $\varphi|_{\mathfrak{l}_3}$, and consequently $\varphi|_{\mathfrak{m}_3}$, are involutive automorphisms, and $\chi(S_1, S_2, S_3; \varphi)$ is not of type 1, see I.3.6 (Definition 19). Then by I.3.8 (Theorem 8) there exists the derived involutive group $\chi^{(1)}(S_1, S_2, S_3; \varphi)$.

Using, furthermore, a Lie algebra $\mathfrak{l}_3 = \mathfrak{m}_3 \dotplus \mathfrak{l}_0$ and the derived involutive group $\chi^{(1)}(S_1, S_2, S_3; \varphi)$, we repeat the above construction *etc.*. The union of all bases introduced in such a way is evidently an involutive base in $\mathfrak{g}$.

Finally, if $\mathfrak{g}$ is compact semi-simple then in the foregoing consideration we take an orthonormal base (in $\mathfrak{m}_1$ *etc.*). Then since $\mathfrak{m}_\alpha \perp \mathfrak{m}_\beta$ ($\alpha \neq \beta$) it follows that $\mathfrak{m}_\alpha \perp \mathfrak{l}_0$, that is, they are orthogonal with respect to the Cartan metric of $\mathfrak{g}$, and so we obtain an orthonormal involutive base. ∎

I.4.3. We consider, furthermore, a compact semi-simple Lie algebra $\mathfrak{g}$ in an orthonormal involutive base, which is possible because of I.3.13 (Theorem 9) and

16

I.3.16 (Theorem 10). Let an involutive base be invariant with respect to an iso-involutive group $\chi(S_1, S_2, S_3; \varphi)$ and its derived subgroup $\chi(\sigma_1, \sigma_2, \sigma_3; \psi)$ (see the notations of I.3.7 (Definition 20)) if, of course, the latter exists.

Let us use (I.7) and introduce the corresponding iso-involutive sums,

$$
\left.
\begin{aligned}
& \mathfrak{g} = \mathfrak{l}_1 + \mathfrak{l}_2 + \mathfrak{l}_3, \quad \mathfrak{l}_3 = \mathfrak{n}_1 + \mathfrak{n}_2 + \mathfrak{n}_3, \quad \mathfrak{n}_1 = \mathfrak{l}_0 = \mathfrak{l}_\alpha \cap \mathfrak{l}_\beta \ (\alpha \neq \beta), \\
& \mathfrak{n}_0 = \mathfrak{n}_1 \cap \mathfrak{n}_2 = \mathfrak{n}_2 \cap \mathfrak{n}_3 = \mathfrak{n}_1 \cap \mathfrak{n}_3, \quad \mathfrak{l}_\alpha = \mathfrak{m}_\alpha \dotplus \mathfrak{n}_1, \\
& \mathfrak{m}_\alpha = \{\, \zeta \in \mathfrak{g} \mid \sigma_\alpha \zeta = -\zeta \,\}, \\
& \mathfrak{n}_1 = \mathfrak{q}_1 \dotplus \mathfrak{n}_0, \quad \mathfrak{n}_2 = \mathfrak{q}_2 \dotplus \mathfrak{n}_0, \quad \mathfrak{n}_3 = \mathfrak{q}_3 \dotplus \mathfrak{n}_0.
\end{aligned}
\right\} \tag{I.12}
$$

We note that (I.12) is also true when $\chi^{(1)}$ does not exists, in which case we need only to put $\mathfrak{n}_2 = \mathfrak{n}_1 = \mathfrak{n}_0$ (that is, $\mathfrak{q}_1 = \mathfrak{q}_2 = \{0\}$).

By I.3.5 (Theorem 7) we have

$$
\mathfrak{m}_2 = \varphi \mathfrak{m}_1, \qquad \mathfrak{q}_2 = \theta \mathfrak{q}_1, \qquad \theta \mathfrak{n}_0 = \mathfrak{n}_0, \qquad \theta \mathfrak{q}_3 = \mathfrak{q}_3, \qquad \theta^2 = \varphi. \tag{I.13}
$$

Let us introduce the notations for the vectors of orthonormal involutive base:

$$
\left.
\begin{aligned}
& X_{i_1} \ (i = 1, \ldots, n) && \text{-- base in } \mathfrak{m}_1 \\
& X_{i_2} \ (i = 1, \ldots, n) && \text{-- base in } \mathfrak{m}_2 \\
& Y_\alpha
\begin{cases}
Y_{\hat\alpha} \ (\alpha = 1, \ldots, \rho) && \text{-- base in } \mathfrak{q}_1 \\
Y_{\alpha \cdot} \ (\alpha = 1, \ldots, r) && \text{-- base in } \mathfrak{n}_0
\end{cases}
\left.\right\} \ \mathfrak{n}_1 = \mathfrak{l}_0 \\
& V_a
\begin{cases}
V_{\hat a} \ (\alpha = 1, \ldots, \rho) && \text{-- base in } \mathfrak{q}_2 \\
V_{a \cdot} \ (a = 1, \ldots, m) && \text{-- base in } \mathfrak{q}_3
\end{cases}
\left.\right\} \ \mathfrak{m}_3
\end{aligned}
\right\} \tag{I.14}
$$

The basis vectors in $\mathfrak{n}_1$ are denoted Y_α; by construction they are either $Y_{\hat\alpha}$, or $Y_{\alpha \cdot}$; the basis vectors in $\mathfrak{m}_3 = \mathfrak{l}_3 \dot- \mathfrak{n}_1$ are denoted V_a, and by construction they are either $V_{\hat a}$, or $V_{a \cdot}$.

Taking into account the action of the involutive automorphisms S_1, S_2, S_3 we can write the structure of the Lie algebra $\mathfrak{g}$:

$$
\left.
\begin{aligned}
& [X_{s_1} X_{k_1}] = -b^\alpha_{s_1 k_1} Y_\alpha, && [X_{s_2} X_{k_2}] = -b^\alpha_{s_2 k_2} Y_\alpha, \\
& [X_{s_1} X_{k_2}] = q^a_{s_1 k_2} V_a, && [X_{s_1} V_a] = t^{k_2}_{s_1 a} X_{k_2}, \\
& [X_{p_2} V_a] = t^{k_1}_{p_2 a} X_{k_1}, && [V_a V_c] = -b^b_{ac} Y_\alpha, \\
& [X_{s_1} Y_\alpha] = a^{p_1}_{s_1 \alpha} X_{p_1}, && [X_{s_2} Y_\alpha] = a^{p_2}_{s_2 \alpha} X_{p_2}, \\
& [V_c Y_\alpha] = a^b_{c\alpha} V_b, && [Y_\alpha Y_\beta] = c^\gamma_{\alpha\beta} Y_\gamma.
\end{aligned}
\right\} \tag{I.15}
$$

By the construction of the involutive base:

$$\varphi X_{s_1} = X_{s_2}, \quad \varphi X_{s_2} = -X_{s_1}, \quad \varphi Y_\alpha = \varphi_\alpha^\beta Y_\beta$$
$$(\varphi Y_{\hat{\alpha}} = -Y_{\hat{\alpha}}, \quad \varphi Y_{\alpha.} = Y_{\alpha.}),$$
$$\varphi V_a = \varphi_a^b V_b \ (\varphi V_{\hat{a}} = -V_{\hat{a}}, \quad \varphi V_{a.} = V_{a.}).$$
\right\} \quad (I.16)

Since φ is an automorphism then $\varphi[\zeta\,\eta] = [(\varphi\zeta)(\varphi\eta)]$; for the structure (I.15), taking into account (I.16), we obtain

$$-b_{s_2 k_2}^\alpha Y_\alpha = [X_{s_2} X_{k_2}]$$
$$= [(\varphi X_{s_1})(\varphi X_{k_1})] = -b_{s_1 k_1}^\alpha (\varphi Y_\alpha) = -b_{s_1 k_1}^\alpha \varphi_\alpha^\beta Y_\beta,$$
$$b_{s_2 k_2}^\alpha = \varphi_\beta^\alpha b_{s_1 k_1}^\beta. \qquad (I.17.1)$$

$$q_{k_1 s_2}^b V_b = -[X_{s_2} X_{k_1}]$$
$$= [(\varphi X_{s_1})(\varphi X_{k_2})] = q_{s_1 k_2}^a (\varphi V_a) = q_{s_1 k_2}^a \varphi_a^b V_b,$$
$$q_{k_1 s_2}^b = \varphi_a^b q_{s_1 k_2}^a. \qquad (I.17.2)$$

$$t_{s_2 b}^{k_1} \varphi_a^b X_{k_1} = \varphi_a^b [X_{s_2} V_b]$$
$$= [(\varphi X_{s_1})(\varphi V_a)] = t_{s_1 a}^{k_2}(\varphi X_{k_2}) = -t_{s_1 a}^{k_2} X_{k_1},$$
$$t_{s_2 b}^{k_1} \varphi_a^b = -t_{s_1 a}^{k_2}. \qquad (I.17.3)$$

$$a_{s_2 \beta}^{p_2} \varphi_\alpha^\beta X_{p_2} = [X_{s_2} Y_\beta]\varphi_\alpha^\beta$$
$$= [(\varphi X_{s_1})(\varphi Y_\alpha)] = a_{s_1 \alpha}^{p_1}(\varphi X_{p_1}) = a_{s_1 \alpha}^{p_1} X_{p_2},$$
$$a_{s_1 \alpha}^{p_1} = \varphi_\alpha^\beta a_{s_2 \beta}^{p_2}. \qquad (I.17.4)$$

$$-\varphi_a^f \varphi_c^d b_{fd}^\beta Y_\beta = \varphi_a^f \varphi_c^d [V_f V_d]$$
$$= [(\varphi V_a)(\varphi V_c)] = -b_{ac}^\alpha(\varphi Y_\alpha) = -b_{ac}^\alpha \varphi_\alpha^\beta Y_\beta,$$
$$b_{ac}^\alpha \varphi_\alpha^\beta = \varphi_a^f \varphi_c^d b_{fd}^\beta. \qquad (I.17.5)$$

$$\varphi_c^d \, \varphi_\alpha^\beta \, a_{d\beta}^f \, V_f = \varphi_c^d \, \varphi_\alpha^\beta \left[V_d \, Y_\beta \right]$$

$$= \left[(\varphi \, V_c)(\varphi \, Y_\alpha) \right] = a_{c\alpha}^b (\varphi \, V_b) = a_{c\alpha}^b \, \varphi_b^f \, V_f,$$

$$\varphi_c^d \, \varphi_\alpha^\beta \, a_{d\beta}^f = a_{c\alpha}^b \, \varphi_b^f. \tag{I.17.6}$$

$$\varphi_\alpha^\lambda \, \varphi_\beta^\mu \, c_{\lambda\mu}^\nu \, Y_\nu = \varphi_\alpha^\lambda \, \varphi_\beta^\mu \left[Y_\lambda \, Y_\mu \right]$$

$$= \left[(\varphi Y_\alpha)(\varphi Y_\beta) \right] = c_{\alpha\beta}^\gamma (\varphi \, Y_\gamma) = c_{\alpha\beta}^\gamma \, \varphi_\gamma^\nu \, Y_\nu,$$

$$\varphi_\alpha^\lambda \, \varphi_\beta^\mu \, c_{\lambda\mu}^\nu = c_{\alpha\beta}^\gamma \, \varphi_\gamma^\nu. \tag{I.17.7}$$

I.4.4. Now we use the Jacobi identities for the case in which not all basis vectors are from $\mathfrak{l}_1$, $\mathfrak{l}_2$, or $\mathfrak{l}_3$.

$$\left[\left[X_{s_1} \, X_{l_1} \right] X_{k_2} \right] + \left[\left[X_{k_2} \, X_{s_1} \right] X_{l_1} \right] + \left[\left[X_{l_1} \, X_{k_2} \right] X_{s_1} \right] = 0,$$

$$b_{s_1 l_1}^\alpha \left[Y_\alpha \, X_{k_2} \right] + q_{l_1 k_2}^a \left[V_a \, X_{s_1} \right] - q_{s_1 k_2}^a \left[V_a \, X_{l_1} \right] = 0,$$

$$a_{k_2 \alpha}^{p_2} \, b_{s_1 l_1}^\alpha \, X_{p_2} - t_{s_1 a}^{p_2} \, q_{l_1 k_2}^a \, X_{p_2} + t_{l_1 a}^{p_2} \, q_{s_1 k_2}^a \, X_{p_2} = 0,$$

$$a_{k_2 \alpha}^{p_2} \, b_{s_1 l_1}^\alpha = t_{s_1 a}^{p_2} \, q_{l_1 k_2}^a - t_{l_1 a}^{p_2} \, q_{s_1 k_2}^a. \tag{I.18.1}$$

$$\left[X_{s_1} \left[V_a \, V_c \right] \right] + \left[V_c \left[X_{s_1} \, V_a \right] \right] + \left[V_a \left[V_c \, X_{s_1} \right] \right] = 0,$$

$$-b_{ac}^\alpha \left[X_{s_1} \, Y_\alpha \right] + t_{s_1 a}^{l_2} \left[V_c \, X_{l_2} \right] - t_{s_1 c}^{l_2} \left[V_a \, X_{l_2} \right] = 0,$$

$$a_{s_1 a}^{p_1} \, b_{ac}^\alpha \, X_{p_1} - t_{l_2 c}^{p_1} \, t_{s_1 a}^{l_2} \, X_{p_1} + t_{l_2 a}^{p_1} \, t_{s_1 c}^{l_2} \, X_{p_1} = 0,$$

$$b_{ac}^\alpha \, a_{s_1 \alpha}^{p_1} = t_{l_2 a}^{p_1} \, t_{s_1 c}^{l_2} - t_{l_2 c}^{p_1} \, t_{s_1 a}^{l_2}. \tag{I.18.2}$$

$$\left[\left[X_{s_1} \, X_{l_2} \right] V_a \right] + \left[\left[V_a \, X_{s_1} \right] X_{l_2} \right] + \left[\left[X_{l_2} \, V_a \right] X_{s_1} \right] = 0,$$

$$q_{s_1 l_2}^c \left[V_c \, V_a \right] - t_{s_1 a}^{p_2} \left[X_{p_2} \, X_{l_2} \right] + t_{l_2 a}^{p_1} \left[X_{p_1} \, X_{s_1} \right] = 0,$$

$$-b_{ca}^\alpha \, q_{s_1 l_2}^c \, Y_\alpha + b_{p_2 l_2}^\alpha \, t_{s_1 a}^{p_2} \, Y_\alpha - b_{p_1 s_1}^\alpha \, t_{l_2 a}^{p_1} \, Y_\alpha = 0,$$

$$q_{s_1 l_2}^c \, b_{ca}^\alpha = t_{s_1 a}^{p_2} \, b_{p_2 l_2}^\alpha - t_{l_2 a}^{p_1} \, b_{p_1 s_1}^\alpha. \tag{I.18.3}$$

$$\left[\left[X_{s_1} \, X_{l_2} \right] Y_\alpha \right] + \left[\left[Y_\alpha \, X_{s_1} \right] X_{l_2} \right] + \left[\left[X_{l_2} \, Y_\alpha \right] X_{s_1} \right] = 0,$$

$$q_{s_1 l_2}^c \left[V_c \, Y_\alpha \right] + a_{l_2 \alpha}^{p_2} \left[X_{p_2} \, X_{s_1} \right] - a_{s_1 \alpha}^{p_1} \left[X_{p_1} \, X_{l_2} \right] = 0,$$

$$q_{s_1 l_2}^c \, a_{c\alpha}^b \, V_b - a_{l_2 \alpha}^{p_2} \, q_{s_1 p_2}^b \, V_b - a_{s_1 \alpha}^{p_1} \, q_{p_1 l_2}^b \, V_b = 0,$$

$$q_{s_1 l_2}^c \, a_{c\alpha}^b = a_{s_1 \alpha}^{p_1} \, q_{p_1 l_2}^b + a_{l_2 \alpha}^{p_2} \, q_{s_1 p_2}^b. \tag{I.18.4}$$

$$[[X_{s_1} V_c] Y_\alpha] + [[Y_\alpha X_{s_1}] V_c] + [[V_c Y_\alpha] X_{s_1}] = 0,$$
$$t^{p_2}_{s_1 c}[X_{p_2} Y_\alpha] - a^{p_1}_{s_1\alpha}[X_{p_1} V_c] + a^b_{c\alpha}[V_b X_{s_1}] = 0,$$
$$t^{p_2}_{s_1 c} a^{k_2}_{p_2\alpha} X_{k_2} - a^{p_1}_{s_1\alpha} t^{k_2}_{p_1 c} X_{k_2} - a^b_{c\alpha} t^{k_2}_{s_1 b} X_{k_2} = 0,$$

$$a^b_{c\alpha} t^{k_2}_{s_1 b} = t^{p_2}_{s_1 c} a^{k_2}_{p_2\alpha} - a^{p_1}_{s_1\alpha} t^{k_2}_{p_1 c}. \tag{I.18.5}$$

Now we introduce the notations:

$$a^{p_1}_{k_1\alpha} = a^p_{k\alpha}, \quad b^\alpha_{k_1 l_1} = b^\alpha_{kl}, \quad q^a_{k_2 l_1} = q^a_{kl}, \quad t^{p_2}_{s_1 a} = t^p_{sa}. \tag{I.19}$$

The conditions (I.17) then take the form:

$$b^\alpha_{s_2 l_2} = \varphi^\alpha_\beta b^\beta_{sl}, \quad q^b_{sl} = \varphi^b_a q^a_{ls}, \quad t^{l_1}_{s_2 b} = -\varphi^a_b t^l_{sa},$$
$$a^{p_2}_{s_2\alpha} = \varphi^\beta_\alpha a^p_{s\beta}, \quad b^\alpha_{ac} = \varphi^f_a \varphi^d_c \varphi^\alpha_\beta b^\beta_{fd}, \tag{I.20}$$
$$a^b_{c\alpha} = \varphi^d_c \varphi^b_f \varphi^\beta_\alpha a^f_{d\beta}, \quad c^\gamma_{\alpha\beta} = \varphi^\lambda_\alpha \varphi^\mu_\beta \varphi^\gamma_\nu c^\nu_{\lambda\mu}.$$

Owing to the orthonormality of our involutive base, from (I.6) we obtain:

$$a^s_{l\alpha} = b^\alpha_{ls}, \quad q^a_{kl} = t^k_{la}, \quad t^{l_1}_{s_2 b} = -t^s_{lb},$$
$$a^f_{c\alpha} = b^\alpha_{cf}, \quad c^\alpha_{\beta\gamma} = -c^\beta_{\alpha\gamma}. \tag{I.21}$$

In the new notations (I.18.1)–(I.18.5), (I.21) takes the form :

$$a^p_{k\alpha} \varphi^\alpha_\beta b^\beta_{sl} = t^p_{lc} q^c_{ks} - t^p_{sc} q^c_{kl}, \tag{I.22.1}$$

$$-b^\alpha_{ac} a^p_{s\alpha} = t^l_{sc} t^p_{lb} \varphi^b_a - t^l_{sa} t^p_{lb} \varphi^b_c, \tag{I.22.2}$$

$$-q^c_{ls} b^\alpha_{ca} = t^p_{sa} b^\beta_{pl} \varphi^\alpha_\beta + t^p_{lc} b^\alpha_{ps} \varphi^c_a, \tag{I.22.3}$$

$$q^c_{ls} a^b_{c\alpha} = q^b_{lp} a^p_{s\alpha} + q^b_{ps} a^p_{l\beta} \varphi^\beta_\alpha, \tag{I.22.4}$$

$$a^b_{c\alpha} t^k_{sb} = t^p_{sc} a^k_{p\beta} \varphi^\beta_\alpha - t^k_{pc} a^p_{s\alpha}. \tag{I.22.5}$$

However, it is easy to see that owing to (I.21) the relations (I.22.3), (I.22.4), (I.22.5) are equivalent. Taking (I.16) also into account we obtain from (I.20) and (I.21):

$$\left.\begin{array}{c} q^{\hat b}_{sl} = -q^{\hat b}_{ls}, \quad q^{b.}_{sp} = q^{b.}_{ps}, \quad t^s_{l\hat b} = -t^l_{s\hat b}, \\[4pt] t^s_{lb.} = t^l_{sb.}, \quad b^{\hat\alpha}_{\hat a\hat c} = 0, \quad b^{\hat\alpha}_{a.\ c.} = 0, \quad a^{\hat b}_{\hat c\hat\alpha} = 0, \\[4pt] a^{b.}_{c.\ \hat\alpha} = 0, \quad b^{\alpha.}_{\hat a\hat c} = 0, \quad a^{\hat b}_{c.\ \alpha.} = 0, \quad c^{\hat\gamma}_{\hat\alpha\hat\beta} = 0, \\[4pt] c^{\hat\gamma}_{\alpha.\ \beta.} = 0, \quad c^{\alpha.}_{\hat\beta.\ \hat\gamma} = 0. \end{array}\right\} \tag{I.23}$$

Now we may write (I.22.1), (I.22.2), (I.22.5) in a more detailed form:

$$a^{\acute{p}}_{k\alpha}\, b^{\alpha}_{sl} = 2\, a^{p}_{k\hat{\alpha}}\, b^{\hat{\alpha}}_{sl} + t^{p}_{l\hat{\alpha}}\, q^{\hat{\alpha}}_{ks} - t^{p}_{s\hat{\alpha}}\, q^{\hat{\alpha}}_{kl} + t^{p}_{l\,a.}\, q^{a.}_{ks} - t^{p}_{s\,a.}\, q^{a.}_{kl}\,, \tag{I.24.1}$$

$$-b^{\alpha.}_{\hat{a}\hat{c}}\, a^{p}_{s\,\alpha.} = t^{p}_{l\hat{c}}\, t^{l}_{s\hat{a}} - t^{p}_{l\hat{a}}\, t^{l}_{s\hat{c}}\,, \tag{I.24.2}$$

$$-b^{\hat{\alpha}}_{a.\,\hat{c}}\, a^{p}_{s\hat{\alpha}} = t^{p}_{l\,a.}\, t^{l}_{s\hat{c}} + t^{p}_{l\hat{c}}\, t^{l}_{s\,a.}\,, \tag{I.24.3}$$

$$-b^{\alpha.}_{a.\,c.}\, a^{p}_{s\alpha.} = t^{p}_{l\,a.}\, t^{l}_{s\,c.} - t^{p}_{l\,c.}\, t^{l}_{sa.}\,, \tag{I.24.4}$$

$$-a^{b.}_{\hat{c}\hat{\alpha}}\, t^{k}_{sb.} = t^{k}_{p\hat{c}}\, a^{p}_{s\hat{\alpha}} + a^{k}_{p\hat{\alpha}}\, t^{p}_{s\hat{c}}\,, \tag{I.24.5}$$

$$a^{\hat{b}}_{\hat{c}\,\alpha.}\, t^{k}_{s\hat{b}} = a^{k}_{p\,\alpha.}\, t^{p}_{s\hat{c}} - t^{k}_{p\hat{c}}\, a^{p}_{s\alpha.}\,, \tag{I.24.6}$$

$$-a^{\hat{b}}_{c.\,\hat{\alpha}}\, t^{k}_{s\hat{b}} = t^{k}_{p\,c.}\, a^{p}_{s\hat{\alpha}} + a^{k}_{p\hat{\alpha}}\, t^{p}_{s\,c.}\,, \tag{I.24.7}$$

$$a^{b.}_{c.\,\alpha.}\, t^{k}_{sb.} = a^{k}_{p\,\alpha.}\, t^{p}_{s\,c.} - t^{k}_{p\,c.}\, a^{p}_{s\,\alpha.}\,. \tag{I.24.8}$$

I.4.5. We obtain more relations by means of an automorphism $\theta = \varphi^{1/2}$. Clearly $\theta X_{s_1} = \lambda^l_s X_{l_1} + \mu^l_s X_{l_2}$, and from $\theta\varphi = \varphi\theta$ and (I.16) we have

$$\theta X_{s_2} = \varphi\,(\theta X_{s_1}) = -\mu^l_s X_{l_1} + \lambda^l_s X_{l_2}.$$

By I.3.10 (Lemma 1), I.3.11 (Lemma 2), and I.3.13 (Theorem 9) we also have $S_1\theta = \theta^{-1} S_1$, therefore

$$S_1\theta X_{s_1} = \theta^{-1} S_1 X_{s_1}\,, \qquad S_1\theta X_{s_2} = \theta^{-1} S_1 X_{s_2}.$$

Since $S_1 X_{s_1} = X_{s_1}$, $S_1 X_{s_2} = -X_{s_2}$, we arrive at

$$\theta^{-1} X_{s_1} = \lambda^l_s\, X_{l_1} - \mu^l_s\, X_{l_2}\,,$$

$$\theta^{-1} X_{s_2} = \mu^l_s\, X_{l_1} + \lambda^l_s\, X_{l_2}.$$

Furthermore,

$$X_{s_1} = \theta\,\bigl(\theta^{-1} X_{s_1}\bigr) = \lambda^l_s\,(\theta X_{l_1}) - \mu^l_s\,(\theta X_{l_2})$$
$$= \lambda^l_s\,(\lambda^p_l X_{p_1} + \mu^p_l X_{p_2}) - \mu^l_s\,(-\mu^p_l X_{p_1} + \lambda^p_l X_{p_2})$$

and

$$\lambda^p_l \lambda^l_s + \mu^p_l \mu^l_s = \delta^p_s\,, \qquad \mu^p_l \lambda^l_s - \lambda^p_l \mu^l_s = 0. \tag{I.25}$$

In addition, from (I.16) and $\theta^2 = \varphi$ we obtain

$$X_{s_2} = \varphi X_{s_1} = \theta\,(\theta X_{s_1}) = \lambda^l_s\,(\theta X_{l_1}) + \mu^l_s\,(\theta X_{l_2})$$
$$= \lambda^l_s\,(\lambda^p_l X_{p_1} + \mu^p_l X_{p_2}) + \mu^l_s\,(-\mu^p_l X_{p_1} + \lambda^p_l X_{p_2})$$

Thus

$$\lambda_l^p \lambda_s^l - \mu_l^p \mu_s^l = 0, \quad \mu_l^p \lambda_s^l + \lambda_l^p \mu_s^l = \delta_s^p. \tag{I.26}$$

From (I.25) and (I.26) it follows that

$$\lambda_l^p \lambda_s^l = \mu_l^p \mu_s^l = \mu_l^p \lambda_s^l = \lambda_l^p \mu_s^l = \frac{1}{2} \delta_s^p,$$

and consequently $\lambda_s^k = \mu_s^k = (1/\sqrt{2})\,\theta_s^k$, where $\theta_l^p \theta_s^l = \delta_s^p$.

But in addition to that the automorphism θ preserves the metric (I.4), which in an orthonormal iso-involutive base implies:

$$\begin{aligned}
\delta_{kl} &= (X_{k_1} X_{l_1}) = ((\theta X_{k_1})(\theta X_{l_1})) \\
&= \frac{1}{2} \theta_k^p \theta_l^s ((X_{p_1} + X_{p_2})(X_{s_1} + X_{s_2})) \\
&= \frac{1}{2} \theta_k^p \theta_l^s (2\,\delta_{ps}) = \sum_{p=1}^{n} \theta_k^p \theta_l^p.
\end{aligned}$$

Thus θ_l^k is an orthogonal matrix, and consequently in an orthogonal involutive base we have $\theta_l^k = \theta_k^l$ (that is, it is symmetric).

As a result:

$$\begin{aligned}
\theta X_{k_1} &= \frac{1}{\sqrt{2}} \theta_k^l (X_{l_1} + X_{l_2}), \\
\theta X_{k_1} &= \frac{1}{\sqrt{2}} \theta_k^l (-X_{l_1} + X_{l_2}), \\
\theta_k^p \theta_s^k &= \delta_s^p, \quad \theta_l^k = \theta_k^l.
\end{aligned} \tag{I.27}$$

By the construction of an iso-involutive sum we have in $\mathfrak{l}_3$, after an appropriate choice of the base,

$$\begin{aligned}
\theta Y_{\hat{\alpha}} = V_{\hat{\alpha}}, \quad \theta V_{\hat{a}} = -Y_{\hat{a}}, \quad \theta Y_{a.} = \theta_{a.}^{\beta.} Y_{\beta.}, \quad \theta V_{a.} = \theta_{a.}^{c.} V_{c.}, \\
\theta_{\alpha.}^{\lambda} \theta_{\mu.}^{\alpha.} = \delta_{\mu.}^{\lambda}, \quad \theta_{a.}^{c.} \theta_{b.}^{a.} = \delta_{b.}^{c.}, \quad \theta_{\alpha.}^{\beta.} = \theta_{\beta.}^{\alpha.}, \quad \theta_{a.}^{c.} = \theta_{c.}^{a.}.
\end{aligned} \tag{I.28}$$

I.4.6. But θ is an automorphism of the structure (I.15), and with the help of (I.27) and (I.28) we obtain

$$[(\theta X_{s_1})(\theta X_{l_1})] = -b_{sl}^{\alpha} (\theta Y_\alpha),$$

$$\begin{aligned}
-b_{sl}^{\hat{\alpha}} V_{\hat{\alpha}} &- b_{sl}^{\alpha.} \theta_{\alpha.}^{\beta.} Y_{\beta.} \\
&= \frac{1}{2} \theta_s^p \theta_l^k \left[(X_{p_1} + X_{p_2})(X_{k_1} + X_{k_2}) \right] \\
&= \frac{1}{2} \theta_s^p \theta_l^k ([X_{p_1} X_{k_1}] + [X_{p_2} X_{k_2}] + [X_{p_2} X_{k_1}] - [X_{k_2} X_{p_1}]) \\
&= \frac{1}{2} \theta_s^p \theta_l^k (-2 b_{pk}^{\beta.} Y_{\beta.} + 2 q_{pk}^{\hat{\alpha}} V_{\hat{\alpha}}),
\end{aligned}$$

$$b_{sl}^{\alpha.} \theta_{\alpha.}^{\beta.} = \theta_s^p \theta_l^k b_{pk}^{\beta.}, \quad b_{sl}^{\hat{\alpha}} = -\theta_s^p \theta_l^k q_{pk}^{\hat{\alpha}}. \tag{I.29.1}$$

$$\left[(\theta X_{s_2})\, (\theta X_{l_1}) \right] = q_{sl}^a\, (\theta V_a),$$

$$-q_{sl}^{\hat{a}} Y_{\hat{a}} + q_{sl}^{a\cdot}\, \theta_{a\cdot}^{b\cdot} V_b.$$

$$= \frac{1}{2}\, \theta_s^p\, \theta_l^k \left[(-X_{p_1} + X_{p_2})\, (X_{k_1} + X_{k_2}) \right]$$

$$= \frac{1}{2}\, \theta_s^p\, \theta_l^k \left(-\left[X_{p_1} X_{k_1} \right] + \left[X_{p_2} X_{k_2} \right] + \left[X_{p_2} X_{k_1} \right] + + \left[X_{k_2} X_{p_1} \right] \right)$$

$$= \frac{1}{2}\, \theta_s^p\, \theta_l^k \left(2\, b_{pk}^{\hat{a}}\, Y_{\hat{a}} + 2\, q_{pk}^{b\cdot}\, V_b. \right),$$

$$q_{sl}^{a\cdot}\, \theta_{a\cdot}^{b\cdot} = \theta_s^p\, \theta_l^k\, q_{pk}^{b\cdot}\,. \tag{I.29.2}$$

$$\left[(\theta X_{l_1})\, (\theta V_{\hat{a}}) \right] = t_{l\hat{a}}^p\, (\theta X_{p_2}),$$

$$-\frac{1}{\sqrt{2}}\, \theta_l^k \left[(X_{k_1} + X_{k_2})\, Y_{\hat{a}} \right] = -\frac{1}{\sqrt{2}}\, \theta_l^k \left(a_{k\hat{a}}^s X_{s_1} - a_{k\hat{a}}^s X_{s_2} \right)$$

$$= \frac{1}{\sqrt{2}}\, t_{l\hat{a}}^p\, \theta_p^s \left(-X_{s_1} + X_{s_2} \right),$$

$$t_{l\hat{a}}^p = \theta_s^p\, \theta_l^k\, a_{k\hat{a}}^s. \tag{I.29.3}$$

$$\left[(\theta X_{l_1})\, (\theta V_{a\cdot}) \right] = t_{la\cdot}^p\, (\theta X_{p_2}),$$

$$\frac{1}{\sqrt{2}}\, \theta_l^k\, \theta_{a\cdot}^{b\cdot} \left[(X_{k_1} + X_{k_2})\, V_b. \right] = \frac{1}{\sqrt{2}}\, \theta_l^k\, \theta_{a\cdot}^{b\cdot} \left(t_{kb\cdot}^s X_{s_2} - t_{kb\cdot}^s X_{s_1} \right)$$

$$= \frac{1}{\sqrt{2}}\, t_{la\cdot}^p\, \theta_p^s \left(-X_{s_1} + X_{s_2} \right),$$

$$\theta_k^l\, \theta_p^s\, t_{la\cdot}^p = \theta_{a\cdot}^{b\cdot}\, t_{kb\cdot}^s\,. \tag{I.29.4}$$

Furthermore, $\left[(\theta X_{p_2})\, (\theta V_{\hat{a}}) \right] = t_{p\hat{a}}^k\, (\theta X_{k_1})$, but applying an automorphism φ to both sides and using $\varphi\theta = \theta\varphi$, we see that it is equivalent to (I.29.3). Analogously $\left[(\theta X_{p_2})\, (\theta V_{a\cdot}) \right] = -t_{pa\cdot}^k\, (\theta X_{k_1})$ is equivalent to (I.29.4).

$$\left[(\theta X_{l_1})\, (\theta Y_{\hat{a}}) \right] = a_{l\hat{a}}^p\, (\theta X_{p_1}),$$

$$\frac{1}{\sqrt{2}}\, a_{l\hat{a}}^p\, \theta_p^k (X_{k_1} + X_{k_2}) = \frac{1}{\sqrt{2}}\, \theta_l^s \left[(X_{s_1} + X_{s_2})\, V_{\hat{a}} \right]$$

$$= \frac{1}{\sqrt{2}}\, \theta_l^s \left(t_{s\hat{a}}^k X_{k_2} + t_{s\hat{a}}^k X_{k_1} \right),$$

which is satisfied by virtue of (I.29.3).

$\left[(\theta X_{l_2}) \, (\theta Y_{\hat{\alpha}}) \right] = -a^p_{l\hat{\alpha}} \, (\theta X_{p_2})$ is equivalent to the above if we take into account $\varphi \, \theta = \theta \, \varphi$.

$$\left[(\theta X_{l_1}) \, (\theta Y_{\alpha.}) \right] = a^p_{l\,\alpha.} \, (\theta X_{p_1}),$$

$$\frac{1}{\sqrt{2}} \, \theta^k_l \, \theta^\beta_{\alpha.} \left[(X_{k_1} + X_{k_2}) \, Y_{\beta.} \right]$$

$$= \frac{1}{\sqrt{2}} \, \theta^k_l \, \theta^\beta_{\alpha.} \, (a^s_{k\,\beta.} \, X_{s_1} + a^s_{k\,\beta.} \, X_{s_2})$$

$$= \frac{1}{\sqrt{2}} \, a^p_{l\,\alpha.} \, \theta^s_p \, (X_{s_1} + X_{s_2}),$$

$$\theta^k_l \, \theta^p_s \, a^s_{k\beta.} = \theta^\alpha_{\beta.} \, a^p_{l\alpha.} \; . \tag{I.29.5}$$

$\left[(\theta X_{l_2}) \, (\theta Y_{\alpha.}) \right] = a^p_{l\alpha.} \, (\theta X_{p_2})$ is equivalent to the preceding if we use $\varphi \, \theta = \theta \, \varphi$.

$$\left[(\theta V_{\hat{\alpha}}) \, (\theta V_{\hat{\beta}}) \right] = -b^\sigma_{\hat{\alpha}\hat{\beta}} \, (\theta Y_{\sigma.}),$$

$$-b^\sigma_{\hat{\alpha}\hat{\beta}} \, \theta^\lambda_{\sigma.} \, Y_\lambda = \left[Y_{\hat{\alpha}} \, Y_{\hat{\beta}} \right] = c^\lambda_{\hat{\alpha}\hat{\beta}} \, Y_\lambda \, ,$$

$$c^\lambda_{\hat{\alpha}\hat{\beta}} = -b^\sigma_{\hat{\alpha}\hat{\beta}} \, \theta^\lambda_{\sigma.} \; . \tag{I.29.6}$$

$$\left[(\theta V_{a.}) \, (\theta V_{\hat{\beta}}) \right] = -b^{\hat{\sigma}}_{a.\,\hat{\beta}} \, (\theta Y_{\hat{\sigma}}),$$

$$-b^{\hat{\sigma}}_{a.\,\hat{\beta}} \, V_{\hat{\sigma}} = -\theta^c_{a.} \left[V_{c.} \, Y_{\hat{\beta}} \right] = -\theta^c_{a.} \, a^{\hat{\sigma}}_{c.\,\hat{\beta}} \, V_{\hat{\sigma}},$$

$$b^{\hat{\sigma}}_{a.\,\hat{\beta}} = \theta^c_{a.} \, a^{\hat{\sigma}}_{c.\,\hat{\beta}} \; . \tag{I.29.7}$$

$$\left[(\theta V_{a.}) \, (\theta V_{c.}) \right] = -b^{\alpha.}_{a.\,c.} \, (\theta Y_{\alpha.}),$$

$$-b^{\alpha.}_{a.\,c.} \, \theta^\beta_{\alpha.} \, Y_{\beta.} = \theta^e_{a.} \, \theta^d_{c.} \left[V_{e.} \, V_{d.} \right] = -\theta^e_{a.} \, \theta^d_{c.} \, b^\beta_{e.\,d.} \, Y_{\beta.} \, ,$$

$$\theta^e_{a.} \, \theta^d_{c.} \, b^\beta_{e.\,d.} = b^{\alpha.}_{a.\,c.} \, \theta^\beta_{\alpha.} \; . \tag{I.29.8}$$

$$\left[(\theta V_{\hat{c}}) \, (\theta Y_{\hat{\alpha}}) \right] = a^b_{\hat{c}\hat{\alpha}} \, (\theta V_{b.}),$$

$$a^b_{\hat{c}\hat{\alpha}} \, \theta^e_{b.} \, V_{e.} = \left[V_{\hat{\alpha}} \, Y_{\hat{c}} \right] = a^e_{\hat{\alpha}\hat{c}} \, V_{e.} \, ,$$

$$a^e_{\hat{\alpha}\hat{c}} = \theta^e_{b.} \, a^b_{\hat{c}\hat{\alpha}} \; . \tag{I.29.9}$$

$$\left[(\theta V_{\hat{e}})\, (\theta Y_{\alpha.})\right] = a_{\hat{e}\alpha.}^{\hat{b}}\, (\theta V_{\hat{b}}),$$

$$-a_{\hat{e}\,\alpha.}^{\hat{b}}\, Y_{\hat{b}} = -\theta_{\alpha.}^{\beta.}\left[Y_{\hat{e}}\, Y_{\beta.}\right] = -\theta_{\alpha.}^{\beta.}\, c_{\hat{e}\beta.}^{\hat{b}}\, Y_{\hat{b}},$$

$$a_{\hat{e}\,\alpha.}^{\hat{b}} = \theta_{\alpha.}^{\beta.}\, c_{\hat{e}\beta.}^{\hat{b}}\, . \tag{I.29.10}$$

$$\left[(\theta V_{e.})\, (\theta Y_{\hat{\alpha}})\right] = a_{e.\,\hat{\alpha}}^{\hat{b}}\, (\theta V_{\hat{b}}),$$

$$-a_{e.\,\hat{\alpha}}^{\hat{b}}\, Y_{\hat{b}} = \theta_{e.}^{c.}\left[V_{c.}\, V_{\hat{\alpha}}\right] = -\theta_{e.}^{c.}\, b_{c.\,\hat{\alpha}}^{\hat{b}}\, Y_{\hat{b}},$$

which is true owing to (I.29.7).

$$\left[(\theta V_{c.})\, (\theta Y_{\alpha.})\right] = a_{c.\,\alpha.}^{b.}\, (\theta V_{b.}),$$

$$a_{c.\,\alpha.}^{b.}\, \theta_{b.}^{e.}\, V_{e.} = \theta_{c.}^{d.}\, \theta_{\alpha.}^{\beta.}\left[V_{d.}\, Y_{\beta.}\right] = \theta_{c.}^{d.}\, \theta_{\alpha.}^{\beta.}\, a_{d.\,\beta.}^{e.}\, V_{e.},$$

$$a_{c.\,\alpha.}^{b.}\, \theta_{b.}^{e.} = \theta_{c.}^{d.}\, \theta_{\alpha.}^{\beta.}\, a_{d.\,\beta.}^{e.}\, . \tag{I.29.11}$$

$$\left[(\theta Y_{\hat{\alpha}})\, (\theta Y_{\hat{\beta}})\right] = c_{\hat{\alpha}\hat{\beta}}^{\sigma.}\, (\theta Y_{\sigma.}),$$

$$c_{\hat{\alpha}\hat{\beta}}^{\sigma.}\, \theta_{\sigma.}^{\lambda}\, Y_{\lambda} = \left[V_{\hat{\alpha}}\, V_{\hat{\beta}}\right] = -b_{\hat{\alpha}\hat{\beta}}^{\lambda}\, Y_{\lambda},$$

which is valid owing to (I.29.6).

$$\left[(\theta Y_{\hat{\alpha}})\, (\theta Y_{\beta.})\right] = c_{\hat{\alpha}\,\beta.}^{\hat{\sigma}}\, (\theta Y_{\hat{\sigma}}),$$

$$c_{\hat{\alpha}\,\nu.}^{\hat{\sigma}}\, V_{\hat{\sigma}} = \theta_{\nu.}^{\lambda}\left[V_{\hat{\alpha}}\, Y_{\lambda}\right] = \theta_{\nu.}^{\lambda}\, a_{\hat{\alpha}\lambda}^{\hat{\sigma}}\, V_{\hat{\sigma}},$$

which is satisfied because of (I.29.10).

$$\left[(\theta Y_{\alpha.})\, (\theta Y_{\beta.})\right] = c_{\alpha.\,\beta.}^{\lambda}\, (\theta Y_{\lambda}),$$

$$c_{\alpha.\,\beta.}^{\lambda}\, \theta_{\lambda}^{\mu.}\, Y_{\mu.} = \theta_{\alpha.}^{\sigma.}\, \theta_{\beta.}^{\nu.}\left[Y_{\sigma.}\, Y_{\nu.}\right] = \theta_{\alpha.}^{\sigma.}\, \theta_{\beta.}^{\nu.}\, c_{\sigma.\,\nu.}^{\mu.}\, Y_{\mu.},$$

$$c_{\alpha.\,\beta.}^{\lambda}\, \theta_{\lambda}^{\mu.} = \theta_{\alpha.}^{\sigma.}\, \theta_{\beta.}^{\nu.}\, c_{\sigma.\,\nu.}^{\mu.}\, . \tag{I.29.12}$$

However, by virtue of (I.21) and (I.28) we take into consideration only (I.29.3), (I.29.4), (I.29.5), (I.29.6), (I.29.7), (I.29.9), (I.29.10), (I.29.11), (I.29.12), since the others are only their consequences. By virtue of (I.29) some relations in (I.24) are redundant, and in (I.24) we may regard only (I.24.1), (I.24.2), (I.24.4), (I.24.5), (I.24.6), (I.24.7), (I.24.8).

Now we are interested in the Jacobi identities for the basis vectors $X_{i_1}, Y_{\alpha.}, Y_{\hat{\beta}}$ which have not yet been considered. We obtain

$$a^s_{k\gamma}\, b^\gamma_{pl} + a^s_{l\gamma}\, b^\gamma_{kp} + a^s_{p\gamma}\, b^\gamma_{lk} = 0, \tag{I.30.1}$$

$$b^\beta_{pl}\, a^p_{k\alpha} + b^\beta_{kp}\, a^p_{l\alpha} = c^\beta_{\gamma\alpha}\, b^\gamma_{kl}, \tag{I.30.2}$$

$$a^l_{p\beta}\, a^p_{s\alpha} - a^l_{p\alpha}\, a^p_{s\beta} = c^\gamma_{\alpha\beta}\, a^l_{s\gamma}, \tag{I.30.3}$$

$$c^\mu_{\lambda\alpha}\, c^\lambda_{\beta\gamma} + c^\mu_{\lambda\gamma}\, c^\lambda_{\alpha\beta} + c^\mu_{\lambda\beta}\, c^\lambda_{\gamma\alpha} = 0. \tag{I.30.4}$$

By (I.21) the relations (I.30.2) are a consequence of (I.30.3), and (I.30.1) are satisfied by virtue of (I.24.1), (I.23), (I.29.3). Thus only (I.30.3) and (I.30.4) have to be regarded. Let us write (27.3) in more detailed form:

$$a^l_{p\hat{\beta}}\, a^p_{s\hat{\alpha}} - a^l_{p\hat{\alpha}}\, a^p_{s\hat{\beta}} = c^{\sigma.}_{\hat{\alpha}\hat{\beta}}\, a^l_{s\sigma.}\ , \tag{I.31.1}$$

$$a^l_{p\hat{\beta}}\, a^p_{s\alpha.} - a^l_{p\ \alpha.}\, a^p_{s\hat{\beta}} = -c^{\hat{\sigma}}_{\hat{\beta}\alpha.}\, a^l_{s\hat{\sigma}}\ , \tag{I.31.2}$$

$$a^l_{p\ \beta.}\, a^p_{s\ \alpha.} - a^l_{p\ \alpha.}\, a^p_{s\ \beta.} = c^{\sigma.}_{\alpha.\ \beta.}\, a^l_{s\ \sigma.}\ . \tag{I.31.3}$$

It is now easily verified that (I.24.2) follows from (I.31.1) and (I.29.6), analogously (I.24.6) follows from (I.31.2), (I.29.10).

The Jacobi identities for $X_{p_2}, Y_{\alpha.}, Y_{\hat{\beta}}$ follow from the Jacobi identities for X_{p_1}, $Y_{\alpha.}, Y_{\hat{\beta}}$ by virtue of the action of the automorphism φ. Finally, since $\mathfrak{g}/\mathfrak{l}_1$ is an elementary involutive pair the Jacobi identities for V_a, Y_α are the consequences of the relations already obtained, since for the elementary pair $\mathfrak{g}/\mathfrak{l}_1$ the restriction $\mathrm{Int}_\mathfrak{g}(\mathfrak{l}_3)$ of the adjoint group $\mathrm{Int}(\mathfrak{g})$ from $\mathfrak{g}$ to $\mathfrak{l}_3$ has a faithful linear representation on $\mathfrak{m}_1 \dotplus \mathfrak{m}_2 = \mathfrak{g} - \mathfrak{l}_3$, whose structure has already been taken into account by the relations written before. In particular, for this reason (I.30.4) follows from the other relations.

We are now going to give some account of what has been obtained above. We have considered a compact semi-simple Lie algebra $\mathfrak{g} = \mathfrak{l}_1 + \mathfrak{l}_2 + \mathfrak{l}_3$, where $\mathfrak{l}_1$, $\mathfrak{l}_2, \mathfrak{l}_3$ are involutive algebras of an iso-involutive decomposition, with respect to an involutive base (I.14) and with the structure (I.15). An automorphism φ has acted on $\mathfrak{g}$ according to (I.16), which has implied (I.20). The orthonormality of the iso-involutive base has implied (I.21) and (I.23). Moreover, the automorphism $\theta = \varphi^{1/2}$ (see (I.27)) has generated additional relations (see (I.29)) for the structure (I.15):

$$t_{l\hat{\alpha}}^{p} = \theta_{s}^{p}\,\theta_{l}^{k}\,a_{k\hat{\alpha}}^{s}\,, \tag{I.32.1}$$

$$\theta_{k}^{l}\,\theta_{p}^{s}\,t_{la.}^{p} = \theta_{a.}^{b.}\,t_{k\,b.}^{s}\,, \tag{I.32.2}$$

$$\theta_{l}^{k}\,\theta_{s}^{p}\,a_{k\ \beta.}^{s} = \theta_{\beta.}^{\alpha.}\,a_{l\ \alpha.}^{p}\,, \tag{I.32.3}$$

$$c_{\hat{\alpha}\hat{\beta}}^{\lambda} = -b_{\hat{\alpha}\hat{\beta}}^{\sigma.}\,\theta_{\sigma.}^{\lambda}\,, \tag{I.32.4}$$

$$b_{a.\ \beta}^{\hat{\sigma}} = \theta_{a.}^{c.}\,a_{c.\ \beta}^{\hat{\sigma}}\,, \tag{I.32.5}$$

$$a_{\hat{\alpha}\hat{c}}^{e.} = \theta_{a.}^{e.}\,a_{\hat{c}\hat{\alpha}}^{a.}\,, \tag{I.32.6}$$

$$a_{\hat{e}\,\alpha.}^{\hat{b}} = \theta_{\alpha.}^{\beta.}\,c_{\hat{e}\hat{\beta}.}^{\hat{b}}\,, \tag{I.32.7}$$

$$a_{c.\ \alpha.}^{a.}\,\theta_{a.}^{e.} = \theta_{c.}^{d.}\,\theta_{\alpha.}^{\beta.}\,a_{d.\ \beta.}^{e.}\,, \tag{I.32.8}$$

$$c_{\alpha.\ \beta.}^{\lambda}\,\theta_{\lambda}^{\mu.} = \theta_{\alpha.}^{\sigma.}\,\theta_{\beta.}^{\nu.}\,c_{\sigma.\ \nu.}^{\mu.}\,. \tag{I.32.9}$$

Finally, the Jacobi identities for (I.15) in an iso-involutive base (I.14) give us (I.24), (I.30), from which, in the case of an elementary involutive pair $\mathfrak{g}/\mathfrak{l}_1$, there are essential and equivalent to the Jacobi identities only the following relations

$$a_{q\alpha}^{p}\,b_{sl}^{\alpha} = 2\,a_{q\hat{\alpha}}^{p}\,b_{sl}^{\hat{\alpha}} + t_{l\hat{\alpha}}^{p}\,q_{qs}^{\hat{\alpha}} - t_{s\hat{\alpha}}^{p}\,q_{ql}^{\hat{\alpha}} + t_{l\,a.}^{p}\,q_{qs}^{a.} - t_{s\,a.}^{p}\,q_{ql}^{a.}\,, \tag{I.33.1}$$

$$-b_{a.\ c.}^{\alpha.}\,a_{s\alpha.}^{p} = t_{l\,a.}^{p}\,t_{s\,c.}^{l} - t_{l\,c.}^{p}\,t_{s\,a.}^{l}\,, \tag{I.33.2}$$

$$-a_{\hat{c}\hat{\alpha}}^{b.}\,t_{s\,b.}^{k} = t_{p\hat{c}}^{k}\,a_{s\hat{\alpha}}^{p} + a_{p\hat{\alpha}}^{k}\,t_{s\hat{c}}^{p}\,, \tag{I.33.3}$$

$$-a_{c.\ \hat{\alpha}}^{\hat{b}}\,t_{s\hat{b}}^{k} = t_{p\,c.}^{k}\,a_{s\hat{\alpha}}^{p} + a_{p\hat{\alpha}}^{k}\,t_{sc.}^{p}\,, \tag{I.33.4}$$

$$a_{c.\ \alpha.}^{b.}\,t_{s\,b.}^{k} = a_{p\,\alpha.}^{k}\,t_{s\,c.}^{p} - t_{p\,c.}^{k}\,a_{s\,\alpha.}^{p}\,, \tag{I.33.5}$$

$$c_{\hat{\alpha}\hat{\beta}}^{\sigma.}\,a_{k\,\sigma.}^{l} = a_{s\hat{\beta}}^{l}\,a_{k\hat{\alpha}}^{s} - a_{s\hat{\alpha}}^{l}\,a_{k\hat{\beta}}^{s}\,, \tag{I.33.6}$$

$$-c_{\hat{\beta}\,\alpha.}^{\hat{\sigma}}\,a_{k\hat{\sigma}}^{l} = a_{s\hat{\beta}}^{l}\,a_{k\alpha.}^{s} - a_{s\,\alpha.}^{l}\,a_{k\hat{\beta}}^{s}\,, \tag{I.33.7}$$

$$c_{\alpha.\ \beta.}^{\sigma.}\,a_{k\,\sigma.}^{l} = a_{s\,\beta.}^{l}\,a_{k\,\alpha.}^{s} - a_{s\,\alpha.}^{l}\,a_{k\,\beta.}^{s}\,. \tag{I.33.8}$$

I.4.7. Definition 23. An iso-involutive group $\chi(S_1, S_2, S_3\,;\,\varphi)$ and the corresponding iso-involutive sum not of type 1 is said to be of type 2 if

$$\theta|_{\mathfrak{g}\dot{-}\mathfrak{l}_3} = \pm\frac{1}{\sqrt{2}}\,(\mathrm{Id} + \varphi|_{(\mathfrak{g}\dot{-}\mathfrak{l}_3)}),$$

and of type 3 otherwise.

I.4.8. Remark. For an iso-involutive sum of type 2 by (I.27) we have

$$\theta_{s}^{p} = \pm\delta_{s}^{p} \quad (p, s = 1, \dots, n).$$

CHAPTER I.5

ISO-INVOLUTIVE SUMS OF TYPES 1 AND 2

I.5.1. Let us consider first an iso-involutive sum of type 1 for a compact semi-simple Lie algebra $\mathfrak{g}$. There are no vectors $Y_{\hat{a}}$, $Y_{\hat{a}}$ in this case, thus we may write Y_α instead of $Y_{\alpha.}$ and Y_a instead of $Y_{a.}$.

Then (I.30) takes the form

$$a_{k\alpha}^p \, b_{ij}^\alpha = t_{ja}^p \, q_{ki}^a - t_{ia}^p \, q_{kj}^a, \tag{I.34.1}$$

$$-b_{ac}^\alpha \, a_{i\alpha}^p = t_{ja}^p \, t_{ic}^j - t_{jc}^p \, t_{ia}^j, \tag{I.34.2}$$

$$a_{c\alpha}^b \, t_{ib}^k = a_{p\alpha}^k \, t_{ic}^p - t_{pc}^k \, a_{i\alpha}^p, \tag{I.34.3}$$

$$c_{\alpha\beta}^\sigma \, a_{j\sigma}^l = a_{i\beta}^l \, a_{j\alpha}^i - a_{i\alpha}^l \, a_{j\beta}^i. \tag{I.34.4}$$

Contracting p and j in (I.34.1) and taking into account (I.23) and (I.21), we obtain

$$\sum_{\alpha,\,p} a_{k\alpha}^p \, a_{i\alpha}^p = t_{pa}^p \, q_{ki}^a - \sum_{p,\,a} t_{ia}^p \, t_{ka}^p,$$

which means that the vector $t_a = a_{pa}^p$ is different from 0.

But then, contracting k and i in (I.34.3), we have $a_{c\alpha}^b t_b = 0$, which means that t_b generates the non-trivial centre in $\mathfrak{l}_3$. We have obtained:

I.5.2. Theorem 12. *Let $\mathfrak{g} = \mathfrak{l}_1 + \mathfrak{l}_2 + \mathfrak{l}_3$ be an iso-involutive decomposition of type 1 for a compact semi-simple Lie algebra $\mathfrak{g}$. Then the involutive algebra $\mathfrak{l}_3$ possesses a non-trivial centre $\mathfrak{z} \subset \mathfrak{m}_3 = \mathfrak{l}_3 \dot- \mathfrak{l}_0$.*

If, furthermore, the involutive pair $\mathfrak{g}/\mathfrak{l}_1$ is elementary then any non-zero $\zeta = \zeta^a V_a \in \mathfrak{z}$ generates $t_j^i = \zeta^a t_{ja}^i$, and (I.34.2), (I.34.3) imply commutativity of t_j^i with t_{qa}^p and $a_{q\alpha}^p$. Thus $t_j^i = t\delta_j^i$, otherwise $\mathrm{Int}_{\mathfrak{g}}^{(\mathfrak{g}\dot-\mathfrak{l}_3)}(\mathfrak{l}_3)$ and $\mathrm{Int}_{\mathfrak{g}}^{(\mathfrak{l}_1\dot-\mathfrak{l}_0)}(\mathfrak{l}_0)$ are reducible, which is impossible for an elementary involutive pair $\mathfrak{g}/\mathfrak{l}_1$.

Thus

$$\mathrm{ad}_{\mathfrak{g}}^{(\mathfrak{g}\dot-\mathfrak{l}_3)}(\mathfrak{z}) = \left\{ t\left(\varphi|_{(\mathfrak{g}\dot-\mathfrak{l}_3)}\right) \right\}.$$

Since in our case $\mathrm{ad}_{\mathfrak{g}}^{(\mathfrak{g}\dot-\mathfrak{l}_3)}(\mathfrak{l}_3)$ is a faithful representation for $\mathfrak{l}_3$ then we have $\dim \mathfrak{z} = 1$ because $\mathrm{ad}_{\mathfrak{g}}^{(\mathfrak{g}\dot-\mathfrak{l}_3)}(\mathfrak{z})$ is one-dimensional.

I.5.3. Theorem 13. *Let* $\mathfrak{g} = \mathfrak{l}_1 + \mathfrak{l}_2 + \mathfrak{l}_3$ *be an iso-involutive sum of type 1. If* $\mathfrak{g}$ *is compact and* $\mathfrak{g}/\mathfrak{l}_1$ *is an elementary involutive pair, then* $\mathfrak{l}_3$ *has the unique one-dimensional centre* $\mathfrak{z} \subset \mathfrak{l}_3 \dot{-} \mathfrak{l}_0$. *Moreover*

$$\mathrm{ad}_{\mathfrak{g}}^{(\mathfrak{g} \dot{-} \mathfrak{l}_3)}(\mathfrak{z}) = \{t\,(\varphi|_{(\mathfrak{g} \dot{-} \mathfrak{l}_3)})\}_{t \in \mathbb{R}}.$$

I.5.4. Theorem 14. *Let* $\mathfrak{g}$ *be a compact Lie algebra,* $\mathfrak{g}/\mathfrak{l}_1$ *be an elementary involutive pair, and let* $\mathfrak{g} = \mathfrak{l}_1 + \mathfrak{l}_2 + \mathfrak{l}_3$ *be an iso-involutive sum of lower index 1 and of type 1, then*

$$\mathfrak{g}/\mathfrak{l}_1 \cong so(n+1)/so(n), \qquad \mathfrak{g}/\mathfrak{l}_2 \cong so(n+1)/so(n),$$

$$\mathfrak{g}/\mathfrak{l}_3 \cong so(n+1)/so(n-1) \oplus so(2),$$

$$\mathfrak{l}_1/\mathfrak{l}_0 \cong so(n)/so(n-1), \qquad \mathfrak{l}_2/\mathfrak{l}_0 \cong so(n)/so(n-1),$$

$$\mathfrak{l}_3/\mathfrak{l}_0 \cong so(2)/\{0\} + so(n-1)/so(n-1)$$

with the natural embeddings.

Proof. By I.5.2 (Theorem 12) $\dim(\mathfrak{l}_3 \dot{-} \mathfrak{l}_0) = 1$. By I.5.3 (Theorem 13) from (I.34) only the following relations are essential:

$$a_{k\alpha}^p\, b_{ij}^\alpha = \rho^2 \left(\delta_j^p\, \delta_{ki} - \delta_i^p\, \delta_{kj} \right),$$

$$a_{i\beta}^l\, a_{j\alpha}^i - a_{i\alpha}^l\, a_{j\beta}^i = c_{\alpha\beta}^\sigma\, a_{j\sigma}^l. \tag{I.34$'$}$$

Furthermore, using (I.34$'$) and the irreducibility of $\mathrm{ad}_{\mathfrak{g}}^{(\mathfrak{g} \dot{-} \mathfrak{l}_1)}(\mathfrak{l}_1)$ we verify that the curvature tensor for $\mathfrak{g}/\mathfrak{l}_1$ has the form indicated in I.2.5 (Theorem 2). Thus

$$\mathfrak{g}/\mathfrak{l}_1 \cong so(n+1)/so(n)$$

with the natural embedding.

Analogously, (I.34$'$) implies that for $\mathfrak{l}_1/\mathfrak{l}_0$ the curvature tensor has the form indicated in I.2.5 (Theorem 2) Thus

$$\mathfrak{l}_1/\mathfrak{l}_0 \cong so(n)/so(n-1)$$

with the natural embedding.

The other relations are true because $\mathfrak{l}_1$ and $\mathfrak{l}_2$ are conjugate in $\mathfrak{g}$. ∎

I.5.5. Corollary 2. *Under the assumptions of I.5.4 (Theorem 14) the Lie algebra* $\mathfrak{g}$ *is not simple only if* $\mathfrak{g}/\mathfrak{l}_1 \cong so(4)/so(3)$.

This follows from the well known result that $so(n)$, $n \geq 2$, is not simple only for $n = 4$.

I.5.6. Now we consider involutive sums of type 2. As before, we consider only elementary compact involutive pairs $\mathfrak{g}/\mathfrak{l}_1$ and compact algebras $\mathfrak{g}$. From (I.32.2) and (I.32.3) it follows that $\theta_{a.}^{b.} = \delta_{a.}^{b.}$, $\theta_{\alpha.}^{\nu.} = \delta_{\alpha.}^{\nu.}$; otherwise $\mathfrak{l}_3$ contains an ideal of $\mathfrak{g}$, which is impossible for an elementary $\mathfrak{g}/\mathfrak{l}_1$.

The conditions (I.32) and (I.33) take now a simpler form. Let us consider (I.33.3):

$$-a_{\hat{c}\hat{\alpha}}^{b.}\, t_{i\ b.}^{k} = a_{p\hat{c}}^{k}\, a_{i\hat{\alpha}}^{p} + a_{p\hat{\alpha}}^{k}\, a_{i\hat{c}}^{p}.$$

Contracting in it k, i and considering $\hat{c} = \hat{\alpha}$ we obtain

$$-a_{\hat{\alpha}\hat{\alpha}}^{b.}\, t_{k\ b.}^{k} = 2\sum_{k,p} \left(a_{p\hat{\alpha}}^{k}\right)^2,$$

where $t_{b.} = t_{k\ b.}^{k}$ is a non-zero vector; otherwise $a_{p\hat{\alpha}}^{k} = 0$ for all $k, p, \hat{\alpha}$, leading to the existence of an ideal of $\mathfrak{g}$ inside of $\mathfrak{l}_3$, which is impossible for an elementary involutive pair. And, furthermore, (I.33.5) implies $a_{c.\ \alpha.}^{b.}\, t_{b.} = 0$, which means that $\mathfrak{n}_3$ possesses a non-trivial central ideal $\mathfrak{z} \subset \mathfrak{q}_3 = \mathfrak{n}_3 - \mathfrak{n}_0$ (see (I.12)).

We have proved the theorem:

I.5.7. Theorem 15. *If a compact Lie algebra $\mathfrak{g}$ has an iso-involutive decomposition $\mathfrak{g} = \mathfrak{l}_1 + \mathfrak{l}_2 + \mathfrak{l}_3$ of type 2 and $\mathfrak{g}/\mathfrak{l}_1$ is elementary then the maximal subalgebra of elements immobile under the action of a conjugate automorphism φ has a non-trivial central ideal*

$$\mathfrak{z} \subset \mathfrak{q}_3 = \mathfrak{n}_3 - \mathfrak{n}_0 \quad (\mathfrak{n}_0 = \mathfrak{n}_3 \cap \mathfrak{l}_1).$$

I.5.8. Now let the iso-involutive sum $\mathfrak{g} = \mathfrak{l}_1 + \mathfrak{l}_2 + \mathfrak{l}_3$ be of lower index 1, then by I.5.7 (Theorem 15) $\mathfrak{q}_3$ is one-dimensional of lower index 1 and type 1. From this it follows easily that $a_{c.\ \hat{\alpha}}^{b.} = \lambda \delta_{\hat{\alpha}}^{b.}$ (since $\mathfrak{l}_0 + \mathfrak{n}_2 + \mathfrak{n}_3 = \mathfrak{l}_3$, where $\tilde{\mathfrak{l}}_3/\tilde{\mathfrak{l}}_0$ is elementary, and because of I.5.3 (Theorem 13)).

Since $\mathfrak{q}_3$ is one-dimensional there is the only one $V_{a.}$ and instead of $t_{j\ a.}^{i}$ we may write simply t_j^i. Then the relations (I.32) take the form:

$$t_{i\hat{\alpha}}^{p} = a_{i\hat{\alpha}}^{p}, \quad c_{\hat{\alpha}\hat{\beta}}^{\lambda.} = -b_{\hat{\alpha}\hat{\beta}}^{\lambda.}, \quad b_{a.\ \hat{\beta}}^{\hat{\sigma}} = a_{a.\ \hat{\beta}}^{\hat{\sigma}},$$

$$a_{\hat{\alpha}\hat{c}}^{e.} = a_{\hat{c}\hat{\alpha}}^{e.}, \quad a_{\hat{e}\ \alpha.}^{\hat{b}} = c_{\hat{e}\ \alpha.}^{\hat{b}}.$$

(I.35)

$$a_{k\alpha}^{p}\, b_{sl}^{\alpha} = 2\, a_{k\hat{\alpha}}^{p}\, b_{sl}^{\hat{\alpha}} + a_{l\hat{\alpha}}^{p}\, b_{sk}^{\hat{\alpha}} - a_{s\hat{\alpha}}^{p}\, b_{lk}^{\hat{\alpha}} + t_{l}^{p}\, q_{ks} - t_{s}^{p}\, q_{kl},$$

(I.36.1)

$$\lambda \delta_{\hat{c}\hat{\alpha}}\, t_{s}^{k} = a_{p\hat{c}}^{k}\, a_{s\hat{\alpha}}^{p} + a_{p\hat{\alpha}}^{k}\, a_{s\hat{c}}^{p},$$

(I.36.2)

$$-\lambda\, a_{s\hat{\alpha}}^{k} = t_{p}^{k}\, a_{s\hat{\alpha}}^{p} + a_{p\hat{\alpha}}^{k}\, t_{s}^{p},$$

(I.36.3)

$$a_{p\,\alpha.}^{k}\, t_{s}^{p} - t_{p}^{k}\, a_{s\,\alpha.}^{p} = 0,$$

(I.36.4)

$$c^{\sigma.}_{\hat{\alpha}\hat{\beta}}\, a^l_{p\sigma.} = a^l_{s\hat{\beta}}\, a^s_{p\hat{\alpha}} - a^l_{s\hat{\alpha}}\, a^s_{p\hat{\beta}}\,, \tag{I.36.5}$$

$$-c^{\hat{\sigma}}_{\hat{\beta}\,\alpha.}\, a^l_{p\hat{\sigma}} = a^l_{s\hat{\beta}}\, a^s_{p\,\alpha.} - a^l_{s\,\alpha.}\, a^s_{p\hat{\beta}}\,, \tag{(33.6)}$$

$$c^{\sigma.}_{\alpha.\,\beta.}\, a^l_{p\sigma.} = a^l_{s\,\beta.}\, a^s_{p\,\alpha.} - a^l_{s\,\alpha.}\, a^s_{p\,\beta.}\,. \tag{I.36.7}$$

For an elementary involutive pair it follows easily from (I.35.2), (I.35.3), (I.35.4) that $t^k_i = -(\lambda/2)\,\delta^k_i$, that is,

$$\mathrm{ad}^{(\mathfrak{g}\dot- \mathfrak{l}_3)}_{\mathfrak{g}}\,(\{t\,\zeta\}) = \{t\,(\varphi|_{(\mathfrak{g}\dot- \mathfrak{l}_3)})\}.$$

Thus we have obtained:

I.5.9. Theorem 16. *Let $\mathfrak{g}$ be compact, $\mathfrak{g}/\mathfrak{l}_1$ be an elementary involutive pair, and let $\mathfrak{g} = \mathfrak{l}_1 + \mathfrak{l}_2 + \mathfrak{l}_3$ be the iso-involutive sum of an iso-involutive group $\chi(S_1, S_2, S_3; \varphi)$ of lower index 1 and of type 2. Then*

$$\mathrm{ad}^{(\mathfrak{g}\dot- \mathfrak{l}_3)}_{\mathfrak{g}}\,(\{t\,\zeta\}) = \{t\,(\varphi|_{(\mathfrak{g}\dot- \mathfrak{l}_3)})\}.$$

I.5.10. Because of what has been proved above, instead of (I.36), we have

$$a^p_{k\alpha}\, b^{\alpha}_{ij} = 2\,a^p_{k\hat{\alpha}}\, b^{\hat{\alpha}}_{ij} + a^p_{j\hat{\alpha}}\, b^{\hat{\alpha}}_{ik} - a^p_{i\hat{\alpha}}\, b^{\hat{\alpha}}_{jk} + + \frac{\lambda^2}{4}\left(\delta^p_j\,\delta_{ki} - \delta^p_i\,\delta_{kj}\right), \tag{I.37.1}$$

$$-\frac{\lambda^2}{2}\,\delta_{\hat{c}\hat{\alpha}}\,\delta^k_i = a^k_{p\hat{c}}\, a^p_{i\hat{\alpha}} + a^k_{p\hat{\alpha}}\, a^p_{i\hat{c}}\,, \tag{I.37.2}$$

$$c^{\sigma.}_{\hat{\alpha}\hat{\beta}}\, a^l_{j\,\sigma.} = a^l_{i\hat{\beta}}\, a^i_{j\hat{\alpha}} - a^l_{i\hat{\alpha}}\, a^i_{j\hat{\beta}}\,, \tag{I.37.3}$$

$$-c^{\hat{\sigma}}_{\hat{\beta}\,\alpha.}\, a^l_{j\hat{\sigma}} = a^l_{i\hat{\beta}}\, a^i_{j\,\alpha.} - a^l_{i\,\alpha.}\, a^i_{j\hat{\beta}}\,, \tag{I.37.4}$$

$$c^{\sigma.}_{\alpha.\,\beta.}\, a^l_{j\,\sigma.} = a^l_{i\,\beta.}\, a^i_{j\,\alpha.} - a^l_{i\,\alpha.}\, a^i_{j\,\beta.}\,. \tag{I.37.5}$$

Furthermore, one can examine the relations (I.37) in general. However, we indicate only two important particular subcases (which almost solves the problem in the general case).

I.5.11. Let $\dim(\mathfrak{l}_3 \dot- \mathfrak{l}_0) = 2$, then $\mathfrak{m}_0 = \mathfrak{l}_0 \dot- \mathfrak{n}_0$ is one-dimensional and there are only one $Y_{\hat{\alpha}}$ and only one $V_{\hat{a}}$.
Thus we write a^p_k instead of $a^p_{k\hat{\alpha}}$ and b_{ij} instead of $b^{\hat{\alpha}}_{ij}$.
Then the relations (I.37) take the form

$$a^p_{k\alpha}\, b^{\alpha}_{ij} = 2\,a^p_k\, b_{ij} + a^p_j\, b_{ik} - a^p_i\, b_{jk} + \frac{\lambda^2}{4}\left(\delta^p_j\,\delta_{ki} - \delta^p_i\,\delta_{kj}\right), \tag{I.38.1}$$

$$a_p^k \, a_i^p = -\frac{\lambda^2}{2} \, \delta_i^k, \qquad\qquad (\text{I.38.2})$$

$$a_i^l \, a_{j\,\alpha}^i - a_{i\,\alpha}^l \, a_j^i = 0, \qquad\qquad (\text{I.38.3})$$

$$a_{i\,\beta}^l \, a_{j\,\alpha}^i - a_{i\,\alpha}^l \, a_{j\,\beta}^i = c_{\alpha.\ \beta.}^{\sigma.} \, a_{j\,\sigma}^l \, . \qquad\qquad (\text{I.38.4})$$

The relation (I.38.1) gives us the form of the curvature tensor of the involutive pairs $\mathfrak{l}_1/\mathfrak{l}_0$ and $\mathfrak{l}_2/\mathfrak{l}_0$ which coincides with the form presented in I.2.6 (Theorem 3) (after some re-normalization of the base).

Furthermore, $\mathfrak{l}_3/\mathfrak{l}_0 = \tilde{\mathfrak{l}}_3/\tilde{\mathfrak{l}}_0 + \tilde{\tilde{\mathfrak{l}}}_0/\tilde{\mathfrak{l}}_0$, where $\tilde{\mathfrak{l}}_3/\tilde{\mathfrak{l}}_0$ is elementary. This follows from the property that $\mathfrak{l}_3/\mathfrak{l}_0$ is an involutive pair of lower index 1. Since $\mathfrak{l}_3$ is an iso-involutive sum of index 1, $\tilde{\mathfrak{l}}_3$ is an iso-involutive sum of index 1 as well, and by I.5.4 (Theorem 14) $\tilde{\mathfrak{l}}_3/\tilde{\mathfrak{l}}_0 \cong so(n+1)/so(n)$. But $\tilde{\mathfrak{l}}_3/\tilde{\mathfrak{l}}_0 \cong so(3)/so(2) \cong su(2)/u(1)$ because $\dim(\tilde{\mathfrak{l}}_3 \dot{-} \tilde{\mathfrak{l}}_0) = 2$, and consequently $\mathfrak{l}_3/\mathfrak{l}_0$ has the form of the curvature tensor indicated in I.2.6 (Theorem 3).

Having used, furthermore, (I.35), (I.38) and the irreducibility of $\mathrm{ad}_{\mathfrak{g}}^{(\mathfrak{g}\dot{-}\mathfrak{l}_1)}(\mathfrak{l}_1)$, by a straightforward verification one obtains that the curvature tensor for $\mathfrak{g}/\mathfrak{l}_1$ has the form indicated in I.2.6 (Theorem 3), whence it follows that $\mathfrak{g}/\mathfrak{l}_1 \cong su(n+1)/su(n) \oplus u(1)$ with a natural embedding. The same is true for $\mathfrak{g}/\mathfrak{l}_2$, since $\mathfrak{l}_1$ and $\mathfrak{l}_2$ are conjugated in $\mathrm{Int}(\mathfrak{g})$.

Furthermore, by (I.38.1) and I.2.6 (Theorem 3) we obtain

$$\mathfrak{l}_1/\mathfrak{l}_0 \cong su(n)/su(n-1) \oplus u(1) + u(1)/u(1)$$

and by the conjugacy of $\mathfrak{l}_1$ and $\mathfrak{l}_2$,

$$\mathfrak{l}_2/\mathfrak{l}_0 \cong su(n)/su(n-1) \oplus u(1) + u(1)/u(1),$$

all with the natural embeddings.

Thus we obtain the theorem:

I.5.12. Theorem 17. *Let $\mathfrak{g}$ be a compact Lie algebra, $\mathfrak{g}/\mathfrak{l}_1$ be an elementary involutive pair, and let $\mathfrak{g} = \mathfrak{l}_1 + \mathfrak{l}_2 + \mathfrak{l}_3$ be an iso-involutive sum of type 2 and of lower index 1, and $\dim(\mathfrak{l}_3 \dot{-} \mathfrak{l}_0) = 2$. Then*

$$\mathfrak{g}/\mathfrak{l}_1 \cong su(n+1)/su(n) \oplus u(1),$$

$$\mathfrak{g}/\mathfrak{l}_2 \cong su(n+1)/su(n) \oplus u(1),$$

$$\mathfrak{g}/\mathfrak{l}_3 \cong su(n+1)/su(n-1) \oplus su(2) \oplus u(1),$$

$$\mathfrak{l}_1/\mathfrak{l}_0 \cong su(n)/su(n-1) \oplus u(1) + u(1)/u(1),$$

$$\mathfrak{l}_2/\mathfrak{l}_0 \cong su(n)/su(n-1) \oplus u(1) + u(1)/u(1),$$

$$\mathfrak{l}_3/\mathfrak{l}_0 \cong su(2)/u(1) + su(n-1) \oplus u(1)/su(n-1) \oplus u(1)$$

with the natural embeddings.

I.5.13. Now let $\dim(\mathfrak{l}_3 \dot{-} \mathfrak{l}_0) = 4$, then $\dim \mathfrak{q}_1 = \dim \mathfrak{q}_2 = 3$. The relation (I.37.1) gives us the form of the curvature tensor of the involutive pairs $\mathfrak{l}_1/\mathfrak{l}_0$, $\mathfrak{l}_2/\mathfrak{l}_0$ coinciding with the form indicated in I.2.7 (Theorem 4) (after some renormalization of a base).

Furthermore, $\mathfrak{l}_3/\mathfrak{l}_0 = \tilde{\mathfrak{l}}_3/\tilde{\mathfrak{l}}_0 + \tilde{\tilde{\mathfrak{l}}}_0/\tilde{\tilde{\mathfrak{l}}}_0$, where $\tilde{\mathfrak{l}}_3/\tilde{\mathfrak{l}}_0$ is elementary. This is because $\mathfrak{l}_3/\mathfrak{l}_0$ is an involutive pair of lower index 1. Since $\mathfrak{l}_3$ is an involutive sum of type 1 $\tilde{\mathfrak{l}}_3$ is an iso-involutive sum of type 1, as well.

Then by I.5.4 (Theorem 14) $\tilde{\mathfrak{l}}_3/\tilde{\mathfrak{l}}_0 \cong so(n+1)/so(n)$. But $\dim(\mathfrak{l}_3 \dot{-} \mathfrak{l}_0) = 4$, which implies

$$\tilde{\mathfrak{l}}_3/\tilde{\mathfrak{l}}_0 \cong so(5)/so(4) \cong sp(2)/sp(1) \oplus sp(1).$$

And $\mathfrak{l}_3/\mathfrak{l}_0$ has the form of the curvature tensor indicated in I.2.7 (Theorem 4).

Having used, furthermore, (I.35), (I.37) and the irreducibility of $\mathrm{ad}_{\mathfrak{g}}^{(\mathfrak{g} \dot{-} \mathfrak{l}_1)}(\mathfrak{l}_1)$, in a straightforward way we verify that for $\mathfrak{g}/\mathfrak{l}_1$ the curvature tensor has the form indicated in I.2.7 (Theorem 4). Whence it follows that

$$\mathfrak{g}/\mathfrak{l}_1 \cong sp(n+1)/sp(n) \oplus sp(1)$$

with the natural embedding.

The same is true for $\mathfrak{g}/\mathfrak{l}_2$ by the conjugacy of $\mathfrak{l}_1$ and $\mathfrak{l}_2$ in $\mathrm{Int}(\mathfrak{g})$. Furthermore, by virtue of I.2.7 (Theorem 4)

$$\mathfrak{l}_1/\mathfrak{l}_0 \cong sp(n)/sp(n-1) \oplus sp(1) + sp(1)/sp(1)$$

and, by the conjugacy of $\mathfrak{l}_1$ and $\mathfrak{l}_2$

$$\mathfrak{l}_2/\mathfrak{l}_0 \cong sp(n)/sp(n-1) \oplus sp(1) + sp(1)/sp(1)$$

(all with the natural embeddings.)

Thus we have:

I.5.14. Theorem 18. *Let $\mathfrak{g}$ be a compact Lie algebra, $\mathfrak{g}/\mathfrak{l}_1$ be an elementary involutive pair, and let $\mathfrak{g} = \mathfrak{l}_1 + \mathfrak{l}_2 + \mathfrak{l}_3$ be an involutive sum of type 2 and of lower index 1, and $\dim(\mathfrak{l}_3 \dot{-} \mathfrak{l}_0) = 4$. Then*

$$\mathfrak{g}/\mathfrak{l}_1 \cong sp(n+1)/sp(n) \oplus sp(1),$$

$$\mathfrak{g}/\mathfrak{l}_2 \cong sp(n+1)/sp(n) \oplus sp(1),$$

$$\mathfrak{g}/\mathfrak{l}_3 \cong sp(n+1)/sp(n-1) \oplus sp(2),$$

$$\mathfrak{l}_1/\mathfrak{l}_0 \cong sp(n)/sp(n-1) \oplus sp(1) + sp(1)/sp(1),$$

$$\mathfrak{l}_2/\mathfrak{l}_0 \cong sp(n)/sp(n-1) \oplus sp(1) + sp(1)/sp(1),$$

$$\mathfrak{l}_3/\mathfrak{l}_0 \cong sp(2)/sp(1) \oplus sp(1) + sp(n-1)/sp(n-1)$$

with the natural embeddings.

CHAPTER I.6

ISO-INVOLUTIVE SUMS OF LOWER INDEX 1

Let us prove first the theorem:

I.6.1. Theorem 19. *Let $\mathfrak{g}$ be a simple compact Lie algebra, and $\mathfrak{g} = \mathfrak{l}_1 + \mathfrak{l}_2 + \mathfrak{l}_3$ be an iso-involutive sum of lower index 1 and not of type 1 generated by an iso-involutive group $\chi(S_1, S_2, S_3; \varphi)$. Then $\chi^{(1)}(S_1, S_2, S_3; \varphi)$ and its corresponding iso-involutive sum $\mathfrak{l}_3 = \mathfrak{l}_0 + \mathfrak{n}_2 + \mathfrak{n}_3$ is of lower index 1 and of type 1 (with the conjugating isomorphism $\theta \in \mathrm{Int}_\mathfrak{g}(\{t\zeta\})$, $\theta^2 = \varphi$).*

Proof. Let us consider the sequence of automorphisms from $\mathrm{Int}_\mathfrak{g}(\{t\zeta\})$:

$$S_3 = \theta_{-2}, \qquad \varphi = \theta_{-1}, \qquad \theta = \theta_0, \theta_1, \theta_2, \ldots, \theta_k, \ldots$$

such that $(\theta_{k+1})^2 = \theta_k$. We denote the subspace of all elements of $\mathfrak{g}$ immobile under the action of θ_k by $\mathfrak{p}_k$. Then

$$\mathfrak{l}_3 = \mathfrak{p}_{-2} \supset \mathfrak{n}_3 = \mathfrak{p}_{-1} \supset \mathfrak{p}_0 \supset \mathfrak{p}_1 \supset \ldots \supset \mathfrak{p}_n = \mathfrak{p}_{n+1}.$$

Evidently all these subspaces are invariant under the action of S_1; thus we have the following involutive pairs of the involutive automorphism S_1:

$$\mathfrak{l}_3/\mathfrak{l}_0, \ \mathfrak{n}/\mathfrak{n}_0, \ \mathfrak{p}_0/\mathfrak{t}_0, \ \mathfrak{p}_1/\mathfrak{t}_1, \ \ldots, \mathfrak{p}_k/\mathfrak{t}_k, \ \ldots, \mathfrak{p}_n/\mathfrak{t}_n,$$

where

$$\mathfrak{l}_0 = \mathfrak{t}_{-2} \supset \mathfrak{n}_0 = \mathfrak{t}_{-1} \supset \mathfrak{t}_0 \supset \ldots \supset \mathfrak{t}_n = \mathfrak{t}_{n+1}.$$

Since $\zeta \in \mathfrak{t}_k$ and $S_1 \zeta = -\zeta$ all these involutive pairs are of lower index 1. Let us consider the involutive pair $\mathfrak{p}_{n-1}/\mathfrak{t}_{n-1}$ of the involutive automorphism S_1. The restriction of θ_{n-1} to $\mathfrak{p}_{n-1}$ is the identity automorphism and the restriction of θ_n to $\mathfrak{p}_{n-1}$ is an involutive automorphism of $\mathfrak{p}_{n-1}$ with the involutive algebra $\mathfrak{p}_n$. But $(\theta_{n+1})^2 = \theta_n$ and $\theta_{n+1} \in \mathrm{Int}_\mathfrak{g}(\{t\zeta\})$; then the restriction of θ_{n+1} to $\mathfrak{p}_{n-1}$ is the conjugate automorphism of the iso-involutive sum $\mathfrak{p}_{n-1} = \mathfrak{t}_{n-1} + \mathfrak{t}'_{n-1} + \mathfrak{p}_n$, and the restriction of θ_{n+1} to $\mathfrak{p}_n$ is the identity automorphism since, by construction, $\mathfrak{p}_n = \mathfrak{p}_{n+1}$. Consequently the preceding iso-involutive sum is of type 1 and of lower index 1, as it has been pointed out earlier. Hence, $\mathfrak{p}_n = \mathfrak{t}_n \oplus \{t\zeta\}$, see I.5.2 (Theorem 12). The linear operator $\mu(x) = [\zeta[\zeta x]]$ is self-dual, whence all its proper values are reals.

Note that $\mathfrak{k}_{n-1} = \mathfrak{t}_{n-1} \dot{-} \mathfrak{t}_n$ and $\mathfrak{k}'_{n-1} = \mathfrak{t}'_{n-1} \dot{-} \mathfrak{t}_n$ are invariant under the action of μ.

Let us take $a \in \mathfrak{k}_{n-1}$ such that $\mu(a) = \lambda a$ $(\lambda \in \mathbb{R})$, and $b = \left[\zeta a \right] \in \mathfrak{k}'_{n-1}$. Then $\left[ab \right] = \sigma \zeta$, $\left[\zeta b \right] = \lambda a$, $\left[\zeta a \right] = b$, and $b \neq 0$ (otherwise $a \in \mathfrak{p}_n$, which is impossible), and $\lambda \neq 0$ (otherwise $b \in \mathfrak{p}_n$, which is impossible). But a, b, ζ are pair-wise orthogonal in the Cartan metric of $\mathfrak{g}$, thus $\sigma \neq 0$, and we obtain the simple compact algebra $\mathfrak{b} = \{ \alpha a + \beta b + \gamma \zeta \}$. Owing to our construction, $\theta_{n+1} a = \tau b$, $\theta_{n+1} b = \nu a$, $\theta_{n+1} \zeta = \zeta$, whence the restriction of $\theta_n = (\theta_{n+1})^2$ to $\mathfrak{b}$ is an involutive automorphism $J \neq \mathrm{Id}$.

Indeed,

$$\theta_{n+1} \mathfrak{k}_{n-1} = \mathfrak{k}'_{n-1}, \qquad \theta_{n+1} \mathfrak{k}'_{n-1} = \mathfrak{k}_{n-1}, \qquad \theta_{n+1} \in \mathrm{Int}\,_{\mathfrak{g}} (\{ t \zeta \}).$$

For this reason $\theta_{n+1} \mathfrak{b} = \mathfrak{b}$. But then

$$\theta_{n+1} (\mathfrak{k}_{n-1} \cap \mathfrak{b}) = \mathfrak{k}'_{n-1} \cap \mathfrak{b}, \qquad \theta_{n+1} (\mathfrak{k}'_{n-1} \cap \mathfrak{b}) = \mathfrak{k}_{n-1} \cap \mathfrak{b}.$$

We have also

$$\mathfrak{k}_{n-1} \cap \mathfrak{b} = \{ t a \}, \qquad \mathfrak{k}'_{n-1} \cap \mathfrak{b} = \{ t b \}.$$

But $\mathrm{Int}_{\mathfrak{g}}(\mathfrak{b})$ is isomorphic to either $SO(3)$ or $SU(2)$, consequently the natural morphism

$$\mathrm{Int}_{\mathfrak{g}}(\mathfrak{b}) \longrightarrow \mathrm{Int}_{\mathfrak{g}}^{\mathfrak{b}}(\mathfrak{b}) = \mathrm{Int}(\mathfrak{b}) \cong SO(3)$$

maps θ_n into a non-trivial involutive automorphism from $SO(3)$, whence $(\theta_n)^2 = \theta_{n-1}$ is mapped into the identity involutive automorphism from $SO(3)$. This means that either $\theta_{n-1} = \mathrm{Id}$ or θ_{n-1} is a non-trivial involutive automorphism. In the first case $\theta_n = S_3$, $\mathfrak{l}_3 = \mathfrak{n}_3 = \mathfrak{l}_0 \oplus \{ t \zeta \}$, and then $\mathfrak{g} = \mathfrak{l}_1 + \mathfrak{l}_2 + \mathfrak{l}_3$ is an iso-involutive sum of type 1 and of lower index 1, which contradicts the conditions of the theorem. Thus the only possibility is that θ_{n-1} is non-trivial, then $\theta_{n-1} = S_3$, $\theta_n = \varphi$, $\theta_{n+1} = \theta$, $\mathfrak{l}_3 \supset \mathfrak{n}_3 = \mathfrak{p}_{-1}$. Furthermore, $\mathfrak{n}_3 = \mathfrak{n}_0 \oplus \{ t \zeta \}$, thus $\mathfrak{l}_3 = \mathfrak{l}_0 + \mathfrak{n}_2 + \mathfrak{n}_3$ is an iso-involutive sum of type 1 and lower index 1 with the conjugate automorphism $\theta \in \mathrm{Int}\,_{\mathfrak{g}} (\{ t \zeta \})$. ∎

I.6.2. Theorem 20. *Let $\mathfrak{g}$ be a simple compact Lie algebra, $\mathfrak{g} = \mathfrak{l}_1 + \mathfrak{l}_2 + \mathfrak{l}_3$ be an iso-involutive sum of lower index 1 and not of type 1. Then it is of type 2.*

Proof. Let us consider $\theta \in \mathrm{Int}\,_{\mathfrak{g}} (\{ t \zeta \})$. From I.6.1 (Theorem 19) it follows that the restriction of θ to $\mathfrak{n}_3$ is the identity automorphism,

$$\theta^{d.}_{c.} = \delta^{d.}_{c.}, \qquad \theta^{\alpha.}_{\nu.} = \delta^{\alpha.}_{\nu.}. \tag{I.39}$$

Since by I.6.1 (Theorem 19) $\mathfrak{l}_3 = \mathfrak{l}_0 + \mathfrak{n}_2 + \mathfrak{n}_3$ is an iso-involutive sum of lower index 1, and not of type 1, with the conjugating automorphism $\theta \in \mathrm{Int}\,_{\mathfrak{g}} (\{ t \zeta \})$, by I.5.3 (Theorem 13) and because of the canonical decomposition for $\mathfrak{l}_3 / \mathfrak{l}_0$ there follow, in an iso-involutive base,

$$\zeta = \mu V_{1.}, \qquad a^{1.}_{\hat{a}\,\hat{\alpha}} = \lambda \delta_{\hat{a}\,\hat{\alpha}}, \qquad a^{c.}_{\hat{a}\,\hat{\alpha}} = 0 \quad (c. \neq 1.), \quad \lambda \neq 0. \tag{I.40}$$

We suppose that $\theta^i_j \neq \pm \delta^i_j$ (i.e., that $\mathfrak{l}_1 + \mathfrak{l}_2 + \mathfrak{l}_3 = \mathfrak{g}$ is not of type 2) and take an iso-involutive base in such a way that θ^i_j has the diagonal form:

$$\theta^{\tilde{i}}_{\tilde{j}\tilde{i}} = \delta_{\tilde{j}\tilde{i}}, \qquad \theta^{\tilde{i}}_{\tilde{j}\tilde{i}} = -\delta_{\tilde{j}\tilde{i}}, \qquad \theta_{\tilde{j}\tilde{i}} = \theta_{\tilde{j}\tilde{i}} = 0,$$

$$\tilde{i} = \tilde{1}, \ldots, \tilde{n}; \quad \tilde{i} = \tilde{1}, \ldots, \tilde{n}; \quad \tilde{n} + \tilde{n} = n; \quad \tilde{n} \neq 0; \quad \tilde{n} \neq 0. \tag{I.41}$$

Furthermore, we use such an involutive base. We note that $t^i_{j\,1.}$ is a non-degenerate symmetric matrix. Indeed, if $t^i_{j\,1.}\,q^j = 0$, where not all q^j are zero, then

$$\left[\left(X_{j_1} q^j \right) V_{1.} \right] = 0$$

and we have

$$\mathrm{Int}_{\mathfrak{g}} \left(\{\,t\,V_{1.}\,\} \right) \left(X_{j_1} q^j \right) = X_{j_1} q^j,$$

whence $S_3 \left(X_{j_1} q^j \right) = X_{j_1} q^j$ and then $X_{j_1} q^j \in \mathfrak{l}_3$ (which is impossible). From (I.32) it follows, furthermore, that

$$t^{\tilde{p}}_{\tilde{i}\,\hat{\alpha}} = a^{\tilde{p}}_{\tilde{i}\,\hat{\alpha}}, \qquad t^{\tilde{p}}_{\tilde{i}\,\hat{\alpha}} = -a^{\tilde{p}}_{\tilde{i}\,\hat{\alpha}}, \qquad t^{\tilde{p}}_{\tilde{i}\,\hat{\alpha}} = a^{\tilde{p}}_{\tilde{i}\,\hat{\alpha}},$$

$$t^{\tilde{p}}_{\tilde{i}\,1.} = 0, \qquad a^{\tilde{j}}_{\tilde{i}\,\alpha.} = 0. \tag{I.42}$$

In particular, it shows that an iso-involutive base can be chosen in such a way that (I.39)–(I.42) are satisfied and, moreover, $t^i_{j\,1.}$ has a diagonal form. Furthermore, we use such choice of a base. Thus

$$t^i_{j\,1.} = 0 \quad (i \neq j), \qquad t^i_{j\,1.} \neq 0 \quad (i = j). \tag{I.43}$$

From (I.33) we obtain, taking into account (I.34)–(I.43):

$$-\lambda\,\delta_{\hat{c}\hat{\alpha}}\,t^k_{i\,1.} = t^k_{p\,\hat{c}}\,a^p_{i\,\hat{\alpha}} + a^k_{p\,\hat{\alpha}}\,t^p_{i\,\hat{c}}, \tag{I.44}$$

$$\lambda\,t^k_{i\,\hat{\alpha}} = t^k_{p\,1.}\,a^p_{i\,\hat{\alpha}} + a^k_{p\,\hat{\alpha}}\,t^p_{i\,1.}, \tag{I.45}$$

$$0 = a^k_{p\,\alpha.}\,t^p_{i\,1.} - t^k_{p\,1.}\,a^p_{i\,\alpha.}. \tag{I.46}$$

From (I.33.1) we have also, taking into account (I.34)–(I.43), and (I.21), (I.23):

$$a^{\tilde{p}}_{\tilde{k}\,\alpha.}\,b^{\alpha.}_{\tilde{s}\,\tilde{q}} = 0 = \sum_{\hat{\alpha}} a^{\tilde{p}}_{\tilde{k}\,\hat{\alpha}}\,a^{\tilde{q}}_{\tilde{s}\,\hat{\alpha}} + \sum_{\tilde{\alpha}} a^{\tilde{p}}_{\tilde{q}\,\hat{\alpha}}\,a^{\tilde{k}}_{\tilde{s}\,\hat{\alpha}} - \sum_{\hat{\alpha}} a^{\tilde{p}}_{\tilde{s}\,\hat{\alpha}}\,a^{\tilde{k}}_{\tilde{q}\,\hat{\alpha}} - t^{\tilde{p}}_{\tilde{s}\,1.}\,t^{\tilde{k}}_{\tilde{q}\,1.},$$

or

$$t^{\tilde{p}}_{\tilde{s}\,1.}\,t^{\tilde{k}}_{\tilde{q}\,1.} = -\sum_{\hat{\alpha}=\hat{1}}^{\hat{r}} \left(a^{\tilde{p}}_{\tilde{k}\,\hat{\alpha}}\,a^{\tilde{s}}_{\tilde{q}\,\hat{\alpha}} + a^{\tilde{p}}_{\tilde{q}\,\hat{\alpha}}\,a^{\tilde{s}}_{\tilde{k}\,\hat{\alpha}} \right) - \sum_{\hat{\alpha}=\hat{1}}^{\hat{r}} a^{\tilde{p}}_{\tilde{s}\,\hat{\alpha}}\,a^{\tilde{k}}_{\tilde{q}\,\hat{\alpha}},$$

whence by the symmetry of the left hand side we have:

$$\sum_{\hat{\alpha}=\hat{1}}^{\hat{r}} a^{\tilde{p}}_{\tilde{s}\,\hat{\alpha}}\,a^{\tilde{k}}_{\tilde{q}\,\hat{\alpha}} = 0, \tag{I.47}$$

$$t^{\tilde{p}}_{\tilde{s}1.}\, t^{\tilde{k}}_{\tilde{q}1.} = -\sum_{\hat{\alpha}=\hat{1}}^{\hat{r}} (a^{\tilde{p}}_{\tilde{k}\hat{\alpha}}\, a^{\tilde{s}}_{\tilde{q}\hat{\alpha}} + a^{\tilde{p}}_{\tilde{q}\hat{\alpha}}\, a^{\tilde{s}}_{\tilde{k}\hat{\alpha}}). \qquad (I.48)$$

From (I.48) we obtain

$$t^{\tilde{p}}_{\tilde{p}1.}\, t^{\tilde{k}}_{\tilde{q}1.} = -2 \sum_{\hat{\alpha}=\hat{1}}^{\hat{r}} a^{\tilde{p}}_{\tilde{k}\hat{\alpha}}\, a^{\tilde{p}}_{\tilde{q}\hat{\alpha}} \quad \text{(no summation over } \tilde{p}).$$

$$t^{\tilde{k}}_{\tilde{k}1.}\, t^{\tilde{p}}_{\tilde{s}1.} = -2 \sum_{\hat{\alpha}=\hat{1}}^{\hat{r}} a^{\tilde{p}}_{\tilde{k}\hat{\alpha}}\, a^{\tilde{s}}_{\tilde{k}\hat{\alpha}} \quad \text{(no summation over } \tilde{k}).$$

$$t^{\tilde{p}}_{\tilde{p}1.}\, t^{\tilde{k}}_{\tilde{k}1.} = -2 \sum_{\hat{\alpha}=\hat{1}}^{\hat{r}} (a^{\tilde{p}}_{\tilde{k}\hat{\alpha}})^2 \quad \text{(no summation over } \tilde{p},\ \tilde{k}).$$

Thus each of the matrices $t^{\tilde{k}}_{\tilde{q}1.}$ and $t^{\tilde{p}}_{\tilde{s}1.}$ has proper values of the same sign, whereas the proper values of one of them have an opposite sign of the proper values of the other,

$$t^{\tilde{p}}_{\tilde{p}1.}\, t^{\tilde{k}}_{\tilde{k}1.} < 0 \quad \text{(no summation over } \tilde{p} \text{ and } \tilde{k}). \qquad (I.49)$$

If also $\tilde{n} > \hat{r}$ then for fixed $\tilde{p}$ there is $\zeta^{\tilde{q}} = \zeta_{\tilde{q}}$ such that $\zeta^{\tilde{q}} a^{\tilde{p}}_{\tilde{q}\hat{\alpha}} = 0$, but then $t^{\tilde{p}}_{\tilde{p}1.}\, t^{\tilde{k}}_{\tilde{q}1.}\, \zeta^{\tilde{q}} = 0$, which is impossible since $t^{\tilde{k}}_{\tilde{q}1.}$ is non-degenerate. Analogously, $\tilde{\tilde{n}} > \hat{r}$ is impossible as well. Thus we have:

$$\tilde{n} \le \hat{r}, \qquad \tilde{\tilde{n}} \le \hat{r}. \qquad (I.50)$$

Introducing the notations

$$\frac{1}{\lambda} t^{k}_{s1.} = \tau^{k}_{s}, \qquad \frac{1}{\lambda} t^{k}_{s\hat{c}} = \tau^{k}_{s\hat{c}}, \qquad \frac{1}{\lambda} a^{k}_{s\hat{\beta}} = \alpha^{k}_{s\hat{\beta}}, \qquad (I.51)$$

we have

$$\tau^{\tilde{p}}_{\tilde{s}\hat{\beta}} = \alpha^{\tilde{p}}_{\tilde{s}\hat{\beta}}, \qquad \tau^{\tilde{p}}_{\tilde{s}\hat{\beta}} = -\alpha^{\tilde{p}}_{\tilde{s}\hat{\beta}}, \qquad \tau^{\tilde{p}}_{\tilde{s}\hat{\beta}} = \alpha^{\tilde{p}}_{\tilde{s}\hat{\beta}}, \qquad \tau^{\tilde{p}}_{\tilde{s}} = 0, \qquad \alpha^{\tilde{p}}_{\tilde{s}\alpha.} = 0, \quad (I.42')$$

$$\tau^{s}_{l} = \begin{cases} 0, & s \ne l, \\ \tau(s), & s = l, \end{cases} \quad \tau(s) \ne 0, \qquad (I.43')$$

$$\tau^{k}_{p\hat{c}}\, \alpha^{p}_{s\hat{\beta}} + \alpha^{k}_{p\hat{\beta}}\, \tau^{p}_{s\hat{c}} = -\delta_{\hat{c}\hat{\beta}}\, \tau^{k}_{s}, \qquad (I.44')$$

$$\tau^{k}_{p}\, \alpha^{p}_{s\hat{\beta}} + \alpha^{k}_{p\hat{\beta}}\, \tau^{p}_{s} = \tau^{k}_{s\hat{\beta}}, \qquad (I.45')$$

$$\alpha^{k}_{p\alpha.}\, \tau^{p}_{s} - \tau^{k}_{p}\, \alpha^{p}_{s\alpha.} = 0, \qquad (I.46')$$

$$\sum_{\hat{\beta}=\hat{1}}^{\hat{r}} \alpha_{\tilde{s}\hat{\beta}}^{\tilde{p}} \alpha_{\tilde{l}\hat{\beta}}^{\tilde{k}} = 0, \tag{I.47'}$$

$$\tau_{\tilde{s}}^{\tilde{p}} \tau_{\tilde{l}}^{\tilde{k}} = -\sum_{\hat{\beta}=\hat{1}}^{\hat{r}} (\alpha_{\tilde{k}\hat{\beta}}^{\tilde{p}} \alpha_{\tilde{l}\hat{\beta}}^{\tilde{s}} + \alpha_{\tilde{l}\hat{\beta}}^{\tilde{p}} \alpha_{\tilde{k}\hat{\beta}}^{\tilde{s}}). \tag{I.48'}$$

Evidently each of the matrices $\tau_{\tilde{s}}^{\tilde{p}}$ and $\tau_{\tilde{l}}^{\tilde{k}}$ has proper values of the same sign, whereas the proper values of one of them have an opposite sign of the proper values of the other, that is,

$$\tau_{\tilde{p}}^{\tilde{p}} \tau_{\tilde{k}}^{\tilde{k}} < 0 \quad \text{(no summation over } \tilde{p} \text{ and } \tilde{k}),$$

or, denoting $\tau_{\tilde{p}}^{\tilde{p}} = \tau(\tilde{p})$, $\tau_{\tilde{k}}^{\tilde{k}} = \tau(\tilde{k})$,

$$\tau(\tilde{p})\,\tau(\tilde{k}) < 0. \tag{I.49'}$$

Let $\tau(\tilde{p}) < 0$, then (I.45') implies

$$(\tau(\tilde{k}) + \tau(\tilde{p}) - 1)\,\alpha_{\tilde{k}\hat{\beta}}^{\tilde{p}} = 0.$$

But

$$(\tau(\tilde{k}) + \tau(\tilde{p}) - 1) < 0,$$

consequently $\alpha_{\tilde{k}\hat{\beta}}^{\tilde{p}} = 0$. But then

$$a_{\tilde{k}\hat{\beta}}^{\tilde{p}} = t_{\tilde{k}\hat{\beta}}^{\tilde{p}} = 0, \tag{I.49''}$$

and from (I.33.1) we obtain

$$a_{\tilde{k}\alpha.}^{\tilde{p}}\, b_{\tilde{s}\tilde{l}}^{\alpha.} = t_{\tilde{l}1.}^{\tilde{p}}\, q_{\tilde{k}\tilde{s}}^{1.} - t_{\tilde{s}1.}^{\tilde{p}}\, q_{\tilde{k}\tilde{l}}^{1.},$$

or

$$\sum_{\alpha.} b_{\tilde{k}\tilde{p}}^{\alpha.}\, b_{\tilde{s}\tilde{l}}^{\alpha.} = q_{\tilde{k}\tilde{s}}^{1.}\, q_{\tilde{l}\tilde{p}}^{1.} - q_{\tilde{k}\tilde{l}}^{1.}\, q_{\tilde{s}\tilde{p}}^{1.}. \tag{I.52}$$

If $t_{\tilde{l}1.}^{\tilde{p}}$ has different proper values then from (I.46) it follows that $(a_{\tilde{l}\alpha.}^{\tilde{s}}\, \zeta^{\tilde{l}}) \perp (\eta^{\tilde{s}})$, where $(\zeta^{\tilde{l}})$ and $(\eta^{\tilde{s}})$ are proper vectors for $t_{\tilde{l}1.}^{\tilde{p}}$ belonging to different proper values (which means $(\zeta^{\tilde{s}}) \perp (\eta^{\tilde{s}})$). This means that $b_{\tilde{k}\tilde{p}}^{\alpha.}\, \zeta^{\tilde{k}} \eta^{\tilde{p}} = 0$. Moreover, $(t_{\tilde{l}1.}^{\tilde{p}}\, \zeta^{\tilde{l}}) \perp (\eta^{\tilde{p}})$, thus $q_{\tilde{p}\tilde{l}}^{1.}\zeta^{\tilde{l}} \eta^{\tilde{p}} = 0$. Now from (I.52) we obtain

$$0 = \sum_{\alpha.} b_{\tilde{k}\tilde{p}}^{\alpha.}\, \zeta^{\tilde{k}} \eta^{\tilde{p}}\, b_{\tilde{s}\tilde{l}}^{\alpha.}\, \zeta^{\tilde{s}} \eta^{\tilde{l}} = (q_{\tilde{k}\tilde{s}}^{1.}\, \zeta^{\tilde{k}} \zeta^{\tilde{s}})(q_{\tilde{l}\tilde{p}}^{1.}\, \eta^{\tilde{l}} \eta^{\tilde{p}}),$$

which is impossible, since for a non-zero vector $\mu^{\tilde{s}}$,

$$q^{1.}_{\tilde{k}\,\tilde{s}}\,\mu^{\tilde{k}}\,\mu^{\tilde{s}} = \sum_{\tilde{s}} t^{\tilde{s}}_{\tilde{k}\,1.}\,\mu^{\tilde{k}}\,\mu^{\tilde{s}} = \lambda \sum_{\tilde{k}=\tilde{1}}^{\tilde{n}} \tau(\tilde{k})(\mu^{\tilde{k}})^2 \neq 0.$$

Thus the only possibility is

$$a^{\tilde{s}}_{\tilde{l}\,\hat{\beta}} = \alpha^{\tilde{s}}_{\tilde{l}\,\hat{\beta}} = 0, \qquad t^{\tilde{s}}_{\tilde{l}\,1.} = \sigma\,\delta^{\tilde{s}}_{\tilde{l}}, \qquad \tau^{\tilde{s}}_{\tilde{l}} = \tilde{\tau}\,\delta^{\tilde{s}}_{\tilde{l}} \quad (\tilde{\tau} < 0). \qquad (I.53)$$

From (I.39′) and (I.42) we have, in addition,

$$\tau^{\tilde{k}}_{\tilde{p}}\,a^{\tilde{p}}_{\tilde{s}\,\hat{\beta}} = -(1+\tilde{\tau})\,a^{\tilde{k}}_{\tilde{s}\,\hat{\beta}}. \qquad (I.54)$$

From (I.48′) we have

$$\tau^{\tilde{s}}_{\tilde{p}}\,\tau^{\tilde{p}}_{\tilde{m}}\,\tau^{\tilde{k}}_{\tilde{l}} = -\sum_{\hat{\beta}}\left[(\tau^{\tilde{s}}_{\tilde{p}}\,\alpha^{\tilde{p}}_{\tilde{k}\,\hat{\beta}})\,\alpha^{\tilde{m}}_{\tilde{l}\,\hat{\beta}} + (\tau^{\tilde{s}}_{\tilde{p}}\,\alpha^{\tilde{p}}_{\tilde{l}\,\hat{\beta}})\,\alpha^{\tilde{m}}_{\tilde{k}\,\hat{\beta}}\right],$$

which together with (I.54) gives

$$\tau^{\tilde{s}}_{\tilde{p}}\,\tau^{\tilde{p}}_{\tilde{m}}\,\tau^{\tilde{k}}_{\tilde{l}} = (1+\tilde{\tau})\sum_{\hat{\beta}}(\alpha^{\tilde{s}}_{\tilde{k}\,\hat{\beta}}\,\alpha^{\tilde{m}}_{\tilde{l}\,\hat{\beta}} + \alpha^{\tilde{s}}_{\tilde{l}\,\hat{\beta}}\,\alpha^{\tilde{m}}_{\tilde{k}\,\hat{\beta}}).$$

Using (I.48′) once more we have

$$\tau^{\tilde{s}}_{\tilde{p}}\,\tau^{\tilde{p}}_{\tilde{m}}\,\tau^{\tilde{k}}_{\tilde{l}} = -(1+\tilde{\tau})\tau^{\tilde{s}}_{\tilde{m}}\,\tau^{\tilde{k}}_{\tilde{l}},$$

whence owing to the non-singularity τ^m_l it follows that:

$$\tau^{\tilde{s}}_{\tilde{m}} = -(1+\tilde{\tau})\,\delta^{\tilde{s}}_{\tilde{m}}. \qquad (I.55)$$

From (I.45′), (I.55), (I.42′) we obtain, furthermore,

$$(3+2\tilde{\tau})\,\alpha^{\tilde{k}}_{\tilde{m}\,\hat{\beta}} = 0. \qquad (I.56)$$

Thus either $\tilde{\tau} = -3/2$ or $\alpha^{\tilde{k}}_{\tilde{m}\,\hat{\beta}} = 0$. From (I.44′) it also follows that

$$2\alpha^{\tilde{k}}_{\tilde{p}\,\hat{\beta}}\,\alpha^{\tilde{p}}_{\tilde{m}\,\hat{\beta}} - 2\alpha^{\tilde{k}}_{\tilde{p}\,\hat{\beta}}\,\alpha^{\tilde{p}}_{\tilde{m}\,\hat{\beta}} = -\tau^{\tilde{k}}_{\tilde{m}}. \qquad (I.57)$$

Therefore if $\alpha^{\tilde{k}}_{\tilde{m}\,\hat{\beta}} = 0$ then $2\sum_{\tilde{p}}\alpha^{\tilde{k}}_{\tilde{p}\,\hat{\beta}}\,\alpha^{\tilde{m}}_{\tilde{p}\,\hat{\beta}} = -\tau^{\tilde{k}}_{\tilde{m}}$, which is impossible since $\tau^{\tilde{k}}_{\tilde{m}}$ is a positive-definite symmetric matrix. Thus we have

$$\tilde{\tau} = -\frac{3}{2}, \qquad \tau^{\tilde{k}}_{\tilde{m}} = -\frac{3}{2}\,\delta^{\tilde{k}}_{\tilde{m}}, \qquad \tau^{\tilde{k}}_{\tilde{m}} = \frac{1}{2}\,\delta^{\tilde{k}}_{\tilde{m}}. \qquad (I.58)$$

Also (I.44′), (I.42), (I.49′) imply

$$-2\,\alpha^{\tilde{k}}_{\tilde{p}\,\hat{\beta}}\,\alpha^{\tilde{p}}_{\tilde{m}\,\hat{\beta}} = -\tau^{\tilde{k}}_{\tilde{m}} = \frac{3}{2}\,\delta^{\tilde{k}}_{\tilde{m}} \tag{I.59}$$

which gives

$$-2\sum_{\hat{\beta}\,\tilde{p}}\alpha^{\tilde{p}}_{\tilde{k}\,\hat{\beta}}\,\alpha^{\tilde{p}}_{\tilde{m}\,\hat{\beta}} = \hat{r}\,\tau^{\tilde{k}}_{\tilde{m}}. \tag{I.60}$$

Contracting $\tilde{\tilde{p}}$ and $\tilde{m}$ in (I.48′) and using (I.58) we obtain

$$\frac{\tilde{\tilde{n}}}{2}\,\tau^{\tilde{k}}_{\tilde{l}} = -2\sum_{\hat{\beta},\,\tilde{p}}\alpha^{\tilde{p}}_{\tilde{k}\,\hat{\beta}}\,\alpha^{\tilde{p}}_{\tilde{l}\,\hat{\beta}}\,. \tag{I.61}$$

From (I.61) and (I.60) there follows

$$\tilde{\tilde{n}} = 2\hat{r}, \tag{I.62}$$

which contradicts (I.50).

Analogously we obtain $\tilde{n} = 2\hat{r}$, considering $\tau(\tilde{\tilde{k}}) < 0$ for (I.49′); but this contradicts (I.50). Finally, $\theta^i_j = \pm\delta^i_j$, which proves our theorem. ∎

I.6.3. Theorem 21. *Let $\mathfrak{g}$ be a compact Lie algebra, $\mathfrak{g} = \mathfrak{l}_1 + \mathfrak{l}_2 + \mathfrak{l}_3$ be the iso-involutive sum of an iso-involutive group $\chi(S_1, S_2, S_3\,;\,\varphi)$, $\mathfrak{g}/\mathfrak{l}_1$ being elementary involutive pair, $\dim(\mathfrak{l}_3 \dot{-} \mathfrak{l}_0) = 1$. Then*

$$\mathfrak{g}/\mathfrak{l}_1 \cong so(n+1)/so(n), \qquad \mathfrak{g}/\mathfrak{l}_2 \cong so(n+1)/so(n),$$

$$\mathfrak{g}/\mathfrak{l}_3 \cong so(n+1)/so(n-1) \oplus so(2),$$

$$\mathfrak{l}_1/\mathfrak{l}_0 \cong so(n)/so(n-1), \qquad \mathfrak{l}_2/\mathfrak{l}_0 \cong so(n)/so(n-1),$$

$$\mathfrak{l}_3/\mathfrak{l}_0 \cong so(2)/\{0\} + so(n-1)/so(n-1)$$

with the natural embeddings.

Proof. Under our assumptions it is obvious that $\mathfrak{g} = \mathfrak{l}_1 + \mathfrak{l}_2 + \mathfrak{l}_3$ is of lower index 1 and type 1. And the theorem follows from I.5.4. (Theorem 14). ∎

I.6.4. Theorem 22. *Let $\mathfrak{g}$ be a simple compact Lie algebra, $\mathfrak{g} = \mathfrak{l}_1 + \mathfrak{l}_2 + \mathfrak{l}_3$ being the iso-involutive sum of the iso-involutive group $\chi(S_1, S_2, S_3\,;\,\varphi)$, $\dim(\mathfrak{l}_3 \dot{-} \mathfrak{l}_0) = 2$. Then*

$$\mathfrak{g}/\mathfrak{l}_1 \cong su(n+1)/su(n) \oplus u(1), \qquad \mathfrak{g}/\mathfrak{l}_2 \cong su(n+1)/su(n) \oplus u(1),$$

$$\mathfrak{g}/\mathfrak{l}_3 \cong su(n+1)/su(n-1) \oplus su(2) \oplus u(1),$$

$$\mathfrak{l}_1/\mathfrak{l}_0 \cong su(n)/su(n-1) \oplus u(1) + u(1)/u(1),$$

$$\mathfrak{l}_2/\mathfrak{l}_0 \cong su(n)/su(n-1) \oplus u(1) + u(1)/u(1),$$

$$\mathfrak{l}_3/\mathfrak{l}_0 \cong su(2)/u(1) + su(n-1) \oplus u(1)/su(n-1) \oplus u(1)$$

with the natural embeddings.

Proof. Let us note that $\mathfrak{l}_3/\mathfrak{l}_0$ is of lower index 1 (otherwise $\left[(\mathfrak{l}_3\dot{-}\mathfrak{l}_0)(\mathfrak{l}_3\dot{-}\mathfrak{l}_0)\right] =$ $\{0\}$, $\mathfrak{l}_3 = \mathfrak{l}_0 \oplus \mathfrak{k}$, where $\mathfrak{l}_3$ is two-dimensional and commutative, which is impossible since for a simple compact Lie algebra $\mathfrak{g}$ the centre of the involutive algebra $\mathfrak{l}_3$ is at most one-dimensional). But then $\mathfrak{g} = \mathfrak{l}_1 + \mathfrak{l}_2 + \mathfrak{l}_3$ is of lower index 1 and type 1 since $\dim(\mathfrak{l}_3\dot{-}\mathfrak{l}_0) \neq 1$.

Consequently by I.6.2 (Theorem 20) $\mathfrak{g} = \mathfrak{l}_1 + \mathfrak{l}_2 + \mathfrak{l}_3$ is of type 2. And the result follows from I.5.12 (Theorem 17). ∎

I.6.5. Theorem 23. *Let $\mathfrak{g}$ be a simple compact Lie algebra, $\mathfrak{g} = \mathfrak{l}_1 + \mathfrak{l}_2 + \mathfrak{l}_3$ be an iso-involutive sum of lower index 1, and $\dim(\mathfrak{l}_3\dot{-}\mathfrak{l}_0) = 4$. Then*

$$\mathfrak{g}/\mathfrak{l}_1 \cong sp(n+1)/sp(n) \oplus sp(1),$$

$$\mathfrak{g}/\mathfrak{l}_2 \cong sp(n+1)/sp(n) \oplus sp(1),$$

$$\mathfrak{g}/\mathfrak{l}_3 \cong sp(n+1)/sp(n-1) \oplus sp(2),$$

$$\mathfrak{l}_1/\mathfrak{l}_0 \cong sp(n)/sp(n-1) \oplus sp(1) + sp(1)/sp(1),$$

$$\mathfrak{l}_2/\mathfrak{l}_0 \cong sp(n)/sp(n-1) \oplus sp(1) + sp(1)/sp(1),$$

$$\mathfrak{l}_3/\mathfrak{l}_0 \cong sp(2)/sp(1) \oplus sp(1) + sp(n-1)/sp(n-1)$$

with the natural embeddings.

Proof. Since $\dim(\mathfrak{l}_3\dot{-}\mathfrak{l}_0) \neq 1$ then $\mathfrak{g} = \mathfrak{l}_1 + \mathfrak{l}_2 + \mathfrak{l}_3$ is not of type 1. Consequently by I.6.2 (Theorem 20) it is of type 2.

The rest of the proof follows from I.5.14 (Theorem 18). ∎

I.6.6. Theorem 24. *Let $\mathfrak{g}$ be a simple compact Lie algebra, $\mathfrak{g} = \mathfrak{l}_1 + \mathfrak{l}_2 + \mathfrak{l}_3$ be an involutive sum of lower index 1 and not of type 1. Then*

$$\mathfrak{l}_3/\mathfrak{l}_0 = \tilde{\mathfrak{l}}_3/\tilde{\mathfrak{l}}_0 + \tilde{\tilde{\mathfrak{l}}}_0/\tilde{\mathfrak{l}}_0,$$

$$\tilde{\mathfrak{l}}_3/\tilde{\mathfrak{l}}_0 \cong so(m+1)/so(m)$$

with the natural embedding, and $\tilde{\mathfrak{l}}_3$ is the special unitary involutive subalgebra of the involutive automorphism S_3.

Proof. Indeed, under our assumptions $\mathfrak{l}_3/\mathfrak{l}_0$ is of lower index 1 and then $\mathfrak{l}_3/\mathfrak{l}_0 = \tilde{\mathfrak{l}}_3/\tilde{\mathfrak{l}}_0 + \tilde{\tilde{\mathfrak{l}}}_0/\tilde{\mathfrak{l}}_0$, where $\tilde{\mathfrak{l}}_3/\tilde{\mathfrak{l}}_0$ is an elementary involutive pair of lower index 1. By I.6.1 (Theorem 19) the iso-involutive group $\chi^{(1)}(S_1, S_2, S_3; \varphi)$ is of lower index 1 and of type 1, as well as the corresponding iso-involutive sum $\tilde{\mathfrak{l}}_3 = \tilde{\mathfrak{l}}_0 + \tilde{\mathfrak{l}}_0' + \tilde{\mathfrak{l}}_0''$.

Using I.5.4 (Theorem 14) we have

$$\tilde{\mathfrak{l}}_3/\tilde{\mathfrak{l}}_0 \cong so(m+1)/so(m)$$

with the natural embedding. Since $S_3 \in \mathrm{Int}_\mathfrak{g}(\{t\zeta\})$ and $\zeta \in \tilde{\mathfrak{l}}_3 \dot{-} \mathfrak{l}_0$ the algebra $\{t\zeta\}$ can be realized as $so(2)$ with the natural embedding into $so(m+1) \cong \tilde{\mathfrak{l}}_3$. Any $so(2)$ can be included into $so(3)$ with the natural embedding into $so(m+1)$. Thus

$$
\begin{array}{ccccc}
\{t\zeta\} & \subset & \mathfrak{b} & \subset & \tilde{\mathfrak{l}}_3 \\
\updownarrow & & \updownarrow & & \updownarrow \\
so(2) & \subset & so(3) & \subset & so(m+1).
\end{array}
$$

Since $\mathrm{Id} \neq S_3 \in \mathrm{Int}_\mathfrak{g}(\{t\zeta\}) \subset \mathrm{Int}_\mathfrak{g}(\mathfrak{b})$ and S_3 commutes with the elements of $\mathrm{Int}_\mathfrak{g}(\mathfrak{b})$ we have $\mathrm{Int}_\mathfrak{g}(\mathfrak{b}) \cong SU(2)$.

Considering, furthermore,

$$
\tilde{\mathfrak{l}}_3/\mathfrak{b} \oplus \mathfrak{n} \cong so(m+1)/so(3) \oplus so(m-2)
$$

we see that $\mathfrak{b} \oplus \mathfrak{n}$ is a principal orthogonal involutive algebra in $\tilde{\mathfrak{l}}_3$ and then $S_3 \in \mathrm{Int}_\mathfrak{g}(\mathfrak{b})$. This means, see I.17 (Definition 8), that $\tilde{\mathfrak{l}}_3$ is the special involutive subalgebra of the involutive automorphism S_3 of type U.   ∎

I.6.7. Theorem 25. *Let $\mathfrak{g}$ be a simple compact Lie algebra, and $\mathfrak{g} = \mathfrak{l}_1 + \mathfrak{l}_2 + \mathfrak{l}_3$ be the iso-involutive sum of lower index 1 generated by an iso-involutive group $\chi(S_1, S_2, S_3; \varphi)$ ($\varphi \in \mathrm{Int}_\mathfrak{g}(\{t\zeta\})$). If $\mathfrak{k}$ is the maximal subalgebra of elements immobile under the action of φ then $\mathfrak{k} \cap (\mathfrak{l}_3 \dot{-} \mathfrak{l}_0) = \{t\zeta\}$ and*

$$
\mathrm{ad}_\mathfrak{g}^{(\mathfrak{g} \dot{-} \mathfrak{l}_3)}(\{t\zeta\}) = \{t(\varphi|_{(\mathfrak{g} \dot{-} \mathfrak{l}_3)})\}.
$$

Proof. From I.6.2 (Theorem 20) it follows that $\mathfrak{g}$ is either of type 1 or of type 2. Using, furthermore, I.5.3 (Theorem 13), I.5.4 (Theorem 14) and I.5.7 (Theorem 15), I.5.9 (Theorem 16), correspondingly with the above two cases, we obtain Theorem 25.   ∎

I.6.8. Theorem 26. *Let $\mathfrak{g}$ be a simple compact Lie algebra, and $\mathfrak{g} = \mathfrak{l}_1 + \mathfrak{l}_2 + \mathfrak{l}_3$ be the iso-involutive sum of lower index 1 with the conjugating automorphism $\varphi \in \mathrm{Int}_\mathfrak{g}(\{t\zeta\})$. Then, for any $x \in \mathfrak{l}_1 \dot{-} \mathfrak{l}_0$,*

$$
\mathfrak{b} = \{\alpha x + \beta \varphi x + \gamma \zeta\}
$$

is a three-dimensional simple compact algebra and $\mathrm{Int}_\mathfrak{g}(\mathfrak{b}) \cong SO(3)$.

Proof. Taking $0 \neq x \in \mathfrak{l}_1 \dot{-} \mathfrak{l}_0$ we consider $[\zeta x] = \mathrm{ad}_\mathfrak{g}^{(\mathfrak{g} \dot{-} \mathfrak{l}_3)}(\zeta)x$. By I.6.7 (Theorem 25) $[\zeta x] = \lambda(\varphi|_{(\mathfrak{g} \dot{-} \mathfrak{l}_3)})x = \lambda \varphi x$. Furthermore,

$$
[\zeta(\varphi x)] = [(\varphi\zeta)(\varphi x)] = \varphi[\zeta x] = \lambda \varphi^2 x = \lambda S_3 x = -\lambda x.
$$

Finally,

$$
\varphi[x(\varphi x)] = [(\varphi x)(\varphi^2 x)] = [(\varphi x)(S_3 x)]
$$

$$
= -[(\varphi x) x] = [x(\varphi x)]
$$

and then $[x(\varphi x)] \in \mathfrak{l}_3 \dot{-} \mathfrak{l}_0$, thus by I.6.7 (Theorem 25) $[x(\varphi x)] = \mu \zeta$.

Consequently $\mathfrak{b} = \{\alpha x + \beta \varphi x + \gamma \zeta\}$ is a three-dimensional compact subalgebra. It is simple since $\lambda \neq 0$ (otherwise $[\zeta x] = [\zeta(\varphi x)] = 0$ and $\mathrm{Int}_\mathfrak{g}(\{t\zeta\})x = x$, whence $S_3 x = x$, which is impossible owing to $0 \neq x \in \mathfrak{l}_1 \dot{-} \mathfrak{l}_0$).

Suppose that $\mathrm{Int}_\mathfrak{g}(\mathfrak{b}) \cong SU(2)$ but $\mathrm{Id} \neq S_3 \in \mathrm{Int}_\mathfrak{g}(\{t\zeta\})$. Then $S_3 \in \mathrm{Int}_\mathfrak{g}(\{tx\})$, hence $S_3 x = x$, which is impossible. Thus $\mathrm{Int}_\mathfrak{g}(\mathfrak{b}) \cong SO(3)$.   ∎

I.6.9. Theorem 27. *Let* $\mathfrak{g}$ *be a compact Lie algebra, and* $\mathfrak{g}/\mathfrak{l}_1$ *be an elementary involutive pair of lower index 1, then all one-dimensional subalgebras* $\{t\,p\} \subset \mathfrak{g}\dot{-}\mathfrak{l}_1$ *are conjugated in* $\mathrm{Int}_{\mathfrak{g}}(\mathfrak{l}_1)$, *and consequently* $\mathfrak{g}/\mathfrak{l}_1$ *is of index 1.*

Proof. Having taken the maximal one-dimensional subalgebra $\{t\,\zeta\} \subset \mathfrak{g}\dot{-}\mathfrak{l}_1$, by I.3.16 (Theorem 10) we construct the iso-involutive sum $\mathfrak{g} = \mathfrak{l}_1 + \mathfrak{l}_2 + \mathfrak{l}_3$ of lower index 1. In order to prove the theorem it is evidently sufficient to prove that any $\{t\,p\} \subset \mathfrak{g}\dot{-}\mathfrak{l}_1$ is conjugated with $\{t\,\zeta\}$ in $\mathrm{Int}_{\mathfrak{g}}(\mathfrak{l}_1)$. If $\mathfrak{g} = \mathfrak{l}_1 + \mathfrak{l}_2 + \mathfrak{l}_3$ is of type 1 then

$$\mathfrak{g}/\mathfrak{l}_1 \cong so(n+1)/so(n),$$

see I.5.4 (Theorem 14), and the assertion of the theorem is evident.

If $\mathfrak{g} = \mathfrak{l}_1 + \mathfrak{l}_2 + \mathfrak{l}_3$ is not of type 1 then it is of type 2 by I.6.2 (Theorem 20). All one-dimensional subalgebras $\{t\,q\} \subset \mathfrak{l}_3\dot{-}\mathfrak{l}_0$ are conjugated in $\mathrm{Int}_{\mathfrak{g}}(\mathfrak{l}_1)$ (moreover, in $\mathrm{Int}_{\mathfrak{g}}(\mathfrak{l}_0)$). Indeed, $\mathfrak{l}_3/\mathfrak{l}_0 = \tilde{\mathfrak{l}}_3/\tilde{\mathfrak{l}}_0 + \tilde{\tilde{\mathfrak{l}}}_0/\tilde{\mathfrak{l}}_0$, where $\tilde{\mathfrak{l}}_3/\tilde{\mathfrak{l}}_0$ is an elementary involutive pair of lower index 1, and the derived involutive group $\chi^{(1)}$, being by I.6.1 (Theorem 19) of type 1, generate the iso-involutive sum $\tilde{\mathfrak{l}}_3 = \tilde{\mathfrak{l}}_0 + \tilde{\mathfrak{l}}_0' + \tilde{\mathfrak{l}}_0''$ of type 1 and lower index 1. Thus

$$\tilde{\mathfrak{l}}_3/\tilde{\mathfrak{l}}_0 \cong so(m+1)/so(m)$$

with the natural embedding.

For this reason all one-dimensional subalgebras $\{\ell\,q\} \subset \mathfrak{l}_3\dot{-}\mathfrak{l}_0 = \tilde{\mathfrak{l}}_3\dot{-}\tilde{\mathfrak{l}}_0$ are conjugated in $\mathrm{Int}_{\mathfrak{g}}(\tilde{\mathfrak{l}}_0)$, and consequently in $\mathrm{Int}_{\mathfrak{g}}(\mathfrak{l}_1)$.

Let us now take an arbitrary $0 \neq p \in \mathfrak{g}\dot{-}\mathfrak{l}_1$, then $p = \alpha p_1 + \beta p_2$, where $p_1 \in \mathfrak{l}_2\dot{-}\mathfrak{l}_0$, $p_2 \in \mathfrak{l}_3\dot{-}\mathfrak{l}_0$. We need to consider $\alpha \neq 0$ (otherwise $p = \beta p_2 \in \mathfrak{l}_3\dot{-}\mathfrak{l}_0$, but it has already been proved that all subalgebras $\{t\,q\} \subset \mathfrak{l}_3\dot{-}\mathfrak{l}_0$ are conjugated in $\mathrm{Int}_{\mathfrak{g}}(\mathfrak{l}_1)$). If $\beta = 0$ then $p \in \mathfrak{l}_2\dot{-}\mathfrak{l}_0$ and, taking $x = -\varphi p \in \mathfrak{l}_1\dot{-}\mathfrak{l}_0$, we have by I.6.8 (Theorem 26) the three-dimensional simple compact algebra $\mathfrak{b} = \{\lambda x + \mu \varphi x + \nu \zeta\}$. Thus $\{t\,p\} = \{t\,\varphi\,x\}$ and $\{t\,\zeta\}$ are conjugated by $\mathrm{Int}_{\mathfrak{g}}(\{t\,x\})$, and consequently in $\mathrm{Int}_{\mathfrak{g}}(\mathfrak{l}_1)$.

The last case to be considered is $p = \alpha p_1 + \beta p_2$, $\alpha \neq 0$, $\beta \neq 0$, $p_1 \in \mathfrak{l}_2\dot{-}\mathfrak{l}_0$, $p_2 \in \mathfrak{l}_3\dot{-}\mathfrak{l}_0$, $p_1 \neq 0$, $p_2 \neq 0$. This case is reduced to the case just considered. Indeed, the same iso-involutive sum

$$\mathfrak{g} = \mathfrak{l}_1 + \mathfrak{l}_2 + \mathfrak{l}_3$$

can be constructed by means of the conjugating one-dimensional group $\mathrm{Int}_{\mathfrak{g}}(\{t\,p_2\})$, since owing to the conjugacy of $\{t\,p_2\}$ and $\{t\,\zeta\}$ we have $S_3 \in \mathrm{Int}_{\mathfrak{g}}(\{t\,p_2\})$. Thus after a proper choice of conjugating one-dimensional subgroup we can put $p_2 = \nu\,\zeta$. In this way $p = \alpha p_1 + \beta\zeta$. But we have already shown that $\mathfrak{b} = \{\lambda p_1 + \mu\varphi p_1 + \sigma\zeta\}$ is a three-dimensional simple compact Lie algebra, thus all one-dimensional subalgebras $\{t\,p\} \subset \{\lambda p_1 + \sigma\zeta\}$ are conjugated by $\mathrm{Int}_{\mathfrak{g}}(\{t\,\varphi\,p_1\})$. In particular, $\{t\,p\} = \{t\,(\alpha p_1 + \beta\zeta)\}$ and $\{t\,\zeta\}$ are conjugated by $\mathrm{Int}_{\mathfrak{g}}(\{t\,\varphi\,p_1\})$, and consequently in $\mathrm{Int}\mathfrak{g}(\mathfrak{l}_1)$. ∎

I.6.10. Theorem 28. *Let* $\mathfrak{g}$ *be a simple compact Lie algebra,* $\mathfrak{g} = \mathfrak{l}_1 + \mathfrak{l}_2 + \mathfrak{l}_3$ *be an iso-involutive sum of lower index 1. Then* $\dim(\mathfrak{l}_3 \dot{-} \mathfrak{l}_0) \neq 3$.

Proof. If $\mathfrak{g} = \mathfrak{l}_1 + \mathfrak{l}_2 + \mathfrak{l}_3$ is of type 1, then by I.5.2 (Theorem 12) $1 = \dim(\mathfrak{l}_3 \dot{-} \mathfrak{l}_0) \neq 3$. Consequently we should consider $\mathfrak{g} = \mathfrak{l}_1 + \mathfrak{l}_2 + \mathfrak{l}_3$ which is not of type 1.

Then by I.6.1 (Theorem 19) we have the iso-involutive sum $\mathfrak{l}_3 = \mathfrak{l}_0 + \mathfrak{n}_2 + \mathfrak{n}_3$ of the iso-involutive group $\chi^{(1)}(S_1, S_2, S_3; \varphi)$ of lower index 1 and of type 1. Hence

$$\mathfrak{l}_3/\mathfrak{l}_0 = \tilde{\mathfrak{l}}_3/\tilde{\mathfrak{l}}_0 + \tilde{\tilde{\mathfrak{l}}}_0/\tilde{\mathfrak{l}}_0,$$

where $\tilde{\mathfrak{l}}_3/\tilde{\mathfrak{l}}_0$ is elementary, and then $\chi^{(1)}$ induces on $\tilde{\mathfrak{l}}_3$ the involutive sum

$$\tilde{\mathfrak{l}}_3 = \tilde{\mathfrak{l}}_0 + \tilde{\mathfrak{l}}_0' + \tilde{\mathfrak{l}}_0''$$

of type 1 and lower index 1. By I.5.4 (Theorem 14) we then have

$$\tilde{\mathfrak{l}}_3/\tilde{\mathfrak{l}}_0 \cong so(m+1)/so(m)$$

with the natural embedding.

If $\dim(\mathfrak{l}_3 \dot{-} \mathfrak{l}_0) = 3$ then $\dim(\tilde{\mathfrak{l}}_3 \dot{-} \tilde{\mathfrak{l}}_0) = 3$. Thus

$$\tilde{\mathfrak{l}}_3/\tilde{\mathfrak{l}}_0 \cong so(4)/so(3)$$

with the natural embedding.

Since $\mathrm{Id} \neq S_3 \in \mathrm{Int}_{\mathfrak{g}}(\{t\zeta\})$, where $\{t\zeta\} \subset \mathfrak{l}_3 \dot{-} \mathfrak{l}_0 = \tilde{\mathfrak{l}}_3 \dot{-} \tilde{\mathfrak{l}}_0$, we have

$$\mathrm{Id} \neq S_3 \in \mathrm{Int}_{\mathfrak{g}}(\tilde{\mathfrak{l}}_3).$$

Since $\tilde{\mathfrak{l}}_3 \cong so(4) = so(3) \oplus so(3)$ and owing to the irreducibility of $\mathfrak{g}/\mathfrak{l}_3$ $\mathrm{Int}_{\mathfrak{g}}(\mathfrak{l}_3)$ contains the unique involutive automorphism $S_3 \neq \mathrm{Id}$ such that $S_3 \eta = \eta$ for $\eta \in \mathfrak{l}_3$, thus we see that $\mathrm{Int}_{\mathfrak{g}}(\tilde{\mathfrak{l}}_3) \cong SO(4)$. For this reason $\mathrm{Int}_{\mathfrak{g}}(\tilde{\mathfrak{l}}_0) \cong SO(3)$ (because that $\tilde{\mathfrak{l}}_0$ is a diagonal of the canonical involutive automorphism $so(3) \leftrightarrow so(3)$ in $\tilde{\mathfrak{l}}_3 \cong so(3) \oplus so(3)$). On the other hand, since $S_3 \in \mathrm{Int}_{\mathfrak{g}}(\{t\zeta\})$ and $\{t\zeta\}$ can be realized as $so(2)$ with the natural embedding into $\tilde{\mathfrak{l}}_3 \cong so(4)$, and since all subalgebras $so(2)$ with the natural embeddings into $so(4)$ are conjugated in $so(4)$, we have $S_3 \in \mathrm{Int}_{\mathfrak{g}}(\tilde{\mathfrak{l}}_0)$. Indeed, $\tilde{\mathfrak{l}}_0$ is $so(3)$ with the natural embedding into $so(4)$, and in $so(3)$ there is $so(2)$ with the natural embedding into $so(4)$; and under the action of $\mathrm{Int}_{\mathfrak{g}}(\mathfrak{l}_3)$ also S_3 is fixed.

As a result $\mathrm{Id} \neq S_3 \in \mathrm{Int}_{\mathfrak{g}}(\tilde{\mathfrak{l}}_0) \cong SO(3)$, which is impossible. Thus our assumption $\dim(\mathfrak{l}_3 \dot{-} \mathfrak{l}_0) = 3$ is incorrect. ∎

CHAPTER I.7

PRINCIPAL CENTRAL INVOLUTIVE
AUTOMORPHISM OF TYPE U

I.7.1. Let $\mathfrak{g}$ be a compact simple Lie algebra, and S be its central principal unitary involutive automorphism with an involutive algebra $\mathfrak{l} = \mathfrak{b} \oplus \mathfrak{z} \oplus \tilde{\mathfrak{l}}$, where $\mathrm{Int}_\mathfrak{g}(\mathfrak{b}) \cong SU(2)$, $\mathfrak{z}$ being the centre in $\mathfrak{l}$. By I.1. the centre $\mathfrak{z}$ is one-dimensional, $\mathrm{Int}_\mathfrak{g}(\mathfrak{z})$ is closed in $\mathrm{Int}(\mathfrak{g})$, and consequently compact, thus $\mathrm{Int}_\mathfrak{g}(\mathfrak{z}) \cong U(1)$. Owing to the irreducibility of $\mathrm{Int}_\mathfrak{g}^{(\mathfrak{g}-\mathfrak{l})}(\mathfrak{l})$ we have $S \in \mathrm{Int}_\mathfrak{g}(\mathfrak{z})$, in addition, $S \in \mathrm{Int}_\mathfrak{g}(\mathfrak{b})$. Let us take $a \in \mathrm{Int}_\mathfrak{g}(\mathfrak{b})$ such that $a^2 = S$ and $b \in \mathrm{Int}_\mathfrak{g}(\mathfrak{z})$ such that $b^2 = S$. Evidently

$$ab = ba, \qquad aS = Sa, \qquad bS = Sb.$$

We introduce $S_1 = ab$, then

$$(S_1)^2 = (ab)^2 = \mathrm{Id}, \qquad S_1 \neq \mathrm{Id}$$

(otherwise $a = b^{-1}$ commutes with $\mathrm{Int}_\mathfrak{g}(\mathfrak{b})$, which is possible only if either $a = S$ or $a = \mathrm{Id}$, contradicting $a^2 = S$) and $S S_1 = S_1 S$.

Let us introduce also $S_2 = S_1 S$ and denote the involutive algebras of the involutive automorphisms S_1 and S_2 by $\mathfrak{l}_1$ and $\mathfrak{l}_2$, respectively. Then we obtain the iso-involutive decomposition $\mathfrak{g} = \mathfrak{l}_1 + \mathfrak{l}_2 + \mathfrak{l}$. Let us also clarify the structure of

$$\mathfrak{l}_0 = \mathfrak{l}_1 \cap \mathfrak{l}_2 = \mathfrak{l}_1 \cap \mathfrak{l} = \mathfrak{l} \cap \mathfrak{l}_2.$$

In $\mathfrak{l}_0$ there are those and only those elements of $\mathfrak{l}$ which are immobile under the action of $S_1 = ab$. Thus $\mathfrak{z} \oplus \tilde{\mathfrak{l}} \subset \mathfrak{l}_0$.

Furthermore, $S_1 \mathfrak{b} = \mathfrak{b}$ and $S_1 x = ax$, if $x \in \mathfrak{b}$, and a generates in $\mathfrak{b}$ an inner automorphism, thus $\mathfrak{b} = \mathfrak{q} + \mathfrak{e}$, where $\mathfrak{q}$ is an one-dimensional subalgebra, $\mathfrak{e}$ is a two-dimensional subspace and $S_1 x = x$ for $x \in \mathfrak{q}$, $S_1 x = -x$ for $x \in \mathfrak{e}$. As a result $\mathfrak{l}_0 = \mathfrak{q} \oplus \mathfrak{z} \oplus \tilde{\mathfrak{l}}$. Let us now take $\zeta \in \mathfrak{e}$ and consider the one-dimensional subgroup $\mathrm{Int}_\mathfrak{g}(\{t\zeta\}) \subset \mathrm{Int}_\mathfrak{g}(\mathfrak{b}) \cong SU(2)$. Then obviously $S \in \mathrm{Int}_\mathfrak{g}(\{t\zeta\})$ and the involutive sum $\mathfrak{g} = \mathfrak{l}_1 + \mathfrak{l}_2 + \mathfrak{l}$ constructed above is iso-involutive. In addition, $\dim(\mathfrak{l} - \mathfrak{l}_0) = \dim \mathfrak{e} = 2$.

Now all conditions of I.6.4 (Theorem 22) are satisfied and we have:

I.7.2. Theorem 29. *Let $\mathfrak{g}$ be a simple compact Lie algebra, and let $\mathfrak{l} = \mathfrak{b} \oplus \mathfrak{z} \oplus \tilde{\mathfrak{l}}$ ($\mathrm{Int}_\mathfrak{g}(\mathfrak{b}) \cong SU(2)$, $\{0\} \neq \mathfrak{z}$ being the center in $\mathfrak{l}$) be a principal unitary central involutive algebra of a principal unitary central involutive automorphism S. Then*

$$\mathfrak{g}/\mathfrak{l} \cong su(n)/su(2) \oplus u(1) \oplus su(n-2)$$

with the natural embedding.

45

CHAPTER I.8

PRINCIPAL UNITARY INVOLUTIVE
AUTOMORPHISM OF INDEX 1

I.8.1. Theorem 30. *Let $\mathfrak{g}$ be a simple compact Lie algebra, and $\mathfrak{g}/\mathfrak{l}$ be an involutive pair of index 1,*

$$\mathfrak{l} = \mathfrak{m} \oplus \mathfrak{n},$$

where $\mathfrak{m} \neq \{0\}$, $\mathfrak{n} \neq \{0\}$.
  Then either

$$\mathfrak{g}/\mathfrak{l} \cong su(n+1)/su(n) \oplus u(1)$$

or

$$\mathfrak{g}/\mathfrak{l} \cong sp(n+1)/sp(n) \oplus sp(1)$$

with the natural embeddings.

Proof. Let us construct an iso-involutive sum $\mathfrak{g} = \mathfrak{l}_1 + \mathfrak{l}_2 + \mathfrak{l}_3$, where $\mathfrak{l}_1 = \mathfrak{l}$, of index 1. This is possible by I.3.16 (Theorem 10). If this sum is of type 1 then by I.5.4 (Theorem 14)

$$\mathfrak{g}/\mathfrak{l} = \mathfrak{g}/\mathfrak{l}_1 \cong so(n+1)/so(n)$$

with the natural embeddings.
  But $so(n) \cong \mathfrak{l}$ is not simple only when $n = 4$. Thus

$$\mathfrak{g}/\mathfrak{l} \cong so(5)/so(4) \cong sp(2)/sp(1) \oplus sp(1).$$

Furthermore, we consider the case in which $\mathfrak{g} = \mathfrak{l}_1 + \mathfrak{l}_2 + \mathfrak{l}_3$ is of index 1 and not of type 1, i.e., of type 2 by I.6.2 (Theorem 20). We can set $\dim(\mathfrak{l}_3 \dot{-} \mathfrak{l}_0) > 4$, otherwise our theorem is valid by I.6.4, I.6.5, I.6.10 (Theorems 22, 23, 28).
  Since $\mathfrak{g}/\mathfrak{l}_1$ is of index 1 we have that $\mathfrak{l}_3/\mathfrak{l}_0$, $\mathfrak{l}_2/\mathfrak{l}_0$ and as well (by the conjugacy of $\mathfrak{l}_1$ and $\mathfrak{l}_2$) $\mathfrak{l}_1/\mathfrak{l}_0$ are of index 1.
  Because $\dim(\mathfrak{l}_1 \dot{-} \mathfrak{l}_0) \neq 1$ (otherwise $\left[\mathfrak{l}_0 (\mathfrak{g} \dot{-} \mathfrak{l}_3)\right] = \{0\}$, which is impossible) we obtain

$$\mathfrak{l}_1/\mathfrak{l}_0 = \tilde{\mathfrak{l}}_1/\tilde{\mathfrak{l}}_0 + \tilde{\tilde{\mathfrak{l}}}_0/\tilde{\tilde{\mathfrak{l}}}_0,$$

where $\tilde{\mathfrak{l}}_1/\tilde{\mathfrak{l}}_0$ is an elementary involutive pair of index 1.
  If $\tilde{\tilde{\mathfrak{l}}}_0 = \{0\}$ then $\mathfrak{m} \subset \tilde{\mathfrak{l}}_1$, $\mathfrak{n} \subset \tilde{\mathfrak{l}}_1$, thus $\mathfrak{l}_1/\mathfrak{l}_0$ is elementary and $\mathfrak{m}, \mathfrak{n}$ are isomorphic to the simple non-one-dimensional algebra $\mathfrak{l}_0$. But then $\mathfrak{l}_3/\mathfrak{l}_0$, by I.6.1 (Theorem 19), is an elementary involutive pair of index 1 and of type 1 and by I.5.4 (Theorem 14)

$$\mathfrak{l}_3/\mathfrak{l}_0 \cong so(n+1)/so(n).$$

But since $\mathfrak{l}_1/\mathfrak{l}_0$ is of index 1 and $\mathfrak{l}_0$ is a diagonal of the canonical symmetry $\mathfrak{m} \leftrightarrow \mathfrak{n}$ then $\mathfrak{l}_0$ is of rank 1, that is, $\mathfrak{l}_0 \cong so(3)$. And

$$\mathfrak{l}_3/\mathfrak{l}_0 \cong so(4)/so(3)$$

with the natural embedding.

But then $\dim(\mathfrak{l}_3 \dot{-} \mathfrak{l}_0) = 3$, which is impossible by I.6.10 (Theorem 28).

Thus we should assume $\tilde{\tilde{\mathfrak{l}}}_0 \neq \{0\}$. Then $\mathfrak{q} = (\varphi \tilde{\mathfrak{l}}_0) \cap \tilde{\tilde{\mathfrak{l}}}_0 = \{0\}$ (otherwise $\mathfrak{q} \subset \mathfrak{l}_0$ is a non-trivial ideal in $\mathfrak{l}_1$, but because of $\varphi \mathfrak{q} = \mathfrak{q}$, $\varphi \mathfrak{l}_1 = \mathfrak{l}_2$, $\mathfrak{q}$ is a non-trivial ideal in $\mathfrak{l}_2$, and consequently in $\mathfrak{l}_3$, whence $\mathfrak{q}$ is a non-trivial ideal in $\mathfrak{g} = \mathfrak{l}_1 + \mathfrak{l}_2 + \mathfrak{l}_3$, which is impossible) and $\varphi \tilde{\mathfrak{l}}_0 \subset \tilde{\mathfrak{l}}_0$. Thus

$$\tilde{\mathfrak{l}}_0 = (\varphi \tilde{\mathfrak{l}}_0) \oplus \mathfrak{k},$$

or

$$\mathfrak{l}_0 = \tilde{\tilde{\mathfrak{l}}}_0 \oplus (\varphi \tilde{\mathfrak{l}}_0) \oplus \mathfrak{k},$$

where $(\varphi \tilde{\mathfrak{l}}_0)$ is an ideal of $\mathfrak{l}_2$.

Furthermore, $\tilde{\tilde{\mathfrak{l}}}_0$ and $(\varphi \tilde{\mathfrak{l}}_0)$ act on $\mathfrak{l}_3 \dot{-} \mathfrak{l}_0$ in a non-trivial way, that is,

$$\left[\tilde{\tilde{\mathfrak{l}}}_0 (\mathfrak{l}_3 \dot{-} \mathfrak{l}_0) \right] \neq \{0\}, \qquad \left[(\varphi \tilde{\mathfrak{l}}_0)(\mathfrak{l}_3 \dot{-} \mathfrak{l}_0) \right] \neq \{0\}.$$

Indeed, if $\left[\tilde{\tilde{\mathfrak{l}}}_0 (\mathfrak{l}_3 \dot{-} \mathfrak{l}_0) \right] = \{0\}$ then, together with $\left[\tilde{\tilde{\mathfrak{l}}}_0 (\mathfrak{l}_1 \dot{-} \mathfrak{l}_0) \right] = \{0\}$, this gives $\left[\tilde{\tilde{\mathfrak{l}}}_0 (\mathfrak{g} \dot{-} \mathfrak{l}_2) \right] = \{0\}$, meaning that $\tilde{\tilde{\mathfrak{l}}}_0$ is an ideal of $\mathfrak{g}$, which is impossible.

Analogously, $\left[(\varphi \tilde{\mathfrak{l}}_0)(\mathfrak{l}_3 \dot{-} \mathfrak{l}_0) \right] = \{0\}$ together with $\left[(\varphi \tilde{\mathfrak{l}}_0)(\mathfrak{l}_2 \dot{-} \mathfrak{l}_0) \right] = \{0\}$ give us $\left[(\varphi \tilde{\mathfrak{l}}_0)(\mathfrak{g} \dot{-} \mathfrak{l}_1) \right] = \{0\}$, meaning that $(\varphi \tilde{\mathfrak{l}}_0)$ is an ideal of $\mathfrak{g}$, which is impossible.

Now we consider

$$\mathfrak{l}_3/\mathfrak{l}_0 = \tilde{\mathfrak{l}}_3/\mathfrak{t} + \mathfrak{p}/\mathfrak{p},$$

where

$$\tilde{\mathfrak{l}}_3/\mathfrak{t} \cong so(m+1)/so(m)$$

with the natural embeddings (as follows from I.6.6 (Theorem 24)).

Since $\dim(\mathfrak{l}_3 \dot{-} \mathfrak{l}_0) > 4$ we have $m \geq 5$ and $so(m) \cong \mathfrak{t}$ is simple and non-one-dimensional. For $\mathfrak{l}_0$ we have

$$\mathfrak{l}_0 = \tilde{\tilde{\mathfrak{l}}}_0 \oplus (\varphi \tilde{\mathfrak{l}}_0) \oplus \mathfrak{k} = \mathfrak{t} \oplus \mathfrak{p}.$$

If $\tilde{\tilde{\mathfrak{l}}}_0 \cap \mathfrak{t} = \{0\}$ or $(\varphi \tilde{\mathfrak{l}}_0) \cap \mathfrak{t} = \{0\}$ then $\tilde{\tilde{\mathfrak{l}}}_0 \subset \mathfrak{p}$ and, respectively, $(\varphi \tilde{\mathfrak{l}}_0) \subset \mathfrak{p}$. But then $\tilde{\tilde{\mathfrak{l}}}_0$ and $(\varphi \tilde{\mathfrak{l}}_0)$ act trivially on $\mathfrak{l}_3 \dot{-} \mathfrak{l}_0 = \tilde{\mathfrak{l}}_3 \dot{-} \mathfrak{t}$, which is impossible. Thus $\tilde{\tilde{\mathfrak{l}}}_0 \cap \mathfrak{t} \neq \{0\}$, $(\varphi \tilde{\mathfrak{l}}_0) \cap \mathfrak{p} \neq \{0\}$, meaning that $\mathfrak{t}$ is not simple, which is impossible for $\mathfrak{t} \cong so(m)$, $m \geq 5$. ∎

I.8.2. Theorem 31. *Let* $\mathfrak{g}$ *be a compact Lie algebra, and* $\mathfrak{g}/\mathfrak{l}$ *be an elementary principal involutive pair,* $\dim \mathfrak{l} = 3$. *Then* $\mathfrak{g}/\mathfrak{l}$ *is a principal orthogonal involutive pair.*

Proof. Let us assume that $\mathrm{Int}_{\mathfrak{g}}(\mathfrak{l}) \cong SU(2)$ and construct an iso-involutive decomposition $\mathfrak{g} = \mathfrak{l}_1 + \mathfrak{l}_2 + \mathfrak{l}_3$, $\mathfrak{l}_1 = \mathfrak{l}$. Then $\dim \mathfrak{l}_0$ is either 2 or 1. But $\dim \mathfrak{l}_0 \neq 2$, otherwise $\mathfrak{l} = \mathfrak{l}_1$ is not simple (indeed, $\mathfrak{l}_0$ is then commutative and $\mathfrak{l}_1/\mathfrak{l}_0$ is reducible).

Thus $\dim \mathfrak{l}_0 = 1$. By our assumption the non-trivial involutive automorphism $S_1 \in \mathrm{Int}_{\mathfrak{g}}(\mathfrak{l}_0) \subset \mathrm{Int}_{\mathfrak{g}}(\mathfrak{l}_1)$. But the conjugating automorphism φ maps $\mathfrak{l}_0$ into itself. Consequently it transforms S_1 into itself. On the other hand, it should transform S_1 into $S_2 \neq S_1$. Thus our assumption is wrong and $\mathrm{Int}_{\mathfrak{g}}(L) \cong SO(3)$, which proves our theorem. ∎

I.8.3. Theorem 32. *Let* $\mathfrak{g}$ *be a compact simple Lie algebra, and* $\mathfrak{g}/\mathfrak{l}$ *be a principal central involutive pair of index 1 for an involutive automorphism* S. *Then* S *is a principal involutive automorphism of type* U *and*

$$\mathfrak{g}/\mathfrak{l} \cong su(3)/su(2) \oplus u(1)$$

with the natural embedding.

Proof. The proof follows immediately from I.8.1 (Theorem 30). ∎

I.8.4. Theorem 33. *Let* $\mathfrak{g}$ *be a simple compact Lie algebra, and* $\mathfrak{g}/\mathfrak{l}$ *be a principal non-central involutive pair of type* U *and of index 1. Then*

$$\mathfrak{g}/\mathfrak{l} \cong sp(n+1)/sp(n) \oplus sp(1)$$

with the natural embedding.

Proof. The proof follows immediately from I.8.1, I.8.2 (Theorems 30, 31). ∎

PART TWO

Я - Свет, Я - Сумрак, Я - Солярис,
Я - Дхармы Сын и Кармы Блик.
Я - трансцендентный Ирреалис,
Я преднамеренно возник.

Квантасмагор

CHAPTER II.1

HYPER-INVOLUTIVE DECOMPOSITION
OF A SIMPLE COMPACT LIE ALGEBRA

II.1.1. Let $\mathfrak{g}$ be a compact simple Lie algebra, $\mathfrak{b}$ be its three-dimensional sub-algebra, and let $\mathrm{Int}_{\mathfrak{g}}\,\mathfrak{b}$ be isomorphic to $SO(3)$. In $SO(3)$ we may choose the elements

$$\bar{\bar{\varphi}}_1 = \begin{pmatrix} 1 & 0 & 0 \\ 0 & 0 & -1 \\ 0 & 1 & 0 \end{pmatrix}, \; \bar{\bar{\varphi}}_2 = \begin{pmatrix} 0 & 0 & 1 \\ 0 & 1 & 0 \\ -1 & 0 & 0 \end{pmatrix}, \; \bar{\bar{\varphi}}_3 = \begin{pmatrix} 0 & -1 & 0 \\ 1 & 0 & 0 \\ 0 & 0 & 1 \end{pmatrix},$$

$$\left. \begin{array}{c} \bar{\bar{p}} = \begin{pmatrix} 0 & 0 & 1 \\ 1 & 0 & 0 \\ 0 & 1 & 0 \end{pmatrix}, \\[2em] \bar{\bar{S}}_1 = \begin{pmatrix} 1 & 0 & 0 \\ 0 & -1 & 0 \\ 0 & 0 & -1 \end{pmatrix}, \; \bar{\bar{S}}_2 = \begin{pmatrix} -1 & 0 & 0 \\ 0 & 1 & 0 \\ 0 & 0 & -1 \end{pmatrix}, \; \bar{\bar{S}}_3 = \begin{pmatrix} -1 & 0 & 0 \\ 0 & -1 & 0 \\ 0 & 0 & 1 \end{pmatrix}. \end{array} \right\} \quad \text{(II.1)}$$

It is easily verified that

$$\left. \begin{array}{c} \bar{\bar{\mathrm{Id}}} = \left(\bar{\bar{p}}\right)^3 = \left(\bar{\bar{S}}_\rho\right)^2, \qquad \bar{\bar{S}}_\rho \bar{\bar{S}}_\mu = \bar{\bar{S}}_\mu \bar{\bar{S}}_\rho, \\[1em] \bar{\bar{S}}_1 \bar{\bar{S}}_2 = \bar{\bar{S}}_3, \qquad \bar{\bar{S}}_2 \bar{\bar{S}}_3 = \bar{\bar{S}}_1, \qquad \bar{\bar{S}}_3 \bar{\bar{S}}_1 = \bar{\bar{S}}_2, \\[1em] \bar{\bar{p}} \bar{\bar{S}}_1 \bar{\bar{p}}^{-1} = \bar{\bar{S}}_2, \qquad \bar{\bar{p}} \bar{\bar{S}}_2 \bar{\bar{p}}^{-1} = \bar{\bar{S}}_3, \qquad \bar{\bar{p}} \bar{\bar{S}}_3 \bar{\bar{p}}^{-1} = \bar{\bar{S}}_1, \\[1em] \left(\bar{\bar{\varphi}}_1 \bar{\bar{\varphi}}_2\right) = \left(\bar{\bar{\varphi}}_2 \bar{\bar{\varphi}}_3\right) = \left(\bar{\bar{\varphi}}_3 \bar{\bar{\varphi}}_1\right) = \bar{\bar{p}}, \\[1em] \left(\bar{\bar{\varphi}}_\rho\right)^2 = \bar{\bar{S}}_\rho, \qquad \rho, \mu = 1, 2, 3. \end{array} \right\} \quad \text{(II.2)}$$

Thus owing to the isomorphism of $SO(3)$ and $\mathrm{Int}_{\mathfrak{g}}(\mathfrak{b})$ we have in $\mathrm{Aut}(\mathfrak{g})$ the automorphisms S_1, S_2, S_3, p, and, in addition, there are $\varphi_\rho \in \mathrm{Int}_{\mathfrak{g}}(\mathfrak{b})$ ($\rho = 1, 2, 3$)

such that

$$(p)^3 = (S_\rho)^2 = \text{Id}, \qquad S_\rho \neq \text{Id}, \qquad S_\rho S_\mu = S_\mu S_\rho$$

$$(\rho, \mu = 1, 2, 3),$$

$$S_1 S_2 = S_3, \qquad S_2 S_3 = S_1, \qquad S_3 S_1 = S_2,$$

$$p S_1 = S_2 p, \qquad p S_2 = S_3 p, \qquad p S_3 = S_1 p,$$

$$(\varphi_\rho)^2 = S_\rho \quad (\rho = 1, 2, 3), \qquad \varphi_1 \varphi_2 = \varphi_2 \varphi_3 = \varphi_3 \varphi_1 = p. \qquad \text{(II.3)}$$

Our further construction depends only on (II.3).

Let $\mathfrak{l}_1$, $\mathfrak{l}_2$, $\mathfrak{l}_3$ be the involutive subalgebras of the involutive automorphisms S_1, S_2, S_3, respectively, and $\mathfrak{l}_0 = \mathfrak{l}_1 \cap \mathfrak{l}_2$. From (II.3) there follows

$$\mathfrak{g} = \mathfrak{l}_1 + \mathfrak{l}_2 + \mathfrak{l}_3, \qquad \mathfrak{l}_1 \cap \mathfrak{l}_2 = \mathfrak{l}_2 \cap \mathfrak{l}_3 = \mathfrak{l}_3 \cap \mathfrak{l}_1 = \mathfrak{l}_0,$$

$$\mathfrak{l}_\alpha = \mathfrak{m}_\alpha \dotplus \mathfrak{l}_0, \qquad \mathfrak{m}_\alpha \perp \mathfrak{l}_0, \qquad \text{(II.4)}$$

$$p \mathfrak{m}_1 = \mathfrak{m}_2, \qquad p \mathfrak{m}_2 = \mathfrak{m}_3, \qquad p \mathfrak{m}_3 = \mathfrak{m}_1, \qquad p \mathfrak{l}_0 = \mathfrak{l}_0,$$

$\mathfrak{l}_\alpha / \mathfrak{l}_0$ $(\alpha = 1, 2, 3)$ are involutive pairs.

If in $\text{Aut}(\mathfrak{g})$ there are automorphisms satisfying (II.3) then the decomposition (II.4) is evidently hyper-involutive, see 1.27 (Definition 17). Then our previous construction results in the following lemma:

II.1.2. Lemma 1. *Let $\mathfrak{b}$ be a three-dimensional simple compact Lie subalgebra in a simple compact Lie algebra $\mathfrak{g}$, and $\text{Int}_\mathfrak{g}(\mathfrak{b}) \cong SO(3)$, then $\mathfrak{g}$ has a hyper-involutive decomposition (II.4) such that automorphisms p, S_1, S_2, S_3 belong to $\text{Int}_\mathfrak{g}(\mathfrak{b})$.*

II.1.3. Definition 24. A base in a Lie algebra $\mathfrak{g}$ is called a hyper-involutive base (hyper-invobase) of a hyper-involutive group $\chi(S_1, S_2, S_3; p)$, and of the corresponding hyper-involutive sum $\mathfrak{g} = \mathfrak{l}_1 + \mathfrak{l}_2 + \mathfrak{l}_3$, if its restriction to $\mathfrak{m}_1 \dotplus \mathfrak{m}_2 \dotplus \mathfrak{m}_3$ is invariant under the action of p, and S_1, S_2, S_3 have diagonal forms in this base.

In the case of a semi-simple Lie algebra $\mathfrak{g}$, by a hyper-involutive base we mean an orthonormal hyper-involutive base only.

II.1.4. Lemma 2. *If $\mathfrak{g}$ is a simple compact Lie algebra and $\chi(S_1, S_2, S_3; p)$ is any of its hyper-involutive group of automorphisms then there exists a hyper-involutive base for $\chi(S_1, S_2, S_3; p)$.*

The proof is based on the following construction. Let us take the orthonormal bases in $\mathfrak{m}_1$ and $\mathfrak{l}_0$, X_{I_1} $(I = 1, \ldots, n)$ and Y_A $(A = 1, \ldots, r)$, respectively, and define an orthonormal base in $\mathfrak{m}_2$ by $p X_{I_1} = X_2$ $(I = 1, \ldots, n)$ and an orthonormal base in $\mathfrak{m}_3$ by $p^2 X_{I_1} = p X_{I_2} = X_{I_3}$ $(I = 1, \ldots, n)$.

The union of all the bases considered above, namely,

$$\left.\begin{aligned}
X_{I_1} &\in \mathfrak{m}_1 \qquad (I = 1, \ldots, n), \\
X_{I_2} &\in \mathfrak{m}_2 \qquad (I = 1, \ldots, n), \\
X_{I_3} &\in \mathfrak{m}_3 \qquad (I = 1, \ldots, n), \\
Y_A &\in \mathfrak{l}_0 \qquad (A = 1, \ldots, r),
\end{aligned}\right\} \tag{II.5}$$

gives us an orthonormal hyper-involutive base in $\mathfrak{g}$.

Because of the construction we obtain

$$\left.\begin{aligned}
p X_{I_1} = X_{I_2}, \qquad p X_{I_2} &= X_{I_3}, \qquad p X_{I_3} = X_{I_1}, \\
p Y_A &= p_A^B Y_B, \\
S_\rho X_{I_\rho} = X_{I_\rho}, \qquad S_\rho Y_A &= Y_A, \qquad S_\rho X_{I_\mu} = -X_{I_\mu}, \\
(\rho, \mu = 1,2,3; \ \rho \neq \mu), \qquad &\left(p^3\right)_A^B = \delta_A^B,
\end{aligned}\right\} \tag{II.6}$$

which is just the condition for being hyper-involutive. It is obvious that any hyper-involutive base can be obtained in the way described above.

II.1.5. Taking into account (II.6) we can write the structure equations for $\mathfrak{g}$ in the form

$$[X_{I_\rho}, X_{J_\rho}] = C_{I_\rho J_\rho}^A Y_A, \qquad [X_{I_\rho}, X_{J_\mu}] = C_{I_\rho J_\mu}^{K_\lambda} X_{K_\lambda},$$

$$[X_{I_\rho}, Y_A] = C_{I_\rho A}^{K_\rho} X_{K_\rho}, \qquad [Y_A Y_B] = C_{AB}^F Y_F, \tag{II.7}$$

$$\lambda, \mu, \rho = 1,2,3; \quad \lambda \neq \mu, \quad \mu \neq \rho, \quad \rho \neq \lambda.$$

Taking into consideration the action of the automorphism p, see (II.6) and (II.7), we obtain

$$C_{I_2 J_3}^{K_1} = C_{I_1 J_2}^{K_3} = C_{I_3 J_1}^{K_2} = t_{IJ}^K, \quad C_{I_1 J_1}^B = C_{IJ}^B,$$

$$C_{I_2 J_2}^B = C_{IJ}^A p_A^B,$$

$$C_{I_3 J_3}^B = C_{IJ}^A \left(p^2\right)_A^B, \quad C_{I_1 A}^{K_1} = C_{IA}^K, \quad C_{I_2 A}^{K_2} = C_{IB}^K \left(p^2\right)_A^B, \tag{II.8}$$

$$C_{I_3 A}^{K_3} = C_{IB}^K p_A^B, \quad C_{AB}^F p_F^D \left(p^2\right)_G^A \left(p^2\right)_H^B = C_{GH}^D.$$

By the orthonormality of the base (II.5) we obtain the skew-symmetry for the structure constants of (II.7) with respect to any pair of indices and the orthogonality of the matrix p_F^D. Together with (II.8), (II.6) this gives us

$$t_{IJ}^K = t_{JK}^I, \quad C_{IJ}^B = -C_{IB}^J, \quad \left(p^2\right)_B^A = p_A^B,$$

$$C_{AB}^F = -C_{FB}^A. \tag{II.9}$$

From (II.9) we obtain also

$$t_{IJ}^R = K_{IJ}^R + E_{IJ}^R, \qquad K_{IJ}^R = -K_{JI}^R = -K_{RJ}^I,$$
$$E_{IJ}^R = E_{JI}^R = E_{RJ}^I. \tag{II.10}$$

II.1.6. Now we intend to consider the Jacobi identities for (II.7) taking into account (II.8), (II.9).

From

$$\left[X_{I_1} \left[X_{J_2} X_{K_3} \right] \right] + \left[X_{K_3} \left[X_{I_1} X_{J_2} \right] \right] + \left[X_{J_2} \left[X_{K_3} X_{I_1} \right] \right] = 0$$

we obtain

$$t_{JK}^P C_{IP}^A + t_{IJ}^P C_{KP}^B \left(p^2 \right)_B^A + t_{KI}^P C_{JP}^B p_B^A = 0. \tag{II.11.1}$$

$$\left[\left[X_{I_1} X_{J_2} \right] Y_A \right] + \left[\left[Y_A X_{I_1} \right] X_{J_2} \right] + \left[\left[X_{J_2} Y_A \right] X_{I_1} \right] = 0,$$

which gives us

$$t_{IJ}^K C_{KB}^P p_A^B - t_{KJ}^P C_{IA}^K - t_{IK}^P C_{JB}^K \left(p^2 \right)_A^B = 0. \tag{II.11.2}$$

$$\left[\left[X_{I_1} X_{J_1} \right] X_{K_2} \right] + \left[\left[X_{K_2} X_{I_1} \right] X_{J_1} \right] + \left[\left[X_{J_1} X_{K_2} \right] X_{I_1} \right] = 0,$$

which gives us

$$t_{JK}^P t_{PI}^R - t_{IK}^P t_{PJ}^R = C_{KB}^R \left(p^2 \right)_A^B C_{IJ}^A. \tag{II.11.3}$$

$$\left[\left[X_{I_1} X_{J_1} \right] X_{K_3} \right] + \left[\left[X_{K_3} X_{I_1} \right] X_{J_1} \right] + \left[\left[X_{J_1} X_{K_3} \right] X_{I_1} \right] = 0,$$

which gives us

$$t_{KJ}^P t_{IP}^R - t_{KI}^P t_{JP}^R = C_{KB}^R \left(p \right)_A^B C_{IJ}^A. \tag{II.11.4}$$

$$\left[X_{I_1} \left[X_{J_1} X_{K_1} \right] \right] + \left[X_{K_1} \left[X_{I_1} X_{J_1} \right] \right] + \left[X_{J_1} \left[X_{K_1} X_{I_1} \right] \right] = 0.$$

which gives us

$$C_{IA}^R C_{JK}^A + C_{KA}^R C_{IJ}^A + C_{JA}^R C_{KI}^A = 0. \tag{II.11.5}$$

$$\left[\left[X_{I_1} X_{J_1} \right] Y_A \right] + \left[\left[Y_A X_{I_1} \right] X_{J_1} \right] + \left[\left[X_{J_1} Y_A \right] X_{I_1} \right] = 0,$$

which gives us

$$C_{IJ}^B C_{BA}^E + C_{JA}^P C_{PI}^E - C_{IA}^P C_{PJ}^E = 0. \tag{II.11.6}$$

$$\left[X_{I_1} \left[Y_A Y_B \right] \right] + \left[Y_B \left[X_{I_1} Y_A \right] \right] + \left[Y_A \left[Y_B X_{I_1} \right] \right] = 0,$$

which gives us

$$C_{AB}^E C_{IE}^R + C_{IB}^P C_{PA}^R - C_{IA}^P C_{PB}^R = 0. \tag{II.11.7}$$

$$[[Y_A Y_B]Y_F] + [[Y_F Y_A]Y_B] + [[Y_B Y_F]Y_A] = 0,$$

which gives us

$$C_{AB}^E C_{EF}^G + C_{FA}^E C_{EB}^G + C_{BF}^E C_{EA}^G = 0. \tag{II.11.8}$$

The other Jacobi identities are consequences of those written above owing to the conditions (II.5) on the automorphism p. Moreover, (II.11.6) follows from (II.11.7), and (II.11.1) follows from (II.11.2) owing to (II.9).

Thus the essential relations are only

$$\left.\begin{aligned}
& t_{IJ}^K C_{KB}^P p_A^B - t_{KJ}^P C_{IA}^K - t_{IK}^P C_{JB}^K \left(p^2\right)_A^B = 0, \\
& t_{JK}^P t_{PI}^R - t_{IK}^P t_{PJ}^R = C_{KB}^R \left(p^2\right)_A^B C_{IJ}^A, \\
& t_{KJ}^P t_{IP}^R - t_{KI}^P t_{JP}^R = C_{KB}^R \left(p\right)_A^B C_{IJ}^A, \\
& C_{IA}^R C_{JK}^A + C_{KA}^R C_{IJ}^A + C_{JA}^R C_{KI}^A = 0, \\
& C_{PB}^R C_{IA}^P - C_{PA}^R C_{IB}^P = C_{AB}^E C_{IE}^R, \\
& C_{AB}^E C_{EF}^G + C_{FA}^E C_{EB}^G + C_{BF}^E C_{EA}^G = 0.
\end{aligned}\right\} \tag{II.12}$$

II.1.7. Definition 25. We say that the hyper-involutive decomposition (II.4) is prime if the restriction of the automorphism p to $\mathfrak{l}_0$ is the identity automorphism, and is non-prime otherwise.

II.1.8. Let us now consider additional relations which appear if we take into account (II.3). From (II.3) we have

$$p\,\varphi_1 = \varphi_2\,p, \qquad p\,\varphi_2 = \varphi_3\,p, \qquad p\,\varphi_3 = \varphi_1\,p,$$
$$\varphi_1\,S_1 = S_1\,\varphi_1, \qquad \varphi_1\,S_2 = S_3\,\varphi_1, \qquad \varphi_2\,S_2 = S_2\,\varphi_2, \tag{II.13}$$
$$\varphi_2\,S_3 = S_1\,\varphi_2, \qquad \varphi_3\,S_3 = S_3\,\varphi_3, \qquad \varphi_3\,S_1 = S_2\,\varphi_3,$$

whence

$$\varphi_\rho\,\mathfrak{l}_\rho = \mathfrak{l}_\rho \quad (\rho = 1,2,3), \qquad \varphi_1\,\mathfrak{l}_2 = \mathfrak{l}_3, \qquad \varphi_1\,\mathfrak{l}_3 = \mathfrak{l}_2,$$
$$\varphi_2\,\mathfrak{l}_1 = \mathfrak{l}_3, \qquad \varphi_2\,\mathfrak{l}_3 = \mathfrak{l}_1, \qquad \varphi_3\,\mathfrak{l}_1 = \mathfrak{l}_2, \qquad \varphi_3\,\mathfrak{l}_2 = \mathfrak{l}_1, \tag{II.14}$$
$$\varphi_\rho\,\mathfrak{l}_0 = \mathfrak{l}_0, \quad \rho = 1,2,3.$$

By virtue of the preceding we obtain:

$$\varphi_1\,X_{I_1} = \varphi_I^K X_{K_1}, \qquad \varphi_1\,X_{I_2} = \theta_I^K X_{K_3}, \qquad \varphi_1\,X_{I_3} = \lambda_I^K X_{K_2}.$$

Using $p\,\varphi_1 = \varphi_2\,p$ and (II.6) we have also

$$\varphi_2\,X_{I_2} = \varphi_I^K X_{K_2}, \qquad \varphi_2\,X_{I_3} = \theta_I^K X_{K_1}, \qquad \varphi_2\,X_{I_1} = \lambda_I^K X_{K_3}.$$

But $\left(\varphi_1\right)^2 = S_1$, implying

$$X_{I_1} = S_1 X_{I_1} = \varphi_1 \left(\varphi_1 X_{I_1} \right) = \varphi_I^K \left(\varphi_1 X_{K_1} \right) = \varphi_I^K \varphi_K^R X_{R_1},$$

whence $\varphi_K^R \varphi_I^K = \delta_I^R$. As well

$$-X_{I_2} = S_1 X_{I_2} = \varphi_1 \left(\varphi_1 X_{I_2} \right) = \theta_I^K \left(\varphi_1 X_{K_3} \right) = \theta_I^K \lambda_K^R X_{R_2},$$

whence $\lambda_K^R \theta_I^K = -\delta_I^R$. Moreover, (II.3) and (II.6) imply

$$X_{I_1} = p X_{I_3} = \varphi_1 \left(\varphi_2 X_{I_3} \right) = \theta_I^K \left(\varphi_1 X_{K_1} \right) = \theta_I^K \varphi_K^R X_{R_1},$$

that is, $\varphi_K^R \theta_I^K = \delta_I^R$. Thus

$$\varphi_K^R \varphi_I^K = \delta_I^R, \qquad \varphi_K^R \theta_I^K = \delta_I^R, \qquad \lambda_K^R \theta_I^K = -\delta_I^R,$$

and consequently

$$\varphi_I^R = \theta_I^R = -\lambda_I^R.$$

Taking into consideration $p \varphi_2 = \varphi_3 p$ we have, finally,

$$\varphi_1 X_{I_1} = \varphi_I^K X_{K_1}, \qquad \varphi_1 X_{I_2} = \varphi_I^K X_{K_3}, \qquad \varphi_1 X_{I_3} = -\varphi_I^K X_{K_2},$$

$$\varphi_2 X_{I_2} = \varphi_I^K X_{K_2}, \qquad \varphi_2 X_{I_3} = \varphi_I^K X_{K_1}, \qquad \varphi_2 X_{I_1} = -\varphi_I^K X_{K_3},$$

$$\varphi_3 X_{I_3} = \varphi_I^K X_{K_3}, \qquad \varphi_3 X_{I_1} = \varphi_I^K X_{K_2}, \qquad \varphi_3 X_{I_2} = -\varphi_I^K X_{K_1}, \tag{II.15}$$

$$\varphi_K^I \varphi_J^K = \delta_J^I.$$

By (II.15) and the orthonormality of base we obtain

$$\varphi_K^I = \varphi_I^K. \tag{II.16}$$

In addition, the conditions (II.3), (II.13), (II.14) are to be considered for $\mathfrak{l}_0$ taking into account the orthonormality of the base. Then

$$(\varphi_\rho) Y_A = (\varphi_\rho)_A^B Y_B,$$

$$(\varphi_1)_A^B (\varphi_2)_C^A = (\varphi_2)_A^B (\varphi_3)_C^A = (\varphi_3)_A^B (\varphi_1)_C^A = p_C^B, \tag{II.17}$$

$$(\varphi_\rho)_A^B (\varphi_\rho)_C^A = \delta_C^B, \qquad (\varphi_\rho)_A^B = (\varphi_\rho)_B^A \quad (\rho = 1, 2, 3).$$

Let us now use that φ_ρ $(\rho = 1, 2, 3)$ are automorphisms of the structure (II.7). Then with the help of (II.15), (II.17), (II.8) we obtain

$$\varphi_I^R \varphi_J^Q t_{QR}^S = t_{IJ}^K \varphi_K^S, \qquad \varphi_I^R C_{RB}^K (\varphi_1)_A^B = \varphi_J^K C_{IA}^J,$$

$$(\varphi_1)_B^A (\varphi_1)_D^E C_{AE}^F = (\varphi_1)_G^F C_{BD}^G. \tag{II.18}$$

The other conditions are the consequences of (II.6)–(II.18).

CHAPTER II.2

SOME AUXILIARY RESULTS

II.2.1. Theorem 1. Let $\mathfrak{g}/\mathfrak{l}$ be an involutive pair of an involutive automor-phism S, and q be an automorphism of $\mathfrak{g}$ such that $qS = Sq$ and $q\eta = \eta$ for any $\eta \in \mathfrak{l}$. Then

$$q\mathfrak{m} = \mathfrak{m} \quad (\mathfrak{m} = \mathfrak{g} \dot{-} \mathfrak{l}), \qquad q^{-1}\mathfrak{m} = \mathfrak{m},$$

$$\left(q \pm q^{-1} \right) \mathrm{ad}_{\mathfrak{g}}^{\mathfrak{g} - \mathfrak{l}}(y) = \mathrm{ad}_{\mathfrak{g}}^{\mathfrak{g} - \mathfrak{l}}(y) \left(q \pm q^{-1} \right)$$

for any $y \in \mathfrak{l}$, and $\left(q - q^{-1} \right)$ is a differentiation of the Lie algebra $\mathfrak{g}$.

Proof. Evidently q^{-1} is an automorphism of $\mathfrak{g}$ and $q^{-1}\eta = \eta$ for any $\eta \in \mathfrak{l}$. Let us consider $\mathfrak{g} = \mathfrak{m} \dot{+} \mathfrak{l}$ ($S\eta = \eta$ for $\eta \in \mathfrak{l}$, $S\eta = -\eta$ for $\eta \in \mathfrak{m}$), then $[\mathfrak{m}\,\mathfrak{l}] \subset \mathfrak{m}$, $[\mathfrak{m}\,\mathfrak{m}] \subset \mathfrak{l}$, $[\mathfrak{l}\,\mathfrak{l}] \subset \mathfrak{l}$. Because of $qS = Sq$ it is evident that $q\mathfrak{m} = \mathfrak{m}$, $q^{-1}\mathfrak{m} = \mathfrak{m}$. If $x \in \mathfrak{m}$, $y \in \mathfrak{l}$, then

$$\left(q \pm q^{-1} \right) [xy] = q [xy] \pm q^{-1} [xy]$$

$$= \left[(qx)(qy) \right] \pm \left[\left(q^{-1}x \right) \left(q^{-1}y \right) \right]$$

$$= \left[(qx)\, y \right] \pm \left[\left(q^{-1}x \right) y \right]$$

$$= \left[\left(\left(q \pm q^{-1} \right) x \right) y \right],$$

which proves the first part of the theorem.

In order to prove the second part we consider

$$\left[\left(\left(q - q^{-1} \right) \zeta \right) \eta \right] + \left[\zeta \left(\left(q - q^{-1} \right) \eta \right) \right]$$

$$= \left[(q\zeta)\eta \right] - \left[\left(q^{-1}\zeta \right) \eta \right] + \left[\zeta (q\eta) \right] - \left[\zeta \left(q^{-1}\eta \right) \right]$$

$$= \left[(q\zeta)\,\eta \right] - q^{-1} \left[\zeta (q\eta) \right] + \left[\zeta (q\eta) \right] - q^{-1} \left[(q\zeta)\,\eta \right]$$

$$= \left(\mathrm{Id} - q^{-1} \right) \left[(q\zeta)\,\eta \right] + \left(\mathrm{Id} - q^{-1} \right) \left[\zeta (q\eta) \right]$$

$$= \left(\mathrm{Id} - q^{-1} \right) \left(\left[(q\zeta)\,\eta \right] + \left[\zeta (q\eta) \right] \right)$$

and also

$$\left(q - q^{-1} \right) [\zeta\eta] = \left(\mathrm{Id} - q^{-1} \right) \left(\mathrm{Id} + q \right) [\zeta\eta]$$

$$= \left(\mathrm{Id} - q^{-1} \right) \left(\left[(q\zeta)(q\eta) \right] + [\zeta\eta] \right).$$

Subtracting term from term we have

$$[((q-q^{-1})\,\zeta)\,\eta] + [\zeta\,((q-q^{-1})\,\eta)] - (q-q^{-1})\,[\zeta\eta]$$
$$= (\mathrm{Id} - q^{-1})\,([(q\zeta)\,\eta] + [\zeta\,(q\eta)]$$
$$- [(q\zeta)\,(q\eta)] - [\zeta\eta])$$
$$= (\mathrm{Id} - q^{-1})\,\xi, \tag{II.9}$$

where $\xi = [(q\zeta)\,\eta] + [\zeta\,(q\eta)] - 2[\zeta\eta]$.

If both ζ and η are from $\mathfrak{m}$ then $\xi \in \mathfrak{l}$, and consequently $(\mathrm{Id} - q^{-1})\,\xi = 0$. If $\eta \in \mathfrak{l}$ then $\xi = 0$, and again $(\mathrm{Id} - q^{-1})\,\xi = 0$. By (II.19) we conclude that $(q - q^{-1})$ is a differentiation. ∎

II.2.2. Theorem 2. *If $\mathfrak{g}$ is a semi-simple compact Lie algebra, $\mathfrak{g}/\mathfrak{l}$ an irreducible (in particular, elementary) involutive pair of an involutive automorphism S, and q an automorphism of $\mathfrak{g}$ such that $q \neq \mathrm{Id}$, $q \neq S$, $q\eta = \eta$ for $\eta \in \mathfrak{l}$, then $q \in \mathrm{Int}_{\mathfrak{g}}(\mathfrak{z})$, where $\mathfrak{z}$ is a non-trivial centre in $\mathfrak{l}$.*

Proof. Since $\mathfrak{g}$ is semi-simple and compact then $\mathfrak{m} = \mathfrak{g} \dot{-} \mathfrak{l}$ is mapped onto itself under the action of q as the orthogonal complement to $\mathfrak{l}$ in $\mathfrak{g}$ with respect to the Cartan metric.

But $S = -\mathrm{Id}$ on $\mathfrak{m}$, whence $qS = Sq$ on $\mathfrak{g}$.

Let $q \neq \mathrm{Id}$, $q \neq S$, then $(q - q^{-1}) = K \neq 0$ is a differentiation of $\mathfrak{g}$ by II.2.1. (Theorem 1), and moreover, is an inner differentiation since $\mathfrak{g}$ is semi-simple, $K(x) = [k\,x] = (\mathrm{ad}_{\mathfrak{g}}k)x$. But $K(x) = 0$ for any $x \in \mathfrak{l}$, thus $[k\,\mathfrak{l}] = \{0\}$.

Let $k = k_{\mathfrak{m}} + k_{\mathfrak{l}}$, where $k_{\mathfrak{m}} \in \mathfrak{m}$, $k_{\mathfrak{l}} \in \mathfrak{l}$, then $[k_{\mathfrak{m}}\,y] + [k_{\mathfrak{l}}\,y] = [k\,y] = 0$ ($y \in \mathfrak{l}$). However, $[k_{\mathfrak{m}}\,\mathfrak{l}] \subset \mathfrak{m}$, $[k_{\mathfrak{l}}\,\mathfrak{l}] \subset \mathfrak{l}$, consequently $[k_{\mathfrak{m}}\,\mathfrak{l}] = \{0\}$. This means that $\mathrm{ad}_{\mathfrak{g}}^{\mathfrak{g}\dot{-}\mathfrak{l}}(\mathfrak{l})\,k_{\mathfrak{m}} = \{0\}$, which, owing to the irreducibility of $\mathrm{ad}_{\mathfrak{g}}^{\mathfrak{g}\dot{-}\mathfrak{l}}(\mathfrak{l})$, is possible only if $k_{\mathfrak{m}} = 0$. Thus $k \in \mathfrak{l}$, $[k\,\mathfrak{l}] = \{0\}$, that is, k belongs to the center $\mathfrak{z}$ of $\mathfrak{l}$. The automorphism $q' = q^{-1}$ is adjoint to q with respect to the Cartan metric. Thus $\frac{1}{2}(q + q^{-1})$ and $\left(\frac{1}{2}(q - q^{-1})\right)^2$ are self-adjoint endomorphisms commuting on $\mathfrak{m}$ with $\mathrm{ad}_{\mathfrak{g}}^{\mathfrak{g}\dot{-}\mathfrak{l}}(\mathfrak{l})$ by II.2.1. (Theorem 1).

Since $\mathrm{ad}_{\mathfrak{g}}^{\mathfrak{g}\dot{-}\mathfrak{l}}(\mathfrak{l})$ is irreducible we have on $\mathfrak{m}$:

$$\frac{1}{2}\left(q|_{\mathfrak{m}} + q^{-1}|_{\mathfrak{m}}\right) = \lambda\,\mathrm{Id}, \qquad \left[\frac{1}{2}\left(q|_{\mathfrak{m}} - q^{-1}|_{\mathfrak{m}}\right)\right]^2 = -\mu^2\,\mathrm{Id}.$$

But

$$q = \frac{1}{2}\left(q + q^{-1}\right) + \frac{1}{2}\left(q - q^{-1}\right) = \frac{1}{2}\left(q + q^{-1}\right) + \frac{1}{2}\,\mathrm{ad}_{\mathfrak{g}}k, \tag{II.20}$$

hence

$$q|_{\mathfrak{m}} = \lambda\,\mathrm{Id} + \mu E \quad (\mu \neq 0), \qquad E^2 = -\mathrm{Id}, \qquad E' = -E,$$
$$q|_{\mathfrak{l}} = \mathrm{Id}. \tag{II.21}$$

Furthermore

$$\mathrm{Id} = (q^{-1}|_{\mathfrak{m}})\,(q|_{\mathfrak{m}}) = (q'|_{\mathfrak{m}})\,(q|_{\mathfrak{m}})$$

$$= (\lambda\,\mathrm{Id} - \mu\,E)\,(\lambda\,\mathrm{Id} + \mu\,E) = (\lambda^2 + \mu^2)\,\mathrm{Id},$$

whence

$$q|_{\mathfrak{m}} = (\cos\varphi)\mathrm{Id} + (\sin\varphi)E = e^{\varphi E}, \qquad E^2 = -\mathrm{Id}, \qquad E' = -E,$$
$$q|_{\mathfrak{l}} = \mathrm{Id}. \tag{II.22}$$

Since $\mathrm{ad}_{\mathfrak{g}}^{\mathfrak{g}^{-\mathfrak{l}}}(\mathfrak{k}) = 2(\sin\varphi)E$, $\mathrm{ad}_{\mathfrak{g}}^{\mathfrak{l}}(\mathfrak{k}) = 0$, see (II.20)–(II.22), we obtain $q = e^{\psi\,\mathrm{ad}_{\mathfrak{g}}\mathfrak{k}}$, where $\psi = \varphi/2\sin\varphi$. Consequently $q \in \mathrm{Int}_{\mathfrak{g}}(t\,\mathfrak{k})$. But $\mathfrak{k} \in \mathfrak{z}$, where $\mathfrak{z}$ is the centre of $\mathfrak{l}$. Thus $q \in \mathrm{Int}_{\mathfrak{g}}(\mathfrak{z})$. ∎

II.2.3. Theorem 3. *Let $\mathfrak{g}$ be a semi-simple compact Lie algebra, and $\mathfrak{l}$ be its involutive algebra of an involutive automorhism S. If $q \in \mathrm{Aut}\,\mathfrak{g}$ $(q \neq \mathrm{Id})$, $q\eta = \eta$ for any $\eta \in \mathfrak{l}$, and $q + \mathrm{Id}$ is invertible (in particular, $q^{2n+1} = \mathrm{Id}$) then $q \in \mathrm{Int}_{\mathfrak{g}}(\mathfrak{z})$, where $\mathfrak{z}$ is a non-trivial centre of $\mathfrak{l}$.*

Proof. For any semi-simple compact Lie algebra $\mathfrak{g}$ with an involutive algebra $\mathfrak{l}$ we have $\mathfrak{g}/\mathfrak{l} = \mathfrak{g}_1/\mathfrak{l}_1 + \ldots + \mathfrak{g}_m/\mathfrak{g}_m$, where $\mathfrak{g}_k/\mathfrak{l}_k$ $(k = 1,\ldots,m)$ are elementary involutive pairs, see, for example, [S. Helgason 62,78].

Since $q\eta = \eta$ for $\eta \in \mathfrak{l}$, we have $q\mathfrak{g}_k = \mathfrak{g}_k$ $(k = 1,\ldots,m)$. Let us consider involutive pairs $\mathfrak{g}_s/\mathfrak{l}_s$ for which $q \neq \mathrm{Id}$. Owing to the invertibility of $q + \mathrm{Id}$ on $\mathfrak{g}$ we have $q \neq S$ on $\mathfrak{g}_k$. Then by II.2.2 (Theorem 2) we have $q|_{\mathfrak{g}_s} \in \mathrm{Int}_{\mathfrak{g}_s}(\mathfrak{z}_s)$, where $\mathfrak{z}_s$ is the (non-trivial) centre of $\mathfrak{l}_s$. But

$$\mathrm{Int}_{\mathfrak{g}}(\mathfrak{z}) = \mathrm{Int}_{\mathfrak{g}}(\mathfrak{z}_1)\cdots\mathrm{Int}_{\mathfrak{g}}(\mathfrak{z}_m),$$

where $\mathfrak{z} = \mathfrak{z}_1 \oplus \ldots \oplus \mathfrak{z}_m$ is the (non-trivial) centre of $\mathfrak{l} = \mathfrak{l}_1 \oplus \ldots \oplus \mathfrak{l}_m$. (Note that for some k $\mathfrak{z}_k = \{0\}$.)

Introducing $q_s = q|_{\mathfrak{g}_s}$ on $\mathfrak{g}_s$, $q_s = \mathrm{Id}$ on $\mathfrak{g}_k$ $(k \neq s)$ we see that $q = q_1 \cdots q_m$ and $q_s \in \mathrm{Int}_{\mathfrak{g}}(\mathfrak{z}_s)$. Consequently $q \in \mathrm{Int}_{\mathfrak{g}}(\mathfrak{z}_1) \cdots \mathrm{Int}_{\mathfrak{g}}(\mathfrak{z}_m) = \mathrm{Int}_{\mathfrak{g}}(\mathfrak{z})$. ∎

II.2.4. Theorem 4. *Let $\mathfrak{g}$ be a compact Lie algebra, $\mathfrak{l} \subset \mathfrak{g}$ be its involutive algebra (of an involutive automorphism S) such that the centre $\mathfrak{g}_0$ of $\mathfrak{g}$ belongs to $\mathfrak{l}$, $\mathfrak{g}_0 \in \mathfrak{l}$. If $q \in \mathrm{Aut}\,\mathfrak{g}$ $(q \neq \mathrm{Id})$, $q\eta = \eta$ for $\eta \in \mathfrak{l}$, and $q + \mathrm{Id}$ is invertible (in particular, $q^{2n+1} = \mathrm{Id}$) then $q \in \mathrm{Int}_{\mathfrak{g}}(\mathfrak{z})$, where $\mathfrak{z}$ is a (non-trivial) centre of $\mathfrak{l}$.*

Proof. For any compact Lie algebra $\mathfrak{g}$ and its involutive algebra $\mathfrak{l}$ we have $\mathfrak{g}/\mathfrak{l} = \mathfrak{g}_0/\mathfrak{l}_0 + \tilde{\mathfrak{g}}/\tilde{\mathfrak{l}}$, where $\tilde{\mathfrak{g}}$ is semi-simple (and compact) and $\mathfrak{g}_0$ is the center in $\mathfrak{g}$. By the conditions $\mathfrak{g}_0 = \mathfrak{l}_0$. Evidently $q\,\mathfrak{g}_0 = \mathfrak{g}_0$, $q\,\tilde{\mathfrak{g}} = \tilde{\mathfrak{g}}$. But $q\eta = \eta$, for $\eta \in \mathfrak{l}$, implies $q = \mathrm{Id}$ on $\mathfrak{g}_0 \subset \mathfrak{l}$, consequently $q|_{\tilde{\mathfrak{g}}} \neq \mathrm{Id}$. Thus $q|_{\tilde{\mathfrak{g}}}$ satisfies on $\tilde{\mathfrak{g}}$ the conditions of II.2.3 (Theorem 3). And $q|_{\tilde{\mathfrak{g}}} \in \mathrm{Int}_{\tilde{\mathfrak{g}}}(\tilde{\mathfrak{z}})$, where $\tilde{\mathfrak{z}}$ is the (non-trivial) centre of $\tilde{\mathfrak{l}}$. But $\mathrm{Int}_{\mathfrak{g}}(\mathfrak{z}) = \mathrm{Int}_{\mathfrak{g}}(\mathfrak{g}_0) \cdot \mathrm{Int}_{\mathfrak{g}}(\tilde{\mathfrak{z}}) = \mathrm{Int}_{\mathfrak{g}}(\tilde{\mathfrak{z}})$, where $\mathfrak{z} = \mathfrak{g}_0 \oplus \tilde{\mathfrak{z}}$, being the non-trivial centre of $\mathfrak{l} = \mathfrak{l}_0 \oplus \tilde{\mathfrak{l}} = \mathfrak{g}_0 \oplus \tilde{\mathfrak{l}}$. (Note that $\mathrm{Int}_{\mathfrak{g}_0}(\mathfrak{g}_0) = \{\mathrm{Id}\}$.) Introducing $\tilde{q} = q|_{\tilde{\mathfrak{g}}}$ on $\tilde{\mathfrak{g}}$, $\tilde{q} = \mathrm{Id}$ on $\mathfrak{g}_0$, we see that $q = \tilde{q}$ and $\tilde{q} \in \mathrm{Int}_{\mathfrak{g}}(\tilde{\mathfrak{z}}) = \mathrm{Int}_{\mathfrak{g}}(\mathfrak{g}_0) \cdot \mathrm{Int}_{\mathfrak{g}}(\tilde{\mathfrak{z}})$. Thus $q \in \mathrm{Int}_{\mathfrak{g}}(\mathfrak{z})$. ∎

CHAPTER II.3

PRINCIPAL INVOLUTIVE AUTOMORPHISMS OF TYPE O

II.3.1. In this Chapter we prove the following theorem: in order for a simple compact Lie algebra $\mathfrak{g}$ to be isomorphic to $so(n)$ or $su(3)$ it is necessary and sufficient that $\mathfrak{g}$ has a principal involutive automorphism of type O.

II.3.2. Let $\mathfrak{g}$ be a simple compact Lie algebra, $\mathfrak{l}$ being its principal involutive algebra of a principal involutive automorphism S of type O. Then $\mathfrak{l} = \mathfrak{b} \oplus \tilde{\mathfrak{l}}$ and $\mathrm{Int}_{\mathfrak{g}}(\mathfrak{b})$ is isomorphic to $SO(3)$.

By II.1.2 (Lemma 1) we can construct a hyper-involutive decomposition of Lie algebra $\mathfrak{g}$ such that in $\mathrm{Int}_{\mathfrak{g}}(\mathfrak{b})$ there are S_1, S_2, S_3, p with the properties:

$$\left.\begin{aligned}
(p)^3 = (S_\rho)^2 &= \mathrm{Id}, \qquad S_\rho = \mathrm{Id}, \qquad S_\rho S_\mu = S_\mu S_\rho \\
&(\rho, \mu = 1, 2, 3), \\
S_1 S_2 = S_3, \qquad & S_2 S_3 = S_1, \qquad S_3 S_1 = S_2, \\
p S_1 = S_2\, p, \qquad & p S_2 = S_3\, p, \qquad p S_3 = S_1\, p.
\end{aligned}\right\} \tag{II.23}$$

And if $\mathfrak{l}_1$, $\mathfrak{l}_2$, $\mathfrak{l}_3$ are involutive algebras of involutive automorphisms S_1, S_2, S_3, respectively, then

$$\left.\begin{aligned}
\mathfrak{g} = \mathfrak{l}_1 + \mathfrak{l}_2 + \mathfrak{l}_3, \qquad & \mathfrak{l}_1 \cap \mathfrak{l}_2 = \mathfrak{l}_2 \cap \mathfrak{l}_3 = \mathfrak{l}_3 \cap \mathfrak{l}_1 = \mathfrak{l}_0, \\
\mathfrak{l}_\alpha = \mathfrak{m}_\alpha &\dotplus \mathfrak{l}_0, \qquad \mathfrak{m}_\alpha \perp \mathfrak{l}_0, \\
p\mathfrak{m}_1 = \mathfrak{m}_2, \qquad p\mathfrak{m}_2 = \mathfrak{m}_3, \qquad & p\mathfrak{m}_3 = \mathfrak{m}_1, \qquad p\mathfrak{l}_0 = \mathfrak{l}_0,
\end{aligned}\right\} \tag{II.24}$$

and $\mathfrak{l}_\alpha/\mathfrak{l}_0$ ($\alpha = 1, 2, 3$) are involutive pairs. (See (II.3), (II.4).)

Under our assumptions we have also

$$\left.\begin{aligned}
p S = S\, p, \qquad S S_\rho &= S_\rho S \quad (\rho = 1, 2, 3), \\
p\eta &= \eta, \quad \eta \in \tilde{\mathfrak{l}}, \\
\tilde{\mathfrak{l}} \subset \mathfrak{l}_0, \qquad & \mathfrak{b} \cap \mathfrak{l}_0 = \{0\},
\end{aligned}\right\} \tag{II.25}$$

and $\mathfrak{l}_0/\tilde{\mathfrak{l}}$ are involutive pairs.

60

II.3.3. Definition 26. A hyper-involutive decomposition of a Lie algebra $\mathfrak{g}$ is said to be basis for a principal involutive automorphism S of type O with the corresponding involutive algebra $\mathfrak{l} = \mathfrak{b} \oplus \tilde{\mathfrak{l}}$ if the involutive automorphisms S_1, S_2, S_3 of the hyper-involutive decomposition belong to $\mathrm{Int}_{\mathfrak{g}}(\mathfrak{b}) \cong SO(3)$, $pS = Sp$, and $p\eta = \eta$ for $\eta \in \tilde{\mathfrak{l}}$.

Evidently (II.25) is valid for any basis hyper-involutive decomposition.

Let us consider the involutive pair $\mathfrak{l}_0/\tilde{\mathfrak{l}}$, then the centre $\mathfrak{z}$ of $\mathfrak{l}_0$ either belongs to $\tilde{\mathfrak{l}}$ or $(\mathfrak{l}_0\dot{-}\tilde{\mathfrak{l}}) \cap \mathfrak{z} \neq \{0\}$.

Let us suppose that the latter is true, and denote by $\mathfrak{q}$ the maximal subalgebra of elements from $\mathfrak{g}$ commuting with $\tilde{\mathfrak{l}}$. Obviously $\mathfrak{b} \subset \mathfrak{q}$, $(\mathfrak{l}_0\dot{-}\tilde{\mathfrak{l}}) \cap \mathfrak{z} \subset \mathfrak{q}$, and $(\mathfrak{g}\dot{-}\mathfrak{l}) \cap \mathfrak{q} \neq \{0\}$.

But then $\mathrm{ad}\,_{\mathfrak{g}}^{\mathfrak{g}-\mathfrak{l}}(\mathfrak{l})$ maps $(\mathfrak{g}\dot{-}\mathfrak{l}) \cap \mathfrak{q}$ into itself, which, owing to the irreducibility of $\mathrm{ad}\,_{\mathfrak{g}}^{\mathfrak{g}-\mathfrak{l}}(\mathfrak{l})$, is possible only if $(\mathfrak{g}\dot{-}\mathfrak{l}) \cap \mathfrak{q} = \mathfrak{g}\dot{-}\mathfrak{l}$. That implies $\mathfrak{g}\dot{-}\mathfrak{l} \subset \mathfrak{q}$, but then $\tilde{\mathfrak{l}}$ is an ideal in the simple Lie algebra $\mathfrak{g}$, that is, $\tilde{\mathfrak{l}} = \{0\}$.

Thus if $\tilde{\mathfrak{l}} \neq \{0\}$ we obtain that the centre of $\mathfrak{l}_0$ belongs to $\tilde{\mathfrak{l}}$.

Assume, furthermore, that the restriction of the automorphism p from (II.3) to $\mathfrak{l}_0$ is not the identity automorphism of $\mathfrak{l}_0$. Then for the Lie algebra $\mathfrak{l}_0$ with an involutive algebra $\tilde{\mathfrak{l}}$ and an automorphism p ($p^3 = \mathrm{Id}$) II.2.4 (Theorem 4) is valid. Thus $p \in \mathrm{Int}_{\mathfrak{l}_0}(\mathfrak{z}) = \mathrm{Int}_{\mathfrak{g}}^{\mathfrak{l}_0}(\mathfrak{z})$ on $\mathfrak{l}_0$, where $\mathfrak{z}$ is the non-trivial centre of $\tilde{\mathfrak{l}}$. But then there exists $q \in \mathrm{Int}_{\mathfrak{g}}(\mathfrak{z})$ such that p and q coincide on $\mathfrak{l}_0$.

Since $\mathfrak{l} = \mathfrak{b} \oplus \tilde{\mathfrak{l}}$ the centre of $\tilde{\mathfrak{l}}$ is also the centre of $\mathfrak{l}$, hence $q\eta = \eta$ for $\eta \in \mathfrak{l}$, and q commutes on $\mathfrak{g}\dot{-}\mathfrak{l}$ with $\mathrm{ad}\,_{\mathfrak{g}}^{\mathfrak{g}-\mathfrak{l}}(\mathfrak{l})$. But then, also, q^3 commutes with $\mathrm{ad}_{\mathfrak{g}}^{\mathfrak{g}-\mathfrak{l}}(\mathfrak{l})$ on $\mathfrak{g}\dot{-}\mathfrak{l}$ and $q^3 = \mathrm{Id}$ on $\mathfrak{l}_0 \cap (\mathfrak{g}\dot{-}\mathfrak{l}) \neq \{0\}$ since there q coincides with p.

By virtue of the irreducibility of $\mathrm{ad}\,_{\mathfrak{g}}^{\mathfrak{g}-\mathfrak{l}}(\mathfrak{l})$ we have $q^3 = \mathrm{Id}$ on $\mathfrak{g}\dot{-}\mathfrak{l}$. Moreover, $q^3 = \mathrm{Id}$ on $\mathfrak{l}$. Thus $q^3 = \mathrm{Id}$ on $\mathfrak{g}$.

Finally, q commutes with S_1, S_2, S_3, p since $q \in \mathrm{Int}_{\mathfrak{g}}(\tilde{\mathfrak{l}})$, and S_1, S_2, S_3, p belong to $\mathrm{Int}_{\mathfrak{g}}(\mathfrak{b})$.

Now we consider $\tilde{p} = pq^{-1}$. By construction $\tilde{p}$ is the identity automorphism on $\mathfrak{l}_0$ and from what has been obtained above and the conditions (II.23) (1), (II.25) (15) there follow:

$$
\left.
\begin{array}{c}
(\tilde{p})^3 = (S_\rho)^2 = \mathrm{Id}, \qquad S_\rho \neq \mathrm{Id}, \qquad S_\rho S_\mu = S_\mu S_\rho \\[4pt]
(\rho,\mu = 1,2,3), \\[4pt]
S_1 S_2 = S_3, \qquad S_2 S_3 = S_1, \qquad S_3 S_1 = S_2, \\[4pt]
\tilde{p} S_1 = S_2 \tilde{p}, \qquad \tilde{p} S_2 = S_3 \tilde{p}, \qquad \tilde{p} S_3 = S_1 \tilde{p}, \\[4pt]
\tilde{p} S = S \tilde{p}, \qquad S S_\rho = S_\rho S \quad (\rho = 1,2,3), \\[4pt]
\tilde{p}\eta = \eta, \quad \eta \in \mathfrak{l}_0.
\end{array}
\right\}
\qquad \text{(II.26)}
$$

Thus we arrive at the theorem:

II.3.4. Theorem 5. *Let* $\mathfrak{g}$ *be a simple compact Lie algebra, and* $\mathfrak{l} = \mathfrak{b} \oplus \tilde{\mathfrak{l}}$ *be its principal involutive algebra of an involutive automorphism* S *of type* O.

If $\tilde{\mathfrak{l}} \neq \{0\}$ *then* $\mathfrak{g}$ *possesses a prime hyper-involutive decomposition basis for* S.

II.3.5. We now consider the case $\tilde{\mathfrak{l}} = \{0\}$, that is, $\mathfrak{l} = \mathfrak{b}$. Then $\mathfrak{l}_0$ (see the decomposition (II.4)) is a commutative subalgebra, since $\mathfrak{l}_0 \subset \mathfrak{m} = \mathfrak{g} - \mathfrak{l}$, $\mathfrak{q}_\alpha = \mathfrak{l}_\alpha \cap \mathfrak{b}$ ($\alpha = 1, 2, 3$) are one-dimensional, and $\mathfrak{l}_\alpha / \mathfrak{q}_\alpha$ ($\alpha = 1, 2, 3$) are the involutive pairs of involutive automorphism S.

Being an involutive algebra of a simple compact Lie algebra $\mathfrak{g}$, $\mathfrak{l}_1$ has at most a one-dimensional centre $\mathfrak{z}$ and $\mathfrak{q}_1$ is one-dimensional.

For this reason we obtain a canonical decomposition $\mathfrak{l}_1 / \mathfrak{q}_1 = \mathfrak{z} / \{0\} + \mathfrak{p}_1 / \mathfrak{q}_1$, where $\mathfrak{z}$ is either one-dimensional or $\mathfrak{z} = \{0\}$. And $\mathfrak{p}_1 / \mathfrak{q}_1$ is an elementary involutive pair. But this is possible only if $\mathfrak{p}_1$ is three-dimensional and simple. But $\mathfrak{l}_0$ is commutative and $\mathfrak{l}_0 \subset \mathfrak{z} + (\mathfrak{p}_1 - \mathfrak{q}_1)$, which implies that $\mathfrak{l}_0$ is at most two-dimensional. However, $\mathfrak{l}_0 \neq \{0\}$, otherwise $\mathfrak{l}_1$ is commutative (since $\mathfrak{l}_1 / \mathfrak{l}_0 = \mathfrak{l}_1 / \{0\}$) and $\mathfrak{g} = \mathfrak{l}$.

Thus $\mathfrak{l}_0$ is commutative and one-dimensional or two-dimensional.

But if $\mathfrak{l}_0$ is one-dimensional then the restriction of the automorphism p (see II.23) to $\mathfrak{l}_0$ is the identity automorphism and the hyper-involutive decomposition (II.24) is prime.

Thus $\mathfrak{l}_0$ is commutative and two-dimensional. Since $S_1 \in \mathrm{Int}_{\mathfrak{g}}(\mathfrak{b})$, $\mathfrak{l}_1$ is the involutive algebra of S_1, and $\mathfrak{l}_1 \cap \mathfrak{b} = \mathfrak{q}_1$ we have $S_1 \in \mathrm{Int}_{\mathfrak{g}}(\mathfrak{q}_1) \subset \mathrm{Int}_{\mathfrak{g}}(\mathfrak{p}_1)$. Hence $\mathrm{Int}_{\mathfrak{g}}(\mathfrak{p}_1)$ is isomorphic to $SU(2)$ and S_1 is a central principal involutive automorphism of the type U.

From what has been presented above and from II.3.4 (Theorem 5) we have the theorem:

II.3.6. Theorem 6. *If* $\mathfrak{g}$ *is a simple compact Lie algebra,* $\mathfrak{l} = \mathfrak{b} \oplus \tilde{\mathfrak{l}}$ ($\mathfrak{b} \cong so(3)$) *being a principal involutive algebra of an involutive automorphism* S *of type* O, *then* $\mathfrak{g}$ *has:*

a) *either a prime hyper-involutive decomposition basis for* S;

b) *or a non-prime hyper-involutive decomposition basis for* S. *In this case*

$$\mathfrak{l} = \mathfrak{b}, \qquad \mathfrak{l}_\alpha = \mathfrak{p}_\alpha \oplus \mathfrak{z}_\alpha \quad (\alpha = 1, 2, 3),$$

where $\mathfrak{p}_\alpha$ *are simple and three-dimensional,* $\mathfrak{z}_\alpha$ *are one-dimensional, and* $\mathfrak{l}_0 = \mathfrak{z}_\alpha \oplus \mathfrak{w}_\alpha$, $\mathfrak{w}_\alpha$ *are one-dimensional, moreover*

$$\mathfrak{w}_\alpha \subset \mathfrak{p}_\alpha, \qquad \mathfrak{l}_\alpha \cap \mathfrak{l} = \mathfrak{p}_\alpha \cap \mathfrak{l} = \mathfrak{q}_\alpha \subset \mathfrak{p}_\alpha \dot{-} \mathfrak{w}_\alpha$$

(where $\mathfrak{q}_\alpha$ *are one-dimensional),* $\dim \mathfrak{g} = 8$, *and* S_α *are principal central involutive automorphisms of type* U.

Thus the further exploration splits into the consideration of two possibilities: the prime decomposition (the main case) and the non-prime decomposition (a singular case).

II.3.7. First we consider the main case. Then the relations (II,23), (II.24), (II.25) are satisfied, $p\eta = \eta$ for $\eta \in \mathfrak{l}_0$, and $\mathfrak{b} \cap \mathfrak{l}_\alpha = \mathfrak{q}_\alpha$ ($\alpha = 1, 2, 3$) are one-dimensional. Since $S\,S_\alpha = S_\alpha S$ we have $S\,\mathfrak{m}_1 = \mathfrak{m}_1$.

We choose an orthonormal base of $\mathfrak{m}_1$ in such a way that the automorphism S has a diagonal form. Then

$$S X_{0_1} = X_{0_1}, \qquad S X_{i_1} = -X_{i_1}, \quad i = 1, \ldots, n, \qquad X_{0_1} \in \mathfrak{b}. \qquad (\text{II.27})$$

Furthermore (as in II.1) we take an orthonormal base in $\mathfrak{l}_0$ and define orthonormal bases in $\mathfrak{m}_2$ and $\mathfrak{m}_3$ by

$$p X_{I_1} = X_{I_2}, \qquad p X_{I_2} = X_{I_3} \quad (I = 0, 1, \ldots, n),$$

see (II.24). The union of all the above bases constitutes an orthonormal base in $\mathfrak{g}$ (see II.5) which is hyper-involutive,

$$\left.\begin{array}{ll} X_{I_1} \in \mathfrak{m}_1 & (I = 0, 1, \ldots, n) \\[4pt] X_{I_2} \in \mathfrak{m}_2 & (I = 0, 1, \ldots, n) \\[4pt] X_{I_3} \in \mathfrak{m}_3 & (I = 0, 1, \ldots, n) \\[4pt] Y_A \in \mathfrak{l}_0 & (A = 1, \ldots, r). \end{array}\right\} \qquad (\text{II.28})$$

Now the relations (II.6)–(II.11) take the form:

$$\left.\begin{array}{llll} p X_{I_1} = X_{I_2}, & p X_{I_2} = X_{I_3}, & p X_{I_3} = X_{I_1}, & p Y_A = Y_A, \\[4pt] S_\rho X_{I_\rho} = X_{I_\rho}, & S_\rho Y_A = Y_A, & S_\rho X_{I_\mu} = -X_{I_\mu}, & \\[4pt] & (\rho, \mu = 1, 2, 3, \ \rho \neq \mu). & & \end{array}\right\} \qquad (\text{II.29})$$

$$\begin{array}{ll} [X_{I_\rho} X_{J_\rho}] = C^A_{I_\rho J_\rho} Y_A, & [X_{I_\rho} X_{J_\mu}] = C^{K_\lambda}_{I_\rho J_\mu} X_{K_\lambda}, \\[6pt] [X_{I_\rho} Y_A] = C^{K_\rho}_{I_\rho A} X_{K_\rho}, & [Y_A Y_B] = C^F_{AB} Y_F, \end{array} \qquad (\text{II.30})$$

$$\lambda, \mu, \rho = 1, 2, 3; \quad \lambda \neq \mu, \quad \mu \neq \rho, \quad \rho \neq \lambda.$$

$$\begin{array}{ll} C^{K_1}_{I_2 J_3} = C^{K_3}_{I_1 J_2} = C^{K_2}_{I_3 J_1} = t^K_{I J}, & C^B_{I_\rho J_\rho} = C^B_{I J}, \\[6pt] C^{K_\rho}_{I_\rho A} = C^K_{I A} \quad (\rho = 1, 2, 3) & \end{array} \qquad (\text{II.31})$$

$$t^K_{I J} = t^I_{J K}, \qquad C^B_{I J} = -C^J_{I B}, \qquad C^F_{A B} = -C^A_{F B}, \qquad (\text{II.32})$$

$$t^R_{I J} = K^R_{I J} + E^R_{I J}, \qquad K^R_{I J} = -K^R_{J I} = -K^I_{R J},$$

$$E^R_{I J} = E^R_{J I} = E^I_{R J}. \qquad (\text{II.33})$$

$$t^K_{IJ} C^P_{KA} - t^P_{KJ} C^K_{IA} - t^P_{IK} C^K_{JA} = 0, \qquad (\text{II.34.1})$$

$$t^P_{JK} t^R_{PI} - t^P_{IK} t^R_{PJ} = C^R_{KB} C^B_{IJ}, \qquad (\text{II.34.2})$$

$$t^P_{KJ} t^R_{IP} - t^P_{KI} t^R_{JP} = C^R_{KB} C^B_{IJ}, \qquad (\text{II.34.3})$$

$$C^R_{IA} C^A_{JK} + C^R_{KA} C^A_{IJ} + C^R_{JA} C^A_{KI} = 0, \qquad (\text{II.34.4})$$

$$C^R_{PB} C^P_{IA} - C^R_{PA} C^P_{IB} = C^E_{AB} C^R_{IE}, \qquad (\text{II.34.5})$$

$$C^E_{AB} C^G_{EF} + C^E_{FA} C^G_{EB} + C^E_{BF} C^G_{EA} = 0. \qquad (\text{II.34.6})$$

$$(\text{II.34})$$

For the involutive automorphism S we also have from (II.27)

$$t^i_{jk} = 0, \qquad t^i_{00} = t^0_{i0} = t^0_{0i} = 0, \quad i = 1, 2, \dots, n. \qquad (\text{II.35})$$

From (II.34.1) and (II.35) we have

$$t^0_{ij} C^p_{0A} - t^p_{0j} C^0_{iA} - t^p_{i0} C^0_{jA} = 0$$

and also

$$t^0_{00} C^p_{0A} - t^p_{k0} C^k_{0A} - t^p_{0k} C^k_{0A} = 0,$$

hence by means of (II.32), (II.33) we obtain:

$$t^0_{ij} C^0_{pA} + t^0_{jp} C^0_{iA} = t^0_{pi} C^0_{jA} = 0, \qquad (\text{II.36})$$

$$t^0_{00} C^0_{pA} = t^k_{0p} C^0_{kA} + t^k_{p0} C^0_{kA} = 2E^k_{p0} C^0_{kA}. \qquad (\text{II.37})$$

We note that $t^0_{00} \neq 0$ (otherwise $\mathfrak{b}$ is commutative, which is impossible).

If $E^0_{ij} = 0$ then from (II.37) it follows that $C^0_{pA} = 0$. Assume that not all E^0_{ij} are equal to zero, then there are ζ^s such that

$$t^0_{ij} \zeta^i \zeta^j = E^0_{ij} \zeta^i \zeta^j \neq 0.$$

By (II.36) we have, contracting over $\zeta^i \zeta^j$:

$$\left(E^0_{ij} \zeta^i \zeta^j \right) C^0_{pA} + 2 \left(E^0_{pj} \zeta^j \right) \left(C^0_{iA} \zeta^i \right) = 0.$$

Contracting the preceding equation once more with ζ^p we obtain

$$3 \left(E^0_{ij} \zeta^i \zeta^j \right) \left(C^0_{pA} \zeta^p \right) = 0,$$

whence $C^0_{pA} \zeta^p = 0$.

But then we have $\left(E^0_{ij} \zeta^i \zeta^j \right) C^0_{pA} = 0$. Then again $C^0_{pA} = 0$. Thus in the main case, taking into account (II.32), we have:

$$C^A_{P0} = C^0_{PA} = C^P_{0A} = 0, \qquad t^0_{00} = \tau \neq 0. \qquad (\text{II.38})$$

Thus we obtain the theorem:

II.3.8. Theorem 7. *Let $\mathfrak{g}$ be a simple compact Lie algebra, and $\mathfrak{l} = \mathfrak{b} \oplus \tilde{\mathfrak{l}}$ be its principal involutive algebra of an involutive automorphism S of type O. If $\mathfrak{g}$ has a prime hyper-involutive decomposition $\mathfrak{g} = \mathfrak{l}_1 + \mathfrak{l}_2 + \mathfrak{l}_3$ basis for S then*

$$\tilde{\mathfrak{l}} = \mathfrak{l}_0 = \mathfrak{l}_1 \cap \mathfrak{l}_2 = \mathfrak{l}_2 \cap \mathfrak{l}_3 = \mathfrak{l}_3 \cap \mathfrak{l}_1,$$

and $\mathfrak{b} \cap \mathfrak{l}_\alpha$ is a one-dimensional centre of $\mathfrak{l}_\alpha$ ($\alpha = 1,2,3$).

II.3.9. Let us now consider the involutive automorphisms $\tilde{S}_1 = S S_1$, $\tilde{S}_2 = S S_2$, $\tilde{S}_3 = S_3$. They are commuting and $\tilde{S}_1 \tilde{S}_2 = \tilde{S}_3$. Thus we may construct the involutive sum $\mathfrak{g} = \tilde{\mathfrak{l}}_1 + \tilde{\mathfrak{l}}_2 + \tilde{\mathfrak{l}}_3$, where $\tilde{\mathfrak{l}}_1$, $\tilde{\mathfrak{l}}_2$, $\tilde{\mathfrak{l}}_3$ are the involutive algebras of the involutive automorphisms $\tilde{S}_1, \tilde{S}_2, \tilde{S}_3$, respectively. Using, furthermore, the commutativity of S, S_1, S_2, S_3 we easily see that

$$\tilde{\mathfrak{l}}_1 = \tilde{\mathfrak{m}}_2 \dot{+} \tilde{\mathfrak{m}}_3 \dot{+} \tilde{\mathfrak{l}} \dot{+} \mathfrak{q}_1, \qquad \tilde{\mathfrak{l}}_2 = \tilde{\mathfrak{m}}_1 \dot{+} \tilde{\mathfrak{m}}_3 \dot{+} \tilde{\mathfrak{l}} \dot{+} \mathfrak{q}_2,$$

$$\tilde{\mathfrak{l}}_3 = \tilde{\mathfrak{m}}_3 \dot{+} \tilde{\mathfrak{l}} \dot{+} \mathfrak{q}_3, \qquad \mathfrak{m}_\alpha = \tilde{\mathfrak{m}}_\alpha \dot{+} \mathfrak{q}_\alpha, \qquad \tilde{\mathfrak{l}}_0 = \tilde{\mathfrak{m}}_3 \dot{+} \tilde{\mathfrak{l}},$$

$\tilde{\mathfrak{m}}_3 = \tilde{\mathfrak{l}}_3 \dot{-} \tilde{\mathfrak{l}}_0 = \mathfrak{q}_3$ is one-dimensional, where

$$\tilde{\mathfrak{l}}_0 = \tilde{\mathfrak{l}}_1 \cap \tilde{\mathfrak{l}}_2 = \tilde{\mathfrak{l}}_2 \cap \tilde{\mathfrak{l}}_3 = \tilde{\mathfrak{l}}_3 \cap \tilde{\mathfrak{l}}_1,$$

$$\mathfrak{l}_\alpha = \tilde{\mathfrak{m}}_\alpha \dot{+} \tilde{\mathfrak{l}} \quad (\alpha = 1,2,3).$$

Moreover, $\tilde{S}_3 = S_3 \in \mathrm{Int}_{\mathfrak{g}}(\mathfrak{q}_3)$, as it was obtained before.

Consequently $\mathfrak{g} = \tilde{\mathfrak{l}}_1 + \tilde{\mathfrak{l}}_2 + \tilde{\mathfrak{l}}_3$ is an iso-involutive sum of type 1 and of lower index 1 and by I.5.4 (Theorem 14) we have the isomorphisms

$$\mathfrak{g} \cong so(m+1), \qquad \tilde{\mathfrak{l}}_1 \cong so(m), \qquad \tilde{\mathfrak{l}}_2 \cong so(m),$$

$$\tilde{\mathfrak{l}}_0 \cong so(m-1) \quad (3 \neq m > 1).$$

with the natural embeddings.

Furthermore, we note that the maximal subalgebra of $\mathfrak{g}$ commuting with $\mathfrak{q}_1$ is $\mathfrak{l}_1$ (since $S_1 \in \mathrm{Int}_{\mathfrak{g}}(\mathfrak{q}_1)$), whence it follows that the maximal subalgebra commuting with $\mathfrak{q}_1$ in $\tilde{\mathfrak{l}}_0$ is $\tilde{\mathfrak{l}}$.

Thus $\tilde{\mathfrak{l}}$ is a maximal subalgebra in $\tilde{\mathfrak{l}}_0$ commuting with one-dimensional $\mathfrak{q}_1 \subset \tilde{\mathfrak{l}}_1$. Consequently since $\tilde{\mathfrak{l}}_0 \cong so(m-1)$ we obtain $\tilde{\mathfrak{l}} \cong so(m-2)$ with the natural embedding.

We see, furthermore, that if $m > 4$ then $\mathfrak{b}$ is a maximal subalgebra commuting with the semi-simple subalgebra $\tilde{\mathfrak{l}} \cong so(m-2)$. This means that $\mathfrak{b} \cong so(3)$ with the natural embedding into $\mathfrak{g} \cong so(m+1)$).

If $m = 4$ then $\tilde{\mathfrak{l}} \cong so(2)$, that is, one-dimensional, and consequently $\mathfrak{l} = \mathfrak{b} \oplus \tilde{\mathfrak{l}}$ is a maximal subalgebra of $\mathfrak{g}$ commuting with $\tilde{\mathfrak{l}}$. This means again that $\mathfrak{b} \cong so(3)$ with the natural embedding.

If $m < 4$ then $\tilde{\mathfrak{l}} = \{0\}$ and, in addition, $m \neq 3$, $m \neq 1$ (since $so(4)$ is not simple and $so(2)$ is not semi-simple), moreover, $m \neq 2$ (otherwise $\mathfrak{g} \cong so(3)$, and the involutive automorphism S is the identity automorphism, which is impossible).

Thus we have:

II.3.10. Theorem 8. *Let $\mathfrak{g}$ be a simple compact Lie algebra, and $\mathfrak{l} = \mathfrak{b} \oplus \tilde{\mathfrak{l}}$ be its principal involutive algebra of an involutive automorphism S of type O.*

If $\mathfrak{g}$ has a prime hyper-involutive decomposition $\mathfrak{g} = \mathfrak{l}_1 + \mathfrak{l}_2 + \mathfrak{l}_3$ basis for S then

$$\mathfrak{g}/\mathfrak{l} \cong so(m)/\left(so(m-3) \oplus so(3) \right)$$
$$\mathfrak{g}/\mathfrak{l}_\alpha \cong so(m)/\left(so(m-2) \oplus so(2) \right)$$
$$(\alpha = 1, 2, 3; \; m > 4)$$

with the natural embeddings.

Together with II.3.4 (Theorem 5), II.3.10 (Theorem 8) gives us:

II.3.11. Theorem 9. *Let $\mathfrak{g}$ be a simple compact Lie algebra, $\mathfrak{l} = \mathfrak{b} \oplus \tilde{\mathfrak{l}}$ be its principal involutive algebra of an involutive automorphism S of type O, and let $\tilde{\mathfrak{l}} \neq \{0\}$. Then*

$$\mathfrak{g}/\mathfrak{l} \cong so(m)/\left(so(m-3) \oplus so(3) \right), \quad m > 4,$$

with the natural embedding.

II.3.12. The considerations of this section are not essential for understanding the other sections. But it is reasonable to finalize the previous studies to a certain extent.

Let $\mathfrak{g}$ be a simple compact Lie algebra, and $\mathfrak{g} = \mathfrak{l}_1 + \mathfrak{l}_2 + \mathfrak{l}_3$ be a prime hyper-involutive decomposition basis for a principal automorphism S of type O.

We now take $p, \varphi_1, \varphi_2, \varphi_3 \in \text{Int}_{\mathfrak{g}}(\mathfrak{b}) \cong so(3)$ according to (II.3). Since $\mathfrak{b}$ commutes with $\mathfrak{l}_0 = \tilde{\mathfrak{l}}$ the restriction of p to $\mathfrak{l}_0$ is the identity automorphism.

Furthermore, $\varphi_\alpha \in \text{Int}_{\mathfrak{g}}(\mathfrak{b})$ and $\varphi_\alpha \in \text{Int}_{\mathfrak{g}}(\mathfrak{l}_\alpha)$ as well, whence $\varphi_\alpha \in \text{Int}_{\mathfrak{g}}(\mathfrak{l}_\alpha \cap \mathfrak{b})$. But $\mathfrak{l}_\alpha \cap \mathfrak{b}$ is a one-dimensional centre in $\mathfrak{l}_\alpha$, as has been shown. Consequently the restriction of φ_α to $\mathfrak{l}_\alpha$ is the identity automorphism.

From (II.15) it follows that $\varphi_J^I = \delta_J^I$ and also (II.18) gives $t_{JI}^P = t_{IJ}^P$, or, what is the same, $K_{IJ}^P = 0$. Thus

$$t_{0J}^I = t_{J0}^I = t_{I0}^J = t_{IJ}^0 = t_{JI}^0. \tag{II.39}$$

Since $X_{0_1} \in \mathfrak{b} \cap \mathfrak{l}_1$, $\mathfrak{b} \cap \mathfrak{l}_1$ being the centre of $\mathfrak{l}_1$, and $\mathfrak{g}$ is simple $\text{ad}_{\mathfrak{g}}^{\mathfrak{g}-\mathfrak{l}_1}(X_{0_1})$ commutes with $\text{ad}_{\mathfrak{g}}^{\mathfrak{g}-\mathfrak{l}_1}(\mathfrak{l}_1)$ which is irreducible.

Thus we have

$$\left[\text{ad}_{\mathfrak{g}}^{\mathfrak{g}-\mathfrak{l}_1}(X_{0_1}) \right]^2 = \nu \, \text{Id}.$$

Using the base (II.28) we obtain

$$C_{\tilde{\alpha} \, 0_1}^{\tilde{J}} \, C_{\tilde{I} \, 0_1}^{\tilde{K}} = \nu \delta_{\tilde{I}}^{\tilde{J}}; \quad \tilde{I} = I_2, I_3; \quad \tilde{J} = J_2, J_3; \quad \tilde{K} = K_2, K_3$$

and, furthermore,

$$C_{K_3 \, 0_1}^{J_2} \, C_{I_2 \, 0_1}^{K_3} = \nu \delta_{I_2}^{J_2},$$

which together with (II.31) gives us

$$t^J_{K0} t^K_{0I} = -\nu \delta^J_I.$$

When $J = I = 0$ we have, in particular, $\left(t^0_{00}\right)^2 = -\nu$, thus $\nu = -\tau^2$.

From the orthonormality of the base (II.28) in the Cartan metric it follows that the scalar square of X_{0_1} is 1, $C^{\tilde{I}}_{0_1 \tilde{K}} C^{\tilde{K}}_{\tilde{I} 0_1} = 1$.

Hence

$$C^{I_3}_{0_1 K_2} C^{K_2}_{I_3 0_1} + C^{I_2}_{0_1 K_3} C^{K_3}_{I_2 0_1} = 1$$

and, finally, $2t^I_{0K} t^K_{I0} = 1.$

The latter implies

$$2\tau^2 \delta^I_I = 2\tau^2 (n+1) = 1,$$

or

$$\tau = \pm(1/\sqrt{2(n+1)}).$$

But the case of negative τ is reduced to $\tau > 0$ by the change of base: X_{0_α} by $-X_{0_\alpha}$, X_{i_α} by X_{i_α} ($\alpha = 1, 2, 3$). Thus we may regard $\tau > 0$.

As a result

$$t^J_{K0} t^K_{0I} = \tau^2 \delta^J_I, \qquad \tau = \frac{1}{\sqrt{2(n+1)}}. \qquad (\text{II}.40)$$

From (II.29.2) and (II.33) there follows, furthermore,

$$t^R_{P0} t^P_{J0} - t^0_{00} t^R_{0J} = 0,$$

$$t^R_{P0} t^P_{J0} = \tau t^R_{0J}. \qquad (\text{II}.41)$$

And then from (II.39), (II.40), (II.41) we obtain

$$t^R_{0J} = \tau \delta^R_J, \qquad \tau = \frac{1}{\sqrt{2(n+1)}}. \qquad (\text{II}.42)$$

The relations (II.34) turn into

$$\frac{1}{2(n+1)} \left(\delta^r_i \delta_{kj} - \delta^r_j \delta_{ki} \right) = C^r_{kB} C^B_{ij},$$

$$C^r_{pB} C^p_{iA} - C^r_{pA} C^p_{iB} = C^E_{AB} C^r_{iE}, \qquad (\text{II}.43)$$

$$C^E_{AB} C^G_{EF} + C^E_{FA} C^G_{EB} + C^E_{BF} C^G_{EA} = 0.$$

From the latter it is easily seen that the algebra generated by the matrices $\overline{\overline{C}}_B = (C^r_{pB})$ is the algebra of all skew-symmetric matrices of an order n.

II.3.13. We have not yet considered the case of non-prime hyper-involutive decomposition basis for involutive automorphism S of type O. (See II.3.6 (Theorem 6).) Here $\dim \mathfrak{g} = 8$. However, II.3.6 (Theorem 6) implies in this case that $\mathfrak{g}$ has a central principal involutive automorphism of type U. This has been considered in Chapter I.6, see I.6.4 (Theorem 22). From these results it follows that $\mathfrak{g} \cong su(3)$ for $\dim \mathfrak{g} = 8$. Owing to the conjugacy of all one-dimensional Lie subalgebras of $\mathfrak{p}_\alpha \cong su(2)$ in $\mathrm{Int}_{\mathfrak{g}}(\mathfrak{p}_\alpha)$ ($\alpha = 1, 2, 3$) we can regard $\mathfrak{q}_1$ as $\mathfrak{q}_1 \cong so(2)$ with the natural embeddings into $\mathfrak{p}_1 \cong su(2)$ and $\mathfrak{g} \cong su(3)$.

Now let us note that, under this natural realization, in $\mathfrak{p}_2 - \mathfrak{w}_2$ there exists $\mathfrak{q}'_2 \cong so(2)$ with the natural embedding into $\mathfrak{g} \cong su(3)$. But $\mathfrak{q}_2$ and $\mathfrak{q}'_2$ are conjugated in $\mathrm{Int}_{\mathfrak{g}}(\mathfrak{z}_1)$. Indeed, if $[\mathfrak{z}_1\, x] = \{0\}$ then $x \in \mathfrak{l}_1$, since $\mathfrak{l}_1$, being an involutive algebra, is the maximal subalgebra. Then $\left[\mathfrak{z}_1\, (\mathfrak{p}_2 - \mathfrak{w}_2)\right] \neq \{0\}$ and in the two-dimensional subspace $\mathfrak{p}_2 - \mathfrak{w}_2$ all one-dimensional subspaces are conjugated in $\mathrm{Int}_{\mathfrak{g}}(\mathfrak{z}_1)$.

Moreover, $\mathrm{Int}_{\mathfrak{g}}(\mathfrak{z}_1) = \{\mathrm{Id}\}$, thus after some transformation of $\mathrm{Int}_{\mathfrak{g}}(\mathfrak{z}_1)$ we can regard

$$\mathfrak{q}_1 \cong so(2), \qquad \mathfrak{q}_2 \cong so(2)$$

with the natural embeddings into $\mathfrak{g} \cong su(3)$.

But then

$$[\mathfrak{q}_1\, \mathfrak{q}_2] = \mathfrak{q}_3 \cong so(2)$$

with the natural embedding into $\mathfrak{g} \cong su(3)$, and

$$\mathfrak{b} = \mathfrak{q}_1 \dot{+} \mathfrak{q}_2 \dot{+} \mathfrak{q}_3 \cong so(3)$$

with the natural embedding into $\mathfrak{g} \cong su(3)$.

Thus we obtain the theorem:

II.3.14. Theorem 10. *Let $\mathfrak{g}$ be a simple compact Lie algebra, $\mathfrak{l} = \mathfrak{b} \oplus \tilde{\mathfrak{l}}$ be the principal involutive algebra of an involutive automorphism S of type O, and let $\mathfrak{g}$ have no prime hyper-involutive decomposition basis for S.*

Then $\mathfrak{b} = \mathfrak{l}$, $\dim \mathfrak{g} = 8$, and

$$\mathfrak{g}/\mathfrak{l} \cong su(3)/so(3)$$

with the natural embedding.

CHAPTER II.4

FUNDAMENTAL THEOREM

In this Chapter we prove that any simple semi-simple compact Lie algebra has a principal involutive automorphism.

II.4.1. Definition 27. Let $\mathfrak{g}$ be a compact Lie algebra, $\mathfrak{l} = \mathfrak{b} \oplus \tilde{\mathfrak{l}}$, $\mathfrak{b} \cong so(3)$, be an involutive algebra of a principal involutive automorphism A of type O, $\mathfrak{r} = \mathfrak{p} \oplus \mathfrak{q} \oplus \tilde{\mathfrak{l}}$, $\mathfrak{p} \cong \mathfrak{q} \cong so(3)$, be an involutive algebra of a di-unitary involutive automorphism J, $JA = AJ$, and let $\mathfrak{b}$ be the diagonal in $\mathfrak{p} \oplus \mathfrak{q}$ of the canonical involutive automorphism $\mathfrak{p} \leftrightarrows \mathfrak{q}$. Then we say that J is an associated involutive automorphism of A.

Respectively, we say that $\mathfrak{r}$ is an associated involutive algebra of $\mathfrak{l}$, and $\mathfrak{g}/\mathfrak{r}$ is an associated involutive pair of $\mathfrak{g}/\mathfrak{l}$.

II.4.2. Theorem 11. *If $\mathfrak{g}$ is a simple compact Lie algebra, $\mathfrak{g}/\mathfrak{b} \oplus \tilde{\mathfrak{l}}$ ($\mathfrak{b} \cong so(3)$) is a principal involutive pair of type O of an involutive automorphism A, $\tilde{\mathfrak{l}} \neq \{0\}$, then there exists in $\mathfrak{g}$ an involutive automorphism J associated with A.*

Proof. Owing to II.3.11 (Theorem 9) we have

$$\mathfrak{g}/\mathfrak{b} \oplus \tilde{\mathfrak{l}} \cong so(n)/(\,so(n-3) \oplus so(3)\,)\,, \ n > 4.$$

with the natural embeddings.

We consider this principal involutive pair $\mathfrak{g}/\mathfrak{b} \oplus \tilde{\mathfrak{l}}$ of the type O of the involutive automorphism A.

In this case the involutive automorphism A acts as

$$A(\hat{b}) = \hat{A}\,\hat{b}\,\hat{A}, \qquad \hat{b} \in so(n), \qquad \hat{A} = \mathrm{diag}\,[-1,-1,-1,1,\ldots,1].$$

Let us introduce, also, an involutive automorphism J,

$$J\hat{b} = \hat{J}\,\hat{b}\,\hat{J}, \qquad \hat{b} \in so(n), \qquad \hat{J} = \mathrm{diag}\,[-1,-1,-1,-1,1\ldots,1].$$

The involutive automorphism J commutes with A and generates the involutive pair

$$\mathfrak{g}/\mathfrak{p} \oplus \mathfrak{q} \oplus \tilde{\mathfrak{l}} \cong so(n)/(\,su(2) \oplus su(2) \oplus so(n-4)\,)$$

and $J \in \mathrm{Int}_{\mathfrak{g}}(\mathfrak{p}) \cong SU(2)$, $J \in \mathrm{Int}_{\mathfrak{g}}(\mathfrak{q}) \cong SU(2)$, $\mathrm{Int}_{\mathfrak{g}}(\mathfrak{p} \oplus \mathfrak{q}) \cong SO(4)$. Thus J is a principal di-unitary involutive automorphism. Moreover, $\mathfrak{b} \subset \mathfrak{p} \oplus \mathfrak{q}$ and is a diagonal in $\mathfrak{p} \oplus \mathfrak{q}$ (that is, the subalgebra of all elements immobile under the action of some canonical involutive automorphism $\mathfrak{p} \leftrightarrows \mathfrak{q}$). ∎

II.4.3. Theorem 12. *If a simple compact Lie algebra* $\mathfrak{g}$ *has an orthogonal principal involutive automorphism* $A \neq \mathrm{Id}$ *then* $\mathfrak{g}$ *has a unitary principal involutive automorphism* J *such that* $JA = AJ$.

Proof. Indeed, in this case we can use the existence of a hyper-involutive decomposition basis for A. See II.3.6 (Theorem 6).

If $\mathfrak{g}$ has a prime hyper-involutive decomposition basis for A then by II.3.10 (Theorem 8) $\mathfrak{g} \cong so(m)$, $m > 4$, and the involutive algebra of the involutive automorphism A is $\mathfrak{l} = \mathfrak{b} \oplus \tilde{\mathfrak{l}}$, $\tilde{\mathfrak{l}} \neq \{0\}$. By II.4.2 (Theorem 11) there exists a di-unitary involutive automorphism J associated with A and $JA = AJ$.

If $\mathfrak{g}$ has a non-prime hyper-involutive decomposition basis for A then by II.3.6 (Theorem 6) $\mathfrak{g}$ has the principal central involutive automorphism J of type U and $JA = AJ$. ■

II.4.4. Theorem 13. *Let* $\mathfrak{g}$ *be a simple compact Lie algebra, and let its involutive algebra* $\mathfrak{l}$ *of an involutive automorphism* A *($A \neq \mathrm{Id}$) have a principal involutive automorphism* ψ, *then* $\mathfrak{g}$ *has a unitary special involutive automorphism.*

Proof. Let us consider the involutive algebra $\mathfrak{l}$ of an involutive automorphism A. By the conditions of theorem, $\mathfrak{l}$ has a principal involutive algebra $\mathfrak{b} \oplus \tilde{\mathfrak{l}} \subset \mathfrak{l}$ ($\mathfrak{b} \cong so(3)$) of the involutive automorphism ψ.

By I.1.20 the involutive pair $\mathfrak{l}/\mathfrak{b} \oplus \tilde{\mathfrak{l}}$ has a unique canonical decomposition:

$$\mathfrak{l}/\mathfrak{b} \oplus \tilde{\mathfrak{l}} = \mathfrak{l}_0/\mathfrak{p}_0 + \sum_{\alpha=1}^{m} (\mathfrak{l}_\alpha/\mathfrak{p}_\alpha),$$

where $\mathfrak{l}_0$ is the centre of $\mathfrak{l}$, and $\mathfrak{l}_\alpha/\mathfrak{p}_\alpha$ ($\alpha = 1, \ldots, m$) are elementary involutive pairs. Owing to $\mathfrak{p}_0 \oplus \mathfrak{p}_1 \oplus \cdots \oplus \mathfrak{p}_m = \mathfrak{b} \oplus \tilde{\mathfrak{l}}$ we have $\mathfrak{b} \subset \mathfrak{p}_{\alpha_0}$, thus after the renumbering we can put $\mathfrak{b} \subset \mathfrak{p}_1 \subset \mathfrak{l}_1$ and $\mathfrak{p}_1 = \mathfrak{b} \oplus \tilde{\mathfrak{p}}$. If $\mathfrak{l}_1$ is semi-simple but not simple then $\mathfrak{l}_1 = \mathfrak{q} \oplus \mathfrak{t}$, where $\mathfrak{q}$ and $\mathfrak{t}$ are simple and isomorphic, $\mathfrak{p}_1$ being the diagonal of the canonical isomorphism $\mathfrak{q} \leftrightarrows \mathfrak{t}$. Therefore $\tilde{\mathfrak{p}} = \{0\}$, that is, $\mathfrak{p}_1 = \mathfrak{b}$, $\mathfrak{q}$ and $\mathfrak{t}$ are three-dimensional and then $\mathfrak{l} = \mathfrak{l}_0 \oplus \mathfrak{q} \oplus \mathfrak{t} \oplus \cdots \oplus \mathfrak{l}_m$ is a principal involutive algebra of $\mathfrak{g}$. Then either A is a unitary principal involutive automorphism or $A \neq \mathrm{Id}$ is a principal involutive automorphism of the type O. In the latter case according to II.4.3 (Theorem 12) $\mathfrak{g}$ has an unitary principal involutive automorphism, which is a particular subcase of a special unitary involutive automorphism.

Thus we should consider only the case in which $\mathfrak{l}_1$ is simple. If $\mathfrak{p}_1 = \mathfrak{l}_1$ then $\tilde{\mathfrak{p}} = \{0\}$, $\mathfrak{l}_1 = \mathfrak{b}$. Thus $\mathfrak{l} = \mathfrak{l}_0 \oplus \mathfrak{b} \oplus \cdots \oplus \mathfrak{l}_m$ is a principal involutive algebra, and either A is a principal unitary involutive automorphism or, by II.4.3 (Theorem 12), $\mathfrak{g}$ has a unitary principal involutive automorphism, which is a particular subcase of a special unitary involutive automorphism.

Thus we have to consider the case $\mathfrak{p}_1 \subset \mathfrak{l}_1$, $\mathfrak{p}_1 \neq \mathfrak{l}_1$. Here $\tilde{\psi}$, the restriction of the automorphism ψ of $\mathfrak{l}$ to $\mathfrak{l}_1$, is a principal involutive automorphism of $\mathfrak{l}_1$, different from the identity automorphism. Furthermore, $\tilde{\psi}$ is either a principal unitary or principal orthogonal, which implies, by II.4.3 (Theorem 12), the existence of a principal unitary involutive automorphism of $\mathfrak{l}_1$.

Thus $\mathfrak{l}_1$ has a unitary principal involutive automorphism $\tilde{J}$ with the involutive algebra $\mathfrak{q}_3 \oplus \mathfrak{q}$, $\mathfrak{q}_3$ being three-dimensional and simple, $\tilde{J} \in \mathrm{Int}_{\mathfrak{l}_1}(\mathfrak{q}_3) \cong SU(2)$. But $\mathrm{Int}_{\mathfrak{l}_1}(\mathfrak{q}_3) = \mathrm{Int}_{\mathfrak{g}}^{\mathfrak{l}_1}(\mathfrak{q}_3)$, and there exists the natural morphism

$$\mathrm{Int}_{\mathfrak{g}}(\mathfrak{q}_3) \longrightarrow \mathrm{Int}_{\mathfrak{g}}^{\mathfrak{l}_1}(\mathfrak{q}_3) = \mathrm{Int}_{\mathfrak{l}_1}(\mathfrak{q}) \cong SU(2),$$

implying that it is an isomorphism, since $\mathrm{Int}_{\mathfrak{g}}(\mathfrak{q}_3) \cong SO(3)$ or $\mathrm{Int}_{\mathfrak{g}}(\mathfrak{q}_3) \cong SU(2)$, but the first possibility is impossible. The inverse image of $\tilde{J} \in \mathrm{Int}_{\mathfrak{l}_1}(\mathfrak{q}_3)$ is an involutive automorphism, $J \in \mathrm{Int}_{\mathfrak{g}}(\mathfrak{q}_3)$, $J \neq \mathrm{Id}$. Because $\mathfrak{q}_3 \subset \mathfrak{l}$ we have $JA = AJ$.

Let $\mathfrak{i}$ be the involutive algebra of the involutive automorphism J, then $\mathfrak{i} \cap \mathfrak{l} = \mathfrak{l}_0 \oplus \mathfrak{q}_3 \oplus \mathfrak{q} \oplus \mathfrak{l}_2 \oplus \cdots \oplus \mathfrak{l}_m$ is a principal involutive algebra of $\mathfrak{i}$, generated by A. The restriction of J to $\mathfrak{i}$ is the identity map, therefore we have a strict morphism

$$SU(2) \cong \mathrm{Int}_{\mathfrak{g}}(\mathfrak{q}_3) \longrightarrow \mathrm{Int}_{\mathfrak{g}}^{\mathfrak{i}}(\mathfrak{q}_3) = \mathrm{Int}_{\mathfrak{i}}(\mathfrak{q}_3) \cong SO(3).$$

Consequently $\mathfrak{i}$ is a special unitary involutive algebra. ∎

II.4.5. Furthermore, we clarify the structure of special unitary involutive automorphisms of a simple compact Lie algebra.

Let $\mathfrak{g}$ be a simple compact Lie algebra, and $\mathfrak{l}$ be its special involutive algebra. Then $\mathfrak{l}$ has an involutive algebra $\mathfrak{b} \oplus \tilde{\mathfrak{l}}$ of the involutive automorphism ψ and $\mathrm{Int}_{\mathfrak{l}}(\mathfrak{b}) \cong SO(3)$, $\mathrm{Int}_{\mathfrak{g}}(\mathfrak{b}) \cong SU(2)$. As we know, there is a unique decomposition

$$\mathfrak{l}/\mathfrak{b} \oplus \tilde{\mathfrak{l}} = \mathfrak{l}_0/\mathfrak{p}_0 + \mathfrak{l}_1/\mathfrak{p}_1 + \cdots + \mathfrak{l}_m/\mathfrak{p}_m,$$

where $\mathfrak{l}_0$ is the centre of $\mathfrak{l}$, and $\mathfrak{l}_\alpha/\mathfrak{p}_\alpha$ ($\alpha = 1, 2, 3$) are elementary involutive pairs. From $\mathfrak{b} \oplus \tilde{\mathfrak{l}} = \mathfrak{p}_0 \oplus \mathfrak{p}_1 \oplus \cdots \oplus \mathfrak{p}_m$ it follows that $\mathfrak{b}$ belongs to one of the $\mathfrak{p}_1, \ldots, \mathfrak{p}_m$, thus after some re-numbering we can put $\mathfrak{b} \subset \mathfrak{p}_1$ and $\mathfrak{p}_1 = \mathfrak{b} \oplus \tilde{\mathfrak{p}}$. If $\mathfrak{l}_1$ is semi-simple and not simple then $\mathfrak{l}_1 = \mathfrak{q} \oplus \mathfrak{t}$, where $\mathfrak{q}$ and $\mathfrak{t}$ are simple and isomorphic to each other, and $\mathfrak{p}_1$ is the diagonal of the canonical involutive automorphism $\mathfrak{q} \leftrightarrows \mathfrak{t}$. Then $\tilde{\mathfrak{p}} = \{0\}$, $\mathfrak{p}_1 = \mathfrak{b}$, $\mathfrak{q}$ and $\mathfrak{t}$ are three-dimensional and simple. Thus $\mathfrak{l} = \mathfrak{l}_0 \oplus \mathfrak{q} \oplus \mathfrak{t} \oplus \cdots \oplus \mathfrak{l}_m$ is a principal involutive algebra.

If $\mathfrak{l}_1$ is simple and $\mathfrak{l}_1 = \mathfrak{p}_1$ then again $\tilde{\mathfrak{p}} = \{0\}$, that is, $\mathfrak{l}_1 = \mathfrak{b}$ and $\mathfrak{l} = \mathfrak{l}_0 \oplus \mathfrak{b}_3 \oplus \cdots \oplus \mathfrak{l}_m$ is a principal involutive algebra.

We have to consider only the subcase when $\mathfrak{p}_1 \subset \mathfrak{l}_1$, $\mathfrak{p}_1 \neq \mathfrak{l}_1$, $\mathfrak{l}_1$ being simple. The morphism

$$\mathrm{Int}_{\mathfrak{l}_1}(\mathfrak{b}) = \mathrm{Int}_{\mathfrak{l}}^{\mathfrak{l}_1}(\mathfrak{b}) \longleftarrow \mathrm{Int}_{\mathfrak{l}}(\mathfrak{b}) \cong SO(3),$$

means that it is an isomorphism, thus $\mathrm{Int}_{\mathfrak{l}_1}(\mathfrak{b}) \cong SO(3)$, and consequently $\mathfrak{l}_1/\mathfrak{b}_3 \oplus \tilde{\mathfrak{p}}$ is a principal orthogonal involutive pair of some involutive automorphism S.

Owing to II.3.6 (Theorem 6), for $\mathfrak{l}_1/\mathfrak{b}_3 \oplus \tilde{\mathfrak{p}}$ there are then two possibilities. Let us suppose first that there is a non-prime hyper-involutive decomposition basis for S. Then $\tilde{\mathfrak{p}} = \{0\}$, $\mathfrak{l}_1 = \mathfrak{n}_1 + \mathfrak{n}_2 + \mathfrak{n}_3$, where $\mathfrak{n}_\alpha$ ($\alpha = 1, 2, 3$) are involutive algebras of invomorphisms S_1, S_2, S_3 generated by the above hyper-involutive decomposition. Also $\mathfrak{n}_\alpha = \mathfrak{t}_\alpha \oplus \mathfrak{z}_\alpha$, where $\mathfrak{t}_\alpha$ ($\alpha = 1, 2, 3$) are three-dimensional simple and $\mathfrak{z}_\alpha$ ($\alpha = 1, 2, 3$) are one-dimensional.

Furthermore, if $n_0 = n_1 \cap n_2 = n_2 \cap n_3 = n_3 \cap n_1$ then $n_0 = \mathfrak{z}_\alpha \oplus \mathfrak{w}_\alpha$, where $\mathfrak{w}_\alpha$ $(\alpha = 1, 2, 3)$ are one-dimensional and $\mathfrak{w}_\alpha \subset \mathfrak{t}_\alpha$, $\mathfrak{w}_\alpha \neq \mathfrak{t}_\alpha$ Moreover, $n_\alpha \cap \mathfrak{b} = \mathfrak{t}_\alpha \cap \mathfrak{b} = \mathfrak{q}_\alpha$ $(\alpha = 1, 2, 3)$ are one-dimensional and S_α $(\alpha = 1, 2, 3)$ are central principal involutive automorphisms of type U. Therefore $\mathrm{Int}_{\mathfrak{l}_1}(\mathfrak{t}_1) \cong SU(2)$, but then the natural morphism $\mathrm{Int}_{\mathfrak{l}_1}(\mathfrak{t}_1) = \mathrm{Int}_\mathfrak{g}^{\mathfrak{l}_1}(\mathfrak{t}_1) \longleftarrow \mathrm{Int}_\mathfrak{g}(\mathfrak{t}_1)$ is an isomorphism.

Since $\mathfrak{q}_1 \subset \mathfrak{t}_1$, $\mathfrak{q}_1 \neq \mathfrak{t}_1$, then, in particular, we have the isomorphism $\mathrm{Int}_{\mathfrak{l}_1}(\mathfrak{q}_1) \leftrightarrows \mathrm{Int}_\mathfrak{g}(\mathfrak{q}_1)$. On the other hand, since $\mathfrak{l}$ is a unitary special involutive algebra there is a strict morphism $SU(2) \cong \mathrm{Int}_\mathfrak{g}(\mathfrak{b}) \longrightarrow \mathrm{Int}_{\mathfrak{l}_1}(\mathfrak{b}) \cong SO(3)$.

But it is easy to see that any morphism $SU(2) \longrightarrow SO(3)$ generates a strict morphism of the corresponding one-dimensional subgroups. Thus for $\mathfrak{q}_1 \subset \mathfrak{b}$ $(\mathfrak{q}_1 \neq \mathfrak{b})$ we have a strict morphism $\mathrm{Int}_\mathfrak{g}(\mathfrak{q}_1) \longrightarrow \mathrm{Int}_{\mathfrak{l}_1}(\mathfrak{q}_1)$. Thus we have the contradiction.

Consequently by II.3.6 (Theorem 6) $\mathfrak{l}_1/\mathfrak{b} \oplus \tilde{\mathfrak{p}}$ has a prime hyper-involutive decomposition only. But then, by II.3.10 (Theorem 8) and II.3.11 (Theorem 9), we obtain $\mathfrak{l}_1/\mathfrak{b} \oplus \tilde{\mathfrak{p}} \cong so(m)/(so(3) \oplus so(m-3))$, $m > 4$.

Thus we have proved the theorem:

II.4.6. Theorem 14. *Let $\mathfrak{g}$ be a simple compact Lie algebra, $\mathfrak{l}$ be its unitary special non-principal involutive algebra, and $\mathfrak{b} \oplus \mathfrak{v}$ be a principal orthogonal involutive algebra of $\mathfrak{l}$, $\mathrm{Int}_\mathfrak{g}(\mathfrak{b}) \cong SU(2)$, $\mathrm{Int}_\mathfrak{l}(\mathfrak{b}) \cong SO(3)$. Then $\mathfrak{l} = \mathfrak{k} \oplus \tilde{\mathfrak{l}}$, $\mathfrak{k}$ has a principal involutive algebra $\mathfrak{b} \oplus \mathfrak{p}$, $\mathfrak{p} \subset \mathfrak{v}$, and*

$$\mathfrak{k}/\mathfrak{b} \oplus \mathfrak{p} \cong so(m)/(so(3) \oplus so(m-3)), \quad m > 4,$$

with the natural embeddings.

Repeating the proof of II.4.6 (Theorem 14) we also obtain the more precise result:

II.4.7. Theorem 15. *If $\mathfrak{g}$ is a simple compact Lie algebra, $\mathfrak{k}$ being its special unitary involutive subalgebra of an involutive automorphism S, then $\mathfrak{k}$ contains a simple ideal $\mathfrak{q} \cong so(m)$ $(m > 2)$ and $\mathfrak{q}$ is a special unitary involutive subalgebra of the involutive automorphism S.*

II.4.8. Theorem 16. *If a simple compact Lie algebra $\mathfrak{g}$ has a unitary special involutive algebra $\mathfrak{l} \ncong so(7)$ then it also has a principal involutive algebra.*

Proof. If $\mathfrak{l}$ is a principal involutive algebra then the result is evident. Thus let $\mathfrak{l}$ be non-principal. By II.46 (Theorem 14) we have $\mathfrak{l} = \mathfrak{i} \oplus \tilde{\mathfrak{l}}$, $\mathfrak{i} \cong so(m)$, $\mathfrak{b} \oplus \mathfrak{r}$ is an involutive algebra in $\mathfrak{i}$ and

$$\mathfrak{i}/\mathfrak{b} \oplus \mathfrak{r} \cong so(m)/(so(3) \oplus so(m-3))$$

with the natural embeddings.

Let us consider a di-unitary principal involutive algebra $\mathfrak{p} \oplus \mathfrak{q} \oplus \mathfrak{r} \cong so(4) \oplus so(m-4)$ with the natural embedding into $\mathfrak{i}$, such that $\mathfrak{b}$ is a diagonal of a canonical involutive automorfism $\mathfrak{p} \leftrightarrows \mathfrak{q}$ (that is, $\mathfrak{p} \oplus \mathfrak{q} \oplus \mathfrak{r}$ is associated with $\mathfrak{b} \oplus \mathfrak{p}$ involutive algebra).

We consider, furthermore, $\mathrm{Int}_\mathfrak{g}(\mathfrak{p} \oplus \mathfrak{q})$.

Since
$$\mathrm{Int}_{\mathfrak{g}}^{\mathfrak{l}}(\mathfrak{p}) = \mathrm{Int}_{\mathfrak{l}}(\mathfrak{p}) \cong \mathrm{Int}_{\mathfrak{i}}(\mathfrak{p}) \cong SU(2)$$
and
$$\mathrm{Int}_{\mathfrak{g}}^{\mathfrak{l}}(\mathfrak{q}) = \mathrm{Int}_{\mathfrak{l}}(\mathfrak{q}) \cong \mathrm{Int}_{\mathfrak{i}}(\mathfrak{q}) \cong SU(2)$$

we have, owing to the morphism $\mathrm{Int}_{\mathfrak{g}}(\mathfrak{p}) \longrightarrow \mathrm{Int}_{\mathfrak{g}}^{\mathfrak{l}}(\mathfrak{p})$, that $\mathrm{Int}_{\mathfrak{g}}(\mathfrak{p}) \cong SU(2)$ and analogously $\mathrm{Int}_{\mathfrak{g}}(\mathfrak{q}) \cong SU(2)$.

In addition, $\mathrm{Int}_{\mathfrak{g}}(\mathfrak{b}) \cong SU(2)$, which implies

$$\mathrm{Int}_{\mathfrak{g}}(\mathfrak{p} \oplus \mathfrak{q}) = \mathrm{Int}_{\mathfrak{g}}(\mathfrak{p}) \cdot \mathrm{Int}_{\mathfrak{g}}(\mathfrak{q}) \cong SO(4),$$

$\mathrm{Int}_{\mathfrak{g}}(\mathfrak{b})$ being the diagonal of a canonical involutive automorphism of this product. Let us consider now $S_1 \in \mathrm{Int}_{\mathfrak{g}}(\mathfrak{p})$, $S_2 \in \mathrm{Int}_{\mathfrak{g}}(\mathfrak{q})$, $S_3 \in \mathrm{Int}_{\mathfrak{g}}(\mathfrak{b})$, such that $S_\mu \neq \mathrm{Id}$, $(S_\mu)^2 = \mathrm{Id}$ ($\mu = 1,2,3$), then $S_1 S_2 = S_2 S_1 = S_3$. By construction S_3 is an involutive automorphism with involutive algebra $\mathfrak{l}$. If $\mathfrak{l}_1$ and $\mathfrak{l}_2$ are the involutive algebras of the involutive automorphisms S_1 and S_2, respectively, then we have the involutive sum $\mathfrak{g} = \mathfrak{l}_1 + \mathfrak{l}_2 + \mathfrak{l}$, where

$$\mathfrak{l}_0 = \mathfrak{l}_1 \cap \mathfrak{l}_2 = \mathfrak{l}_2 \cap \mathfrak{l} = \mathfrak{l} \cap \mathfrak{l}_1 = \mathfrak{p} \oplus \mathfrak{q} \oplus \mathfrak{n} \oplus \tilde{\mathfrak{l}},$$

whence it follows that $\mathfrak{l}_1$ and $\mathfrak{l}_2$ are unitary special involutive algebras in $\mathfrak{g}$ (since $\mathfrak{l}_0$ is a principal involutive algebra in $\mathfrak{l}_1$ and in $\mathfrak{l}_2$, $S_1 \in \mathrm{Int}_{\mathfrak{g}}(\mathfrak{p})$, $S_2 \in \mathrm{Int}_{\mathfrak{g}}(\mathfrak{q})$).

Assuming that $\mathfrak{l}_1$ is not a principal involutive algebra in $\mathfrak{g}$, we have by II.46. (Theorem 14)

$$\mathfrak{l}_1 = \tilde{\mathfrak{i}} \oplus \mathfrak{k}, \qquad \tilde{\mathfrak{i}}/\mathfrak{p} \oplus \mathfrak{t} \cong so(m)/(so(3) \oplus so(m-3)), \qquad m > 4,$$

with the natural embeddings, and $\mathfrak{t}$ is an ideal of $\mathfrak{q} \oplus \mathfrak{n} \oplus \tilde{\mathfrak{l}}$.

Moreover, evidently

$$\mathfrak{l}_1/\mathfrak{l}_0 = \tilde{\mathfrak{i}}/(\mathfrak{p} \oplus \mathfrak{t}) + \mathfrak{k}/\mathfrak{n}, \qquad \mathfrak{n} \oplus \mathfrak{t} = \mathfrak{q} \oplus \mathfrak{n} \oplus \tilde{\mathfrak{l}}.$$

Let $\mathfrak{q} \not\subset \mathfrak{t}$, then $\mathfrak{q} \subset \mathfrak{k}$ and $\mathfrak{q}$ commutes with $\tilde{\mathfrak{i}}$, but then the restriction of $S_2 \in \mathrm{Int}_{\mathfrak{g}}(\mathfrak{q})$ to $\tilde{\mathfrak{i}}$ is the identity automorphism, and therefore $\tilde{\mathfrak{i}} \subset \mathfrak{l}_2$. By the choice $\tilde{\mathfrak{i}} \subset \mathfrak{l}_1$, thus $\tilde{\mathfrak{i}} \subset \mathfrak{l}_0$, whence $\tilde{\mathfrak{i}} = \mathfrak{p} \oplus \mathfrak{t}$ and $\tilde{\mathfrak{i}} = \mathfrak{p}$, since $\tilde{\mathfrak{i}}$ is simple. And $\mathfrak{l}_1 = \mathfrak{p} \oplus \mathfrak{k}$ is a principal involutive algebra, which contradicts the assumption.

Thus we should assume that $\mathfrak{q} \subset \mathfrak{t}$. Since $\mathfrak{q}$ and $\mathfrak{t}$ are ideals in $\mathfrak{l}_0$ we have $\mathfrak{t} = \mathfrak{q} \oplus \mathfrak{f}$. But $\mathfrak{t} \cong so(m-3)$ ($m > 4$), and either $m = 6$ or $m = 7$.

If $m = 6$ then

$$\mathfrak{t} \cong so(3) \cong \mathfrak{q}, \qquad \tilde{\mathfrak{i}}/\mathfrak{p} \oplus \mathfrak{q} \cong so(6)/(so(3) \oplus so(3)),$$

with the natural embedding. Hence $\mathrm{Int}_{\mathfrak{g}}^{\mathfrak{l}_1}(\mathfrak{q}) \cong \mathrm{Int}_{\mathfrak{l}_1}(\mathfrak{q}) \cong SO(3)$ and then $S_2 = S_1$, which is impossible.

Thus the only possibility is $m = 7$. Then

$$\mathfrak{t} \cong so(4), \qquad \mathfrak{t} = \mathfrak{q} \oplus \mathfrak{f}, \qquad \tilde{\mathfrak{i}}/\mathfrak{p} \oplus \mathfrak{q} \oplus \mathfrak{f} \cong so(7)/(so(3) \oplus so(4)).$$

Since $\mathfrak{k}$ commutes with $\mathfrak{q}$ and $\mathfrak{p}$, $S_1 \in \mathrm{Int}_\mathfrak{g}(\mathfrak{p})$ and $S_2 \in \mathrm{Int}_\mathfrak{g}(\mathfrak{q})$ act on $\mathfrak{k}$ as the identity map. Thus $\mathfrak{k} \subset \mathfrak{l}_1$, $\mathfrak{k} \subset \mathfrak{l}_2$, that is, $\mathfrak{k} \subset \mathfrak{l}_0$. Therefore $\mathfrak{l}_0 = \mathfrak{p} \oplus \mathfrak{q} \oplus \mathfrak{f} \oplus \mathfrak{k}$.

On the other hand, $\mathfrak{l}_0 = \mathfrak{p} \oplus \mathfrak{q} \oplus \mathfrak{r} \oplus \tilde{\mathfrak{l}}$, consequently $\mathfrak{f} \oplus \mathfrak{k} = \mathfrak{n} \oplus \tilde{\mathfrak{l}}$. But $\mathfrak{f} \not\subset \tilde{\mathfrak{l}}$, otherwise $\mathfrak{f}$ is a three-dimensional simple ideal in $\mathfrak{l} = \mathfrak{i} \oplus \tilde{\mathfrak{l}}$, and $\mathfrak{l}$ is a principal involutive algebra, which is a contradiction. Thus $\mathfrak{f} \subset \mathfrak{n} \cong so(m-4)$, which means either $m = 7$ or $m = 8$.

If $m = 8$ then $\mathfrak{n} = \mathfrak{f} \oplus \mathfrak{h} \cong so(4)$ and from $\mathfrak{f} \oplus \mathfrak{k} = \mathfrak{n} \oplus \tilde{\mathfrak{l}}$ it follows that $\mathfrak{k} = \mathfrak{h} \oplus \tilde{\mathfrak{l}}$, that is, $\mathfrak{l}_1 = \tilde{\mathfrak{i}} \oplus \mathfrak{k} = \tilde{\mathfrak{i}} \oplus \tilde{\mathfrak{l}} \oplus \mathfrak{h}$ is a principal involutive algebra, being a contradiction.

Finally, if $m = 7$ then $\mathfrak{n} = \mathfrak{f}$, $\mathfrak{k} = \tilde{\mathfrak{l}}$. Thus $\tilde{\mathfrak{l}} \subset \mathfrak{l}_0$ and $\tilde{\mathfrak{l}}$ is an ideal in $\mathfrak{l}$ and in $\mathfrak{l}_1$. If $\mathfrak{l}_2$ is not a principal involutive algebra then analogously we can show that $\tilde{\mathfrak{l}}$ is an ideal in $\mathfrak{l}_2$,

As a result $\tilde{\mathfrak{l}}$ is an ideal in $\mathfrak{g} = \mathfrak{l}_1 + \mathfrak{l}_2 + \mathfrak{l}$.

But $\mathfrak{g}$ is a simple semi-simple Lie algebra, which implies that $\tilde{\mathfrak{l}} = \{0\}$.

Thus if $\mathfrak{l}_1$ and $\mathfrak{l}_2$ are not principal involutive algebras then $\mathfrak{l} = \mathfrak{i} \cong so(7)$. Therefore if $\mathfrak{l} \not\cong so(7)$ then either $\mathfrak{l}_1$ or $\mathfrak{l}_2$ is a principal involutive algebra, which proves the theorem. ∎

In fact, the proof given above contains more than II.4.8 (Theorem 16). Namely, there has been proved:

II.4.9. Theorem 17. *If $\mathfrak{g}$ is a simple compact Lie algebra which has a unitary special involutive algebra $\mathfrak{l}$ of an involutive automorphism S, but has no principal involutive algebras, then $\mathfrak{g} = \mathfrak{l}_1 + \mathfrak{l}_2 + \mathfrak{l}$, where $\mathfrak{l}_1$ and $\mathfrak{l}_2$ are unitary special involutive algebras of the involutive automorphisms S_1 and S_2, respectively, $S_1 S_2 = S_2 S_1 = S$, $\mathfrak{l} \cong so(7)$, $\mathfrak{l}_1 \cong so(7)$, $\mathfrak{l}_2 \cong so(7)$, $\mathfrak{l}_0 = \mathfrak{l}_1 \cap \mathfrak{l}_2 = \mathfrak{l}_1 \cap \mathfrak{l} = \mathfrak{l}_2 \cap \mathfrak{l} = \mathfrak{p} \oplus \mathfrak{q} \oplus \mathfrak{f}$, where $\mathfrak{p}, \mathfrak{q}, \mathfrak{f}$ are simple three-dimensional Lie algebras,*

$$\mathfrak{l}/\mathfrak{l}_0 \cong so(7)/(so(3) \oplus so(4)),$$

$$\mathfrak{l}_1/\mathfrak{l}_0 \cong so(7)/(so(3) \oplus so(4)),$$

$$\mathfrak{l}_2/\mathfrak{l}_0 \cong so(7)/(so(3) \oplus so(4))$$

with the natural embeddings; $S_1 \in \mathrm{Int}_\mathfrak{g}(\mathfrak{p})$, $S_2 \in \mathrm{Int}_\mathfrak{g}(\mathfrak{q})$, $S \in \mathrm{Int}_\mathfrak{g}(\mathfrak{b})$, where $\mathfrak{b}$ is a diagonal of a canonical involutive automorphism of $\mathfrak{p} \oplus \mathfrak{q}$.

II.4.10. Let us explore the case described in II.4.9 (Theorem 17) more thoroughly. We realize $\mathfrak{l}$ as $so(7)$, then $\mathfrak{b}$ is $so(3)$ with the natural embedding into $so(7)$. However, since $\mathfrak{f} \cong so(3) \subset so(7) \cong \mathfrak{l}$ with the natural embedding and all subalgebras $so(3)$ with the natural embedding into $so(7)$ are conjugated in $\mathrm{Int}(so(7))$ we have that $\mathfrak{f}$ and $\mathfrak{b}$ are conjugated in $\mathrm{Int}_\mathfrak{g}(\mathfrak{l})$ and consequently in $\mathrm{Int}(\mathfrak{g})$. From this, in particular, it follows that $S \in \mathrm{Int}_\mathfrak{g}(\mathfrak{f})$.

Let us now note that since $\mathfrak{l}_1, \mathfrak{l}_2, \mathfrak{l}_3$ are special unitary involutive algebras and

$$\mathfrak{l}_1/\mathfrak{l}_0 \cong \mathfrak{l}_2/\mathfrak{l}_0 \cong \mathfrak{l}/\mathfrak{l}_0 \cong so(7)/so(4) \oplus so(3)$$

(with the natural embeddings) we have the iso-involutive sums

$$\mathfrak{g} = \mathfrak{l}_1 + \mathfrak{l}_2 + \mathfrak{l}, \qquad \mathfrak{g} = \mathfrak{l} + \mathfrak{l}_1 + \mathfrak{l}_2, \qquad \mathfrak{g} = \mathfrak{l}_2 + \mathfrak{l} + \mathfrak{l}_1$$

with the conjugating automorphisms $\varphi \in \mathrm{Int}_{\mathfrak{g}}(t\zeta)$, $\varphi_2 \in \mathrm{Int}_{\mathfrak{g}}(t\zeta_2)$, $\varphi_1 \in \mathrm{Int}_{\mathfrak{g}}(t\zeta_1)$, respectively, where

$$so(2) \cong \{t\zeta\} \subset \mathfrak{l}\dot{-}\mathfrak{l}_0 \subset \mathfrak{l} \cong so(7),$$

$$so(2) \cong \{t\zeta_2\} \subset \mathfrak{l}_2\dot{-}\mathfrak{l}_0 \subset \mathfrak{l}_2 \cong so(7),$$

$$so(2) \cong \{t\zeta_1\} \subset \mathfrak{l}_1\dot{-}\mathfrak{l}_0 \subset \mathfrak{l}_1 \cong so(7)$$

with the natural embeddings.

For this reason the involutive automorphisms S_1, S_2, S are conjugated in $\mathrm{Int}(\mathfrak{g})$:

$$\varphi^{-1} S_1 \varphi = S_2, \qquad \varphi_2^{-1} S \varphi_2 = S_1, \qquad \varphi_1^{-1} S_2 \varphi_1 = S_1.$$

Also

$$\varphi^{-1} S \varphi = S, \qquad \varphi_2^{-1} S_2 \varphi = S_2, \qquad \varphi_1^{-1} S_1 \varphi_1 = S_1,$$

since $(\varphi)^2 = S$, $(\varphi_1)^2 = S_1$, $(\varphi_2)^2 = S_2$.

As well we have

$$\varphi\,\mathfrak{l}_0 = \mathfrak{l}_0, \qquad \varphi_2\,\mathfrak{l}_0 = \mathfrak{l}_0, \qquad \varphi_1\,\mathfrak{l}_0 = \mathfrak{l}_0$$

which, owing to

$$\mathfrak{l}_0 = \mathfrak{p} \oplus \mathfrak{q} \oplus \mathfrak{f}, \qquad S_1 \in \mathrm{Int}_{\mathfrak{g}}(\mathfrak{p}), \qquad S_2 \in \mathrm{Int}_{\mathfrak{g}}(\mathfrak{q}), \qquad S \in \mathrm{Int}_{\mathfrak{g}}(\mathfrak{f}),$$

imply

$$\varphi\,\mathfrak{p} = \mathfrak{q}, \qquad \varphi_2\,\mathfrak{f} = \mathfrak{p}, \qquad \varphi_1\,\mathfrak{q} = \mathfrak{f}.$$

Thus the subalgebras $\mathfrak{p}$, $\mathfrak{q}$, $\mathfrak{f}$ are conjugated in $\mathrm{Int}(\mathfrak{g})$ and, additionally, $\mathfrak{f}$ is conjugated in $\mathrm{Int}(\mathfrak{g})$ with the diagonal $\mathfrak{b}$ of $\mathfrak{p} \oplus \mathfrak{q}$. We note also that $\varphi\,\mathfrak{b} = \mathfrak{b}$.

Let us show that this is impossible, since $\mathfrak{p} \cong \mathfrak{q} \cong so(3)$.

Indeed, if $\varphi\,\mathfrak{p} = \mathfrak{q}$, then φ acts on $\mathfrak{p} \oplus \mathfrak{q}$ as a canonical involutive automorphism generating the diagonal $\mathfrak{b}$ of $\mathfrak{p} \oplus \mathfrak{q}$ and

$$\tau: \zeta \to \zeta + \varphi\zeta = (\mathrm{Id} + \varphi)\,\zeta, \qquad \zeta \in \mathfrak{p},$$

is the natural isomorphism of $\mathfrak{p}$ onto $\mathfrak{b}$. However, by the conditions $\mathfrak{b}$ and $\mathfrak{p}$ are conjugated in $\mathrm{Int}(\mathfrak{g})$, that is, there exists $\psi \in \mathrm{Int}(\mathfrak{g})$ such that $\psi\,\mathfrak{b} = \mathfrak{p}$. But then $\psi\tau = (\mathrm{Id} + \varphi)$ is an automorphism of $\mathfrak{p} \cong so(3)$. Since all automorphisms of $so(3)$ are inner automorphisms, $\psi\tau = \mathrm{Int}_{\mathfrak{p}}(\eta)$, $\eta \in \mathfrak{p}$.

We consider now $\nu = \mathrm{Int}_{\mathfrak{g}}(\eta)$. Evidently for the restriction of ν to $\mathfrak{p}$ we have $\nu|_{\mathfrak{p}} = \mathrm{Int}_{\mathfrak{g}}^{\mathfrak{p}}(\eta) = \mathrm{Int}_{\mathfrak{p}}(\eta) = \psi\tau$, that is, $\psi^{-1}\nu|_{\mathfrak{p}} = \tau$, or $(\psi^{-1}\nu)|_{\mathfrak{p}} = \tau$. Thus τ is a restriction to $\mathfrak{p}$ of some inner automorphism $\mu = \psi^{-1}\nu \in \mathrm{Int}(\mathfrak{g})$.

If now $a(x,y)$ is the Cartan metric of $\mathfrak{g}$ then it is positive-definite ($a(x,x) > 0$ for $x \neq 0$), and $a(\mu x, \mu y) = a(x,y)$, since μ is an automorphism of $\mathfrak{g}$.

Furthermore, if $0 \neq \zeta \in \mathfrak{p}$ then $\mu\zeta = \mu|_{\mathfrak{p}}\,\zeta = \psi^{-1}\nu|_{\mathfrak{p}}$, and $a(\psi^{-1}\tau\zeta, \psi^{-1}\tau\zeta) = a(\zeta,\zeta)$. Since ψ is an isomorphism we obtain $a(\tau\zeta,\tau\zeta) = a(\zeta,\zeta)$ and, furthermore, owing to $\tau = \mathrm{Id} + \varphi$, $a(\zeta,\zeta) + 2a(\varphi\zeta,\zeta) + a(\varphi\zeta,\varphi\zeta) = a(\zeta,\zeta)$. Since $\varphi \in \mathrm{Int}(\mathfrak{g})$ we have $a(\varphi\zeta,\varphi\zeta) = a(\zeta,\zeta)$ and, finally, $2a(\varphi\zeta,\zeta) + a(\zeta,\zeta) = 0$.

But $\zeta \in \mathfrak{p}$, $\varphi\zeta \in \mathfrak{q}$, which give $a(\varphi\zeta,\zeta) = 0$, since $\mathfrak{p}$ and $\mathfrak{q}$ are simple semi-simple ideals in subalgebra $\mathfrak{p} \oplus \mathfrak{q} \subset \mathfrak{g}$ and $a(x,y)$ is the Cartan metric. As a result $a(\zeta,\zeta) = 0$, contradicting $\zeta \neq 0$.

Thus we have the following theorems:

II.4.11. Theorem 18. *There are no Lie algebras satisfying the conditions of II.4.9 (Theorem 17).*

II.4.12. Theorem 19. *There are no simple compact Lie algebras* $\mathfrak{g}$ *which have a special unitary involutive algebra* $\mathfrak{l} \cong so(7)$, *but have no principal involutive algebras.*

Combining II.4.8 (Theorem 16) and II.4.12 (Theorem 19) we have:

II.4.13. Theorem 20. *If a simple compact Lie algebra* $\mathfrak{g}$ *has a unitary special involutive automorphism (algebra) then it has a principal involutive automorphism (algebra).*

It is now easy to prove the fundamental theorem:

II.4.14. Theorem 21. *A simple semi-simple compact Lie algebra has a principal involutive automorphism.*

Proof. We proceed by the induction. The assertion of the theorem is true for a simple semi-simple compact Lie algebra $\mathfrak{g}$ of the minimal possible dimension 3, in this case the identity (involutive) automorphism is principal.

Suppose that our assertion is true for any simple semi-simple compact Lie algebra of dimension less then r and let us show that it is true for a simple semi-simple compact Lie algebra of the dimension r.

Let us consider any non-identity involutive automorphism of $\mathfrak{g}$ (which evidently exists) and its involutive algebra $\mathfrak{l}$. Then either $\mathfrak{l}$ has a simple semi-simple ideal or $\mathfrak{l}$ is commutative. If $\mathfrak{l}$ is commutative then it is one-dimensional, since the centre of an involutive algebra for a simple semi-simple compact Lie algebra is at most one-dimensional. But then $\mathfrak{g}-\mathfrak{l}$ is two-dimensional since $\mathrm{ad}_{\mathfrak{g}}^{(\mathfrak{g}-\mathfrak{l})}(\mathfrak{l})$ is irreducible on $\mathfrak{g}-\mathfrak{l}$, and $\mathfrak{g}$ is three-dimensional, that is, it has the identity principal involutive automorphism.

Thus we should consider the case $\mathfrak{l} = \mathfrak{p} \oplus \mathfrak{q}$, where $\mathfrak{p}$ is a simple semi-simple compact Lie algebra and $\dim \mathfrak{p} \leq r$. By the inductive assumption $\mathfrak{p}$ has a principal involutive automorphism μ. But then $\mathfrak{l}$ has a principal involutive automorphism as well, since we can define in a unique way the involutive automorphism ν of $\mathfrak{l}$ by $\nu|_{\mathfrak{p}} = \mu$, $\nu|_{\mathfrak{q}} = \mathrm{Id}$. However by II.4.4 (Theorem 13) $\mathfrak{g}$ then has a unitary special involutive automorphism, and consequently by II.4.13 (Theorem 20) $\mathfrak{g}$ has a principal involutive automorphism.

Furthermore, II.4.14 (Theorem 21) and II.4.3 (Theorem 12) imply:

II.4.15. Theorem 22. *If* $\mathfrak{g}$ *is a simple compact non-commutative Lie algebra and* $\dim \mathfrak{g} \neq 3$ *then* $\mathfrak{g}$ *has a unitary principal involutive automorphism.*

This theorem is of great importance, it allows us to characterize simple compact Lie algebras by the type of their principal unitary involutive automorphisms. For example, a central unitary principal involutive automorphism leads us to the Lie algebras of type $su(n)$, a di-unitary principal involutive automorphism leads us to the Lie algebras of type $so(n)$ or g_2 etc..

All this is considered later.

CHAPTER II.5

PRINCIPAL DI-UNITARY INVOLUTIVE AUTOMORPHISM

II.5.1. Let $\mathfrak{g}$ be a simple compact Lie algebra, and $\mathfrak{l}$ be its principal involutive algebra of a principal involutive automorphism S of type $U^{(2)}$. Then $\mathfrak{l} = \mathfrak{p} \oplus \mathfrak{q} \oplus \tilde{\mathfrak{l}}$, $S \in \text{Int}_{\mathfrak{g}}\mathfrak{p} \cong SU(2)$, $S \in \text{Int}_{\mathfrak{g}}(\mathfrak{q}) \cong SU(2)$.

We consider the diagonal $\mathfrak{b}$ in $\mathfrak{p} \oplus \mathfrak{q}$ of the canonical involutive automorphism $\mathfrak{p} \leftrightarrows \mathfrak{q}$. Then evidently $\text{Int}_{\mathfrak{g}}(\mathfrak{b}) \cong SO(3)$.

Consequently owing to II.1.2 (Lemma 1) we can construct a hyper-involutive decomposition. Thus we have $S_1, S_2, S_3, p \in \text{Int}_{\mathfrak{g}}(\mathfrak{b})$ such that

$$
\left.
\begin{aligned}
(p)^3 = (S_\rho)^2 = \text{Id}, \qquad S_\rho \neq \text{Id}, \qquad S_\rho S_\mu = S_\mu S_\rho \\
(\rho, \mu = 1, 2, 3), \\
S_1 S_2 = S_3, \qquad S_2 S_3 = S_1, \qquad S_3 S_1 = S_2, \\
p S_1 = S_2 p, \qquad p S_2 = S_3 p, \qquad p S_3 = S_1 p.
\end{aligned}
\right\}
\tag{II.44}
$$

If $\mathfrak{l}_1, \mathfrak{l}_2, \mathfrak{l}_3$ are the involutive algebras of the involutive automorphisms S_1, S_2, S_3, respectively, then

$$
\left.
\begin{aligned}
\mathfrak{g} = \mathfrak{l}_1 + \mathfrak{l}_2 + \mathfrak{l}_3, \qquad \mathfrak{l}_1 \cap \mathfrak{l}_2 = \mathfrak{l}_2 \cap \mathfrak{l}_3 = \mathfrak{l}_3 \cap \mathfrak{l}_1 = \mathfrak{l}_0, \\
\mathfrak{l}_\alpha/\mathfrak{l}_0 \quad (\alpha = 1, 2, 3), \\
\mathfrak{l}_\alpha = \mathfrak{m}_\alpha + \mathfrak{l}_0, \qquad \mathfrak{m}_\alpha \perp \mathfrak{l}_0, \\
p\mathfrak{m}_1 = \mathfrak{m}_2, \qquad p\mathfrak{m}_2 = \mathfrak{m}_3, \qquad p\mathfrak{m}_3 = \mathfrak{m}_1, \qquad p\mathfrak{l}_0 = \mathfrak{l}_0.
\end{aligned}
\right\}
\tag{II.45}
$$

However, in our assumptions we also have

$$
\left.
\begin{aligned}
pS = Sp, \qquad S_\alpha S = S S_\alpha \quad (\alpha = 1, 2, 3), \\
p\eta = \eta, \qquad \eta \in \tilde{\mathfrak{l}}, \qquad \tilde{\mathfrak{l}} \subset \mathfrak{l}_0, \\
(\mathfrak{p} \oplus \mathfrak{q}) \cap \mathfrak{l}_0 = \{0\},
\end{aligned}
\right\}
\tag{II.46}
$$

where $\mathfrak{l}_0/\tilde{\mathfrak{l}}$ is an involutive pair.

77

II.5.2. Definition 28. A hyper-involutive decomposition of a Lie algebra $\mathfrak{g}$ is called basis for a principal involutive automorphism S of type $U^{(2)}$ with the involutive algebra $\mathfrak{l} = \mathfrak{p} \oplus \mathfrak{q} \oplus \tilde{\mathfrak{l}}$ ($\mathrm{Int}_\mathfrak{g}(\mathfrak{p}) \cong SU(2)$, $\mathrm{Int}_\mathfrak{g}(\mathfrak{q}) \cong SU(2)$) if the involutive automorphisms S_1, S_2, S_3 of this hyper-involutive decomposition belong to $\mathrm{Int}_\mathfrak{g}(\mathfrak{b})$ ($\mathfrak{b}$ being a diagonal in $\mathfrak{p} \oplus \mathfrak{q}$), $pS = Sp$, and $p\eta = \eta$ for $\eta \in \tilde{\mathfrak{l}}$.

For such a hyper-involutive decomposition the conditions (II.46) are evidently satisfied.

II.5.3. Let us consider the involutive pair $\mathfrak{l}_0/\tilde{\mathfrak{l}}$, then either the centre $\mathfrak{z}$ of $\mathfrak{l}_0$ belongs to $\tilde{\mathfrak{l}}$ or else $(\mathfrak{l}_0 \dot- \tilde{\mathfrak{l}}) \cap \mathfrak{z} \neq \{0\}$.

Let us assume the latter possibility. We denote by $\mathfrak{r}$ the maximal subalgebra of vectors from $\mathfrak{g}$ commuting with $\tilde{\mathfrak{l}}$. Evidently $(\mathfrak{p} \oplus \mathfrak{q}) \subset \mathfrak{r}$, $(\mathfrak{l}_0 \dot- \tilde{\mathfrak{l}}) \cap \mathfrak{z} \subset \mathfrak{r}$, and $(\mathfrak{g} \dot- \mathfrak{l}) \cap \mathfrak{r} \neq \{0\}$. But then $\mathrm{ad}_\mathfrak{g}^{\mathfrak{g} \dot- \mathfrak{l}}(\mathfrak{l})$ maps $(\mathfrak{g} \dot- \mathfrak{l}) \cap \mathfrak{r}$ into itself, which, by virtue of the irreducibility of $\mathrm{ad}_\mathfrak{g}^{\mathfrak{g} \dot- \mathfrak{l}}(\mathfrak{l})$, is possible only if $(\mathfrak{g} \dot- \mathfrak{l}) \cap \mathfrak{r} = \mathfrak{g} \dot- \mathfrak{l}$, that is, $\mathfrak{g} \dot- \mathfrak{l} \subset \mathfrak{r}$. But then $\tilde{\mathfrak{l}}$ is an ideal of a simple Lie algebra $\mathfrak{g}$, which means that $\tilde{\mathfrak{l}} = \{0\}$.

If $\tilde{\mathfrak{l}} \neq \{0\}$ then from what has been considered above it follows that the centre $\mathfrak{z}$ of the compact Lie algebra $\mathfrak{l}_0$ belongs to $\tilde{\mathfrak{l}}$, and also that the restriction of p (see (II.44)) to $\mathfrak{l}_0$ is not the identity automorphism.

Then for the Lie algebra $\mathfrak{l}_0$ with the involutive algebra $\tilde{\mathfrak{l}}$ and automorphism p ($p^3 = \mathrm{Id}$) II.2.4 (Theorem 4) is satisfied.

Consequently $p \in \mathrm{Int}_{\mathfrak{l}_0}(\mathfrak{z}) = \mathrm{Int}_\mathfrak{g}^{\mathfrak{l}_0}(\mathfrak{z})$ on $\mathfrak{l}_0$, where $\mathfrak{z}$ is the non-trivial centre of $\tilde{\mathfrak{l}}$. But then there exists $q \in \mathrm{Int}_\mathfrak{g}(\mathfrak{z})$ such that p and q coincide on $\mathfrak{l}_0$.

Since $\mathfrak{l} = \mathfrak{p} \oplus \mathfrak{q} \oplus \tilde{\mathfrak{l}}$ the centre of $\tilde{\mathfrak{l}}$ is the centre of $\mathfrak{l}_0$ as well, whence $q\eta = \eta$ for $\eta \in \mathfrak{l}$ and q commutes with $\mathrm{ad}_\mathfrak{g}^{\mathfrak{g} \dot- \mathfrak{l}}(\mathfrak{l})$ on $\mathfrak{g} \dot- \mathfrak{l}$; moreover, $q^3 = \mathrm{Id}$ on $\mathfrak{l}_0 \cap (\mathfrak{g} \dot- \mathfrak{l}) \neq \{0\}$ since $q = p$ there.

Owing to the irreducibility of $\mathrm{ad}_\mathfrak{g}^{\mathfrak{g} \dot- \mathfrak{l}}(\mathfrak{l})$ we then obtain $q^3 = \mathrm{Id}$ on $\mathfrak{g} \dot- \mathfrak{l}$. In addition, $q^3 = \mathrm{Id}$ on $\mathfrak{l}$, thus $q^3 = \mathrm{Id}$ on $\mathfrak{g}$. Lastly, q commutes with S_1, S_2, S_3, p, since $q \in \mathrm{Int}_\mathfrak{g}(\tilde{\mathfrak{l}})$ and $S_1, S_2, S_3, p \in \mathrm{Int}_\mathfrak{g}(\mathfrak{p} \oplus \mathfrak{q})$.

Now we consider $\tilde{p} = pq^{-1}$. By the construction presented above $\tilde{p}$ is the identity map on $\mathfrak{l}_0$.

Thus what has been presented above and (II.44), (II.46) imply

$$
\left.
\begin{aligned}
&(\tilde{p})^3 = (S_\rho)^2 = \mathrm{Id}, \quad S_\rho \neq \mathrm{Id}, \quad S_\rho S_\mu = S_\mu S_\rho \\
&\qquad (\rho, \mu = 1,2,3), \\
&S_1 S_2 = S_3, \quad S_2 S_3 = S_1, \quad S_3 S_1 = S_2, \\
&\tilde{p} S_1 = S_2 \tilde{p}, \quad \tilde{p} S_2 = S_3 \tilde{p}, \quad \tilde{p} S_3 = S_1 \tilde{p}, \\
&\tilde{p} S = S \tilde{p}, \quad S S_\rho = S_\rho S \ (\rho = 1,2,3), \\
&\tilde{p} \eta = \eta, \quad \eta \in \mathfrak{l}_0.
\end{aligned}
\right\}
\quad \text{(II.47)}
$$

Thus we arrive at the theorem:

II.5.4. Theorem 23. *Let $\mathfrak{g}$ be a simple compact Lie algebra, and $\mathfrak{l} = \mathfrak{p} \oplus \mathfrak{q} \oplus \tilde{\mathfrak{l}}$ be its principal involutive algebra of an involutive automorphism S of type $U^{(2)}$ ($\mathrm{Int}_{\mathfrak{g}}(\mathfrak{p}) \cong SU(2)$, $\mathrm{Int}_{\mathfrak{g}}(\mathfrak{q}) \cong SU(2)$). If $\tilde{\mathfrak{l}} \neq \{0\}$ then $\mathfrak{g}$ has a prime hyper-involutive decomposition basis for S.*

II.5.5. Now we consider the case $\tilde{\mathfrak{l}} = \{0\}$, that is, $\mathfrak{l} = \mathfrak{p} \oplus \mathfrak{q}$. Then $\mathfrak{l}_0$ is commutative, because $\mathfrak{l}_0 \subset \mathfrak{m} = \mathfrak{g} \dot{-} \mathfrak{l}$, and $\mathfrak{l}_\alpha = \mathfrak{l}_\alpha \cap \mathfrak{l}$ ($\alpha = 1, 2, 3$) are two-dimensional, and moreover $\mathfrak{l}_\alpha \cap \mathfrak{p}$, $\mathfrak{l}_\alpha \cap \mathfrak{q}$ are one-dimensional, $\mathfrak{l}_\alpha$ being commutative.

We note that $\mathfrak{l}_\alpha / \mathfrak{l}_\alpha$ ($\alpha = 1, 2, 3$) are the involutive pairs of the involutive automorphism S.

Let us assume that $\mathfrak{l}_1$ has a non-trivial, and consequently one-dimensional centre $\mathfrak{z}_1$ (an involutive algebra of a simple compact Lie algebra has at most a one-dimensional centre). Then either $\mathfrak{z}_1 \subset \mathfrak{l}_1$ or $\mathfrak{z}_1 \subset \mathfrak{l}_1 \dot{-} \mathfrak{l}_1$.

If $\mathfrak{z}_1 \subset \mathfrak{l}_1 \dot{-} \mathfrak{l}_1$ we consider $S_1 = \tilde{S}_1$, $\tilde{S}_2 = S\, S_1$, $\tilde{S}_3 = S$, and their involutive algebras $\tilde{\mathfrak{l}}_1 = \mathfrak{l}_1$, $\tilde{\mathfrak{l}}_2$, $\tilde{\mathfrak{l}}_3 = \mathfrak{l}$.

Since $\tilde{S}_1, \tilde{S}_2, \tilde{S}_3$ commute pair-wise and the product of any two of them gives the third, we have the involutive decomposition $\mathfrak{g} = \tilde{\mathfrak{l}}_1 + \tilde{\mathfrak{l}}_2 + \tilde{\mathfrak{l}}_3$. Moreover, $\tilde{\mathfrak{l}}_0 = \mathfrak{l}_1$. Taking $\zeta \in \mathfrak{p} \dot{-} (\mathfrak{p} \cap \mathfrak{l}_1)$ we have $S = \tilde{S}_3 \in \mathrm{Int}_{\mathfrak{g}}(\mathfrak{p})$, $\tilde{S}_3 \in \mathrm{Int}_{\mathfrak{g}}(\{t\zeta\}) \subset \mathrm{Int}_{\mathfrak{g}}(\mathfrak{p})$, which means that the decomposition written above is iso-involutive. Thus

$$\varphi\, \tilde{\mathfrak{l}}_1 = \tilde{\mathfrak{l}}_2, \qquad \varphi\, \tilde{\mathfrak{l}}_0 = \tilde{\mathfrak{l}}_0, \qquad \varphi^2 = S, \qquad \varphi \in \mathrm{Int}_{\mathfrak{g}}(\{t\zeta\}) \,.$$

Consequently $\tilde{\mathfrak{l}}_2$ has the one-dimensional centre $\varphi\, \mathfrak{z}_1 = \mathfrak{z}_2 \subset \tilde{\mathfrak{l}}_2 \dot{-} \tilde{\mathfrak{l}}_0$. Evidently $S_1 = \tilde{S}_1 \in \mathrm{Int}_{\mathfrak{g}}(\mathfrak{z}_1)$, and taking $\tilde{\varphi} \in \mathrm{Int}_{\mathfrak{g}}(\mathfrak{z}_1)$, $\tilde{\varphi}^2 = S_1$, we obtain, analogously to the preceding, $\tilde{\varphi}\, \tilde{\mathfrak{l}}_2 = \tilde{\mathfrak{l}}_3 = \mathfrak{l}$, whence $\mathfrak{l}$ has the one-dimensional centre $\mathfrak{z} = \tilde{\varphi}\, \mathfrak{z}_2$. This contradicts the semi-simplicity of $\mathfrak{l} = \mathfrak{p} \oplus \mathfrak{q}$.

Thus $\mathfrak{z}_1$ (the centre of $\mathfrak{l}_1$) belongs to $\mathfrak{l}_1$ (in particular, being trivial). Let us use the canonical decomposition for $\mathfrak{l}_1 / \mathfrak{l}_1$. Obviously either $\mathfrak{l}_1 / \mathfrak{l}_1 = \mathfrak{z}_1 / \mathfrak{z}_1 + \mathfrak{r}_1 / \mathfrak{t}_1$, where $\mathfrak{z}_1$, $\mathfrak{t}_1$ are one-dimensional, $\mathfrak{r}_1$ is simple three-dimensional, or $\mathfrak{l}_1 / \mathfrak{l}_1 = \mathfrak{r}_1 / \mathfrak{t}_1 + \mathfrak{w}_1 / \mathfrak{n}_1$, where $\mathfrak{t}_1$, $\mathfrak{n}_1$ are one-dimensional, and $\mathfrak{r}_1$ and $\mathfrak{w}_1$ are simple three-dimensional.

In the first case $\mathfrak{l}_0 \subset \mathfrak{l}_1 \dot{-} \mathfrak{l}_1$ is at most one-dimensional, but then the restriction of the automorphism p of the hyper-involutive decomposition to $\mathfrak{l}_0$ is the identity automorphism and our hyper-involutive decomposition (II.45) is again prime.

In the second case $\mathfrak{l}_0$ is evidently at most two-dimensional. Moreover, if $\mathfrak{l}_0$ is one-dimensional or $\mathfrak{l}_0 = \{0\}$, then the hyper-involutive decomposition (II.45) is again prime.

Thus if $\mathfrak{g}/\mathfrak{l}$ does not have a prime hyper-involutive decomposition basis for S then $\mathfrak{l} = \mathfrak{p} \oplus \mathfrak{q}$, $\mathfrak{l}_\alpha / \mathfrak{l}_\alpha = \mathfrak{r}_\alpha / \mathfrak{t}_\alpha + \mathfrak{w}_\alpha / \mathfrak{n}_\alpha$ ($\alpha = 1, 2, 3$), where $\mathfrak{r}_\alpha$, $\mathfrak{w}_\alpha$ are three-dimensional, $\mathfrak{t}_\alpha$, $\mathfrak{n}_\alpha$ are one-dimensional, $\mathfrak{l}_0 = \mathfrak{l}_1 \cap \mathfrak{l}_2$ is two-dimensional. Hence $\dim \mathfrak{g} = 14$.

Let us also consider $\mathrm{Int}_{\mathfrak{g}}(\mathfrak{r}_\alpha)$ and $\mathrm{Int}_{\mathfrak{g}}(\mathfrak{w}_\alpha)$. If one of them is isomorphic to $SO(3)$ then $\mathfrak{g}/\mathfrak{l}_\alpha$ is a principal involutive pair of type O. Also $\dim \mathfrak{l}_\alpha = 6$, which by II.3 implies $\mathfrak{g}/\mathfrak{l}_\alpha \cong so(6)/(so(3) \oplus so(3))$. Then $\dim \mathfrak{g} = 15$, which is wrong. Thus $\mathrm{Int}_{\mathfrak{g}}(\mathfrak{r}_\alpha) \cong SU(2)$, $\mathrm{Int}_{\mathfrak{g}}(\mathfrak{w}_\alpha) \cong SU(2)$.

Let us also show that $\mathfrak{l}$ and $\mathfrak{l}_\alpha$ are conjugated in $\mathrm{Int}(\mathfrak{g})$. Indeed, the involutive automorphisms $S_1 = \tilde{S}_1$, $\tilde{S}_2 = S_1 S$, $\tilde{S}_3 = S$ generate, as has been shown, the involutive decomposition $\mathfrak{g} = \tilde{\mathfrak{l}}_1 + \tilde{\mathfrak{l}}_2 + \tilde{\mathfrak{l}}_3$, where $\tilde{\mathfrak{l}}_1 = \mathfrak{l}_1$ and $\tilde{\mathfrak{l}}_2$ are conjugated in $\mathrm{Int}(\mathfrak{g})$.

Taking $\eta \in \mathfrak{r}_1 \dot{-} \mathfrak{t}_1$, $\widetilde{S}_1 = S_1 \in \mathrm{Int}_\Gamma\left(\{t\eta\}\right) \subset \mathrm{Int}_\mathfrak{g}(\mathfrak{r}_1)$, and $\tilde{\varphi} \in \mathrm{Int}_\Gamma\left(\{t\eta\}\right)$, such that $\tilde{\varphi}^2 = S_1$, we have $\tilde{\varphi} \mathfrak{l}_2 = \mathfrak{l}$. Thus $\mathfrak{l}_2$ and $\mathfrak{l}$ are conjugated in $\mathrm{Int}(\mathfrak{g})$. Consequently $\mathfrak{l}_1$ and $\mathfrak{l}$ are conjugated in $\mathrm{Int}(\mathfrak{g})$, which implies the conjugacy of S_1, and S in $\mathrm{Int}(\mathfrak{g})$.

We have arrived at the theorem:

II.5.6. Theorem 24. *If $\mathfrak{g}$ is a simple compact Lie algebra and $\mathfrak{l}$ is its principal involutive algebra of an involutive automorphism S of type $U^{(2)}$ then either $\mathfrak{g}$ has a prime hyper-involutive decomposition basis for S or $\mathfrak{g}$ has a non-prime hyper-involutive decomposition basis for S such that $\mathfrak{l} = \mathfrak{p} \oplus \mathfrak{q}$, $\mathfrak{l}_\alpha = \mathfrak{r}_\alpha \oplus \mathfrak{w}_\alpha$, $\mathrm{Int}_\mathfrak{g}(\mathfrak{r}_\alpha) \cong SU(2)$, $\mathrm{Int}_\mathfrak{g}(\mathfrak{w}_\alpha) \cong SU(2)$, $\mathfrak{l}_0 = \mathfrak{t}_\alpha \oplus \mathfrak{n}_\alpha$, where $\mathfrak{t}_\alpha \subset \mathfrak{r}_\alpha$, and $\mathfrak{n}_\alpha \subset \mathfrak{w}_\alpha$, $\mathfrak{t}_\alpha$ and $\mathfrak{n}_\alpha$ are one-dimensional, $\mathfrak{l}_\alpha \cap \mathfrak{l} = \mathfrak{a}_\alpha \oplus \mathfrak{c}_\alpha$, where $\mathfrak{a}_\alpha$ and $\mathfrak{c}_\alpha$ are one-dimensional and $\mathfrak{a}_\alpha \subset \mathfrak{p}_\alpha$, $\mathfrak{c}_\alpha \subset \mathfrak{q}$, $\dim \mathfrak{g} = 14$. The involutive automorphisms S and S_1, S_2, S_3 are conjugated in $\mathrm{Int}(\mathfrak{g})$.*

II.5.7. Thus there are two possibilities for further considerations: the prime decomposition (main case) and the non-prime decomposition (singular case).

We start first with the main case of a prime hyper-involutive decomposition basis for a principal di-unitary automorphism. Then (II.44), (II.45), (II.46) are valid, $p\eta = \eta$ for $\eta \in \mathfrak{l}_0$, and $(\mathfrak{p} \oplus \mathfrak{q}) \cap \mathfrak{l}_\alpha = \mathfrak{a}_\alpha \oplus \mathfrak{c}_\alpha$ ($\alpha = 1, 2, 3$), where $\mathfrak{a}_\alpha$, $\mathfrak{c}_\alpha$ are one-dimensional.

Since $S S_\alpha = S_\alpha S$ we have $S \mathfrak{m}_1 = \mathfrak{m}_1$.

Choosing an orthonormal base $X_{1_1}, \ldots, X_{n_1}$ in $\mathfrak{m}_1 = \mathfrak{g} \dot{-} \mathfrak{l}_1$ in such a way that S has a diagonal matrix on $\mathfrak{m}_1$, we have

$$S X_{a_1} = X_{a_1}, \qquad S X_{i_1} = -X_{i_1}; \quad a = 1, 2; \quad i = 3, \ldots, n;$$
$$X_{a_1} \in \mathfrak{p} \oplus \mathfrak{q}. \tag{II.48}$$

Furthermore (as in II.1) we choose an orthonormal base in $\mathfrak{l}_0$ and define bases in $\mathfrak{m}_1$ and $\mathfrak{m}_2$ by means of the automorphism p see (II.4), (II.5), (II.6).

The union of all the bases introduced constitute a canonical base of the hyper-involutive decomposition

$$\mathfrak{g} = \mathfrak{l}_1 + \mathfrak{l}_2 + \mathfrak{l}_3,$$

or a hyper-involutive base, see (II.5), (II.6),

$$\left.\begin{array}{ll} X_{I_1} \in \mathfrak{m}_1 & (I = 1, \ldots, n), \\ X_{I_2} \in \mathfrak{m}_2 & (I = 1, \ldots, n), \\ X_{I_3} \in \mathfrak{m}_3 & (I = 1, \ldots, n), \\ Y_A \in \mathfrak{l}_0 & (A = 1, \ldots, r). \end{array}\right\} \tag{II.49}$$

Now the relations (II.6)–(II.11) have the form (II.29)–(II.34), which we write down once more.

$$\left.\begin{array}{l} p X_{I_1} = X_{I_2}, \qquad p X_{I_2} = X_{I_3}, \qquad p X_{I_3} = X_{I_1}, \qquad p Y_A = Y_A, \\ S_\rho X_{I_\rho} = X_{I_\rho}, \qquad S_\rho Y_A = Y_A, \qquad S_\rho X_{I_\mu} = -X_{I_\mu} \\ (\rho, \mu = 1, 2, 3, \ \rho \neq \mu). \end{array}\right\} \tag{II.50}$$

$$\left.\begin{array}{ll} \left[X_{I_\rho} X_{J_\rho}\right] = C_{I_\rho J_\rho}^A Y_A, & \left[X_{I_\rho} X_{J_\mu}\right] = C_{I_\rho J_\mu}^{K_\lambda} X_{K_\lambda}, \\[2mm] \left[X_{I_\rho} Y_A\right] = C_{I_\rho A}^{K_\rho} X_{K_\rho}, & \left[Y_A Y_B\right] = C_{AB}^F Y_F, \\[2mm] \lambda, \mu, \rho = 1, 2, 3; \quad \lambda \neq \mu, \quad \mu \neq \rho, \quad \rho \neq \lambda. \end{array}\right\} \quad \text{(II.51)}$$

$$\left.\begin{array}{l} C_{I_2 J_3}^{K_1} = C_{I_1 J_2}^{K_3} = C_{I_3 J_1}^{K_2} = t_{IJ}^K, \qquad C_{I_\rho J_\rho}^B = C_{IJ}^B, \\[2mm] C_{I_\rho A}^{K_\rho} = C_{IA}^K \quad (\rho = 1, 2, 3). \end{array}\right\} \quad \text{(II.52)}$$

$$t_{IJ}^K = t_{JK}^I, \qquad C_{IJ}^B = -C_{IB}^J, \qquad C_{AB}^F = -C_{FB}^A. \quad \text{(II.53)}$$

$$\left.\begin{array}{l} t_{IJ}^R = K_{IJ}^R + E_{IJ}^R, \qquad K_{IJ}^R = -K_{JI}^R = -K_{RJ}^I, \\[2mm] E_{IJ}^R = E_{JI}^R = E_{RJ}^I. \end{array}\right\} \quad \text{(II.54)}$$

$$\left.\begin{array}{ll} t_{IJ}^K C_{KA}^P - t_{KJ}^P C_{IA}^K - t_{IK}^P C_{JA}^K = 0, & \text{(II.55.1)} \\[2mm] t_{JK}^P t_{PI}^R - t_{IK}^P t_{PJ}^R = C_{KB}^R C_{IJ}^B, & \text{(II.55.2)} \\[2mm] t_{KJ}^P t_{IP}^R - t_{KI}^P t_{JP}^R = C_{KB}^R C_{IJ}^B, & \text{(II.55.3)} \\[2mm] C_{IA}^R C_{JK}^A + C_{KA}^R C_{IJ}^A + C_{JA}^R C_{KI}^A = 0, & \text{(II.55.4)} \\[2mm] C_{PB}^R C_{IA}^P - C_{PA}^R C_{IB}^P = C_{AB}^M C_{IM}^R, & \text{(II.55.5)} \\[2mm] C_{AB}^M C_{MF}^G + C_{FA}^M C_{MB}^G + C_{BF}^M C_{MA}^G = 0. & \text{(II.55.6)} \end{array}\right\} \quad \text{(II.55)}$$

For the involutive automorphism S we also obtain from (II.48)

$$\begin{array}{c} t_{jk}^i = 0, \qquad t_{ab}^i = t_{ib}^a = t_{bi}^a = 0 \\[2mm] (a, b = 1, 2; \ i = 3, 4, \ldots, n). \end{array} \quad \text{(II.56)}$$

Before using the above relations we consider the involutive pair $\mathfrak{l}_1/\mathfrak{l}_0$. There is a canonical decomposition

$$\mathfrak{l}_1/\mathfrak{l}_0 = \mathfrak{z}/\tilde{\mathfrak{z}} + \sum_{\alpha=1}^n (\mathfrak{m}_\alpha/\mathfrak{n}_\alpha),$$

where $\mathfrak{z}$ is the center of $\mathfrak{l}_1$ and $\mathfrak{m}_\alpha/\mathfrak{n}_\alpha$ are elementary involutive pairs.

Since $S \in \text{Int}_g(\mathfrak{a}_1 \oplus \mathfrak{c}_1)$ and $(\mathfrak{a}_1 \oplus \mathfrak{c}_1) \subset \mathfrak{l}_1$ the restriction of S onto $\mathfrak{l}_1$ is an inner automorphism of $\mathfrak{l}_1$, and then $S\mathfrak{z} = \mathfrak{z}$, $S\mathfrak{m}_\alpha = \mathfrak{m}_\alpha$ (an inner automorphism maps the center and semi-simple ideals into itself), moreover S maps $\mathfrak{l}_0$ into itself, thus $S\,\mathfrak{n}_\alpha = \mathfrak{n}_\alpha$.

The involutive algebra of the involutive automorphism S of $\mathfrak{l}_1$ is $\tilde{\mathfrak{l}} \oplus \mathfrak{a}_1 \oplus \mathfrak{c}_1$ ($\mathfrak{a}_1$ and $\mathfrak{c}_1$ are one-dimensional, $\tilde{\mathfrak{l}} \subset \mathfrak{l}_0$, $\mathfrak{a}_1 \oplus \mathfrak{c}_1 \subset \mathfrak{l}_1 \dot{-} \mathfrak{l}_0$), consequently $\mathfrak{a}_1 \oplus \mathfrak{c}_1 = \mathfrak{t} = \mathfrak{t}_0 \oplus \mathfrak{t}_1 \oplus \cdots \oplus \mathfrak{t}_n$, where $\mathfrak{t}_0 \subset \mathfrak{z}$, $\mathfrak{t}_\alpha \subset (\mathfrak{m}_\alpha \dot{-} \mathfrak{n}_\alpha)$ $(\alpha = 1, \ldots, n)$ and $\mathfrak{t}_0, \mathfrak{t}_1 \ldots \mathfrak{t}_n$ are at most two-dimensional.

But, in fact, $\mathfrak{t}_\alpha$ $(\alpha = 1, \ldots, n)$ are at most one-dimensional.

Indeed, if $\mathfrak{t}_\alpha \neq \{0\}$ then $S \in \mathrm{Int}_{\mathfrak{m}_\alpha}(\mathfrak{t}_\alpha)$ on $\mathfrak{m}_\alpha$, and the involutive algebra for S is $(\tilde{\mathfrak{l}} \cap \mathfrak{m}_\alpha) \oplus \mathfrak{t}_\alpha \subset \mathfrak{m}_\alpha$. If $\mathfrak{m}_\alpha$ is simple then any its involutive algebra has at most one-dimensional centre, thus $\mathfrak{t}_\alpha$ is one-dimensional. If $\mathfrak{m}_\alpha$ is semi-simple then, since $\mathfrak{m}_\alpha / \mathfrak{n}_\alpha$ $(\alpha = 1, \ldots, n)$ are elementary, $\mathfrak{m}_\alpha = \mathfrak{m}'_\alpha \oplus \mathfrak{m}''_\alpha$, where simple $\mathfrak{m}'_\alpha$ and $\mathfrak{m}''_\alpha$ are isomorphic, and $\mathfrak{n}_\alpha$ is a diagonal subalgebra of the canonical isomorphism $\mathfrak{m}'_\alpha \leftrightarrows \mathfrak{m}''_\alpha$ (being isomorphic to $\mathfrak{m}'_\alpha$, $\mathfrak{m}''_\alpha$). The inner involutive automorphism S generates involutive automorphisms in $\mathfrak{m}'_\alpha$ and $\mathfrak{m}''_\alpha$ with the involutive algebras $\mathfrak{n}'_\alpha$ and $\mathfrak{n}''_\alpha$, respectively, which are conjugated by the canonical involutive automorphism $\mathfrak{m}'_\alpha \leftrightarrows \mathfrak{m}''_\alpha$. Thus $\mathfrak{n}'_\alpha \leftrightarrows \mathfrak{n}''_\alpha$.

Let $\mathfrak{d}_\alpha$ be the diagonal subalgebra of the canonical involutive automorphism $\mathfrak{n}'_\alpha \leftrightarrows \mathfrak{n}''_\alpha$, then $\mathfrak{n}'_\alpha \oplus \mathfrak{n}''_\alpha = \mathfrak{d}_\alpha \dot{+} \mathfrak{k}_\alpha$, $\dim \mathfrak{d}_\alpha = \dim \mathfrak{k}_\alpha$, and $\mathfrak{d}_\alpha$ is isomorphic to $\mathfrak{n}'_\alpha$, $\mathfrak{n}''_\alpha$. Since $\mathfrak{k}_\alpha$ is at most two-dimensional the same is true for $\mathfrak{d}_\alpha$, as well as for $\mathfrak{n}'_\alpha$. For this reason $\mathfrak{n}'_\alpha$ is a commutative involutive algebra in the compact simple algebra $\mathfrak{m}'_\alpha$, which is possible only if $\mathfrak{n}'_\alpha$ is at most one-dimensional. But then $\mathfrak{k}_\alpha$ is at most one-dimensional as well.

The subalgebra $\mathfrak{t}_0 \subset \mathfrak{z}$ also is at most one-dimensional, because $\mathfrak{z}$ is at most one-dimensional (as the centre of the involutive algebra $\mathfrak{l}_1$ in a simple compact Lie algebra $\mathfrak{g}$).

After the re-numbering of the involutive pairs $\mathfrak{m}_\alpha / \mathfrak{n}_\alpha$ we may assume that $\mathfrak{t}_1$ is one-dimensional and, denoting $\mathfrak{m}_1 = \mathfrak{l}'_1$, $\mathfrak{z} \oplus \mathfrak{m}_2 \oplus \cdots \oplus \mathfrak{m}_n = \mathfrak{l}''_1$, $\mathfrak{n}_1 = \mathfrak{l}'_0$, $\mathfrak{z} \oplus \mathfrak{n}_2 \oplus \cdots \oplus \mathfrak{n}_n = \mathfrak{l}''_0$, we have the decomposition

$$\mathfrak{l}_1 / \mathfrak{l}_0 = \mathfrak{l}'_1 / \mathfrak{l}'_0 + \mathfrak{l}''_1 / \mathfrak{l}''_0, \qquad S\,\mathfrak{l}'_1 = \mathfrak{l}'_1, \qquad S\,\mathfrak{l}''_1 = \mathfrak{l}''_1,$$

$$S\,\mathfrak{l}'_0 = \mathfrak{l}'_0, \qquad S\,\mathfrak{l}''_0 = \mathfrak{l}''_0, \tag{II.57}$$

$$\dim\{(\mathfrak{a}_1 \oplus \mathfrak{c}_1) \cap (\mathfrak{l}'_1 \dot{-} \mathfrak{l}'_0)\} = \dim\{(\mathfrak{a}_1 \oplus \mathfrak{c}_1) \cap (\mathfrak{l}''_1 \dot{-} \mathfrak{l}''_0)\} = 1.$$

Let us use (II.57) for the specialization of the choice of base (II.49). From (II.57) there follows $\mathfrak{m}_1 = \mathfrak{m}'_1 \dot{+} \mathfrak{m}''_1$, where $\mathfrak{m}'_1 = \mathfrak{l}'_1 \dot{-} \mathfrak{l}'_0$, $\mathfrak{m}''_1 = \mathfrak{l}''_1 \dot{-} \mathfrak{l}''_0$. Evidently $\mathfrak{m}'_1 \perp \mathfrak{m}''_1$, $\mathfrak{l}'_0 \perp \mathfrak{l}''_0$ with respect to the Cartan metric of $\mathfrak{g}$.

Let us choose an orthonormal base in $\mathfrak{m}'_1$ in such a way that S has a diagonal matrix on $\mathfrak{m}'_1$. Then for the basis vectors X_{1_1}, $X_{3_1} \ldots X_{m_1}$

$$S\,X_{a'_1} = X_{a'_1}, \qquad S\,X_{i'_1} = -X_{i'_1};$$

$$a' = 1; \quad i' = 3, 4, \ldots, m. \tag{II.58}$$

Analogously, let us choose an orthonormal base in $\mathfrak{m}''_1$ in such a way that S has a diagonal matrix on $\mathfrak{m}_1''$. Then for the basis vectors X_{2_1}, $X_{(m+1)_1}$, $\ldots X_{n_1}$

$$S X_{a_1''} = X_{a_1''}, \qquad S X_{i_1''} = -X_{i_1''};$$

$$a'' = 2; \quad i'' = m+1, \ldots n.$$

(II.59)

The union of all the bases constructed above constitutes the base $\{X_{I_1}\}$ in $\mathfrak{m}_1$ ($I = a, i$; $I' = a', i'$; $I'' = a'', i''$).

Furthermore, we choose an orthonormal base $\{Y_A\} = \{Y_{A'}, Y_{A''}\}$, $Y_{A'} \in \mathfrak{l}_0'$, $Y_{A''} \in \mathfrak{l}_0''$ in $\mathfrak{l}_0$ and induce bases in $\mathfrak{m}_2$, $\mathfrak{m}_3$ by means of the automorphism p, see (II.2).

The union of all the above bases gives us the hyper-involutive base which will be used later. In this base we have

$$C_{I'\ J''}^{A} = C_{I'\ J}^{A''} = C_{I''\ J}^{A'} = C_{I''\ A}^{J''} = C_{I'\ A''}^{J} = C_{I''\ A'}^{J} = 0,$$

$$C_{A'\ B''}^{F} = 0.$$

(II.60)

Moreover, we have (II.50)–(II.56).

From (II.55.1), (II.56), (II.60) we obtain

$$t_{i'\ j'}^{a'}\, C_{a'\ A'}^{p'} - t_{a'\ j'}^{p'}\, C_{i'\ A'}^{a'} - t_{i'\ a'}^{p'}\, C_{j'\ A'}^{a'} = 0,$$

$$t_{b'\ b'}^{a'}\, C_{a'\ A'}^{p'} - t_{k'\ b'}^{p'}\, C_{b'\ A'}^{k'} - t_{b'\ k'}^{p'}\, C_{b'\ A'}^{k'} = 0,$$

$$t_{i''\ j''}^{a''}\, C_{a''\ A''}^{p''} - t_{a''\ j''}^{p''}\, C_{i''\ A''}^{a''} - t_{i''\ a''}^{p''}\, C_{j''\ A''}^{a''} = 0,$$

$$t_{b''\ b''}^{a''}\, C_{a''\ A''}^{p''} - t_{k''\ b''}^{p''}\, C_{b''\ A''}^{k''} - t_{b''\ k''}^{p''}\, C_{b''\ A''}^{k''} = 0.$$

Hence taking into account that $a' = 1$, $b'' = 2$ we have

$$t_{i'\ j'}^{1}\, C_{1\ A'}^{p'} - t_{1\ j'}^{p'}\, C_{i'\ A'}^{1} - t_{i'\ 1}^{p'}\, C_{j'\ A'}^{1} = 0,$$

$$t_{1\ 1}^{1}\, C_{1\ A'}^{p'} - t_{k'\ 1}^{p'}\, C_{1\ A'}^{k'} - t_{1\ k'}^{p'}\, C_{1\ A'}^{k'} = 0,$$

$$t_{i''\ j''}^{2}\, C_{2\ A''}^{p''} - t_{2\ j''}^{p''}\, C_{i''\ A''}^{2} - t_{i''\ 2}^{p''}\, C_{j''\ A''}^{2} = 0,$$

$$t_{2\ 2}^{2}\, C_{2\ A''}^{p''} - t_{k''\ 2}^{p''}\, C_{2\ A''}^{k''} - t_{2\ k''}^{p''}\, C_{2\ A''}^{k''} = 0.$$

(II.61)

From (II.61), (II.53), (II.54) we have, furthermore,

$$\left.\begin{aligned} t_{i'\ j'}^{1}\, C_{p'\ A'}^{1} + t_{j'\ p'}^{1}\, C_{i'\ A'}^{1} + t_{p'\ i'}^{1}\, C_{j'\ A'}^{1} &= 0, \\ t_{1\ 1}^{1}\, C_{p'\ A'}^{1} &= 2E_{1\ p'}^{k'}\, C_{k'\ A'}^{1}. \end{aligned}\right\}$$

(II.62′)

$$t^2_{i''\,j''}\,C^2_{p''\,A''} + t^2_{j''\,p''}C^2_{i''\,A''} + t^2_{p''\,i''}\,C^2_{j''\,A''} = 0,$$

$$t^2_{2\,2}\,C^2_{p''\,A''} = 2E^{k''}_{2\,p''}\,C^2_{k''\,A''}.$$

(II.62'')

From (II.55.2) and (II.60) it follows that:

$$t^c_{b''\,h''}\,t^{f'}_{c\,a'} - t^c_{a'\,h''}\,t^{f'}_{c\,b''} = C^{f'}_{h''\,B}\,C^B_{a'\,b''} = 0,$$

that is,

$$t^2_{2\,2}\,t^1_{c\,1} - t^c_{1\,2}\,t^1_{c\,2} = 0 \quad (c = 1,2).$$

(II.63)

Let us suppose that $t^1_{1\,1} = t^2_{2\,2} = 0$. Then (II.63) implies $t^1_{1\,2}\,t^1_{1\,2} + t^2_{1\,2}\,t^1_{2\,2} = 0$, which, together with (II.43), gives us $\left(t^1_{1\,2}\right)^2 + \left(t^2_{1\,2}\right)^2 = 0$, that is, $t^1_{1\,2} = t^2_{1\,2} = 0$.

This means that $t^a_{bc} = 0$, but then $\mathfrak{p} \oplus \mathfrak{q}$ is commutative which contradicts its semi-simplicity.

Consequently either $t^1_{1\,1}$ or $t^2_{2\,2}$ (or both) are different from zero. Without loss of generality we can put $t^1_{1\,1} \neq 0$. The conditions (II.62') coincide, up to notations, with (II.36), (II.37) which have already been examined. Therefore for $t^1_{1\,1} \neq 0$ (II.62') gives $C^1_{j'\,A'} = 0$. Together with (II.60) this implies

$$C^1_{j\,A} = C^A_{1\,j} = C^j_{1\,A} = 0.$$

(II.64)

The above means that $\mathfrak{z}_1 = \{t\,X_{1_1}\}$ is the centre in $\mathfrak{l}_1$, indeed, $t\,X_{1_1}$ commutes owing to (II.64) with $\mathfrak{l}_1$, but the centre in $\mathfrak{l}_1$ is at most one-dimensional. Analogously, $\mathfrak{z}_2 = \{t\,X_{1_2}\}$ is the centre in $\mathfrak{l}_2$, and $\mathfrak{z}_3 = \{t\,X_{1_3}\}$ is the centre in $\mathfrak{l}_3$.

If we assume also that $t^2_{2\,2} \neq 0$, then from (II.62'') we obtain in a similar way that $\{t\,X_{2_1}\}$ is the centre in $\mathfrak{l}_1$, which is impossible, since the basis vectors X_{1_1} and X_{2_1} are linearly independent and the centre in $\mathfrak{l}_1$ is at most one-dimensional. Thus

$$t^1_{1\,1} \neq 0, \qquad t^2_{2\,2} = 0.$$

(II.65)

Let us note that $\mathfrak{z}_\alpha$ is the maximal subset of elements from $\mathfrak{l}_\alpha \dot{-} \mathfrak{l}_0$ commuting with $\mathfrak{l}_0$. Indeed, the maximal subset $\mathfrak{c}_\alpha \subset \mathfrak{l}_\alpha \dot{-} \mathfrak{l}_0$ is a commutative ideal in $\mathfrak{l}_\alpha$, that is, it belongs to the centre $\mathfrak{z}_\alpha$, $\mathfrak{c}_\alpha \subset \mathfrak{z}_\alpha$. However, by the conditions $\mathfrak{z}_\alpha \subset \mathfrak{l}_\alpha \dot{-} \mathfrak{l}_0$ and it commutes with $\mathfrak{l}_0$, thus $\mathfrak{z}_\alpha \subset \mathfrak{c}_\alpha$. As a result $\mathfrak{c}_\alpha = \mathfrak{z}_\alpha$.

We now consider the maximal subset $\mathfrak{w} \subset \mathfrak{m}_1 \dot{+} \mathfrak{m}_2 \dot{+} \mathfrak{m}_3$ of vectors commuting with $\mathfrak{l}_0$. If $\zeta \in \mathfrak{w}$ then $\zeta = \zeta_1 + \zeta_2 + \zeta_3$, where $\zeta_\alpha \in \mathfrak{m}_\alpha$ ($\alpha = 1,2,3$).

Let $\eta \in \mathfrak{l}_0$, then from (II.51) it follows that $[\eta\,\zeta_\alpha] \in \mathfrak{m}_\alpha$. Thus from $0 = [\eta\,\zeta] = [\eta\,\zeta_1] + [\eta\,\zeta_2] + [\eta\,\zeta_3]$ we have $[\eta\,\zeta_\alpha] = 0$, that is, $\zeta_\alpha \in \mathfrak{z}_\alpha$. Conversely, from $[\eta\,\zeta_\alpha] = 0$ ($\alpha = 1,2,3$) we obtain $[\eta\,\zeta] = 0$. Consequently $\mathfrak{w} = \mathfrak{z}_1 \dot{+} \mathfrak{z}_2 \dot{+} \mathfrak{z}_3$.

It is easily verified that $\mathfrak{w}$ is a subalgebra. Indeed, owing to the Jacobi identities, $[\eta\,[\mathfrak{z}_\alpha\,\mathfrak{z}_\beta]] = \{0\}$ for $\eta \in \mathfrak{l}_0$, but $[\mathfrak{z}_\alpha\,\mathfrak{z}_\beta] \subset \mathfrak{m}_\gamma$ ($\alpha \neq \beta$, $\alpha \neq \gamma$, $\beta \neq \gamma$). Thus $[\mathfrak{z}_\alpha\,\mathfrak{z}_\beta] \subset \mathfrak{z}_\gamma$ ($\alpha \neq \beta$, $\beta \neq \gamma$, $\gamma \neq \alpha$; $\alpha,\beta,\gamma = 1,2,3$), whence it follows that $\mathfrak{w}$ is a subalgebra.

Let us also note that $\mathfrak{w}$ is a three-dimensional simple subalgebra (since $t^1_{1\,1} \neq 0$). Furthermore, since $S_\alpha \in \text{Int}_\mathfrak{g}(\{t\,X_{1_\alpha}\})$, $\text{Int}_\mathfrak{g}(\mathfrak{w}) \cong SO(3)$ and, in addition, since $X_{1_\alpha} \in \mathfrak{p} \oplus \mathfrak{q}$ we have $\mathfrak{w} \subset \mathfrak{p} \oplus \mathfrak{q}$.

Evidently $\mathfrak{w} \neq \mathfrak{p}$, $\mathfrak{w} \neq \mathfrak{q}$ (owing to $\mathrm{Int}_{\mathfrak{g}}(\mathfrak{p}) \cong SU(2)$, $\mathrm{Int}_{\mathfrak{g}}(\mathfrak{q}) \cong SU(2)$, $\mathrm{Int}_{\mathfrak{g}}(\mathfrak{w}) \cong SO(3)$), consequently $\mathfrak{w}$ is the diagonal subalgebra of the canonical involutive automorphism $\mathfrak{p} \leftrightarrows \mathfrak{q}$.

Let us show, finally, that $\mathfrak{w} = \mathfrak{b}$, where $\mathfrak{b}$ is the three-dimensional subalgebra of $\mathfrak{p} \oplus \mathfrak{q}$, by means of which we have constructed the hyper-involutive decomposition (see II.5.2 Definition 28).

Indeed, $\mathfrak{w}$ and $\mathfrak{b}$ are diagonals in $\mathfrak{p} \oplus \mathfrak{q}$ and, as is well known, are conjugated in $\mathrm{Int}_{\mathfrak{g}}(\mathfrak{p} \oplus \mathfrak{q})$. But $S_\alpha \in \mathrm{Int}_{\mathfrak{g}}(\mathfrak{b})$, $S_\alpha \in \mathrm{Int}_{\mathfrak{g}}(\mathfrak{w})$, implying that there exists a conjugating automorphism $\tau \in \mathrm{Int}_{\mathfrak{g}}(\mathfrak{p} \oplus \mathfrak{q})$ such that $\tau S_\alpha = S_\alpha \tau$, whence $\tau \, \mathfrak{l}_\alpha = \mathfrak{l}_\alpha$. Since $\{t X_{1_\alpha}\}$ is the centre in $\mathfrak{l}_\alpha$ we have $\tau \{t X_{1_\alpha}\} \subset \{t X_{1_\alpha}\}$, or $\tau (X_{1_\alpha}) = \lambda X_{1_\alpha}$.

But $X_{1_\alpha} \in \mathfrak{w}$, $\tau (X_{1_\alpha}) \in \mathfrak{b}$, consequently $X_{1_\alpha} \in \mathfrak{b}$, implying $\mathfrak{w} = \mathfrak{b}$ (since $X_{1_1}, X_{1_2}, X_{1_3}$ constitute a base in $\mathfrak{w}$). Since $\mathfrak{b}$ is a subalgebra of $\mathfrak{p} \oplus \mathfrak{q}$ we obtain $t_{11}^2 = t_{21}^1 = t_{12}^1 = 0$. The latter, together with (II.56), (II.60), ((II.64), gives us

$$t^1_{1\tilde{I}} = t^{\tilde{I}}_{11} = t^1_{\tilde{I}1} = 0, \qquad t^{\tilde{P}}_{\tilde{J}\tilde{I}} = 0,$$
$$C^1_{\tilde{I}A} = C^{\tilde{I}}_{1A} = C^A_{1\tilde{I}} = 0 \quad (\tilde{I}, \tilde{J}, \tilde{P} = 2, 3, \ldots, n). \tag{II.66}$$

Then from (II.66) and (II.52) it follows that the linear transformation A,

$$A X_{1_\rho} = X_{1_\rho}, \qquad A Y_B = Y_B, \qquad A X_{\tilde{I}_\rho} = -X_{I_\rho} \quad (\rho = 1, 2, 3), \tag{II.67}$$

is an involutive automorphism of $\mathfrak{g}$ with the involutive subalgebra $\mathfrak{b} \oplus \mathfrak{l}_0$.

Since $\mathrm{Int}_{\mathfrak{g}}(\mathfrak{b}) \cong SO(3)$ we see that A is a principal involutive automorphism of type O such that $A S = S A$, $A S_\rho = S_\rho A$ ($\rho = 1, 2, 3$).

The prime hyper-involutive decomposition basis for S, which has been constructed above, is as well a prime hyper-involutive decomposition basis for A.

Thus we have the theorem:

II.5.8. Theorem 25. *Let $\mathfrak{g}$ be a simple compact Lie algebra, $\mathfrak{l} = \mathfrak{p} \oplus \mathfrak{q} \oplus \tilde{\mathfrak{l}}$ be its principal di-unitary involutive algebra of an involutive automorphism S (that is, $\mathrm{Int}_{\mathfrak{g}}(\mathfrak{p}) \cong SU(2)$, $\mathrm{Int}_{\mathfrak{g}}(\mathfrak{q}) \cong SU(2)$), and let $\mathfrak{g}$ have a prime hyper-involutive decomposition basis for S. Then there exists in $\mathfrak{g}$ a principal involutive automorphism A of type O such that S is an associated involutive automorphism for A. Moreover, the prime hyper-involutive decomposition basis for S is also a prime hyper-involutive decomposition basis for A.*

From II.5.8 (Theorem 25) and II.3.10 (Theorem 8) there follows in this case that

$$\mathfrak{g}/\mathfrak{l}_\alpha \cong so(m)/(so(m-2) \oplus so(2)) \quad (\alpha = 1, 2, 3),$$
$$\mathfrak{g}/(\mathfrak{b} \oplus \mathfrak{l}_0) \cong so(m)/(so(3) \oplus so(m-3)) \quad (m > 4)$$

with the natural embeddings.

Thus with such a realization

$$(\mathfrak{l}_1 \dot{-} \{t X_{1_1}\})/\mathfrak{l}_0 \cong so(m-2)/so(m-3),$$

but $\{t\,X_{2_1}\} \subset ([_1 \dot{-} \{t\,X_{1_1}\}) \dot{-} [_0$, and for this reason $\{t\,X_{2_1}\} \cong so(2)$ with the natural embedding into $[_1 \dot{-} \{t\,X_{1_1}\} \cong so(m-2)$ and into $\mathfrak{g} \cong so(m)$.

Thus

$$\{t\,X_{1_1}\} \oplus \{t\,X_{2_1}\} \cong so(2) \oplus so(2)$$

with the natural embedding into $\mathfrak{g} \cong so(m)$.

In addition,

$$\mathfrak{b} = \{t\,X_{1_1}\} + \{t\,X_{1_2}\} + \{t\,X_{1_3}\} \cong so(3)$$

with the natural embedding into $\mathfrak{g} \cong so(m)$.

Let us note, furthermore, that

$$\mathfrak{p} \oplus \mathfrak{q} = \left[\{t\,X_{2_1}\}\,\mathfrak{b}\right] + \mathfrak{b} + \{t\,X_{2_1}\},$$

whence in our realization it follows that

$$\mathfrak{p} \oplus \mathfrak{q} \cong so(4)$$

with the natural embeddings into $\mathfrak{g} \cong so(m)$.

Furthermore, $\tilde{[}$ is the maximal subalgebra of vectors from $\mathfrak{g}$ commuting with $\mathfrak{p} \oplus \mathfrak{q}$ (in our realization, $\tilde{[} \cong so(m-4)$ with the natural embedding into $\mathfrak{g} \cong so(m)$).

As a result we have proved the theorem:

II.5.9.   Theorem 26. *Let* $[= \mathfrak{p} \oplus \mathfrak{q} \oplus \tilde{[}$ *be the principal involutive algebra of an involutive automorphism* S *of type* $U^{(2)}$ *(that is,* $\mathrm{Int}_{\mathfrak{g}}(\mathfrak{p}) \cong SU(2)$, $\mathrm{Int}_{\mathfrak{g}}(\mathfrak{q}) \cong SU(2)$*) of a simple compact Lie algebra* $\mathfrak{g}$ *having a prime hyper-involutive decomposition basis for* S. *Then*

$$\mathfrak{g}/\mathfrak{p} \oplus \mathfrak{q} \oplus \tilde{[} \cong so(m)/(\,so(4) \oplus so(m-4)\,), \quad m > 4,$$

with the natural embedding.

From II.5.6 (Theorem 24) and II.5.9 (Theorem 26) there follows:

II.5.10.   Theorem 27. *If* $\mathfrak{g}$ *is a simple compact Lie algebra,* $[= \mathfrak{p} \oplus \mathfrak{q} \oplus \tilde{[}$ *is its principal involutive algebra of type* $U^{(2)}$ *(that is,* $\mathrm{Int}_{\mathfrak{g}}(\mathfrak{p}) \cong SU(2)$, $\mathrm{Int}_{\mathfrak{g}}(\mathfrak{q}) \cong SU(2)$*), and* $\tilde{[} \neq \{0\}$ *then*

$$\mathfrak{g}/[\cong so(m)/(\,so(4) \oplus so(m-4)\,), \quad m > 5,$$

with the natural embedding.

CHAPTER II.6

SINGULAR PRINCIPAL DI-UNITARY
INVOLUTIVE AUTOMORPHISM

II.6.1. We intend to consider a Lie algebra $\mathfrak{g}$ with a principal di-unitary involutive automorphism S having only a non-prime involutive decomposition basis for S. Then the assertion of II.5.6. (Theorem 24) is valid.

We specialize the choice of a hyper-involutive base in the following way:

$$X_{I_1'} \in \mathfrak{r}_1 \dot{-} \mathfrak{r}_1 \cap \mathfrak{l}_0 \quad (I' = 1', 2'),$$

$$X_{I_1''} \in \mathfrak{w}_1 \dot{-} \mathfrak{w}_1 \cap \mathfrak{l}_0 \quad (I'' = 1'', 2''),$$

$$Y_{A'} \in \mathfrak{r}_1 \cap \mathfrak{l}_0 \quad (A' = 3'), \qquad Y_{A''} \in \mathfrak{w}_1 \cap \mathfrak{l}_0 \quad (A'' = 3''), \qquad \text{(II.68)}$$

$$S X_{a_1} = X_{a_1} \quad (a = 1', 1''), \qquad S X_{i_1} = -X_{i_1} \quad (i = 2', 2''),$$

$$S Y_A = -Y_A.$$

Owing to (II.5) the orthogonal matrix p_B^A satisfies the conditions $\left(p^3\right)_B^A = \delta_B^A$. Therefore we have the two possibilities:

$$p Y_{3'} = -\frac{1}{2} Y_{3'} \pm \frac{\sqrt{3}}{2} Y_{3''}, \qquad p Y_{3''} = \mp \frac{\sqrt{3}}{2} Y_{3'} - \frac{1}{2} Y_{3''}.$$

But the change of base $Y_{3'} \longrightarrow -Y_{3'}$, $Y_{3''} \longrightarrow Y_{3''}$ reduces one of the bases to the other, which does not affect (II.68) since this allows the multiplication of basis vectors by ± 1.

We choose a base (II.68) for the certainty in such a way that

$$p_{3'}^{3'} = p_{3''}^{3''} = -\frac{1}{2}, \qquad p_{3''}^{3'} = \frac{\sqrt{3}}{2}, \qquad p_{3'}^{3''} = -\frac{\sqrt{3}}{2}. \qquad \text{(II.69)}$$

The automorphism S together with (II.68) and (II.8) gives us

$$t_{ai}^b = t_{ab}^i = t_{ib}^a = t_{jk}^i = 0, \qquad C_{bA}^a = C_{ab}^A = C_{jA}^i = C_{ij}^A = 0. \qquad \text{(II.70)}$$

By the commutativity of $\mathfrak{l}_0$ we have also

$$C_{BF}^A = 0. \qquad \text{(II.71)}$$

87

By the commutativity of $\mathfrak{r}_1$ and $\mathfrak{w}_1$ for the base (II.68) there also follows

$$C^{I'}_{J''\,A} = C^A_{I'\,J''} = C^{I'}_{J\,A''} = C^{A''}_{I'\,J} = C^{A'}_{I''\,J} = C^{I''}_{J\,A'} = 0. \tag{II.72}$$

The relations (II.71), (II.72) imply that (II.11.5)–(II.11.8) are satisfied. Moreover, (II.11.2) follows from (II.11.1) by virtue of (II.9), and (II.11.3) follows from (II.11.4) by virtue of (II.15)–(II.18).

Thus the essential relations are only

$$t^P_{J\,K}\,C^A_{I\,P} + t^P_{K\,I}\,C^B_{J\,P}\,p^A_B + t^P_{I\,J}\,C^B_{K\,P}\,(p^2)^A_B = 0, \tag{II.73.1}$$

$$t^P_{K\,J}\,t^R_{I\,P} - t^P_{K\,I}\,t^R_{J\,P} = C^R_{K\,B}\,p^B_A\,C^A_{I\,J}, \tag{II.73.2}$$

$$\left.\begin{array}{c} \varphi^R_I\,\varphi^Q_J\,t^S_{Q\,R} = t^K_{I\,J}\,\varphi^S_K, \qquad \varphi^R_I\,C^K_{R\,B}\,(\varphi_1)^B_A = \varphi^K_J\,C^J_{I\,A}, \\[2mm] \varphi^I_K\,\varphi^K_J = \delta^I_J, \qquad (\varphi_1)^B_A\,(\varphi_1)^A_C = \delta^B_C, \qquad \varphi^I_K = \varphi^K_I, \\[2mm] (\varphi_1)^A_B = (\varphi_1)^B_A. \end{array}\right\} \tag{II.74}$$

II.6.2. Now we clarify the form of the matrices φ^I_J, $(\varphi_1)^A_B$ of the automorphism φ_1 (see (II.3), (II.14), (II.15)). II.5.4 (Theorem 23) implies that $\varphi_1 \in \mathrm{Int}_\mathfrak{g}(\mathfrak{r}_1 \oplus \mathfrak{w}_1)$, thus $\varphi_1 = a \cdot b = b \cdot a$, where $a \in \mathrm{Int}_\mathfrak{g}(\mathfrak{r}_1)$, $b \in \mathrm{Int}_\mathfrak{g}(\mathfrak{w}_1)$. But $(\varphi_1)^2 = S_1$, whence $a^2 \cdot b^2 = S_1$, or $\mathrm{Int}_\mathfrak{g}(\mathfrak{r}_1) \ni S_1(a^2) = (b^{-1})^2 \in \mathrm{Int}_\mathfrak{g}(\mathfrak{w}_1)$. This means that either $S_1(a^2) = (b^{-1})^2 = \mathrm{Id}$ or $S_1(a^2) = (b^{-1})^2 = S_1$.

Thus either $a^2 = \mathrm{Id}$, $b^2 = S_1$ or $b^2 = \mathrm{Id}$, $a^2 = S_1$, but changing the notations to $\mathfrak{w}_1$ instead of $\mathfrak{r}_1$ and $\mathfrak{r}_1$ instead of $\mathfrak{w}_1$ we reduce one possibility to the other.

Thus we may put $a^2 = S_1$, $b^2 = \mathrm{Id}$ and consequently either $b = S_1$ or $b = \mathrm{Id}$.

But then $\varphi_1 = a \cdot b \in \mathrm{Int}_\mathfrak{g}(\mathfrak{r}_1)$. On the other hand $\varphi_1 \in \mathrm{Int}_\mathfrak{g}(\mathfrak{r}_1 \cap (\mathfrak{p} \oplus \mathfrak{q}))$ and in the base (II.68) we have $\varphi_1 \in \mathrm{Int}_\mathfrak{g}(\{t\,X_{1'_1}\})$, $(\varphi_1)^2 = S_1$. The restriction of φ_1 to $\mathfrak{l}_1$ is not the identity automorphism (otherwise, from (II.17) it would follow $(\varphi_3)^A_B = p^A_B$, which is possible only if $p^A_B = \delta^A_B$; but this contradicts the conditions. Consequently the restriction of φ_1 to $\mathfrak{r}_1$ is an involutive automorphism of $\mathfrak{r}_1$.

Therefore we have

$$\varphi^I_{J''} = \delta^{I''}_{J''}, \qquad \varphi^{I'}_{J''} = 0, \qquad \varphi^{i'}_{j'} = -1, \qquad \varphi^{a'}_{b'} = 1, \qquad \varphi^{i'}_{a'} = 0,$$
$$(\varphi_1)^{A'}_{B''} = 0, \qquad (\varphi_1)^{A'}_{B'} = -1, \qquad (\varphi_1)^{A''}_{B''} = 1. \tag{II.75}$$

By means of (II.75), (II.70), (II.72) the conditions (II.74) turn into

$$t^a_{i'\,j'} = t^a_{j'\,i'}, \qquad t^a_{i''\,j''} = t^a_{j''\,i''}, \qquad t^a_{b\,f} = t^a_{f\,b}, \qquad t^a_{i'\,j''} = -t^a_{j''\,i'},$$

However, the first three relations are identically satisfied because $j' = i' = 2'$, $j'' = i'' = 2''$, $\{a, b\} = \{1', 1''\}$, and (II.8).

Thus instead of (II.74) we obtain

$$t^a_{i'\,j''} = -t^a_{j''\,i'}. \tag{II.76}$$

Now we need to satisfy (II.73.1), (II.73.2) under the conditions (II.69), (II.70), (II.72), (II.76), (II.9). The orthonormal base (II.68) is defined up to the multiplication by ± 1, which implies that we may choose it in such a way that

$$t^{a'}_{a'\,a'} \geq 0, \qquad t^{a''}_{a''\,a''} \geq 0, \qquad C^{B'}_{a'\,i'} > 0, \qquad C^{B''}_{a''\,i''} > 0. \tag{II.77}$$

We use such a choice later.

Let us consider first (II.73.1). Taking into account (II.69)–(II.72) we have from (II.73.1)

$$t^{P'}_{J\,K}\,C^{A'}_{I\,P'} + t^{P}_{K\,I}\left(-\frac{1}{2}C^{A'}_{J\,P} - \frac{\sqrt{3}}{2}C^{A''}_{J\,P}\right),$$
$$+ t^{P}_{I\,J}\left(-\frac{1}{2}C^{A'}_{K\,P} + \frac{\sqrt{3}}{2}C^{A''}_{K\,P}\right) = 0. \tag{II.78}$$

$$t^{P''}_{J\,K}\,C^{A''}_{I\,P''} + t^{P}_{K\,I}\left(-\frac{1}{2}C^{A''}_{J\,P} + \frac{\sqrt{3}}{2}C^{A'}_{J\,P}\right)$$
$$+ t^{P}_{I\,J}\left(-\frac{1}{2}C^{A''}_{K\,P} - \frac{\sqrt{3}}{2}C^{A'}_{K\,P}\right) = 0. \tag{II.79}$$

Whence first of all we obtain

$$t^{P'}_{J'\,K'}\,C^{A'}_{I'\,P'} - \frac{1}{2}t^{P'}_{K'\,I'}\,C^{A'}_{J'\,P'} - \frac{1}{2}t^{P'}_{I'\,J'}\,C^{A'}_{K'\,P'} = 0, \tag{II.78.1}$$

$$t^{P''}_{J''\,K''}\,C^{A''}_{I''\,P''} - \frac{1}{2}t^{P''}_{K''\,I''}\,C^{A''}_{J''\,P''} - \frac{1}{2}t^{P''}_{I''\,J''}\,C^{A''}_{K''\,P''} = 0. \tag{II.79.1}$$

When $I' = J' = K'$ the relation (II.78) is identically satisfied.

When $I' = J' \neq K'$ we have

$$t^{K'}_{I'\,K'}\,C^{A'}_{I'\,K'} - \frac{1}{2}t^{K'}_{K'\,I'}\,C^{A'}_{I'\,K'} - \frac{1}{2}t^{I'}_{I'\,I'}\,C^{A'}_{K'\,I'} = 0,$$

where there is no summation over I', K'. Hence

$$t^{K'}_{I'\,K'} = -t^{I'}_{I'\,I'} \quad (I' \neq K'). \tag{II.78.1'}$$

The cases $I' \neq J' = K'$ and $I' = K' \neq J'$ are consequences of (II.78.1').

Analogously, from (II.79), with $I = I''$, $J = J''$, $K = K''$ (II.79.1) follows, whence

$$t^{K''}_{I''\,K''} = -t^{I''}_{I''\,I''} \quad (I'' \neq K''). \tag{II.79.1'}$$

Furthermore, by (II.78) we obtain

$$-\frac{1}{2}t^{P'}_{K'\,I''}\,C^{A'}_{J'\,P'} - \frac{1}{2}t^{P'}_{I''\,J'}\,C^{A'}_{K'\,P'} = 0. \tag{II.78.2}$$

When $J' = K'$ the relation (II.78.2) is satisfied by (II.76) if the latter is rewritten in the equivalent form

$$t_{K'\,I''}^{P'} = -t_{I''\,K'}^{P'} \quad (P' \neq K'), \qquad t_{K''\,I'}^{P''} = -t_{I'\,K''}^{P''} \quad (P'' \neq K''). \quad (\text{II.76}')$$

When $J' \neq K'$ the relation (II.78.2) takes the form

$$t_{K'\,I''}^{K'} C_{J'\,K'}^{A'} + t_{I''\,J''}^{J'} C_{K'\,J'}^{A'} = 0,$$

where there is no summation over K', J', whence

$$t_{K'\,I''}^{K'} = t_{J'\,I''}^{J'}. \quad (\text{II.78.2}')$$

Analogously, from (II.79), when $I = I'$, $J = J''$, $K = K''$ (II.79.2) follows, whence

$$t_{K''\,I'}^{K''} = t_{J''\,I'}^{J''}. \quad (\text{II.79.2}')$$

Furthermore, from (II.78) we have

$$t_{J''\,K'}^{P'} C_{I'\,P'}^{A'} + t_{K'\,I'}^{P''} \left(-\frac{\sqrt{3}}{2} C_{J''\,P''}^{A''} \right) + t_{I'\,J''}^{P'} \left(-\frac{1}{2} C_{K'\,P'}^{A'} \right) = 0. \quad (\text{II.78.3})$$

When $I' = K'$ the relation (II.78.3) gives us

$$\sqrt{3}\, C_{I'\,P'}^{A'} t_{J''\,I'}^{P'} = C_{J''\,P''}^{A''} t_{I'\,I'}^{P''}.$$

When $I' \neq K'$ the relation (II.78.3) gives us

$$\sqrt{3}\, C_{I'\,P'}^{A'} t_{J''\,K'}^{P'} = C_{J''\,P''}^{A''} t_{K'\,I'}^{P''}.$$

Or, more precisely,

$$\sqrt{3}\, C_{i'\,a'}^{A'} t_{j''\,i'}^{a'} = C_{j''\,a''}^{A''} t_{a'\,a'}^{a''},$$

$$\sqrt{3}\, C_{i'\,a'}^{A'} t_{a'\,a'}^{a''} = C_{j''\,a''}^{A''} t_{j''\,i'}^{a'},$$

where there is no summation over repeated indices.

Solving this system we obtain

$$t_{j''\,i'}^{a'} = t_{a'\,a'}^{a''}, \quad (\text{II.78.3}')$$

and into account also (II.78) taking,

$$C_{j''\,a''}^{A''} = \sqrt{3}\, C_{j'\,a'}^{A'}, \quad t_{a'\,a'}^{a''} \neq 0. \quad (\text{II.78.3}'')$$

The cases $A = A'$, $I = I'$, $J = J'$, $K = K''$ for (II.78) do not give new relations.

Furthermore, from (II.78) we obtain.

$$t^{P'}_{J''\,K''}\,C^{A'}_{I'\,P'} + t^{P''}_{K''\,I'}\left(-\frac{\sqrt{3}}{2}\,C^{A''}_{J''\,P''}\right) + t^{P''}_{I'\,J''}\left(\frac{\sqrt{3}}{2}\,C^{A''}_{K''\,P''}\right) = 0. \quad \text{(II.78.4)}$$

This gives, when $J'' = K''$,

$$-t^{P'}_{K''\,K''}\,C^{A'}_{I'\,P'} + t^{P''}_{I'\,K''}\,C^{A''}_{K''\,P''}\,\sqrt{3} = 0.$$

From (II.78.4), when $J'' \neq K''$, we have

$$t^{P'}_{J''\,K''}\,C^{A'}_{I'\,P'} + t^{P''}_{K''\,I'}\,C^{A''}_{P''\,J''}\,\sqrt{3} = 0 \quad (J'' \neq K'').$$

And in more detailed form (II.78.4) is reduced to

$$C^{A'}_{i'\,a'}\,t^{a'}_{a''\,a''} = \sqrt{3}\,C^{A''}_{k''\,a''}\,t^{a''}_{k''\,i'},$$

$$C^{A'}_{i'\,a'}\,t^{a''}_{k''\,i'} = \sqrt{3}\,C^{A''}_{k''\,a''}\,t^{a''}_{a''\,a''}.$$

(no summation over repeating indices).

Solving this system we obtain

$$t^{a''}_{k''\,i'} = t^{a'}_{a''\,a''}, \quad\quad\quad\quad \text{(II.78.4$'$)}$$

and into account also (II.77) taking,

$$C^{A'}_{i'\,a'} = \sqrt{3}\,C^{A''}_{i'\,a''}, \quad t^{a'}_{a''\,a''} \neq 0. \quad\quad \text{(II.78.4$''$)}$$

Further consideration of the consequences of (II.78), (II.79) does not give new relations.

II.6.3. Let us assume that $t^{a''}_{a'\,a'} = t^{a'}_{a''\,a''} = 0$, then $\mathfrak{t} = \{\alpha\,X_{a'_1} + \beta\,X_{a'_2} + \gamma\,X_{a'_3}\}$ is an ideal in $\mathfrak{p}\oplus\mathfrak{q}$, thus either $\mathfrak{t} = \mathfrak{p}$ or $\mathfrak{t} = \mathfrak{q}$. But $S_\rho \in \mathrm{Int}_\mathfrak{g}(\{t\,X_{a'_\rho}\}) \subset \mathrm{Int}_\mathfrak{g}(\mathfrak{t}) \cong SO(3)$ and $\mathrm{Int}_\mathfrak{g}(\mathfrak{p}) \cong \mathrm{Int}_\mathfrak{g}(\mathfrak{q}) \cong SU(2)$, leading to a contradiction. Owing to (II.78.3$''$), II.(78.4$''$) the case $t^{a''}_{a'\,a'} \neq 0$, $t^{a'}_{a''\,a''} \neq 0$ gives us $C^{A'}_{i'\,a'} = C^{A''}_{i''\,a''} = 0$, which is impossible.

Lastly, let $t^{a''}_{a''\,a''} = 0$, $t^{a''}_{a'\,a'} \neq 0$, then $\mathfrak{t} = \{\alpha\,X_{a''_1} + \beta\,X_{a''_2} + \gamma\,X_{a''_3}\}$ is a three-dimensional subalgebra of $\mathfrak{p}\oplus\mathfrak{q}$, with $\mathfrak{t}\neq\mathfrak{p}$, $\mathfrak{t}\neq\mathfrak{q}$. Evidently $\mathfrak{t}$ is not commutative since $\mathfrak{p}\oplus\mathfrak{q}$ contains no three-dimensional commutative subalgebras.

Consequently $\mathfrak{t}$ is simple compact, being the diagonal in $\mathfrak{p}\oplus\mathfrak{q}$ of the same canonical involutive automorphism as $\mathfrak{b}$. Owing to the uniqueness of the diagonal of the canonical involutive automorphism we obtain $\mathfrak{t} = \mathfrak{b}$. But

$$\varphi_1 \in \mathrm{Int}_\mathfrak{g}(\mathfrak{b}\cap(\mathfrak{l}_1\dot{-}\mathfrak{l}_0)) = \mathrm{Int}_\mathfrak{g}(\mathfrak{t}\cap(\mathfrak{l}_1\dot{-}\mathfrak{l}_0)) = \mathrm{Int}_\mathfrak{g}(\{t\,X_{a''_1}\}).$$

Thus the restriction of φ_1 to $\mathfrak{r}_1$ is the identity automorphism, which contradicts to (II.75).

Consequently there is the only possibility $t^{a''}_{a'\,a'} = 0$, $t^{a'}_{a''\,a''} \neq 0$, that is, $\mathfrak{t} = \{\alpha\,X_{a'_1} + \beta\,X_{a'_2} + \gamma\,X_{a'_3}\}$ is the diagonal in $\mathfrak{p}\oplus\mathfrak{q}$ of the same canonical involutive automorphism as $\mathfrak{b}$. Owing to the uniqueness of the diagonal of the canonical involutive automorphism we obtain $\mathfrak{t} = \mathfrak{b}$, that is, $X_{a'_\rho} \in \mathfrak{b}$ $(\rho = 1,2,3)$.

We have thus obtained a more precise version of II.5.6 (Theorem 24):

II.6.4. Theorem 28. *Under the conditions and notations of II.5.6 (Theorem 24), if $\mathfrak{g}$ has only non-prime hyper-involutive decomposition basis for S then one may regard $\mathfrak{b} \cap \mathfrak{l}_\alpha = \mathfrak{b} \cap \mathfrak{r}_\alpha$, where $\mathfrak{b}$ is the diagonal of $\mathfrak{l} = \mathfrak{p} \oplus \mathfrak{q}$ generating the hyper-involutive decomposition basis for S.*

II.6.5. In our consideration the conditions (II.73.1) have been reduced to (II.78.1′), (II.79.1′), (II.78.2′), (II.79.2′), (II.78.3′), (II.78.3″), (II.78.4′), (II.78.4″); in addition, $t^{a''}_{a'\,a'} = 0$, $t^{a'}_{a''\,a''} \neq 0$. Therefore instead of (II.73.1) and (II.76) we have

$$t^{K'}_{I'\,K'} = -t^{I'}_{I'\,I'} \quad (I' \neq K'), \qquad t^{K'}_{I''\,J'} = 0, \qquad t^{K''}_{K''\,I'} = t^{J''}_{J''\,I'},$$

$$t^{P''}_{K''\,I'} = -t^{P''}_{I'\,K''} \quad (P'' \neq K''), \qquad t^{K''}_{I''\,K''} = -t^{I''}_{I''\,I''} \quad (I'' \neq K''), \qquad \text{(II.80)}$$

$$0 \neq t^{a''}_{a'\,a'} = t^{a''}_{k''\,i'}, \qquad C^{A'}_{a'\,i'} = \sqrt{3}\, C^{A''}_{a''\,i''}.$$

Now we consider (II.73.2) which, together with what has been obtained above, gives

$$t^{P'}_{K'\,J'}\, t^{R'}_{I'\,P'} - t^{P'}_{K'\,I'}\, t^{R'}_{J'\,P'} = -\frac{1}{2}\, C^{R'}_{K'\,B'}\, C^{B'}_{I'\,J'}. \qquad \text{(II.81.1)}$$

When $I' = J'$ or $K' = R'$ the relation (II.81.1) is identically satisfied. When $I' \neq J'$, $K' \neq R'$, $I' = K'$ we have $J' = R'$ and (II.81.1) implies

$$\left(t^{I'}_{I'\,I'}\right)^2 + \left(t^{J'}_{J'\,J'}\right)^2 = \frac{1}{4}\left(C^{B'}_{I'\,J'}\right)^2 \quad \text{(no summation over I', J'),}$$

or

$$\left(t^{a'}_{a'\,a'}\right)^2 = \frac{1}{4}\left(C^{B'}_{a'\,j'}\right)^2.$$

Together with (II.77) this gives

$$t^{a'}_{a'\,a'} = \frac{1}{2}\, C^{B'}_{a'\,j'}. \qquad \text{(II.81.1′)}$$

When $I' \neq J'$, $K' \neq R'$, $I' = R'$ we again obtain (II.81.1′).

Furthermore, from (II.73.2) we have

$$t^{P'}_{K''\,J''}\, t^{R'}_{I'\,P'} - t^{P''}_{K''\,I'}\, t^{R'}_{J''\,P''} = 0. \qquad \text{(II.81.2)}$$

When $R' = I' = a'$, $K'' = J'' = a''$ the preceding equation implies

$$t^{a'}_{a''\,a''}\left(t^{a'}_{a'\,a'} - t^{a'}_{a''\,a''}\right) = 0,$$

thus

$$t^{a'}_{a'\,a'} = t^{a'}_{a''\,a''}. \qquad \text{(II.81.2′)}$$

The other choices of indices in (II.81.2) do not give new relations.

Furthermore, from (II.73.2) we have

$$t^P_{K''\,J''}\,t^{R''}_{I''\,P} - t^P_{K''\,I''}\,t^{R''}_{J''\,P} = -\frac{1}{2}\,C^{R''}_{K''\,B''}\,C^{B''}_{I''\,J''}. \tag{II.81.3}$$

For $I'' = R'' = a''$, $J'' = K'' = k''$ we obtain from (II.81.3), using (II.80),

$$\left(t^{a'}_{a''\,a''}\right)^2 - \left(t^{a''}_{a''\,a''}\right)^2 = -\frac{1}{4}\left(C^{B''}_{a''\,k''}\right)^2,$$

or, taking into account (II.80), (II.81.1'), (II.81.2'),

$$\left(t^{a''}_{a''\,a''}\right)^2 = \left(C^{B''}_{a''\,k''}\right)^2,$$

whence using (II.77) we obtain

$$t^{a''}_{a''\,a''} = C^{B''}_{a''\,k''}. \tag{II.81.3'}$$

The other selections of indices in (II.81.3) do not give new relations.

It is easily verified that the other subcases for (II.73.2) give no new relations, except (II.81.1'), (II.81.2'), (II.81.3').

Thus (II.73.2) is equivalent to

$$t^{a'}_{a''\,a''} = t^{a'}_{a'\,a'} = \frac{1}{2}\,C^{B'}_{a'\,j'}, \qquad t^{a''}_{a''\,a''} = C^{B''}_{a''\,k''}. \tag{II.82}$$

Lastly, using the orthonormality of the base in $\mathfrak{g}$ with respect to the Cartan metric and the relations obtained above, we have

$$C^{B'}_{a'\,i'} = \frac{1}{2}. \tag{II.83}$$

And from (II.80), (II.82), (II.83), (II.68)–(II.77), we obtain

$$\left.\begin{aligned}
&t^{a'}_{a'\,a'} = \frac{1}{4}, \qquad t^{a''}_{a''\,a''} = \frac{1}{2\sqrt{3}}, \qquad t^{a'}_{a''\,a''} = t^{a''}_{a''\,a'} = t^{a''}_{a''\,a'} = \frac{1}{4}, \\[2mm]
&t^{a'}_{k''\,k''} = t^{k''}_{a'\,k''} = t^{k''}_{k''\,a'} = \frac{1}{4}, \qquad t^{a'}_{k'\,k'} = t^{k'}_{a'\,k'} = t^{k'}_{k'\,a'} = -\frac{1}{4}, \\[2mm]
&t^{a''}_{k''\,k''} = t^{k''}_{a''\,k''} = t^{k''}_{k''\,a''} = -\frac{1}{2\sqrt{3}}, \\[2mm]
&t^{a''}_{k''\,i'} = t^{i'}_{a''\,k''} = t^{k''}_{i'\,a''} = \frac{1}{4}, \qquad t^{a''}_{i'\,k''} = t^{i'}_{k''\,a''} = t^{k''}_{a''\,i'} = -\frac{1}{4}, \\[2mm]
&C^{B'}_{a'\,k'} = C^{k'}_{B'\,a'} = C^{a'}_{k'\,B'} = \frac{1}{2}, \qquad C^{B'}_{k'\,a'} = C^{a'}_{B'\,k'} = C^{k'}_{a'\,B'} = -\frac{1}{2}, \\[2mm]
&C^{B''}_{a''\,k''} = C^{k''}_{B''\,a''} = C^{a''}_{k''\,B''} = \frac{1}{2\sqrt{3}}, \\[2mm]
&C^{B''}_{k''\,a''} = C^{a''}_{B''\,k''} = C^{k''}_{a''\,B''} = -\frac{1}{2\sqrt{3}}, \qquad p^{A'}_{A'} = p^{B'}_{B'} = -\frac{1}{2}, \\[2mm]
&p^{A''}_{B'} = -p^{B'}_{A''} = \frac{\sqrt{3}}{2}, \quad (a = 1;\ i,k = 2;\ A,B = 3),
\end{aligned}\right\} \tag{II.84}$$

The other constants are equal to zero.

By means of (II.84) we determine the structure equations (II.7) of the Lie algebra $\mathfrak{g}$ in the hyper-involutive base (II.5), (II.68).

Thus we have the theorem:

II.6.6. Theorem 29. *There exists a unique simple compact Lie algebra $\mathfrak{g}$ with a principal involutive automorphism S of type $U^{(2)}$ having only non-prime hyper-involutive decomposition basis for S.*

The embedding of the involutive algebra $\mathfrak{l}$ of the involutive automorphism S into $\mathfrak{g}$ is unique, up to transformations from $\mathrm{Int}(\mathfrak{g})$.

II.6.7. Remark. In the hyper-involutive base (II.5), (II.68) the structure (II.7) of $\mathfrak{g}$ from the preceding theorem is defined by the conditions (II.84).

We denote such an algebra by g_2.

II.6.8. Let us note some useful consequences of what has been obtained above.

From (II.84) it is easy to see that $\mathfrak{r}_1 + \mathfrak{r}_2 + \mathfrak{r}_3$ is a subalgebra of $\mathfrak{g} \cong g_2$, see the notations of II.5.6 (Theorem 24) and (II.68).

Moreover,

$$\mathfrak{l}_0 \subset \mathfrak{r}_1 + \mathfrak{r}_2 + \mathfrak{r}_3, \quad \mathfrak{b} \subset \mathfrak{r}_1 + \mathfrak{r}_2 + \mathfrak{r}_3,$$

$(\mathfrak{r}_1 + \mathfrak{r}_2 + \mathfrak{r}_3)/\mathfrak{b}$ is an involutive pair.

Thus $\dim(\mathfrak{r}_1 + \mathfrak{r}_2 + \mathfrak{r}_3) = 8$, and the hyper-involutive decomposition of $\mathfrak{g} \cong g_2$ basis for S is simultaneously the hyper-involutive decomposition of $\mathfrak{r}_1 + \mathfrak{r}_2 + \mathfrak{r}_3$ with the involutive algebras $\mathfrak{r}_\alpha$ ($\alpha = 1, 2, 3$). Since $\mathrm{Int}_{\mathfrak{g}}(\mathfrak{b}) \cong SO(3)$ we have

$$\mathrm{Int}_{(\mathfrak{r}_1+\mathfrak{r}_2+\mathfrak{r}_3)}(\mathfrak{b}) = \mathrm{Int}_{\mathfrak{g}}^{(\mathfrak{r}_1+\mathfrak{r}_2+\mathfrak{r}_3)}(\mathfrak{b}) \cong SO(3),$$

that is, $(\mathfrak{r}_1 + \mathfrak{r}_2 + \mathfrak{r}_3)/\mathfrak{b}$ is a principal involutive pair of type O.

Since $\mathfrak{r}_\alpha$ ($\alpha = 1, 2, 3$) are simple and compact, we see that $(\mathfrak{r}_1 + \mathfrak{r}_2 + \mathfrak{r}_3)$ is simple and compact as well. Then according to II.3.14 (Theorem 10),

$$(\mathfrak{r}_1 + \mathfrak{r}_2 + \mathfrak{r}_3)/\mathfrak{b} \cong su(3)/so(3)$$

with the natural embedding.

It is easy to see that $\mathfrak{r}_1 + \mathfrak{r}_2 + \mathfrak{r}_3$ is a maximal subalgebra of $\mathfrak{g} \cong g_2$.

Thus we have the theorem:

II.6.9. Theorem 30. *The Lie algebra $su(3)$ can be embedded into g_2 in such a way that $su(3)$ is a maximal subalgebra of g_2, but the pair $(g_2, su(3))$ is not an involutive pair.*

In the notations of II.5.6 (Theorem 24)

$$su(3) \cong (\mathfrak{r}_1 + \mathfrak{r}_2 + \mathfrak{r}_3) \subset \mathfrak{g} \cong g_2.$$

II.6.10. Remark. We note also that the pair $(g_2, su(3))$ locally generates a homogeneous space V_6, $\dim V_6 = 6$, in such a way that the involutive algebras $\mathfrak{l}_\alpha$ ($\alpha = 1, 2, 3$) generate in V_6 three mirrors (see Part III) with trivial intersections.

Thus V_6 is a tri-symmetric, but not symmetric, Riemannian space, see Part III, or [L.V. Sabinin 61].

CHAPTER II.7

MONO-UNITARY NON-CENTRAL
PRINCIPAL INVOLUTIVE AUTOMORPHISM

We prove first some auxiliary theorems.

II.7.1. Theorem 31. *Let* $\mathfrak{g}$ *be a simple compact Lie algebra,* $\mathfrak{l} = \mathfrak{p} \oplus \mathfrak{q} \oplus \tilde{\mathfrak{l}}$ *be its principal di-unitary involutive algebra of an involutive automorphism* S *(*$S \in \mathrm{Int}_\mathfrak{g}(\mathfrak{p})$*,* $S \in \mathrm{Int}_\mathfrak{g}(\mathfrak{q})$*). If* $\tilde{\mathfrak{l}}$ *is a special unitary subalgebra of the involutive automorphism* S *then either* $\mathfrak{g} \cong so(8)$ *or* $\mathfrak{g} \cong so(12)$.

Proof. Since $\tilde{\mathfrak{l}} \neq \{0\}$ we have according to II.5.10 (Theorem 27)

$$\mathfrak{g}/\mathfrak{l} \cong so(n)/(\,so(4) \oplus so(n-4)\,), \quad n > 5,$$

with the natural embedding. Thus $\tilde{\mathfrak{l}} \cong so(n-4)$.

The case $n = 6$ contradicts that $\tilde{\mathfrak{l}}$ is a special unitary subalgebra.

When $n = 7$ we obtain $\tilde{\mathfrak{l}} \cong so(3)$ and

$$\mathfrak{g}/\mathfrak{l} \cong so(7)/(\,so(4) \oplus so(3)\,)$$

with the natural embedding; then $\mathrm{Int}_\mathfrak{g}(\tilde{\mathfrak{l}}) \cong SO(3)$, which again contradicts that $\tilde{\mathfrak{l}}$ is a special unitary subalgebra.

The case $n = 8$ gives us $\tilde{\mathfrak{l}} \cong so(4) = su(2) \oplus su(2)$, and is obviously suitable leading to $\mathfrak{g} \cong so(8)$.

Now we consider $n > 8$. Then $\tilde{\mathfrak{l}}$ is simple and more than three-dimensional. Since $\tilde{\mathfrak{l}}$ is a special algebra of the involutive automorphism S there exists a principal orthogonal involutive pair $\tilde{\mathfrak{l}}/\mathfrak{b} \oplus \tilde{\tilde{\mathfrak{l}}}$, $S \in \mathrm{Int}_\mathfrak{g}(\mathfrak{b})$.

Evidently $\tilde{\tilde{\mathfrak{l}}} \neq \{0\}$ (otherwise, either $\tilde{\mathfrak{l}} = \mathfrak{b}$ or $\dim \tilde{\mathfrak{l}} = 8$ by II.3.10 (Theorem 8) and II.3.11 (Theorem 9), but $\tilde{\mathfrak{l}}$ is more than three-dimensional, and $\dim \tilde{\mathfrak{l}} = \dim so(n-4) = (1/2)(n-4)(n-5) \neq 8$).

Now, by II.3.11 (Theorem 9) we have

$$\tilde{\mathfrak{l}}/\mathfrak{b} \oplus \tilde{\tilde{\mathfrak{l}}} \cong so(n-4)/(\,so(3) \oplus so(n-7)\,)$$

with the natural embedding.

We consider, furthermore, in $\tilde{\mathfrak{l}}$ the principal di-unitary involutive algebra $\mathfrak{f} \oplus \mathfrak{e} \oplus \mathfrak{c} \cong so(4) \oplus so(n-8)$ such that $\mathfrak{b}$ is a diagonal in $\mathfrak{f} \oplus \mathfrak{e} \cong so(4)$.

Since $\mathrm{Int}_{\tilde{\mathfrak{l}}}(\mathfrak{f}) \cong SU(2)$, $\mathrm{Int}_{\tilde{\mathfrak{l}}}(\mathfrak{e}) \cong SU(2)$ we have

$$\mathrm{Int}_\mathfrak{g}(\mathfrak{f}) \cong SU(2), \quad \mathrm{Int}_\mathfrak{g}(\mathfrak{e}) \cong SU(2).$$

We now consider non-trivial involutive automorphisms $S_1 \in \mathrm{Int}_\mathfrak{g}(\mathfrak{f})$, $S_2 \in \mathrm{Int}_\mathfrak{g}(\mathfrak{e})$. Then $S_1 S_2 = S_2 S_1 = S$, because $S \in \mathrm{Int}_\mathfrak{g}(\mathfrak{b})$.

Let $\mathfrak{l}_1$ be the involutive algebra of the involutive automorphism S_1, and $\mathfrak{l}_2$ be the involutive algebra of the involutive automorphism S_2. Being commutative S_1 and S_2 generate the involutive pair $\mathfrak{l}_1/(\mathfrak{p} \oplus \mathfrak{q} \oplus \mathfrak{f} \oplus \mathfrak{e} \oplus \mathfrak{c})$ of the involutive automorphism S, whereas $\mathfrak{c} \cong so(n - 8)$. The restrictions of S and S_2 to $\mathfrak{l}_1$ evidently coincide and are different from the identity automorphism. We denote such restriction by A. Then

$$A \in \mathrm{Int}_{\mathfrak{l}_1}(\mathfrak{p}) \cong SU(2), \qquad A \in \mathrm{Int}_{\mathfrak{l}_1}(\mathfrak{q}) \cong SU(2),$$

$$A \in \mathrm{Int}_{\mathfrak{l}_1}(\mathfrak{e}) \cong SU(2), \qquad \mathrm{Int}_{\mathfrak{l}_1}(\mathfrak{f}) \cong SO(3).$$

Furthermore, we consider the decomposition of the involutive pair

$$\mathfrak{l}_1/(\mathfrak{p} \oplus \mathfrak{q} \oplus \mathfrak{f} \oplus \mathfrak{e} \oplus \mathfrak{c})$$

of the involutive automorphism A into elementary involutive pairs:

$$\mathfrak{l}_1^{(1)}/\mathfrak{t}^{(1)} + \mathfrak{l}_1^{(2)}/\mathfrak{t}^{(2)} + \ldots + \mathfrak{l}_1^{(m)}/\mathfrak{t}^{(m)}.$$

If any two of $\mathfrak{p}$, $\mathfrak{q}$, $\mathfrak{e}$ belong to different $\mathfrak{t}^{(i)}$ then all $\mathfrak{l}_1^{(k)} \subset \mathfrak{l}$ and $\mathfrak{l}_1^{(1)} \oplus \ldots \oplus \mathfrak{l}_1^{(m)} = \mathfrak{l}_1 \subset \mathfrak{l}$, which is impossible.

After re-numbering we may assume $\mathfrak{p} \subset \mathfrak{t}^{(1)}$, $\mathfrak{q} \subset \mathfrak{t}^{(1)}$, $\mathfrak{e} \subset \mathfrak{t}^{(1)}$, and $\mathfrak{l}_1^{(k)} = \mathfrak{t}^{(k)}$ for $(k = 2, \ldots, m)$. Since $\mathfrak{t}^{(1)} = \mathfrak{p} \oplus \mathfrak{q} \oplus \mathfrak{e} \oplus \mathfrak{a}$ we have evidently that for an elementary involutive pair $\mathfrak{l}_1^{(1)}/\mathfrak{t}^{(1)}$ the subalgebra $\mathfrak{l}_1^{(1)}$ is simple.

Moreover, $\mathfrak{p} \oplus \mathfrak{q} \oplus (\mathfrak{e} \oplus \mathfrak{a})$ is the principal di-unitary involutive algebra of the involutive automorphism A. Consequently

$$\mathfrak{l}_1^{(1)}/\mathfrak{p} \oplus \mathfrak{q} \oplus (\mathfrak{e} \oplus \mathfrak{a}) \cong so(m)/(so(4) \oplus so(m - 4)) \quad (m > 5)$$

according to II.5.10 (Theorem 27).

Thus $\mathfrak{e} \oplus \mathfrak{a} \cong so(m - 4)$, whence either $m = 7$ or $m = 8$.

But $m = 7$ implies $\mathfrak{a} = \{0\}$ and $\mathfrak{l}_1^{(1)}/\mathfrak{p} \oplus \mathfrak{q} \oplus \mathfrak{e} \cong so(7)/(so(4) \oplus so(3))$, which is impossible because $\mathrm{Int}_{\mathfrak{l}_1^{(1)}}(\mathfrak{e}) \cong SU(2)$, $\mathrm{Int}_{so(7)}(so(3)) \cong SO(3)$.

Consequently $m = 8$ and $\mathfrak{a}$ is three-dimensional and simple. Since $\mathrm{Int}_{\mathfrak{l}_1^{(1)}}(\mathfrak{a}) \cong SU(2)$ we then have $\mathfrak{a} \neq \mathfrak{f}$.

Furthermore, we use the equality

$$\mathfrak{p} \oplus \mathfrak{q} \oplus \mathfrak{f} \oplus \mathfrak{e} \oplus \mathfrak{c} = \mathfrak{p} \oplus \mathfrak{q} \oplus \mathfrak{e} \oplus \mathfrak{a} \oplus \tilde{\mathfrak{a}},$$

whence $\mathfrak{f} \oplus \mathfrak{c} = \mathfrak{a} \oplus \tilde{\mathfrak{a}}$.

Since $\mathfrak{a} \neq \mathfrak{f}$ we have $\mathfrak{a} \subset \mathfrak{c}$, hence $\mathfrak{c} = \mathfrak{a} \oplus \tilde{\mathfrak{c}}$. But $\mathfrak{c} = so(n - 8)$, consequently either $n = 11$ or $n = 12$.

When $n = 11$ we have $\mathfrak{g}/\mathfrak{l} \cong so(11)/(so(4) \oplus so(7))$ with the natural embeddings. Thus $\tilde{\mathfrak{l}} \cong so(7)$.

But $\operatorname{Int}_{\mathfrak{g}}(\tilde{\mathfrak{l}}) \cong \operatorname{Int}_{so(11)}(so(7)) \cong SO(7)$, having the trivial centre, which contradicts that $\tilde{\mathfrak{l}}$ is the di-unitary special subalgebra of the involutive automorphism S.

Thus the only possibility is $n = 12$, and then

$$\mathfrak{g}/\mathfrak{l} \cong so(12)/(so(4) \oplus so(8))$$

with the natural embedding.

It is now easily verified that $\tilde{\mathfrak{l}} \cong so(8)$ is, indeed, a unitary special subalgebra of $\mathfrak{g}$. For this we consider the involutive pair

$$so(8)/(so(4) \oplus so(4)) \cong \tilde{\mathfrak{l}}/(\mathfrak{p} \oplus \mathfrak{q}) \oplus (\mathfrak{f} \oplus \mathfrak{e})$$

with the natural embedding.

Then $\operatorname{Int}_{\mathfrak{g}}(\mathfrak{p}) \cong SU(2)$, $\operatorname{Int}_{\mathfrak{g}}(\mathfrak{f}) \cong SU(2)$, $\operatorname{Int}_{\mathfrak{g}}(\mathfrak{p} \oplus \mathfrak{f}) \cong SO(4)$. And if $\mathfrak{b}$ is a diagonal of a canonical involutive automorphism of $\mathfrak{p} \oplus \mathfrak{f}$ then $\operatorname{Int}_{\mathfrak{g}}(\mathfrak{b}) \cong SU(2)$. Furthermore, $\operatorname{Int}_{\mathfrak{l}}(\mathfrak{p} \oplus \mathfrak{f}) \cong SO(4)$, whence $\operatorname{Int}_{\mathfrak{l}}(\mathfrak{b}) = \operatorname{Int}_{\mathfrak{g}}^{\mathfrak{l}}(\mathfrak{b}) \cong SO(3)$.

Lastly, $\tilde{\mathfrak{l}}/(\mathfrak{p} \oplus \mathfrak{f} \oplus (\mathfrak{q} \oplus \mathfrak{e})) \cong so(8)/(so(4) \oplus so(4))$ is a principal di-unitary involutive pair, thus the diagonal $\mathfrak{b}$ is isomorphic to $so(3)$ with the natural embedding into $\mathfrak{p} \oplus \mathfrak{f} \cong so(4)$, and therefore there exists the principal orthogonal involutive pair $\tilde{\mathfrak{l}}/\mathfrak{b} \oplus \tilde{\mathfrak{l}} \cong so(8)/(so(3) \oplus so(5))$ (with the natural embedding). Since $\operatorname{Int}_{\mathfrak{g}}(\mathfrak{b}) \cong SU(2)$ that means that $\mathfrak{l}$ is a special unitary subalgebra. ∎

II.7.2. Theorem 32. *If $\mathfrak{g}$ is a simple compact Lie algebra and $\mathfrak{k} \cong so(6)$ is its special unitary subalgebra of an involutive automorphism S then S is a central involutive automorphism.*

Proof. Let us consider in $\mathfrak{k}$ an involutive algebra $\mathfrak{b} \oplus \tilde{\mathfrak{k}}$ such that $\operatorname{Int}_{\mathfrak{g}}(\mathfrak{b}) \cong SU(2)$, $\operatorname{Int}_{\mathfrak{k}}(\mathfrak{b}) \cong SO(3)$, which, by the definition of a special unitary subalgebra, I.1.17 (Definition 8), is possible. Then by II.3.10 (Theorem 8), II.3.11 (Theorem 9), II.3.14 (Theorem 10)

$$\mathfrak{k}/\mathfrak{b} \oplus \tilde{\mathfrak{k}} \cong so(n)/(so(3) \oplus so(n-3))$$

with the natural embedding.

In our case $\mathfrak{k} \cong so(6)$, consequently

$$\mathfrak{k}/\mathfrak{b} \oplus \tilde{\mathfrak{k}} \cong so(6)/(so(3) \oplus so(3))$$

with the natural embedding.

Since $S \in \operatorname{Int}_{\mathfrak{g}}(\mathfrak{b}) \cong SU(2)$ we have $S \in \operatorname{Int}_{\mathfrak{g}}(\{t\zeta\})$, where $\{t\zeta\} \subset \mathfrak{b}$. But $\{t\zeta\} \cong so(2) \subset so(3) \cong \mathfrak{b}$ with the natural embedding.

By virtue of the conjugacy of all subalgebras $so(2)$ naturally embedded into $so(6)$ we have $S \in \operatorname{Int}_{\mathfrak{g}}(\{t\zeta\})$, where $\{t\zeta\}$ is any one-dimensional subalgebra of $\mathfrak{k}$ isomorphic to $so(2)$ with the natural embedding into $so(6) \cong \mathfrak{k}$ in the constructed realization.

Now we consider in $so(6)$ the matrices $\hat{a}^{(i)}_{(j)}$ defined by

$$\hat{a}^{(i)\,p}_{(j)\,q} = \delta^i_q \delta^p_j - \delta^j_q \delta^p_i \,,$$

$$(i, j, p, q = 1, \dots, 6) \,.$$

All matrices $\pm\hat{a}^{(i)}_{(j)}$ are conjugated in $\mathrm{Int}(so(6))$ and all subalgebras $so(2)$, with the natural embedding into $so(6)$, have the form $t\,\hat{a}^{(i)}_{(j)}$.

For this reason $S = \mathrm{Int}_{\mathfrak{g}}\big(t_0\,\hat{a}^{(i)}_{(j)}\big)$ in our realization, where t_0 does not depend on i, j.

Introducing also

$$\hat{g} = \hat{a}^{(2)}_{(1)} + \hat{a}^{(4)}_{(3)} + \hat{a}^{(6)}_{(5)}$$

we obtain

$$\mathrm{Int}_{\mathfrak{g}}\big(t_0\,\hat{g}\big) = \mathrm{Int}_{\mathfrak{g}}\big(t_0\,\hat{a}^{(2)}_{(1)}\big)\,\mathrm{Int}_{\mathfrak{g}}\big(t_0\,\hat{a}^{(4)}_{(3)}\big)\,\mathrm{Int}_{\mathfrak{g}}\big(t_0\,\hat{a}^{(6)}_{(5)}\big) = S^3 = S.$$

Lastly, taking

$$\varphi = \mathrm{Int}_{\mathfrak{g}}\big(\tfrac{1}{2}\,t_0\,\hat{g}\big), \qquad \varphi_1 = \mathrm{Int}_{\mathfrak{g}}\big(\tfrac{1}{2}\,t_0\,\hat{a}^{(2)}_{(1)}\big),$$

$$\varphi_2 = \mathrm{Int}_{\mathfrak{g}}\big(\tfrac{1}{2}\,t_0\,\hat{a}^{(4)}_{(3)}\big), \qquad \varphi_3 = \mathrm{Int}_{\mathfrak{g}}\big(\tfrac{1}{2}\,t_0\,\hat{a}^{(6)}_{(5)}\big)$$

we have

$$(\varphi_1)^2 = (\varphi_2)^2 = (\varphi_3)^2 = \varphi^2 = S, \qquad \varphi_1\,\varphi_2\,\varphi_3 = \varphi.$$

Let us consider, finally, the natural morphism

$$\mathrm{Int}_{\mathfrak{g}}(\mathfrak{k}) \longrightarrow \mathrm{Int}^{\mathfrak{k}}_{\mathfrak{g}}(\mathfrak{k}) \cong \mathrm{Int}(\mathfrak{k}).$$

Since the restriction of S to $\mathfrak{k}$ is the identity involutive automorphism, $S|_{\mathfrak{k}} = \mathrm{Id}$, we have $(\varphi_\alpha|_{\mathfrak{k}})^2 = (\varphi|_{\mathfrak{k}})^2 = \mathrm{Id}$ ($\alpha = 1, 2, 3$), that is, $\varphi_\alpha|_{\mathfrak{k}}$ and $\varphi|_{\mathfrak{k}}$ are involutive automorphisms in $\mathfrak{k} \cong so(6)$. But $\varphi_\alpha|_{\mathfrak{k}}$ ($\alpha = 1, 2, 3$) are non-trivial involutive automorphisms of $\mathfrak{k} \cong so(6)$ (indeed, they are generated by subalgebras isomorphic to $so(2)$ with the natural embedding into $so(6)$, but $so(2)$ can be embedded into $so(3)$ with the natural embedding into $so(6)$; furthermore, $\mathrm{Int}_{\mathfrak{g}}\,(so(3)) \cong SU(2)$, $\mathrm{Int}_{\mathfrak{k}}\,(so(3)) \cong SO(3)$, and the kernel of the morphism $SU(2) \longrightarrow SO(3)$ consists of involutive elements only). Therefore,

$$\varphi_1|_{\mathfrak{k}}\,(\hat{x}) = \mathrm{e}^{-\pi\,\hat{a}^{(2)}_{(1)}}\,\hat{x}\,\mathrm{e}^{\pi\,\hat{a}^{(2)}_{(1)}}, \qquad \varphi_2|_{\mathfrak{k}}\,(\hat{x}) = \mathrm{e}^{-\pi\,\hat{a}^{(4)}_{(3)}}\,\hat{x}\,\mathrm{e}^{\pi\,\hat{a}^{(4)}_{(3)}},$$

$$\varphi_3|_{\mathfrak{k}}\,(\hat{x}) = \mathrm{e}^{-\pi\,\hat{a}^{(6)}_{(5)}}\,\hat{x}\,\mathrm{e}^{\pi\,\hat{a}^{(6)}_{(5)}}.$$

But then

$$\varphi|_{\mathfrak{k}}(\hat{x}) = (\varphi_1 \, \varphi_2 \, \varphi_3)|_{\mathfrak{k}}(\hat{x}) = (\varphi_1|_{\mathfrak{k}})(\varphi_2|_{\mathfrak{k}})(\varphi_3|_{\mathfrak{k}})(\hat{x})$$

$$= e^{-\pi \hat{g}} \, \hat{x} \, e^{\pi \hat{g}} = (-\hat{\mathrm{Id}}) \, \hat{x} \, (-\hat{\mathrm{Id}}) = \hat{x}, \quad \hat{x} \in so(6) \cong \mathfrak{k}.$$

Thus $\varphi|_{\mathfrak{k}} = \mathrm{Id}$, that is, the natural morphism

$$\mathrm{Int}_{\mathfrak{g}}(\mathfrak{k}) \longrightarrow \mathrm{Int}_{\mathfrak{g}}^{\mathfrak{k}}(\mathfrak{k}) \cong \mathrm{Int}(\mathfrak{k})$$

maps φ into Id. Consequently φ belongs to the centre of $\mathrm{Int}_{\mathfrak{g}}(\mathfrak{k})$ and $\varphi \neq \mathrm{Id}$, since $\varphi^2 = S \neq \mathrm{Id}$.

Now we consider the involutive algebra $\mathfrak{l}$ of the involutive automorphism S, then $\mathfrak{l} = \mathfrak{k} \oplus \mathfrak{a}$, whence $\varphi\eta = \eta$ for $\eta \in \mathfrak{l}$. By II.2.3 (Theorem 3) we have in $\mathfrak{l}$ the non-trivial centre $\mathfrak{z}$. ∎

II.7.3. Theorem 33. *Let $\mathfrak{g}$ be a simple compact Lie algebra, and $\mathfrak{l}_1 = \mathfrak{p} \oplus \tilde{\mathfrak{l}}$ be its principal non-central mono-unitary involutive algebra of an involutive automorphism $S_1 \in \mathrm{Int}_{\mathfrak{g}}(\mathfrak{p}) \cong SU(2)$. Then $\tilde{\mathfrak{l}}$ is simple and $\dim \tilde{\mathfrak{l}} > 3$.*

Proof. From the conditions it is evident that either $\tilde{\mathfrak{l}} = \{0\}$ or $\tilde{\mathfrak{l}}$ is semi-simple and has no three-dimensional ideals.

Let us consider first the case $\tilde{\mathfrak{l}} \neq \{0\}$. Then $\tilde{\mathfrak{l}} = \mathfrak{m} \oplus \mathfrak{n}$, where $\mathfrak{m}$ is simple, $\dim \mathfrak{m} > 3$, and $\mathfrak{n}$ is semi-simple. Let us suppose that $\mathfrak{n} \neq \{0\}$. By II.4.15 (Theorem 22) $\mathfrak{m}$ has a principal unitary involutive algebra $\mathfrak{q} \oplus \mathfrak{t}$ with $\mathrm{Int}_{\mathfrak{m}}(\mathfrak{q}) \cong SU(2)$. But then evidently $\mathrm{Int}_{\mathfrak{g}}(\mathfrak{q}) \cong SU(2)$.

Let $S_2 \in \mathrm{Int}_{\mathfrak{g}}(\mathfrak{q})$, then $S_1 \neq S_2$, $S_1 S_2 = S_2 S_1 = S_3$, where S_3 is an involutive automorphism (the assumption $S_1 = S_2$ gives $\mathrm{Int}_{\mathfrak{m}}(\mathfrak{q}) = \mathrm{Int}_{\mathfrak{g}}^{\mathfrak{m}}(\mathfrak{q}) \cong \mathrm{Int}_{\mathfrak{g}}^{\mathfrak{l}_1}(\mathfrak{q}) \cong SO(3)$, which is impossible).

Using S_1, S_2, S_3 we now construct the involutive sum $\mathfrak{g} = \mathfrak{l}_1 + \mathfrak{l}_2 + \mathfrak{l}_3$ (where $\mathfrak{l}_\alpha$ are the involutive algebras of the involutive automorphisms S_α) such that $\mathfrak{l}_0 = \mathfrak{l}_\alpha \cap \mathfrak{l}_\mu$ ($\alpha \neq \mu$) and $\mathfrak{l}_\alpha/\mathfrak{l}_0$ are involutive pairs. Obviously $\mathfrak{l}_0 = \mathfrak{p} \oplus \mathfrak{q} \oplus \mathfrak{n} \oplus \mathfrak{t}$.

Now we consider the decomposition into elementary involutive pairs

$$\mathfrak{l}_3/\mathfrak{l}_0 = \mathfrak{l}_3^{(1)}/\mathfrak{l}_0^{(1)} + \cdots + \mathfrak{l}_3^{(m)}/\mathfrak{l}_0^{(m)}.$$

Since $S_1 S_3 = S_2$, the restrictions of S_1 and S_2 to $\mathfrak{l}_3$ coincide and are different from the identity automorphism. We denote such a restriction by A. Then

$$A \in \mathrm{Int}_{\mathfrak{l}_3}(\mathfrak{p}) \cong SU(2), \qquad A \in \mathrm{Int}_{\mathfrak{l}_3}(\mathfrak{q}) \cong SU(2),$$

and $\mathfrak{l}_3/\mathfrak{l}_0$ is the involutive pair of the di-unitary involutive automorphism A. If $\mathfrak{p}$ and $\mathfrak{q}$ belong to different $\mathfrak{l}_0^{(k)}$ then $\mathfrak{l}_3^{(s)} \subset \mathfrak{l}_0$, for some s, whence $\mathfrak{l}_3 \subset \mathfrak{l}_0 \subset \mathfrak{l}_1$, which is impossible. Therefore after some re-numbering we may assume $\mathfrak{p} \oplus \mathfrak{q} \subset \mathfrak{l}_0^{(1)}$, $\mathfrak{l}_3^{(k)} \subset \mathfrak{l}_0$ ($k = 2, \ldots, m$).

Therefore we have the decomposition $\mathfrak{l}_3/\mathfrak{l}_0 = \mathfrak{l}_3^{(1)}/\mathfrak{l}_0^{(1)} + \mathfrak{a}/\mathfrak{a}$, whence, in particular, it follows that $\mathfrak{l}_3^{(1)}$ is simple and $\mathfrak{l}_3^{(1)}/\mathfrak{l}_0^{(1)}$ is the principal di-unitary involutive pair for the involutive automorphism A.

From

$$\mathfrak{l}_0 = \mathfrak{p} \oplus \mathfrak{q} \oplus \mathfrak{n} \oplus \mathfrak{t} = \mathfrak{l}_0^{(1)} \oplus \mathfrak{a}$$

we have also

$$\mathfrak{l}_0^{(1)} = \mathfrak{p} \oplus \mathfrak{q} \oplus \mathfrak{n}' \oplus \mathfrak{t}',$$

$$\mathfrak{a} = \mathfrak{n}'' \oplus \mathfrak{t}'',$$

where $\mathfrak{n} = \mathfrak{n}' \oplus \mathfrak{n}''$, $\mathfrak{t} = \mathfrak{t}' \oplus \mathfrak{t}''$.

If we suppose that $\mathfrak{n}'' \neq \{0\}$, then

$$\left[\mathfrak{n}'' \left(\mathfrak{l}_1 \dot{-} \mathfrak{l}_0 \right) \right] = \left[\mathfrak{n}'' \left(\mathfrak{l}_3 \dot{-} \mathfrak{l}_0 \right) \right] = \{0\},$$

so that $\mathfrak{n}'' \subset \mathfrak{l}_0 \subset \mathfrak{l}_2$, and $\left[\mathfrak{n}'' \left(\mathfrak{g} \dot{-} \mathfrak{l}_2 \right) \right] = \{0\}$, $\mathfrak{n}''$ having generated a non-trivial ideal in $\mathfrak{g}$, which is impossible. Thus we have $\mathfrak{n}'' = \{0\}$, $\mathfrak{n} = \mathfrak{n}'$, and as result $\mathfrak{l}_3 / \mathfrak{l}_0 = \mathfrak{l}_3^{(1)} / (\mathfrak{p} \oplus \mathfrak{q} \oplus \mathfrak{n} \oplus \mathfrak{t}') + \mathfrak{t}''/\mathfrak{t}''$.

By virtue of II.5.10 (Theorem 27)

$$\mathfrak{l}_3^{(1)} / (\mathfrak{p} \oplus \mathfrak{q} \oplus (\mathfrak{n} \oplus \mathfrak{t}')) \cong so(m) / (so(4) \oplus so(m-4)) \quad (m > 5)$$

with the natural embedding.

But $\mathfrak{n}$ is non-one-dimensional and has no three-dimensional ideals, thus $m > 8$. But then $\mathfrak{n} \oplus \mathfrak{t}' \cong so(m-4)$ is simple.

Consequently $\mathfrak{n} \cong so(m-4)$ and is simple, $\mathfrak{t}' = \{0\}$.

Thus

$$\mathfrak{l}_3^{(1)} / \mathfrak{p} \oplus \mathfrak{q} \oplus \mathfrak{n} \cong so(m) / (so(4) \oplus so(m-4)) \quad (m > 8)$$

with the natural embedding.

Now we consider a diagonal $\mathfrak{b}$ in $\mathfrak{p} \oplus \mathfrak{q} \cong so(4)$. Then $\mathfrak{b} \cong so(3)$ with the natural embedding into $\mathfrak{p} \oplus \mathfrak{q} \cong so(4)$.

Let us also take a subalgebra $\mathfrak{f}$ of $\mathfrak{n}$ such that $\mathfrak{f} \cong so(3)$ with the natural embedding into $\mathfrak{n} \cong so(m-4)$.

All subalgebras of $so(3)$ with the natural embedding into $\mathrm{Int}(so(m))$, $m > 4$, are conjugated in $\mathrm{Int}(so(m))$, therefore $\mathfrak{b}$ and $\mathfrak{f}$ are conjugated in $\mathrm{Int}_{\mathfrak{g}}(\mathfrak{l}_3)$.

But $S_3 \in \mathrm{Int}_{\mathfrak{g}}(\mathfrak{b})$, which implies $S_3 \in \mathrm{Int}_{\mathfrak{g}}(\mathfrak{f}) \subset \mathrm{Int}_{\mathfrak{g}}(\mathfrak{n})$. However, then, owing to $[\mathfrak{n}\,\mathfrak{m}] = \{0\}$, the vectors from $\mathfrak{m}$ are immobile under the action of S_3, that is, $\mathfrak{m} \subset \mathfrak{l}_3$, which is impossible by the construction.

Consequently $\mathfrak{n} = \{0\}$, that is, $\tilde{\mathfrak{l}}$ is simple and $\dim \tilde{\mathfrak{l}} > 3$.

Now we consider the case $\tilde{\mathfrak{l}} = \{0\}$. Then we have a mono-unitary involutive pair $\mathfrak{g}/\mathfrak{p}$, $S_1 \in \mathrm{Int}_{\mathfrak{g}}(\mathfrak{p}) \cong SU(2)$. But according to I.8.2 (Theorem 31) this is impossible. Thus $\tilde{\mathfrak{l}} \neq \{0\}$. ∎

II.7.4. Definition 29. Let $\mathfrak{g}$ be a simple compact Lie algebra, and $\mathfrak{l}_1 = \mathfrak{p} \oplus \tilde{\mathfrak{l}}$ be its principal involutive algebra of an involutive automorphism $S_1 \in \mathrm{Int}_{\mathfrak{g}}(\mathfrak{p})$, $S_1 \neq \mathrm{Id}$. Let, furthermore, $\mathfrak{q} \oplus \tilde{\tilde{\mathfrak{l}}}$ be a principal unitary involutive algebra in $\tilde{\mathfrak{l}}$ of an involutive automorphism $S_2 \in \mathrm{Int}_{\mathfrak{g}}(\mathfrak{q})$; $\mathfrak{l}_2$ being the involutive algebra of the involutive automorphism $S_2 \neq S_1$, $\mathfrak{l}_3$ being the involutive algebra of the involutive automorphism $S_3 = S_1 S_2 = S_2 S_1$. Then the involutive sum $\mathfrak{g} = \mathfrak{l}_1 + \mathfrak{l}_2 + \mathfrak{l}_3$ is called the basis involutive sum for a principal unitary involutive algebra $\mathfrak{l}_1$.

II.7.5. Theorem 34. *Let* $\mathfrak{g}$ *be a simple compact Lie algebra, and* $\mathfrak{l}_1 = \mathfrak{p} \oplus \tilde{\mathfrak{l}}$ *be its principal mono-unitary non-central involutive algebra of an involutive automorphism* $S_1 \in \mathrm{Int}_{\mathfrak{g}}(\mathfrak{p})$. *Then the basis for* $\mathfrak{l}_1$ *involutive sum* $\mathfrak{g} = \mathfrak{l}_1 + \mathfrak{l}_2 + \mathfrak{l}_3$ *of the involutive automorphisms* $S_1, S_2, S_3 = S_1 S_2 = S_2 S_1$ *exists and is iso-involutive with a conjugating automorphism* $\varphi \in \mathrm{Int}_{\mathfrak{g}}(\{t\zeta\})$, $\zeta \in \mathfrak{l}_3 - \mathfrak{l}_0$ ($\mathfrak{l}_0 = \mathfrak{l}_\alpha \cap \mathfrak{l}_\beta$, $\alpha \neq \beta$).

Moreover,

$$\mathfrak{l}_3 / \mathfrak{l}_0 = \mathfrak{k} / \mathfrak{l}_0^{(1)} + \mathfrak{l}_0^{(2)} / \mathfrak{l}_0^{(2)},$$

where $\mathfrak{k}$ *is a simple special unitary subalgebra of the involutive automorphism* $S_3 = \varphi^2$ *and* $\mathfrak{k} / \mathfrak{l}_0^{(1)}$ *is a principal di-unitary involutive pair.*

Proof. According to II.7.3 (Theorem 33) $\tilde{\mathfrak{l}}$ is simple and $\dim \tilde{\mathfrak{l}} > 3$, but then by II.4.15 (Theorem 22) $\tilde{\mathfrak{l}}$ has a unitary principal involutive algebra $\mathfrak{q} \oplus \mathfrak{t}$ such that $\mathrm{Int}_{\tilde{\mathfrak{l}}}(\mathfrak{q}) \cong SU(2)$. Owing to the morphisms

$$\mathrm{Int}_{\mathfrak{g}}(\mathfrak{q}) \longrightarrow \mathrm{Int}_{\mathfrak{g}}^{\mathfrak{l}_1}(\mathfrak{q}) \longrightarrow \mathrm{Int}_{\mathfrak{l}_1}^{\tilde{\mathfrak{l}}}(\mathfrak{q}) \cong \mathrm{Int}_{\tilde{\mathfrak{l}}}(\mathfrak{q}) \cong SU(2),$$

it is evident that $\mathrm{Int}_{\mathfrak{g}}(\mathfrak{q}) \cong SU(2)$.

Now we take an involutive automorphism $\mathrm{Id} \neq S_2 \in \mathrm{Int}_{\mathfrak{g}}(\mathfrak{q})$.

By the commutativity of $\mathfrak{q}$ and $\mathfrak{p}$ we have $S_2 S_1 = S_1 S_2 = S_3 \neq \mathrm{Id}$ (otherwise $S_1 = S_2$, but the restriction of S_1 to $\mathfrak{l}_1$ is the identity automorphism, and the restriction of S_2 to $\mathfrak{l}_1$ is different from the identity automorphism; which is impossible).

Taking $\mathfrak{l}_1, \mathfrak{l}_2, \mathfrak{l}_3$, the involutive algebras of S_1, S_2, S_3, respectively, we obtain $\mathfrak{g} = \mathfrak{l}_1 + \mathfrak{l}_2 + \mathfrak{l}_3$, the involutive sum basis for $\mathfrak{l}_1$.

Now we consider $\mathfrak{l}_0 = \mathfrak{l}_\alpha \cap \mathfrak{l}_\beta$ ($\alpha \neq \beta$). By construction we have $\mathfrak{l}_0 = \mathfrak{p} \oplus \mathfrak{q} \oplus \mathfrak{t}$. Since $S_2 S_1 = S_3$ the restrictions of S_1 and S_2 to $\mathfrak{l}_3$ coincide and give a non-identity involutive automorphism A of $\mathfrak{l}_3$ with the involutive algebra $\mathfrak{l}_0$.

Therefore $\mathfrak{l}_3 / \mathfrak{p} \oplus \mathfrak{q} \oplus \mathfrak{t}$ is the di-unitary principal involutive pair of the involutive automorphism A.

The decomposition

$$\mathfrak{l}_3 / \mathfrak{l}_0 = \mathfrak{l}_3^{(1)} / \mathfrak{l}_0^{(1)} + \ldots + \mathfrak{l}_3^{(m)} / \mathfrak{l}_0^{(m)}$$

into elementary involutive pairs implies that $\mathfrak{p}$ and $\mathfrak{q}$ belong to the same $\mathfrak{l}^{(k)}$. (Otherwise $\mathfrak{l}_3^{(s)} \subset \mathfrak{l}_0$ ($s = 1, \ldots, m$) since $A \in \mathrm{Int}_{\mathfrak{l}_3}(\mathfrak{p})$, $A \in \mathrm{Int}_{\mathfrak{l}_3}(\mathfrak{q})$. But then $\mathfrak{l}_3 \subset \mathfrak{l}_0$, which is impossible).

After re-numbering we may assume $\mathfrak{p} \oplus \mathfrak{q} \subset \mathfrak{l}_0^{(1)} \subset \mathfrak{k} = \mathfrak{l}_3^{(1)}$, then evidently $\mathfrak{l}_3 \dot{-} \mathfrak{k} \subset \mathfrak{l}_0$, and there is the decomposition

$$\mathfrak{l}_3 / \mathfrak{l}_0 = \mathfrak{k} / \mathfrak{p} \oplus \mathfrak{q} \oplus \mathfrak{t}' + \mathfrak{t}'' / \mathfrak{t}'',$$

where $\mathfrak{k} / \mathfrak{p} \oplus \mathfrak{q} \oplus \mathfrak{t}'$ is an elementary involutive pair. It is possible only if $\mathfrak{k}$ is simple.

From the theorems of II.5 and II.6 it follows that either a principal di-unitary involutive pair $\mathfrak{k} / \mathfrak{p} \oplus \mathfrak{q} \oplus \mathfrak{t}'$ is isomorphic to $so(n) / (so(4) \oplus so(n-4))$, $n > 4$, (with the natural embedding) or $\mathfrak{k} \cong g_2$, and then $\mathfrak{t}' = \{0\}$.

Let us show that the latter is impossible. Indeed, using II.5.6 (Theorem 24), II.6.4 (Theorem 28) and their notations (replacing, of course, $\mathfrak{g}$ by $\mathfrak{k}$)) we have $\text{Int}_{\mathfrak{k}}(\mathfrak{r}_1) \cong SU(2)$, but then also $\text{Int}_{\mathfrak{g}}(\mathfrak{r}_1) \cong SU(2)$, $\text{Int}_{\mathfrak{k}}(\mathfrak{b}) \cong SO(3)$ (where $\mathfrak{b}$ is a diagonal in $\mathfrak{p} \oplus \mathfrak{q}$) and $S_3 \in \text{Int}_{\mathfrak{g}}(\mathfrak{b}) \cong SU(2)$.

Now we consider the one-dimensional subalgebra $\mathfrak{b} \cap \mathfrak{r}_1$. Then the isomorphism $\text{Int}_{\mathfrak{g}}(\mathfrak{r}_1) \cong \text{Int}_{\mathfrak{k}}(\mathfrak{r}_1)$ implies the isomorphism $\text{Int}_{\mathfrak{g}}(\mathfrak{b} \cap \mathfrak{r}_1) \cong \text{Int}_{\mathfrak{k}}(\mathfrak{r}_1 \cap \mathfrak{b})$. On the other hand, because of the strict morphism $\text{Int}_{\mathfrak{g}}(\mathfrak{b}) \longrightarrow \text{Int}_{\mathfrak{k}}(\mathfrak{b})$ and that $S_3 \in \text{Int}_{\mathfrak{g}}(\mathfrak{b} \cap \mathfrak{r}_1)$ we have the strict morphism $\text{Int}_{\mathfrak{g}}(\mathfrak{b} \cap \mathfrak{r}_1) \longrightarrow \text{Int}_{\mathfrak{k}}(\mathfrak{b} \cap \mathfrak{r}_1)$, which is a contradiction.

Thus we should consider that

$$\mathfrak{k}/\mathfrak{p} \oplus \mathfrak{q} \oplus \mathfrak{t}' \cong so(m)/(\,so(4) \oplus so(m-4)\,), \quad m > 4,$$

with the natural embedding.

In addition, the diagonal $\mathfrak{b}$ in $\mathfrak{p} \oplus \mathfrak{q}$ is isomorphic to $so(3)$ with the natural embedding into $so(3)$, which determines the involutive algebra $\mathfrak{b} \oplus \mathfrak{n} \cong so(3) \oplus so(m-3)$ with the natural embedding into $\mathfrak{k} \cong so(m)$. But $S_3 \in \text{Int}_{\mathfrak{g}}(\mathfrak{b}) \cong SU(2)$, therefore (see (I.1.17) (Definition 8)) $\mathfrak{k}$ is a unitary special simple subalgebra of the involutive automorphism S_3. Since all subalgebras $so(2)$ with the natural embedding into $su(2)$ are conjugated in $\text{Int}(\mathfrak{k})$ they are conjugated in $\text{Int}_{\mathfrak{g}}(\mathfrak{k})$ as well.

But $S_3 \in \text{Int}_{\mathfrak{g}}(\mathfrak{a})$ if $\mathfrak{a} \cong so(2) \subset so(3) \cong \mathfrak{b}$ with the natural embedding, consequently for any $\mathfrak{c} \cong so(2) \subset so(m) \cong \mathfrak{k}$ with the natural embedding we have $S_3 \in \text{Int}_{\mathfrak{g}}(\mathfrak{c})$.

Taking, furthermore, $\mathfrak{c} \cong so(2) \subset so(m) \dot{-} (\,so(4) \oplus so(m-4)\,)$ with the natural embedding into $so(m)$ (this is obviously possible) and representing $\mathfrak{c}$ as $\{t\,\zeta\}$ we have

$$\zeta \in \mathfrak{k} \dot{-} (\mathfrak{p} \oplus \mathfrak{q} \oplus \mathfrak{t}') = \mathfrak{l}_3 \dot{-} \mathfrak{l}_0, \qquad S_3 \in \text{Int}_{\mathfrak{g}}(\{t\,\zeta\}).$$

Choosing $\varphi \in \text{Int}_{\mathfrak{g}}(\{t\,\zeta\})$ such that $\varphi^2 = S_3$ we obtain the conjugating automorphism of the involutive algebras $\mathfrak{l}_1$ and $\mathfrak{l}_2$. And $\mathfrak{g} = \mathfrak{l}_1 + \mathfrak{l}_2 + \mathfrak{l}_3$ is an iso-involutive sum. ∎

II.7.6. Definition 30. Let $\mathfrak{g}$ be a simple compact Lie algebra, $\mathfrak{l} = \mathfrak{p} \oplus \tilde{\mathfrak{l}}$ being its involutive algebra of a principal non-central mono-unitary automorphism $S \in \text{Int}_{\mathfrak{g}}(\mathfrak{p})$, where S is not of index 1. We call such an involutive automorphism *exceptional principal* (respectively, we speak of an exceptional principal involutive algebra and exceptional involutive pair)

Moreover, we say in this case that $\mathfrak{g}$ is of type:

(1) f_4 if $\tilde{\mathfrak{l}}$ has a principal mono-unitary non-central involutive automorphism of index 1;

(2) e_6 if $\tilde{\mathfrak{l}}$ has a principal central unitary involutive automorphism;

(3) e_7 if $\tilde{\mathfrak{l}}$ has a principal non-central di-unitary involutive automorphism;

(4) e_8 if $\tilde{\mathfrak{l}}$ has only an exceptional principal involutive automorphism.

Since we have used all possible hypotheses about $\tilde{\mathfrak{l}}$ there are no other possibilities.

CHAPTER II.8

EXCEPTIONAL PRINCIPAL INVOLUTIVE
AUTOMORPHISM OF TYPES f AND e

II.8.1. Theorem 35. *Let $\mathfrak{g}$ be a Lie algebra of type f_4, then there exists an iso-involutive decomposition*

$$\mathfrak{g} = \mathfrak{l}_1 + \mathfrak{l}_2 + \mathfrak{l}_3 \qquad (\mathfrak{l}_\alpha \cap \mathfrak{l}_\beta = \mathfrak{l}_0, \quad \alpha \neq \beta)$$

basis for an exceptional involutive algebra $\mathfrak{l}_1$, where

$$\mathfrak{l}_1/\mathfrak{l}_0 \cong sp(3)/(\,sp(2) \oplus sp(1)\,) + sp(1)/sp(1),$$

$$\mathfrak{l}_2/\mathfrak{l}_0 \cong sp(3)/(\,sp(2) \oplus sp(1)\,) + sp(1)/sp(1),$$

$$\mathfrak{l}_3/\mathfrak{l}_0 \cong so(9)/(\,so(4) \oplus so(5)\,)$$

with the natural embeddings, and $\dim \mathfrak{g} = 52$.

Proof. By II.7.6 (Definition 30) $\mathfrak{g}$ has the exceptional principal involutive algebra $\mathfrak{l}_1 = \mathfrak{p} \oplus \tilde{\mathfrak{l}}$ of an involutive automorphism $S_1 \in \mathrm{Int}\,_\mathfrak{g}(\mathfrak{p})$, whereas $\tilde{\mathfrak{l}}$ has a principal mono-unitary non-central involutive algebra $\mathfrak{q} \oplus \mathfrak{t}$ of index 1.

By II.7.3 (Theorem 33) $\tilde{\mathfrak{l}}$ is simple compact, therefore

$$\tilde{\mathfrak{l}}/\mathfrak{q} \oplus \mathfrak{t} \cong sp(m)/(\,sp(1) \oplus sp(m-1)\,), \quad m > 2,$$

with the natural embedding, according to I.8.4 (Theorem 33).

Thus $\mathfrak{t} \cong sp(m-1)$, that is, $\mathfrak{t}$ is simple.

Using the involutive automorphisms

$$S_1 \in \mathrm{Int}\,_\mathfrak{g}(\mathfrak{p}), \qquad S_2 \in \mathrm{Int}\,_\mathfrak{g}(\mathfrak{q}), \qquad S_3 = S_1 S_2 = S_2 S_1,$$

we may construct the corresponding iso-involutive sum $\mathfrak{g} = \mathfrak{l}_1 + \mathfrak{l}_2 + \mathfrak{l}_3$ basis for $\mathfrak{l}_1$, see II.7.4 (Definition 29).

By II.7.5 (Theorem 34) we have, furthermore,

$$\mathfrak{l}_3/\mathfrak{l}_0 = \mathfrak{k}/\mathfrak{l}_0^{(1)} + \mathfrak{l}_0^{(2)}/\mathfrak{l}_0^{(2)},$$

where $\mathfrak{k}$ is a simple special unitary subalgebra of the involutive automorphism S_3, and evidently $\mathfrak{l}_0^{(1)} = \mathfrak{p} \oplus \mathfrak{q} \oplus \mathfrak{t}'$, $\mathfrak{l}_0^{(2)} = \mathfrak{t}''$, $\mathfrak{t} = \mathfrak{t}' \oplus \mathfrak{t}''$.

Since t is simple we have either $t' = \{0\}$ or $t'' = \{0\}$. By II.4.7 (Theorem 15) and II.5.9 (Theorem 26) we obtain

$$\mathfrak{k}/\mathfrak{p} \oplus \mathfrak{q} \oplus t' \cong so(q)/(\,so(4) \oplus so(q-4)\,), \quad q > 4,$$

with the natural embedding.

If $t' = \{0\}$ then $q = 5$ and the involutive pair

$$\mathfrak{k}/\mathfrak{p} \oplus \mathfrak{q} \oplus t' \cong so(5)/so(4)$$

is of index 1. However, $\mathfrak{k}$ is a special unitary subalgebra, whence $\mathfrak{g}/\mathfrak{l}_1$ is of index 1, which contradicts the conditions of the theorem.

Thus we should assume $t' \neq \{0\}$, that is, $t' = t$, $t'' = \{0\}$. Hence $t \cong sp(m-1) \cong so(q-4)$, $m > 2$, which is possible only for $t \cong sp(2) \cong so(5)$. But then

$$\mathfrak{l}_3/\mathfrak{l}_0 = \mathfrak{k}/\mathfrak{p} \oplus \mathfrak{q} \oplus t \cong so(9)/(\,so(4) \oplus so(5)\,),$$

$$\mathfrak{l}_1/\mathfrak{l}_0 = \mathfrak{l}_1/\mathfrak{p} \oplus \mathfrak{q} \oplus t = \tilde{\mathfrak{l}}/\mathfrak{q} \oplus t + \mathfrak{p}/\mathfrak{p}$$

$$\cong sp(3)/(\,sp(2) \oplus sp(1)\,) + sp(1)/sp(1),$$

with the natural embeddings.

Owing to the conjugacy of $\mathfrak{l}_1$ and $\mathfrak{l}_2$ in $\mathrm{Int}_\mathfrak{g}(\mathfrak{l}_3)$ we have also

$$\mathfrak{l}_2/\mathfrak{l}_0 \cong sp(3)/(\,sp(2) \oplus sp(1)\,) + sp(1)/sp(1). \qquad \blacksquare$$

II.8.2. **Theorem 36.** *Let $\mathfrak{g}$ be a Lie algebra of type e_6, then there is an iso-involutive decomposition $\mathfrak{g} = \mathfrak{l}_1 + \mathfrak{l}_2 + \mathfrak{l}_3$ ($\mathfrak{l}_\alpha \cap \mathfrak{l}_\beta = \mathfrak{l}_0$, $\alpha \neq \beta$) basis for an exceptional involutive algebra $\mathfrak{l}_1$, where*

$$\mathfrak{l}_1/\mathfrak{l}_0 \cong su(6)/(\,su(4) \oplus su(2) \oplus u(1)\,) + su(2)/su(2),$$

$$\mathfrak{l}_2/\mathfrak{l}_0 \cong su(6)/(\,su(4) \oplus su(2) \oplus u(1)\,) + su(2)/su(2),$$

$$\mathfrak{l}_3/\mathfrak{l}_0 \cong so(10)/(\,so(4) \oplus so(6)\,) + u(1)/u(1)$$

with the natural embedding, and $\dim \mathfrak{g} = 78$.

Proof. By II.7.6 Definition 30 $\mathfrak{g}$ has the exceptional principal involutive algebra $\mathfrak{l}_1 = \mathfrak{p} \oplus \tilde{\mathfrak{l}}$ of an involutive automorphism $S_1 \in \mathrm{Int}_\mathfrak{g}(\mathfrak{p})$, whereas $\tilde{\mathfrak{l}}$ has a principal central unitary involutive algebra $\mathfrak{q} \oplus t$ ($t = \mathfrak{w} \oplus \mathfrak{z}$, where $\mathfrak{z}$ is the one-dimensional centre in $\mathfrak{w}$ and $\mathfrak{w}$ is either simple and non-one-dimensional or $\mathfrak{w} = \{0\}$).

Indeed, by II.7.3 Theorem 33 $\tilde{\mathfrak{l}}$ is simple compact, consequently

$$\tilde{\mathfrak{l}}/\mathfrak{q} \oplus t \cong su(m)/(su(2) \oplus su(m-2) \oplus u(1))$$

with the natural embedding, according to I.7.2. (Theorem 29).

Thus $\mathfrak{w} \cong su(m-2)$.

Using, furthermore, the involutive automorphisms

$$S_1 \in \operatorname{Int}_{\mathfrak{g}}(\mathfrak{p}), \qquad S_2 \in \operatorname{Int}_{\mathfrak{g}}(\mathfrak{q}), \qquad S_3 = S_1 S_2 = S_2 S_1,$$

we construct the corresponding iso-involutive sum $\mathfrak{g} = \mathfrak{l}_1 + \mathfrak{l}_2 + \mathfrak{l}_3$ basis for the involutive algebra $\mathfrak{l}_1$, see II.7.4 (Definition 29).

By II.7.5 (Theorem 34) we then have

$$\mathfrak{l}_3/\mathfrak{l}_0 = \mathfrak{k}/\mathfrak{l}_0^{(1)} + \mathfrak{l}_0^{(2)}/\mathfrak{l}_0^{(2)},$$

where $\mathfrak{k}$ is a simple special unitary subalgebra of the involutive automorphism S_3, and evidently

$$\mathfrak{l}_0^{(1)} = \mathfrak{p} \oplus \mathfrak{q} \oplus \mathfrak{t}', \qquad \mathfrak{l}_0^{(2)} = \mathfrak{t}'', \qquad \mathfrak{t} = \mathfrak{t}' \oplus \mathfrak{t}''.$$

Since $\mathfrak{t} = \mathfrak{w} \oplus \mathfrak{z}$ and $\mathfrak{w}$ is either simple non-one-dimensional or trivial then either $\mathfrak{t}' = \mathfrak{z}$ or $\mathfrak{t}' = \mathfrak{w}$ or else $\mathfrak{t}' = \{0\}$. From II.4.7 (Theorem 15), II.5.9 (Theorem 26) there follows

$$\mathfrak{k}/\mathfrak{p} \oplus \mathfrak{q} \oplus \mathfrak{t}' \cong so(q)/(so(4) \oplus so(q-4)), \qquad q > 4,$$

with the natural embedding.

If $\mathfrak{t}' = \{0\}$ then $q = 5$ and the involutive pair

$$\mathfrak{k}/\mathfrak{p} \oplus \mathfrak{q} \cong so(5)/so(4)$$

is of index 1.

But $\mathfrak{k}$ is a special unitary subalgebra, which implies that $\mathfrak{g}/\mathfrak{l}_1$ is of index 1, contradicting the conditions of the theorem.

Consequently we should assume that $\mathfrak{t}' \neq \{0\}$.

If $\mathfrak{t}' = \mathfrak{z}$ then $\mathfrak{t}'' = \mathfrak{w} \cong su(m-2)$. In addition $\mathfrak{l}_3 = \mathfrak{w} \oplus \mathfrak{k}$, $\mathfrak{k} \cong so(6)$ is the special unitary subalgebra of involutive automorphism S_3. Then by II.7.2 (Theorem 32) $\mathfrak{l}_3$ is a central involutive algebra, which is impossible.

Thus $\mathfrak{t}' = \mathfrak{w}$, and $\mathfrak{w} \cong su(m-2) \cong so(q-4)$, which gives the two possibilities that either $\mathfrak{w} \cong su(2) \cong so(3)$ or $\mathfrak{w} \cong su(4) \cong so(6)$.

Let us consider the first possibility, then

$$\tilde{\mathfrak{l}}/\mathfrak{q} \oplus \mathfrak{w} \oplus \mathfrak{z} \cong su(4)/(su(2) \oplus su(2) \oplus u(1))$$

$$\cong so(6)/(so(4) \oplus so(2)),$$

and consequently $\operatorname{Int}_{\tilde{\mathfrak{l}}}(\mathfrak{w}) \cong SU(2)$. But then also $\operatorname{Int}_{\mathfrak{g}}(\mathfrak{w}) \cong SU(2)$.

Let us take the non-trivial involutive automorphism $\tilde{S} \in \operatorname{Int}_{\mathfrak{g}}(\mathfrak{w})$. Since

$$\mathfrak{k}/(\mathfrak{p} \oplus \mathfrak{q}) \oplus \mathfrak{w} \cong so(7)/(so(4) \oplus so(3))$$

with the natural embedding then the restriction of $\tilde{S} \neq \operatorname{Id}$ to $\mathfrak{l}_3$ is the identity automorphism, and owing to the maximality of the involutive algebra $\mathfrak{l}_3$ of the simple algebra $\mathfrak{g}$ we have $\widetilde{S} = S_3$. But then $S_1 \in \mathrm{Int}_\mathfrak{g}(\mathfrak{b})$, where $\mathfrak{b}$ is a diagonal of $\mathfrak{q} \oplus \mathfrak{w}$, that is, $\mathfrak{b} \cong so(3)$ with the natural embedding in $\mathfrak{p} \oplus \mathfrak{q} \cong so(4) \subset so(6) \cong \tilde{\mathfrak{l}}$. Therefore $\tilde{\mathfrak{l}} \cong so(6)$ is a special unitary subalgebra of the involutive automorphism S_1. By II.7.2 (Theorem 32) $\mathfrak{l}_1 = \tilde{\mathfrak{l}} \oplus \mathfrak{p}$ is then a central involutive algebra, which is impossible.

Thus there remains only the second possibility $\mathfrak{w} \cong su(4) \cong so(6)$. Then evidently

$$\mathfrak{l}_1/\mathfrak{l}_0 \cong su(6)/(\, su(4) \oplus su(2) \oplus u(1)\,) + su(2)/su(2)$$

with the natural embedding, and similarly

$$\mathfrak{l}_2/\mathfrak{l}_0 \cong su(6)/(\, su(4) \oplus su(2) \oplus u(1)\,) + su(2)/su(2)$$

with the natural embedding.

Also we have

$$\mathfrak{l}_3/\mathfrak{l}_0 \cong so(10)/(\, so(4) \oplus so(6)\,) + u(1)/u(1).$$

This proves the theorem.   ∎

II.8.3. Theorem 37. *Let $\mathfrak{g}$ be a Lie algebra of type e_7. Then there exists an iso-involutive decomposition*

$$\mathfrak{g} = \mathfrak{l}_1 + \mathfrak{l}_2 + \mathfrak{l}_3 \qquad (\, \mathfrak{l}_\alpha \cap \mathfrak{l}_\beta = \mathfrak{l}_0, \quad \alpha \neq \beta\,)$$

basis for an exceptional involutive algebra $\mathfrak{l}_1$, where

$$\mathfrak{l}_\alpha/\mathfrak{l}_0 \cong so(12)/(\, so(4) \oplus so(8)\,) + su(2)/su(2) \quad (\,\alpha = 1,2,3\,)$$

with the natural embedding, and $\dim \mathfrak{g} = 133$.

Proof. By II.7.6 (Definition 30) $\mathfrak{g}$ has the exceptional involutive algebra $\mathfrak{l}_1 = \mathfrak{p} \oplus \tilde{\mathfrak{l}}_1$ of an involutive automorphism $S_1 \in \mathrm{Int}_\mathfrak{g}(\mathfrak{p})$; moreover, $\tilde{\mathfrak{l}}_1$ has a principal di-unitary non-central involutive algebra $\mathfrak{q} \oplus \mathfrak{f} \oplus \mathfrak{t}$.

By II.7.3 (Theorem 33) $\tilde{\mathfrak{l}}_1$ is simple compact, therefore either

$$\tilde{\mathfrak{l}}_1/\mathfrak{q} \oplus \mathfrak{f} \oplus \mathfrak{t} \cong so(m)/(\, so(4) \oplus so(m-4)\,), \quad 4 < m \neq 6,$$

or

$$\mathfrak{t} = \{0\}, \qquad \tilde{\mathfrak{l}}_1/\mathfrak{q} \oplus \mathfrak{f} \oplus \mathfrak{t} \cong g_2/so(4),$$

with the natural embedding, according to II.5 (Theorems 23–27), II.6 (Theorems 28–29).

Using the involutive automorphisms

$$S_1 \in \mathrm{Int}_\mathfrak{g}(\mathfrak{p}), \qquad S_2 \in \mathrm{Int}_\mathfrak{g}(\mathfrak{q}), \qquad S_3 = S_1 S_2 = S_2 S_1,$$

we construct the involutive sum $\mathfrak{g} = \mathfrak{l}_1 + \mathfrak{l}_2 + \mathfrak{l}_3$ basis for $\mathfrak{l}_1$, see II.7.4 Definition 29.

By II.7.5 (Theorem 34) we obtain

$$\mathfrak{l}_3/\mathfrak{l}_0 = \mathfrak{k}/\mathfrak{l}_0^{(1)} + \mathfrak{l}_0^{(2)}/\mathfrak{l}_0^{(2)},$$

where $\mathfrak{k}$ is the simple special unitary subalgebra of the involutive automorphism S_3. Evidently

$$\mathfrak{l}_0^{(1)} = \mathfrak{p} \oplus \mathfrak{q} \oplus \mathfrak{m}', \qquad \mathfrak{l}_0^{(2)} = \mathfrak{m}'', \qquad \mathfrak{f} \oplus \mathfrak{t} = \mathfrak{m}' \oplus \mathfrak{m}''.$$

By the theorems of II.5, II.6 we have, furthermore,

$$\mathfrak{k}/\mathfrak{p} \oplus \mathfrak{q} \oplus \mathfrak{m}' \cong so(q)/(\,so(4) \oplus so(q-4)\,), \quad q > 4,$$

with the natural embedding.

If $\mathfrak{m}' = \{0\}$ then $q = 5$, and the involutive pair

$$\mathfrak{k}/\mathfrak{p} \oplus \mathfrak{q} \oplus \mathfrak{m}' \cong so(5)/so(4)$$

is of index 1. However, $\mathfrak{k}$ is a unitary special subalgebra, this implies that $\mathfrak{g}/\mathfrak{l}_1$ is of index 1 as well, which contradicts the conditions of theorem. Thus $\mathfrak{m}' \neq \{0\}$.

Let us show first that $S_3 \in \mathrm{Int}_\mathfrak{g}(\mathfrak{f})$. Since $\mathrm{Int}_{\tilde{\mathfrak{l}}_1}(\mathfrak{g}) \cong SU(2)$ we have also $\mathrm{Int}_\mathfrak{g}(\mathfrak{f}) \cong SU(2)$. This gives us a non-identity involutive automorphism $\tilde{S} \in \mathrm{Int}_\mathfrak{g}(\mathfrak{f})$. But the restrictions of $\tilde{S}$ and S_2 to $\mathfrak{l}_1$ coincide, whence it follows that the restriction of $\tilde{S} S_2 = S_2 \tilde{S}$ to $\mathfrak{l}_1$ is the identity involutive automorphism.

Thus either $\tilde{S} S_2 = S_1$ or $\tilde{S} S_2 = \mathrm{Id}$.

Let us consider the conjugating automorphism φ of the basis iso-involutive sum. By the construction $\varphi S_2 \varphi^{-1} = S_1$, $\varphi \mathfrak{l}_0 = \mathfrak{l}_0$. Therefore, if $S_2 \in \mathrm{Int}_\mathfrak{g}(\mathfrak{f})$ then $S_1 \in \mathrm{Int}_\mathfrak{g}(\varphi \mathfrak{f})$. In addition, $\varphi \mathfrak{q} = \mathfrak{p}$ by the construction, thus $\varphi \mathfrak{f} \neq \mathfrak{p}$.

Finally, since $\mathfrak{f}$ is an ideal of $\mathfrak{l}_0$ we have that $\varphi \mathfrak{f}$ is an ideal of $\mathfrak{l}_0$ as well, and $\varphi \mathfrak{f} \subset \mathfrak{t} \neq \{0\}$. But then

$$\tilde{\mathfrak{l}}_1/\mathfrak{q} \oplus \mathfrak{f} \oplus \mathfrak{t} \cong so(m)/(\,so(4) \oplus so(m-4)\,),$$

and either $\mathfrak{t} = \varphi \mathfrak{f} \cong so(3)$, or $\mathfrak{t} = (\varphi \mathfrak{f}) \oplus \mathfrak{m}_3 \cong so(4)$.

But the latter is impossible since $\tilde{\mathfrak{l}}_1/(\varphi \mathfrak{f}) \oplus \mathfrak{q} \oplus \mathfrak{f} \oplus \mathfrak{m}_3$ is a principal ortogonal pair, which is possible only when

$$\mathfrak{l}_1/(\varphi \mathfrak{f}) \oplus \mathfrak{q} \oplus \mathfrak{f} \oplus \mathfrak{m} \cong so(7)/(\,so(3) \oplus so(4)\,)$$

with the natural embedding. But then $\mathfrak{q} \oplus \mathfrak{f} \oplus \mathfrak{m} \cong so(4)$, which is wrong.

Thus we should assume $\mathfrak{t} = \varphi \mathfrak{f}$. But then

$$\tilde{\mathfrak{l}}_1/(\varphi \mathfrak{f}) \oplus \mathfrak{q} \oplus \mathfrak{f} \cong so(7)/(\,so(3) \oplus so(4)\,)$$

with the natural embedding.

Owing to the conjugacy of all subalgebras $so(3)$ with the natural embeddings into $so(7)$, a diagonal $\mathfrak{b}$ of $\mathfrak{q} \oplus \mathfrak{f}$ is conjugated with $(\varphi \mathfrak{f})$.

But then $S_1 \in \mathrm{Int}_\mathfrak{g}(\mathfrak{b})$ since $S_1 \in \mathrm{Int}_\mathfrak{g}(\varphi\mathfrak{f})$ and S_1 commutes with $\mathrm{Int}_\mathfrak{g}(\tilde{\mathfrak{l}}_1)$. Thus $\mathrm{Int}_\mathfrak{g}(\mathfrak{b}) \cong SU(2)$. On the other hand, $S_2 \in \mathrm{Int}_\mathfrak{g}(\mathfrak{q})$, $S_2 \in \mathrm{Int}_\mathfrak{g}(\mathfrak{f})$ imply $\mathrm{Int}_\mathfrak{g}(\mathfrak{b}) \cong SO(3)$, which is impossible. Consequently $S_3 \in \mathrm{Int}_\mathfrak{g}(\mathfrak{f})$.

Now, substituting $\mathfrak{f}$ instead of $\mathfrak{q}$, and S_3 instead of S_2, respectively, in the foregoing consideration we see that $\mathfrak{g} = \mathfrak{l}_1 + \mathfrak{l}_2 + \mathfrak{l}_3$ is an iso-involutive sum basis for $\mathfrak{l}_1$ with the conjugating automorphism $\psi \in \mathrm{Int}_\mathfrak{g}(\{t\eta\})$, $\eta \in \mathfrak{l}_2 \dot{-} \mathfrak{l}_0$. Since $\mathfrak{l}_2 = \tilde{\mathfrak{l}}_2 \oplus \mathfrak{q}$ we obtain by II.7.5 (Theorem 34)

$$\tilde{\mathfrak{l}}_2/\mathfrak{p} \oplus \mathfrak{f} \oplus \mathfrak{t} \cong so(q)/(so(4) \oplus so(q-4)) \,.$$

Owing to the conjugacy of $\tilde{\mathfrak{l}}_1$ and $\tilde{\mathfrak{l}}_2$ we have

$$\tilde{\mathfrak{l}}_1/\mathfrak{q} \oplus \mathfrak{f} \oplus \mathfrak{t} \cong so(q)/(so(4) \oplus so(q-4)),$$

and owing to the conjugacy of $\tilde{\mathfrak{l}}_1$ and $\tilde{\mathfrak{l}}_3 = \mathfrak{k}$ we have also

$$\mathfrak{l}_3 = \mathfrak{k} \oplus \mathfrak{f}, \qquad \mathfrak{k}/\mathfrak{p} \oplus \mathfrak{q} \oplus \mathfrak{t} \cong so(q)/(so(4) \oplus so(q-4))$$

with the natural embeddings. In addition, we know that $q > 5$. But $q \neq 6$, otherwise $\mathfrak{t} \cong so(2)$, and $\tilde{\mathfrak{l}}_1/\mathfrak{q} \oplus \mathfrak{f} \oplus \mathfrak{t}$ is then a principal di-unitary central involutive pair, which is excluded by the conditions of the theorem.

Thus $\mathfrak{t} \cong so(q-4)$, $q > 6$.

The diagonal $\mathfrak{b}$ in $\mathfrak{q} \oplus \mathfrak{f}$ is isomorphic to $so(3)$ with the natural embeddings into $so(q) \cong \tilde{\mathfrak{l}}_1$ and $S_1 \in \mathrm{Int}_\mathfrak{g}(\mathfrak{b})$. Owing to the conjugacy of all subalgebras $so(3)$ with the natural embedding into $\tilde{\mathfrak{l}}_1 \cong so(q)$ $(q > 6)$ there exists

$$\mathfrak{a} \cong so(3) \subset so(q-4) \cong \mathfrak{t} \quad (q > 6)$$

with the natural embedding.

But then $S_1 \in \mathrm{Int}_\mathfrak{g}(\mathfrak{a})$, and, moreover, there exists $\mathfrak{c}$ such that

$$\mathfrak{t}/\mathfrak{a} \oplus \mathfrak{c} \cong so(q-4)/(so(3) \oplus so(q-7))$$

is a principal orthogonal involutive pair.

Thus, furthermore, for

$$\mathfrak{k}/\mathfrak{p} \oplus \mathfrak{q} \oplus \mathfrak{t} \cong so(q)/(so(4) \oplus so(q-4))$$

we obtain that $\mathfrak{t} \cong so(q-4)$ is a special unitary subalgebra of the involutive automorphism $\tilde{S}_1$, ($\tilde{S}_1$ being the restriction of S_1 to $\mathfrak{k}$).

By II.7.1 (Theorem 31) we conclude that either $q = 8$ or $q = 12$.

But $q = 8$ implies $\mathfrak{t}/\mathfrak{a} \cong so(4)/so(3)$, and then $su(2) \oplus su(2) \cong \mathfrak{t} = \mathfrak{t}' \oplus \mathfrak{t}''$, $\mathfrak{a}$ being a diagonal in $\mathfrak{t}' \oplus \mathfrak{t}''$.

But then $\tilde{S}_1 \in \mathrm{Int}_\mathfrak{t}(\mathfrak{t}')$, $\tilde{S}_1 \in \mathrm{Int}_\mathfrak{t}(\mathfrak{t}'')$, whence $\mathrm{Int}_\mathfrak{t}(\mathfrak{a}) \cong SO(3)$. On the other hand, $\tilde{S}_1 \in \mathrm{Int}_\mathfrak{t}(\mathfrak{a})$. This contradiction shows that $q \neq 8$.

Thus $q = 12$, which proves the theorem. ∎

II.8.4. Theorem 38. *Let* $\mathfrak{g}$ *be a Lie algebra of type* e_8. *Then there exists an iso-involutive decomposition*

$$\mathfrak{g} = \mathfrak{l}_1 + \mathfrak{l}_2 + \mathfrak{l}_3 \quad (\mathfrak{l}_\alpha \cap \mathfrak{l}_\beta = \mathfrak{l}_0, \quad \alpha \neq \beta)$$

basis for an exceptional involutive algebra $\mathfrak{l}_1$, *where*

$$\mathfrak{l}_1/\mathfrak{l}_0 \cong e_7/(\,so(12) \oplus so(3)\,) + su(2)/su(2),$$

$$\mathfrak{l}_2/\mathfrak{l}_0 \cong e_7/(\,so(12) \oplus so(3)\,) + su(2)/su(2),$$

$$\mathfrak{l}_3/\mathfrak{l}_0 \cong so(16)/(\,so(4) \oplus so(12)\,)$$

with the natural embeddings, and $\dim \mathfrak{g} = 248$.

Proof. By II.7.6 Definition 30 $\mathfrak{g}$ has the exceptional principal involutive algebra $\mathfrak{l}_1 = \mathfrak{p} \oplus \tilde{\mathfrak{l}}$ of an involutive automorphism $S_1 \in \mathrm{Int}_\mathfrak{g}(\mathfrak{p})$; moreover, $\tilde{\mathfrak{l}}$ has a principal exceptional involutive algebra $\mathfrak{q} \oplus \mathfrak{t}$.

By II.7.3 (Theorem 33) $\tilde{\mathfrak{l}}$ is simple and compact and also $\mathfrak{t}$ is simple, $\dim \mathfrak{t} > 3$. Using the involutive automorphisms

$$S_1 \in \mathrm{Int}_\mathfrak{g}(\mathfrak{p}), \qquad S_2 \in \mathrm{Int}_\mathfrak{g}(\mathfrak{q}), \qquad S_3 = S_1 S_2 = S_2 S_1,$$

we construct the iso-involutive sum $\mathfrak{g} = \mathfrak{l}_1 + \mathfrak{l}_2 + \mathfrak{l}_3$ basis for the involutive algebra $\mathfrak{l}_1$, see II.7.4 (Definition 29). By II.7.5 (Theorem 34) we have, furthermore,

$$\mathfrak{l}_3/\mathfrak{l}_0 = \mathfrak{k}/\mathfrak{l}_0^{(1)} + \mathfrak{l}_0^{(2)}/\mathfrak{l}_0^{(2)},$$

where $\mathfrak{k}$ is a simple special unitary subalgebra of the involutive automorphism S_3. Evidently

$$\mathfrak{l}_0^{(1)} = \mathfrak{p} \oplus \mathfrak{q} \oplus \mathfrak{t}', \qquad \mathfrak{l}_0^{(2)} = \mathfrak{t}'', \qquad \mathfrak{t} = \mathfrak{t}' \oplus \mathfrak{t}'',$$

thus either $\mathfrak{t}' = \{0\}$ or $\mathfrak{t}'' = \{0\}$.

By the results of II.5, II.6 we have

$$\mathfrak{k}/\mathfrak{p} \oplus \mathfrak{q} \oplus \mathfrak{t}' \cong so(q)/(\,so(4) \oplus so(q-4)\,), \quad q > 4,$$

with the natural embedding.

If $\mathfrak{t}' = \{0\}$ then $q = 5$, and the involutive pair

$$\mathfrak{k}/\mathfrak{p} \oplus \mathfrak{q} \oplus \mathfrak{t}' \cong so(5)/so(4)$$

is of index 1. But $\mathfrak{k}$ is a special unitary subalgebra, whence $\mathfrak{g}/\mathfrak{l}_1$ is of index 1, contradicting the conditions of the theorem.

Thus the only case is $\mathfrak{t}' \neq \{0\}$, $\mathfrak{t}'' = \{0\}$, that is, $\mathfrak{t}' = \mathfrak{t}$, $\mathfrak{l}_3 = \mathfrak{k}$. But then $\mathfrak{t} \cong so(q-4)$ and, by the simplicity of $\mathfrak{t}$ and $\dim \mathfrak{t} > 3$, we obtain $q > 8$.

If $q = 10$ then the involutive pair $\tilde{\mathfrak{l}}/\mathfrak{q} \oplus \mathfrak{t}$ is of type e_6, and then $\mathfrak{t} \cong so(6)$.

But on the other hand, by II.8.2 (Theorem 36) $t \cong su(6)$, which is a contradiction.

If $q \neq 10$ then the involutive pair $\tilde{l}/q \oplus t$ is of type e_7, whence, by II.8.3 (Theorem 37),

$$\tilde{l}/q \oplus t \cong e_7/(su(2) \oplus so(12)).$$

Consequently $q = 16$ and

$$l_3/\mathfrak{p} \oplus q \oplus t \cong so(16)/(so(4) \oplus so(12)).$$

This proves our theorem. ∎

The following result is true.

II.8.5. Theorem 39. *The simple compact non-one-dimensional Lie algebras are isomorphic to*

$$so(n) \quad (n \neq 4, 2, 1), \qquad g_2, \qquad su(n) \quad (n > 1),$$

$$sp(n) \quad (n > 0), \qquad f_4, \qquad e_6, \qquad e_7, \qquad e_8.$$

All these algebras are pair-wise non-isomorphic and uniquely determined by the type of a principal unitary involutive automorphism, except for the cases

$$so(3) \cong su(2) \cong sp(1), \qquad so(5) \cong sp(2), \qquad so(6) \cong su(4).$$

Outline of the proof. It is sufficient to compare the dimensions of different types:

$$\dim so(n) = \frac{1}{2}n(n-1), \qquad \dim g_2 = 14,$$

$$\dim su(n) = n^2 - 1, \qquad \dim sp(n) = n(2n+1),$$

$$\dim f_4 = 52, \qquad \dim e_6 = 78,$$

$$\dim e_7 = 133, \qquad \dim e_8 = 248,$$

and to compare their ranks (the dimensions of maximal commutative subalgebras). Then the only thing is to be proved that $so(13)$ is non-isomorphic to e_6 and that $sp(6)$ is non-isomorphic to e_6.

These results are well known and we leave them without proof here. ∎

II.8.6. Remark. It is of value to prove in an elementary way that $so(13) \not\cong e_6$, $sp(6) \not\cong e_6$.

CHAPTER II.9

CLASSIFICATION OF SIMPLE
SPECIAL UNITARY SUBALGEBRAS

We are interested in unitary special subalgebras $\mathfrak{k}$ (dim $\mathfrak{k} > 3$) in a simple compact Lie algebra $\mathfrak{g}$ (the case dim $\mathfrak{k} = 3$ gives us principal unitary involutive automorphisms which have already been considered).

First we generalize II.7.5 (Theorem 34).

II.9.1. Theorem 40. *Let $\mathfrak{g}$ be a simple compact Lie algebra, $\mathfrak{l}_1 = \mathfrak{p} \oplus \tilde{\mathfrak{l}}$ be its principal unitary involutive algebra of an involutive automorphism $S_1 \in \mathrm{Int}_\mathfrak{g}(\mathfrak{p})$, and let $\tilde{\mathfrak{l}}$ have a simple ideal $\mathfrak{a}$ such that $\dim \mathfrak{a} > 3$.*

Then the basis for $\mathfrak{l}_1$ involutive sum $\mathfrak{g} = \mathfrak{l}_1 + \mathfrak{l}_2 + \mathfrak{l}_3$ of the involutive automorphisms $S_1, S_2, S_3 = S_1 S_2 = S_2 S_1$ exists, being iso-involutive with the conjugating automorphism $\varphi \in \mathrm{Int}_\mathfrak{g}(\{t\zeta\})$, $\zeta \in \mathfrak{l}_3 \dot{-} \mathfrak{l}_0$ ($\mathfrak{l}_0 = \mathfrak{l}_\alpha \cap \mathfrak{l}_\beta$, $\alpha \neq \beta$).

Furthermore,

$$\mathfrak{l}_3/\mathfrak{l}_0 = \mathfrak{k}/\mathfrak{l}_0^{(1)} + \mathfrak{l}_0^{(2)}/\mathfrak{l}_0^{(2)},$$

where $\mathfrak{k}$ is the simple special unitary subalgebra of the involutive automorphism $S_3 = \varphi^2$, and $\mathfrak{k}/\mathfrak{l}_0^{(1)}$ is a principal unitary involutive pair.

Proof. By II.4.15 (Theorem 22) $\mathfrak{a}$ has a principal unitary involutive algebra, but then $\tilde{\mathfrak{l}}$ also has a principal unitary involutive algebra $\mathfrak{q} \oplus \mathfrak{t}$, and $S_2 \in \mathrm{Int}_\mathfrak{g}(\mathfrak{q})$, $S_2 \neq S_1$.

The rest of the proof coincides with the proof of II.7.5 (Theorem 34). ∎

II.9.2. Theorem 41. *Let $\mathfrak{g}$ be a simple compact Lie algebra, $\mathfrak{l}_1 = \mathfrak{p} \oplus \mathfrak{m} \oplus \tilde{\mathfrak{l}}$ be its principal di-unitary involutive algebra of an involutive automorphism S_1 ($S_1 \in \mathrm{Int}_\mathfrak{g}(\mathfrak{p})$, $S_1 \in \mathrm{Int}_\mathfrak{g}(\mathfrak{m})$), and let $\tilde{\mathfrak{l}}$ have a simple ideal $\mathfrak{a}$, $\dim \mathfrak{a} > 3$.*

Then $\mathfrak{g} \cong so(n)$ ($n > 8$), and there exists the iso-involutive decomposition $\mathfrak{g} = \mathfrak{l}_1 + \mathfrak{l}_2 + \mathfrak{l}_3$ basis for $\mathfrak{l}_1$ ($\mathfrak{l}_0 = \mathfrak{l}_\alpha \cap \mathfrak{l}_\beta$, $\alpha \neq \beta$).

Moreover,

$$\mathfrak{l}_1/\mathfrak{l}_0 \cong so(n-4)/(\,so(n-8) \oplus so(4)\,) + so(4)/so(4),$$

$$\mathfrak{l}_2/\mathfrak{l}_0 \cong so(n-4)/(\,so(n-8) \oplus so(4)\,) + so(4)/so(4),$$

$$\mathfrak{l}_3/\mathfrak{l}_0 \cong so(8)/(\,so(4) \oplus so(4)\,) + so(n-8)/so(n-8)$$

with the natural embeddings into $\mathfrak{g} \cong so(n)$.

Proof. By II.5.10 (Theorem 27) we have

$$\mathfrak{g}/\mathfrak{l}_1 \cong so(n)/(\,so(4) \oplus so(n-4)\,)$$

with the natural embeddings. Thus $\tilde{\mathfrak{l}} \cong so(n-4)$. But $n = 5,6,7,8$ is impossible, contradicting the existence of a simple ideal $\mathfrak{a} \subset \tilde{\mathfrak{l}}$, dim $\mathfrak{a} > 3$.

Consequently $n > 8$, $\tilde{\mathfrak{l}}$ is simple, dim $\tilde{\mathfrak{l}} > 3$. Taking now the di-unitary involutive algebra $\mathfrak{q} \oplus \mathfrak{n} \oplus \tilde{\tilde{\mathfrak{l}}} \subset \tilde{\mathfrak{l}}$ of the involutive automorphism $S_2 \in \text{Int}_\mathfrak{g}(\mathfrak{q})$, we construct by II.9.1 (Theorem 40) the basis for $\mathfrak{l}_1$ involutive sum $\mathfrak{g} = \mathfrak{l}_1 + \mathfrak{l}_2 + \mathfrak{l}_3$.

Moreover,

$$\mathfrak{l}_0 = \mathfrak{p} \oplus \mathfrak{q} \oplus \mathfrak{m} \oplus \mathfrak{n} \oplus \tilde{\tilde{\mathfrak{l}}}, \quad \mathfrak{p} \subset \mathfrak{k}, \quad \mathfrak{q} \subset \mathfrak{k}.$$

But in this consideration we may interchange $\mathfrak{p}$ and $\mathfrak{m}$ then evidently $\mathfrak{m} \subset \mathfrak{k}$. But $\mathfrak{k}/\mathfrak{p} \oplus \mathfrak{q} \oplus \mathfrak{m} \oplus \mathfrak{w}$ is a di-unitary principal involutive pair, and the restrictions of the involutive automorphisms $S_1 \in \text{Int}_\mathfrak{g}(\mathfrak{p})$, $S_2 \in \text{Int}_\mathfrak{g}(\mathfrak{q})$, $S_1 \in \text{Int}_\mathfrak{g}(\mathfrak{m})$ to $\mathfrak{k}$ coincide, being different from the identity automorphism. This is possible only if $\mathfrak{w}$ is simple, dim $\mathfrak{w} = 3$.

Thus $\mathfrak{k} \cong so(8)$ and

$$\mathfrak{l}_3/\mathfrak{l}_0 \cong so(8)/(\,so(4) \oplus so(4)\,) + so(n-8)/so(n-8).$$

with the natural embedding   ∎

II.9.3. Theorem 42. *Let $\mathfrak{g}$ be a simple compact Lie algebra, $\mathfrak{l}_1 = \mathfrak{p} \oplus \tilde{\mathfrak{l}} \oplus \mathfrak{z}$ be its central unitary involutive algebra (dim $\mathfrak{z} = 1$) of an involutive automorphism $S_1 \in \text{Int}_\mathfrak{g}(\mathfrak{p})$, and let $\tilde{\mathfrak{l}}$ have a simple ideal $\mathfrak{a}$, dim $\mathfrak{a} > 3$.*

Then $\mathfrak{g} \cong su(n)$ $(n > 4)$ and there exists the iso-involutive decomposition basis for $\mathfrak{l}_1$, $\mathfrak{g} = \mathfrak{l}_1 + \mathfrak{l}_2 + \mathfrak{l}_3$ $(\mathfrak{l}_0 = \mathfrak{l}_\alpha \cap \mathfrak{l}_\beta, \quad \alpha \neq \beta)$.

Moreover,

$$\mathfrak{l}_1/\mathfrak{l}_0 \cong su(n-2)/(\,su(2) \oplus su(n-4) \oplus u(1)\,) + su(2)/su(2) + u(1)/u(1),$$

$$\mathfrak{l}_2/\mathfrak{l}_0 \cong su(n-2)/(\,su(2) \oplus su(n-4) \oplus u(1)\,) + su(2)/su(2) + u(1)/u(1),$$

$$\mathfrak{l}_3/\mathfrak{l}_0 \cong so(6)/(\,so(4) \oplus so(2)\,) + su(n-4)/su(n-4) + u(1)/u(1)$$

with the natural embeddings into $\mathfrak{g} \cong su(n)$.

Proof. By I.7.2 (Theorem 29) we have

$$\mathfrak{g}/\mathfrak{l}_1 \cong su(n)/(\,su(2) \oplus su(n-2) \oplus u(1)\,), \quad n > 2$$

with the natural embedding. By the existence of a simple ideal $\mathfrak{a}$, dim $\mathfrak{a} > 3$, we obtain $n > 4$. Thus $\tilde{\mathfrak{l}} \cong su(n-2)$ is simple. Taking, furthermore, the unitary central involutive algebra $\mathfrak{q} \oplus \mathfrak{z}' \oplus \tilde{\tilde{\mathfrak{l}}} \subset \tilde{\mathfrak{l}}$ (dim $\mathfrak{z}' = 1$) of an involutive automorphism $S_2 \in \text{Int}_\mathfrak{g}(\mathfrak{q})$, we construct by II.9.1 (Theorem 40) the involutive sum $\mathfrak{g} = \mathfrak{l}_1 + \mathfrak{l}_2 + \mathfrak{l}_3$ basis for $\mathfrak{l}_1$. Moreover, $\mathfrak{l}_0 = \mathfrak{p} \oplus \mathfrak{q} \oplus \tilde{\tilde{\mathfrak{l}}} \oplus \mathfrak{z} \oplus \mathfrak{z}'$ and, by the construction and I.7.2 (Theorem 29),

$$\mathfrak{l}_1/\mathfrak{l}_0 \cong su(n-2)/(\,su(2) \oplus u(1) \oplus su(n-4)\,) + su(2)/su(2) + u(1)/u(1),$$

that is, $\tilde{\tilde{\mathfrak{l}}} \cong su(n-4)$.

Consequently, either $\tilde{\tilde{\mathfrak{l}}}$ is simple, $\dim \tilde{\tilde{\mathfrak{l}}} \geqslant 3$, or $\tilde{\tilde{\mathfrak{l}}} = \{0\}$. Furthermore, we have

$$\mathfrak{l}_3/\mathfrak{l}_0 = \mathfrak{k}/\mathfrak{l}_0^{(1)} + \mathfrak{l}_0^{(2)}/\mathfrak{l}_0^{(2)}.$$

We note that $(\mathfrak{z} \oplus \mathfrak{z}') \not\subseteq \mathfrak{l}_0^{(2)}$, otherwise $\mathfrak{l}_3$ would have a more than one-dimensional centre, which is impossible. But $(\mathfrak{z} \oplus \mathfrak{z}') \not\subseteq \mathfrak{l}_0^{(1)}$ as well, otherwise the involutive algebra $\mathfrak{l}_0^{(1)}$ of a simple compact algebra $\mathfrak{k}$ would have more than a one-dimensional centre, which is impossible. Consequently $\mathfrak{z} \oplus \mathfrak{z}' = \mathfrak{z}_1 \oplus \mathfrak{z}_2$, where $\mathfrak{z}_1$ and $\mathfrak{z}_2$ are one-dimensional, and $\mathfrak{z}_1 \subset \mathfrak{l}_0^{(1)}$, $\mathfrak{z}_2 \subset \mathfrak{l}_0^{(2)}$.

But

$$\mathfrak{k}/\mathfrak{l}_0^{(1)} \cong so(m)/(\,so(4) \oplus so(m-4)\,)$$

with the natural embedding, which implies $m = 6$ (otherwise there is no one-dimensional centre in $\mathfrak{l}_0^{(1)}$).   ∎

II.9.4. Theorem 43. *Let $\mathfrak{g}$ be a simple compact Lie algebra, and $\mathfrak{l}_1 = \mathfrak{p} \oplus \tilde{\mathfrak{l}}$ be its principal mono-unitary non-central involutive algebra of index 1 of an involutive automorphism $S_1 \in \mathrm{Int}_{\mathfrak{g}}(\mathfrak{p})$.*

Then $\mathfrak{g} \cong sp(n)$ and there exists the iso-involutive decomposition

$$\mathfrak{g} = \mathfrak{l}_1 + \mathfrak{l}_2 + \mathfrak{l}_3 \quad (\mathfrak{l}_0 = \mathfrak{l}_\alpha \cap \mathfrak{l}_\beta, \quad \alpha \neq \beta)$$

basis for $\mathfrak{l}_1$. Moreover,

$$\mathfrak{l}_1/\mathfrak{l}_0 \cong sp(n-1)/(\,sp(n-2) \oplus sp(1)\,) + sp(1)/sp(1),$$

$$\mathfrak{l}_2/\mathfrak{l}_0 \cong sp(n-1)/(\,sp(n-2) \oplus sp(1)\,) + sp(1)/sp(1),$$

$$\mathfrak{l}_3/\mathfrak{l}_0 \cong so(5)/so(4) + sp(n-2)/sp(n-2),$$

with the natural embedding into $\mathfrak{g} \cong sp(n)$.

Proof. Using II.7.5 (Theorem 34) we construct the iso-involutive sum $\mathfrak{g} = \mathfrak{l}_1 + \mathfrak{l}_2 + \mathfrak{l}_3$ basis for $\mathfrak{l}_1$. By I.8.4 (Theorem 33) we conclude that $\mathfrak{l}_1/\mathfrak{l}_0$ has the form indicated in the theorem (up to isomorphism).

Owing to the iso-involutivity of $\mathfrak{g} = \mathfrak{l}_1 + \mathfrak{l}_2 + \mathfrak{l}_3$, $\mathfrak{l}_2/\mathfrak{l}_0$ has the analogous form (up to isomorphism). Furthermore,

$$\mathfrak{l}_3/\mathfrak{l}_0 = \mathfrak{k}/\mathfrak{l}_0^{(1)} + \mathfrak{l}_0^{(2)}/\mathfrak{l}_0^{(2)},$$

where $\mathfrak{k}/\mathfrak{l}_0^{(1)}$ is a di-unitary principal involutive pair. But $\mathfrak{g}/\mathfrak{l}_1$ is of index 1, consequently $\mathfrak{l}_3/\mathfrak{l}_0$ and $\mathfrak{k}/\mathfrak{l}_0^{(1)}$ are of index 1. Hence

$$\mathfrak{k}/\mathfrak{l}_0^{(1)} \cong so(5)/so(4)(\cong sp(2)/sp(1) \oplus sp(1)).$$

Therefore

$$\mathfrak{l}_3/\mathfrak{l}_0 \cong so(5)/so(4) + sp(n-2)/sp(n-2),$$

which proves our theorem.   ∎

II.9.5. Theorem 44. *Let* $\mathfrak{g}$ *be a simple compact Lie algebra,* $\mathfrak{k}$ *be its simple special subalgebra of an involutive automorphism* S_3, $\dim \mathfrak{k} > 3$.

Then there exist the principal unitary automorphisms S_1 *and* S_2 *such that* $S_1 S_2 = S_2 S_1 = S_3$, *and there exists the iso-involutive decomposition*

$$\mathfrak{g} = \mathfrak{l}_1 + \mathfrak{l}_2 + \mathfrak{l}_3 \quad (\mathfrak{l}_0 = \mathfrak{l}_\alpha \cap \mathfrak{l}_\beta, \quad \alpha \neq \beta)$$

basis for $\mathfrak{l}_1$, *where* $\mathfrak{l}_1$, $\mathfrak{l}_2$, $\mathfrak{l}_3$ *are the involutive algebras of the involutive automorphisms* S_1, S_2, S_3, *respectively, and*

$$\mathfrak{l}_3/\mathfrak{l}_0 = \mathfrak{k}/\mathfrak{l}_0^{(1)} + \mathfrak{l}_0^{(2)}/\mathfrak{l}_0^{(2)},$$

where $\mathfrak{k}/\mathfrak{l}_0^{(1)}$ *is a principal di-unitary invoutive pair.*

Proof. Since $\mathfrak{k}$ is a special unitary subalgebra of an involutive automorphism S_3 we have $\mathfrak{k} \cong so(m)$. And taking $\mathfrak{b} \cong so(3)$ with the natural embedding into $\mathfrak{k} \cong so(m)$ we have $S_3 \in \mathrm{Int}_\mathfrak{g}(\mathfrak{b})$. We now choose $\mathfrak{p} \oplus \mathfrak{q} \cong so(4)$ with the natural embedding into $\mathfrak{k} \cong so(m)$ in such a way that $\mathfrak{b} \cong so(3)$ has the natural embedding into $\mathfrak{p} \oplus \mathfrak{q} \cong so(4)$. We then see that $\mathfrak{b}$ is a diagonal of $\mathfrak{p} \oplus \mathfrak{q}$.

Moreover, it is evident that if $S_1 \in \mathrm{Int}_\mathfrak{g}(\mathfrak{p})$, $S_2 \in \mathrm{Int}_\mathfrak{g}(\mathfrak{q})$, and S_1, S_2 are the non-identity involutive automorphisms then $S_1 S_2 = S_2 S_1 = S_3$.

Taking the involutive algebras $\mathfrak{l}_1$, $\mathfrak{l}_2$, $\mathfrak{l}_3$ of the involutive automorphisms S_1, S_2, S_3, respectively, we obtain the involutive sum

$$\mathfrak{g} = \mathfrak{l}_1 + \mathfrak{l}_2 + \mathfrak{l}_3, \quad (\mathfrak{l}_0 = \mathfrak{l}_\alpha \cap \mathfrak{l}_\beta, \quad \alpha \neq \beta).$$

By the construction $\mathfrak{l}_3 = \mathfrak{k} \oplus \mathfrak{m}$,

$$\mathfrak{l}_3/\mathfrak{l}_0 = \mathfrak{k}/\mathfrak{p} \oplus \mathfrak{q} \oplus \mathfrak{t} + \mathfrak{m}/\mathfrak{m},$$

thus denoting $\mathfrak{l}_0^{(2)} = \mathfrak{m}$ and $\mathfrak{l}_0^{(1)} = \mathfrak{p} \oplus \mathfrak{q} \oplus \mathfrak{t}$ we see that the assertion of the theorem about $\mathfrak{l}_3/\mathfrak{l}_0$ is true.

Taking $\{t\,\zeta\} \cong so(2 \subset so(m) \cong \mathfrak{k}$ with the natural embedding and in such a way that $\zeta \in \mathfrak{k} \dot{-} \mathfrak{l}_0^{(1)}$ we have, owing to the conjugacy of all subalgebras $so(2)$ with the natural embeddings into $so(m) \cong \mathfrak{k}$, $S_3 \in \mathrm{Int}_\mathfrak{g}(\{t\,\zeta\})$.

Taking $\varphi \in \mathrm{Int}_\mathfrak{g}(\{t\,\zeta\})$, $\varphi^2 = S_3$, we obtain a conjugating automorphism φ of the involutive sum $\mathfrak{g} = \mathfrak{l}_1 + \mathfrak{l}_2 + \mathfrak{l}_3$, which consequently is iso-involutive.

Let us consider

$$\mathfrak{l}_1/\mathfrak{l}_0 = \mathfrak{l}_1/\mathfrak{p} \oplus \mathfrak{q} \oplus \mathfrak{t} \oplus \mathfrak{m}$$

in the decomposition into elementary involutive pairs:

$$\mathfrak{l}_1/\mathfrak{l}_0 = \sum_{\alpha=1}^{m} \mathfrak{l}_1^{(\alpha)}/\mathfrak{l}_0^{(\alpha)}.$$

After some re-numbering we may consider $\mathfrak{q} \subset \mathfrak{l}_0^{(1)}$. But then, by the commutativity of $\mathfrak{q}$ and $\mathfrak{l}_1^{(\alpha)}$ ($\alpha = 2, 3, \ldots, m$), and because $S_2 \in \mathrm{Int}_\mathfrak{g}(\mathfrak{q})$, it follows that vectors of $\mathfrak{l}_1^{(\alpha)}$ ($\alpha = 2, 3, \ldots, m$) are immobile under the action of S_2, that is, $\mathfrak{l}_1^{(\alpha)} \subset \mathfrak{l}_2$ ($\alpha = 2, 3, \ldots, m$), and consequently $\mathfrak{l}_1^{(\alpha)} \subset \mathfrak{l}_0$ ($\alpha = 2, 3, \ldots, m$).

Thus we obtain the decomposition

$$\mathfrak{l}_1/\mathfrak{l}_0 = \tilde{\mathfrak{l}}_1/(\mathfrak{q} \oplus \mathfrak{w}) + \mathfrak{a}/\mathfrak{a}.$$

If $\mathfrak{p} \subset \mathfrak{a}$ then evidently $\mathfrak{l}_1 = \mathfrak{p} \oplus \tilde{\tilde{\mathfrak{l}}}_1$, whence it follows that $\mathfrak{l}_1$ (and owing to the conjugacy $\mathfrak{l}_2$ as well) is a principal unitary involutive algebra, as the theorem asserts.

The only case to be considered, finally, is $\mathfrak{p} \subset \mathfrak{w}$. Then we have

$$\mathfrak{l}_1/\mathfrak{l}_0 = \tilde{\mathfrak{l}}_1/(\mathfrak{q} \oplus \mathfrak{p} \oplus \mathfrak{v}) + \mathfrak{a}/\mathfrak{a}.$$

The involutive pair $\tilde{\mathfrak{l}}_1/(\mathfrak{q} \oplus \mathfrak{p} \oplus \mathfrak{v})$ is elementary, thus it is evident that $\tilde{\mathfrak{l}}_1$ is simple. The restriction of $\mathrm{Int}_\mathfrak{g}(\mathfrak{p}) \cong SU(2)$ onto $\mathfrak{l}_1$ is $SO(3)$, which means that the above involutive pair is principal orthogonal. This is possible only if

$$\tilde{\mathfrak{l}}_1/\mathfrak{p} \oplus \mathfrak{q} \oplus \mathfrak{v} \cong so(7)/(so(3) \oplus so(4)), \qquad \dim \mathfrak{v} = 3,$$

that is, $\mathrm{Int}_{\mathfrak{l}_1}(\mathfrak{v}) \cong SU(2)$.

Now, let us note that $\mathfrak{t} \oplus \mathfrak{m} = \mathfrak{v} \oplus \mathfrak{a}$. Thus

$$\mathfrak{m} \ominus \mathfrak{v} \overset{\text{def}}{=} \{\xi \in \mathfrak{m} \mid \xi \perp \mathfrak{v}\} \subset \mathfrak{a}.$$

But then $\mathfrak{m} \ominus \mathfrak{v}$ commutes with $(\mathfrak{l}_\alpha \dot{-} \mathfrak{l}_0)$ ($\alpha = 1, 2, 3$), that is, $\mathfrak{m} \ominus \mathfrak{v}$ is an ideal of a simple Lie algebra $\mathfrak{g}$. Consequently $\mathfrak{m} \ominus \mathfrak{v} = \{0\}$, but then either $\mathfrak{m} = \{0\}$ or $\mathfrak{m} = \mathfrak{v}$. Since $\mathfrak{t} \cong so(m-4)$ then $\mathfrak{m} = \{0\}$ implies $\mathfrak{v} \oplus \mathfrak{a} \cong so(m-4)$, which is possible only for $m = 7$ and $m = 8$.

But $m = 7$ gives us $\mathfrak{l}_3 = \mathfrak{k} \cong so(7)$, $\mathfrak{l}_1 \cong so(7)$, $\mathfrak{l}_2 \cong so(7)$, and then $\mathfrak{g}$ satisfies the conditions of II.4.9 (Theorem 17). But in the proof of II.4.11 (Theorem 18) it has been shown that such an algebra $\mathfrak{g}$ does not exist.

Consequently $m = 8$ and then $\dim \mathfrak{a} = 3$,

$$\mathfrak{k}/\mathfrak{p} \oplus \mathfrak{q} \oplus \mathfrak{v} \oplus \mathfrak{a} \cong so(8)/so(4) \oplus so(4).$$

Since $S_1 \in \mathrm{Int}_\mathfrak{g}(\mathfrak{p})$, $S_2 \in \mathrm{Int}_\mathfrak{g}(\mathfrak{q})$ we evidently have either $S_2 \in \mathrm{Int}_\mathfrak{g}(\mathfrak{a})$ or $S_1 \in \mathrm{Int}_\mathfrak{g}(\mathfrak{a})$. But $\mathfrak{l}_1 = \tilde{\mathfrak{l}}_1 \oplus \mathfrak{a}$, $S_2 \notin \mathrm{Int}_\mathfrak{g}(\mathfrak{a})$, consequently $S_1 \in \mathrm{Int}_\mathfrak{g}(\mathfrak{a})$, meaning that $\mathfrak{l}_1$ is a principal mono-unitary non-central involutive algebra, $\mathfrak{l}_1 \cong so(7) \oplus so(3)$.

By I.8.4 (Theorem 33), II.8.4 (Theorem 35), II.8.2 (Theorem 36), II.8.3 (Theorem 37), II.8.4 (Theorem 38) we conclude that such a principal mono-unitary non-central involutive algebra does not exist.

As a result we should assume $\mathfrak{m} = \mathfrak{v}$ ($\dim \mathfrak{v} = 3$), $\mathfrak{t} = \mathfrak{a}$. Then evidently $S_3 \in \mathrm{Int}_\mathfrak{g}(\mathfrak{v})$ and $\mathfrak{l}_3 = \mathfrak{m} \oplus \mathfrak{k} \cong so(3) \oplus so(m)$ is a non-central mono-unitary involutive algebra.

Moreover, $\mathfrak{g}/\mathfrak{l}_3$ is not of index 1, otherwise $\mathfrak{l}_1/\mathfrak{l}_0$ is of index 1, but

$$\mathfrak{l}_1/\mathfrak{l}_0 = \mathfrak{n}_1/\mathfrak{n}_0 + \mathfrak{t}/\mathfrak{t},$$

$$\mathfrak{n}_1/\mathfrak{n}_0 \cong so(7)/(so(3) \oplus so(4))$$

and $so(7)/(so(3) \oplus so(4))$ is not of index 1.

Consequently $\mathfrak{l}_3 = \mathfrak{m} \oplus \mathfrak{k} \cong so(3) \oplus so(m)$ is an exceptional principal involutive algebra and is evidently of type e_7. Then $\mathfrak{k} \cong so(12)$.

But $S_1 \in \mathrm{Int}_{\mathfrak{g}}(\mathfrak{p})$, $S_2 \in \mathrm{Int}_{\mathfrak{g}}(\mathfrak{q})$, $S_3 \in \mathrm{Int}_{\mathfrak{g}}(\mathfrak{v})$, thus $\tilde{\mathfrak{l}}_1 \cong so(7)$ is a special unitary involutive algebra of an involutive automorphism S_1.

Therefore $\mathfrak{g} = \mathfrak{l}_1 + \mathfrak{l}_2 + \mathfrak{l}_3$ is an iso-involutive decomposition basis for $\mathfrak{l}_3$.

But then by II.8.3 (Theorem 37) for $\mathfrak{g} \cong e_7$ we have $\mathfrak{l}_1 \cong so(12) \oplus so(3)$. However, by our assumptions $\mathfrak{l}_1 \cong \mathfrak{c} \oplus \mathfrak{t}$, $\mathfrak{c} \cong so(7)$. This contradiction shows that $\mathfrak{p}$ is an ideal of $\mathfrak{l}_1$, which proves the theorem. ∎

II.9.6. Remark. II.9.5 (Theorem 44) gives us the way to determine all simple special unitary subalgebras of simple compact Lie algebras, since the problem is reduced to the consideration of basis involutive decompositions for principal unitary involutive algebras. The latter problem is solved by II.8.1 (Theorem 35), II.8.2 (Theorem 36), II.8.3 (Theorem 37), II.8.4 (Theorem 38), II.9.2 (Theorem 41), II.9.3 (Theorem 42), II.9.4 (Theorem 43).

In such a way we have arrived at a peculiar involutive principle of duality for principal unitary and special unitary involutive automorphisms.

Thus the following theorem is true.

II.9.10. Theorem 45. *Let $\mathfrak{g}$ be a simple compact Lie algebra, $\mathfrak{k}$ be its special unitary simple subalgebra, $\dim \mathfrak{k} > 3$, and let $\mathfrak{l} = \mathfrak{k} \oplus \mathfrak{m}$ be its special unitary involutive algebra.*

Then $\mathfrak{k} \cong so(5)$, $so(6)$, $so(8)$, $so(9)$, $so(10)$, $so(12)$, $so(16)$ and, respectively,

$$\mathfrak{g}/\mathfrak{l} \cong sp(n)/(sp(2) \oplus sp(n-2)) \quad (n > 2),$$

$$\cong su(n)/(su(4) \oplus su(n-4) \oplus u(1)) \quad (n > 4),$$

$$\cong so(n)/(so(8) \oplus so(n-8)) \quad (n > 8),$$

$$\cong f_4/so(9),$$

$$\cong e_6/(so(10) \oplus so(2)),$$

$$\cong e_7/(so(12) \oplus so(3)),$$

$$\cong e_8/so(16)$$

with the natural embeddings.

II.9.11. Remark. Under the natural embeddings in the last four cases we understand the embeddings described by II.8.1 (Theorem 35), II.8.2 (Theorem 36), II.8.3 (Theorem 37), II.8.4 (Theorem 38).

CHAPTER II.10

HYPER-INVOLUTIVE RECONSTRUCTIONS
OF BASIS INVOLUTIVE DECOMPOSITIONS

II.10.1. Let $\mathfrak{g}$ be a simple compact Lie algebra of type f or e, and $\mathfrak{g} = \mathfrak{l}_1 + \mathfrak{l}_2 + \mathfrak{l}_3$ be an iso-involutive sum basis for the exceptional principal involutive algebra $\mathfrak{l}_1 = \mathfrak{p} \oplus \tilde{\mathfrak{l}}$ of type $U^{(1)}$ (see II.7.4 (Definition 29), II.7.6 (Definition 30).

By II.7.5 (Theorem 34) we have

$$\mathfrak{l}_3/\mathfrak{l}_0 = \mathfrak{k}/\mathfrak{l}_0^{(1)} + \mathfrak{l}_0^{(2)}/\mathfrak{l}_0^{(2)},$$

where $\mathfrak{k}$ is a simple special subalgebra of type U, $\mathfrak{l}_0^{(1)} = \mathfrak{p} \oplus \mathfrak{q} \oplus \mathfrak{r}$,

$$\mathfrak{k}/\mathfrak{l}_0^{(1)} \cong so(m)/so(4) \oplus so(m-4)$$

with the natural embedding.

Moreover, $m > 5$ (otherwise $\mathfrak{k}/\mathfrak{l}_0^{(1)}$, and also $\mathfrak{g}/\mathfrak{l}_1$ are of index 1, contradicting that $\mathfrak{l}_1$ is an exceptional principal involutive algebra). By II.8.1 (Theorem 35), II.8.2 (Theorem 36), II.8.3 (Theorem 37), II.8.4 (Theorem 38) it follows that in this case $m = 9, 10, 12, 16$. Respectively, then $\mathfrak{l}_0^{(2)} \cong \{0\}$, $u(1)$, $su(2)$, $\{0\}$.

We note also that $\mathfrak{l}_0^{(2)} \subset \tilde{\mathfrak{l}}$ and $\mathfrak{r} \cong so(5)$, $so(6)$, $so(8)$, $so(12)$, respectively.

Let us consider the chain of subalgebras

$$\Delta(so(2) \oplus so(2)) \subset \Delta(so(3) \oplus so(3)) \subset su(3) \subset u(3) \subset so(6) \subset so(m),$$

with the natural embeddings, where $\Delta(\mathfrak{m} \oplus \mathfrak{m}')$ means a diagonal of the canonical symmetry $\mathfrak{m} \leftrightarrows \mathfrak{m}'$, and its corresponding (by virtue of isomorphism $\mathfrak{k} \cong so(m)$) chain of subalgebras $\mathfrak{p}' \subset \mathfrak{b} = \Delta(\mathfrak{m} \oplus \mathfrak{m}') \subset \mathfrak{n} \subset \mathfrak{t} \subset \mathfrak{f} \subset \mathfrak{k}$.

Since $\mathfrak{p} \oplus \mathfrak{q} \cong so(4)$ (owing to the isomorphism $\mathfrak{k} \cong so(m)$) any one-dimensional subalgebra of $\mathfrak{p}$ is isomorphic to $\Delta(so(2) \oplus so(2))$ with the natural embedding into $\mathfrak{k} \cong so(m)$. Therefore we may start the construction of the above chain with $\mathfrak{p}' \subset \mathfrak{p}$.

Since, by the definition of a special subalgebra $S_3 \in \text{Int}_\mathfrak{g}(\mathfrak{m}') \cong SU(2)$, $S_3 \in \text{Int}_\mathfrak{g}(\mathfrak{m}) \cong SU(2)$ we obtain $\text{Int}_\mathfrak{g}(\mathfrak{b}) \cong SO(3)$.

Finally, the chain may be constructed in such a way that $\mathfrak{p} \subset \mathfrak{n}$ with the natural embedding (that is, as $su(2) \subset su(3)$).

Indeed, for this we have to begin the construction in such a way that $\mathfrak{p} \oplus \mathfrak{q} \cong so(4)$ is naturally embedded into $\mathfrak{f} \cong so(6)$ as well as into $\mathfrak{k} \cong so(m)$.

In what follows we consider such a choice of the chain. With this choice $S_1 \in \text{Int}_\mathfrak{g}(\mathfrak{b}) \cong SO(3)$, the algebra $\mathfrak{g}$ has a hyper-involutive decomposition with the involutive automorphisms $\widetilde{S}_1 = S_1$, $\widetilde{S}_2$, $\widetilde{S}_3$, according to II.1.2 (Lemma 1). But all non-identity involutive automorphisms of $\mathrm{Int}_{\mathfrak{g}}(B_3) \cong SO(3)$ are conjugated and $\widetilde{S}_1 = S_1$ is a principal exceptional involutive automorphism; consequently, $\widetilde{S}_1$, $\widetilde{S}_2$, $\widetilde{S}_3$ are principal exceptional involutive automorphisms.

By the construction the involutive automorphisms $\widetilde{S}_1$, $\widetilde{S}_2$, $\widetilde{S}_3$ generate a hyper-involutive decomposition in $\mathfrak{n} \cong su(3)$. Thus

$$\mathfrak{n} = \mathfrak{n}_1 \oplus \mathfrak{n}_2 \oplus \mathfrak{n}_3, \qquad \mathfrak{n}_0 = \mathfrak{n}_\alpha \underset{\alpha \neq \beta}{\cap} \mathfrak{n}_\beta = \mathfrak{z} \oplus \mathfrak{w},$$

$$\dim \mathfrak{z} = \dim \mathfrak{w} = 1, \qquad \mathfrak{n}_\alpha/\mathfrak{n}_0 = \mathfrak{p}^{(\alpha)}/\mathfrak{z}^{(\alpha)} + \mathfrak{w}^{(\alpha)}/\mathfrak{w}^{(\alpha)},$$

$$\mathfrak{p}^{(\alpha)} \cong su(2), \qquad \mathfrak{z}^{(\alpha)} \oplus \mathfrak{w}^{(\alpha)} = \mathfrak{z} \oplus \mathfrak{w}, \qquad \dim \mathfrak{z}^{(\alpha)} = \dim \mathfrak{w}^{(\alpha)} = 1.$$

This follows from II.3.8 (Theorem 7), II.3.10 (Theorem 8), II.3.11 (Theorem 9), II.3.14 (Theorem 10) if one applies them to a principal involutive pair of type O, $\mathfrak{n}/\mathfrak{b}$). Lastly, by the construction of $\mathfrak{n}$ and a choice of $\widetilde{S}_1 = S_1$ we have $\mathfrak{p} = \mathfrak{p}^{(1)}$. Note also that the conjugating automorphism p ($p^3 = \mathrm{Id}$) acts on $\mathfrak{z} \oplus \mathfrak{w}$ in a non-trivial way.

As a result we have the hyper-involutive sum $\mathfrak{g} = \tilde{\mathfrak{l}}_1 + \tilde{\mathfrak{l}}_2 + \tilde{\mathfrak{l}}_3$ of the principal involutive automorphisms $\widetilde{S}_1$, $\widetilde{S}_2$, $\widetilde{S}_3$; moreover,

$$\tilde{\mathfrak{l}}_\alpha/\tilde{\mathfrak{l}}_0 = \mathfrak{l}^{(\alpha)}/\mathfrak{l}_0^{(\alpha)} + \mathfrak{p}^{(\alpha)}/\mathfrak{z}^{(\alpha)},$$

$$\mathfrak{p}^{(\alpha)}/\mathfrak{z}^{(\alpha)} \cong su(2)/u(1),$$

so that $\tilde{\mathfrak{l}}_0 = \mathfrak{l}_0^{(\alpha)} \oplus \mathfrak{z}^{(\alpha)}$. Therefore the maximal commutative ideal of $\tilde{\mathfrak{l}}_0$ is not $\{0\}$. However, $p\mathfrak{z}^{(\alpha)} \cap \mathfrak{z}^{(\alpha)} = \{0\}$ (since p is non-trivial on $\mathfrak{z} \oplus \mathfrak{w}$), and consequently the maximal commutative ideal of $\tilde{\mathfrak{l}}_0$ is at least two-dimensional and $\mathfrak{l}_0^{(\alpha)} = {}'\mathfrak{l}_0^{(\alpha)} \oplus \mathfrak{w}^{(\alpha)}$.

By II.7.3 (Theorem 33) $\mathfrak{l}^{(\alpha)}$ is simple and more than three-dimensional.

Consequently $\mathfrak{l}_0^\alpha$ has at most a one-dimensional centre as an involutive algebra of the simple algebra $\mathfrak{l}_0^{(\alpha)}$. As a result ${}'\mathfrak{l}_0^{(\alpha)}$ is semi-simple and ${}'\mathfrak{l}_0^{(\alpha)} = {}'\mathfrak{l}_0^{(\beta)} = {}'\mathfrak{l}_0$. Now it is evident that $p = \mathrm{Id}$ on ${}'\mathfrak{l}_0$.

We have proved the theorem:

II.10.2. **Theorem 46.** *Let $\mathfrak{g}$ be a simple compact Lie algebra, and let $\tilde{\mathfrak{l}}_1 = \mathfrak{p}^{(1)} \oplus \tilde{\mathfrak{l}}_1$ be its exceptional principal involutive algebra of an involutive automorphism $\widetilde{S}_1 \in \mathrm{Int}_{\mathfrak{g}}(\mathfrak{p}^{(1)}) \cong SU(2)$.*

Then there exists the hyper-involutive decomposition $\mathfrak{g} = \tilde{\mathfrak{l}}_1 + \tilde{\mathfrak{l}}_2 + \tilde{\mathfrak{l}}_3$ of the exceptional principal involutive automorphisms $\widetilde{S}_1, \widetilde{S}_2, \widetilde{S}_3$ such that

$$\tilde{\mathfrak{l}}_\alpha/\tilde{\mathfrak{l}}_0 = \mathfrak{l}^{(\alpha)}/{}'\mathfrak{l}_0 \oplus \mathfrak{w}^{(\alpha)} + \mathfrak{p}^{(\alpha)}/\mathfrak{z}^{(\alpha)}, \qquad \dim \mathfrak{w}^{(\alpha)} = \dim \mathfrak{z}^{(\alpha)} = 1,$$

the conjugating automorphism $p = \text{Id}$ *on* $'\mathfrak{l}_0$ *and is non-trivial on* $\mathfrak{w}^{(\alpha)} \oplus \mathfrak{z}^{(\alpha)} = \mathfrak{t}$, $p\mathfrak{t} = \mathfrak{t}$.

Moreover,

$$p\mathfrak{p}^{(1)} = \mathfrak{p}^{(2)}, \qquad p\mathfrak{p}^{(2)} = \mathfrak{p}^{(3)}, \qquad p\mathfrak{p}^{(3)} = \mathfrak{p}^{(1)},$$

and

$$\tilde{S}_\alpha \in \text{Int}_\mathfrak{g}(\mathfrak{b} \cap \mathfrak{p}^{(\alpha)}) \subset \text{Int}_\mathfrak{g}(\mathfrak{b}) \cong SO(3), \qquad p \in \text{Int}_\mathfrak{g}(\mathfrak{b}),$$

where

$$\mathfrak{p}^{(1)} + \mathfrak{p}^{(2)} + \mathfrak{p}^{(3)} = \mathfrak{n} \cong su(3), \qquad \mathfrak{n}/\mathfrak{b} \cong su(3)/so(3)$$

with the natural embedding.

II.10.3. Corollary. *Under the assumptions and notations of II.10.2 (Theorem 46),* $'\mathfrak{l}_0 \neq \{0\}$ *is the maximal subalgebra of elements of* $\mathfrak{g}$ *commuting with* $\mathfrak{n}$, *and* $\mathfrak{n}$ *is the maximal subalgebra of elements of* $\mathfrak{g}$ *commuting with* $'\mathfrak{l}_0$.

II.10.4. Now let us clarify how to find $'\mathfrak{l}_0 \oplus \mathfrak{w}^{(\alpha)} = \mathfrak{l}_0^{(\alpha)}$ and an involutive pair $\mathfrak{l}^{(\alpha)}/\mathfrak{l}_0^\alpha$ for any type of exceptional Lie algebras.

For this we note that by construction

$$(\mathfrak{p} \oplus \mathfrak{q}) \oplus \mathfrak{r} \cong so(4) \oplus so(m-4)$$

is invariant under the action of $\tilde{S}_2$ (and $\tilde{S}_3$ as well).

Therefore, $\tilde{S}_2$ generates the involutive pair

$$\mathfrak{p} \oplus \mathfrak{q} \oplus \mathfrak{r}/\mathfrak{p}' \oplus \mathfrak{q}_1 \oplus \mathfrak{r}_1 \oplus \tilde{\mathfrak{r}} = \mathfrak{p}/\mathfrak{p}' + \mathfrak{q}/\mathfrak{q}_1 + \mathfrak{r}/\mathfrak{r}_1 \oplus \tilde{\mathfrak{r}}$$

$$\cong su(2)/u(1) + su(2)/u(1) + so(n-4)/so(2) \oplus so(m-6).$$

Furthermore, we note that $\tilde{S}_2$ and S_2 commute, therefore if we consider the restrictions of $\tilde{S}_2$ and S_2 to $\mathfrak{l}^{(1)}$ and denote the involutive algebra of the restriction $\tilde{S}_2$ to $\mathfrak{l}^{(1)}$ by $\mathfrak{d}$ then S_2 generates the involutive pair

$$\mathfrak{d}/\mathfrak{q}_1 \oplus \mathfrak{r}_1 \oplus \tilde{\mathfrak{r}} \oplus \mathfrak{a},$$

$$\mathfrak{q}_1 \oplus \mathfrak{r}_1 \oplus \tilde{\mathfrak{r}} \cong so(2) \oplus so(2) \oplus so(m-6).$$

However,

$$\mathfrak{d} \oplus \mathfrak{p}' = \tilde{\mathfrak{l}}_0 = \mathfrak{l}_0^{(1)} \oplus \mathfrak{z}^{(1)} \qquad (\mathfrak{z}^{(1)} = \mathfrak{p}^{(1)} = \mathfrak{p}').$$

Thus $\mathfrak{d} = \mathfrak{l}_0^{(1)} = {'\mathfrak{l}_0} \oplus \mathfrak{w}^{(1)}$, where $'\mathfrak{l}_0$ is semi-simple (as we have obtained before).

Therefore

$$\mathfrak{d}/\mathfrak{q}_1 \oplus \mathfrak{r}_1 \oplus \tilde{\mathfrak{r}} \oplus \mathfrak{a} = {'\mathfrak{l}_0}/\mathfrak{c} \oplus \mathfrak{a} + \mathfrak{n}^{(1)}\mathfrak{n}^{(1)},$$

$$\mathfrak{c} \cong so(2) \oplus so(m-6),$$

and
$$'\mathfrak{l}_0/\mathfrak{t} \oplus \mathfrak{v} \oplus \mathfrak{a}, \qquad \mathfrak{t} \oplus \mathfrak{v} \cong so(2) \oplus so(m-6),$$

is a special involutive pair with the special involutive subalgebra $\mathfrak{v} \cong so(m-6)$ (indeed, $so(m-6)$ is naturally embedded into $so(m) \cong \mathfrak{k}$, and $\mathfrak{k}$ is a special subalgebra in $\mathfrak{g}$). Thus $\widetilde{S}_2$, the restriction of the involutive automorphism S_2 to $'\mathfrak{l}_0$, belongs to $\mathrm{Int}\,'\mathfrak{l}_0(\mathfrak{v})$, but $\mathfrak{t} \cong so(2)$ is naturally embedded into $\mathfrak{k} \cong so(m)$, therefore $S_2|'\mathfrak{l}_0 \in \mathrm{Int}\,'\mathfrak{l}_0(\mathfrak{t})$, and consequently $\mathfrak{t}$ and $\mathfrak{v}$ can not belong to different ideals of $'\mathfrak{l}_0$.

In the cases of types f_4, e_8 we have $'\mathfrak{l}_0 = \{0\}$ by II.8.1 (Theorem 35), II.8.4 (Theorem 38) and then $\mathfrak{v}$ is a simple special subalgebra in $'\mathfrak{l}_0$, therefore $'\mathfrak{l}_0$ is simple.

Then by II.8.1 (Theorem 35), II.8.4 (Theorem 38) we obtain for $\mathfrak{t} \oplus \mathfrak{v}$, up to isomorphisms, respectively,

$$so(2) \oplus so(3) \cong u(1) \oplus su(2), \qquad so(2) \oplus so(10).$$

And from the classification of principal and special unitary involutive automorphisms for $'\mathfrak{l}_0/\mathfrak{t} \oplus \mathfrak{v}$, up to isomorphisms, we have in the cases f_4, e_8, respectively, $su(3)/u(2)$, $e_6/so(2) \oplus so(10)$ with the natural embeddings (see II.9).

In the case e_7 we have $\mathfrak{a} \cong so(3)$ and $S_2 \in \mathrm{Int}_\mathfrak{g}(\mathfrak{a})$, thus $'\mathfrak{l}_0/\mathfrak{t} \oplus \mathfrak{v} \oplus \mathfrak{a}$ is a principal unitary central involutive pair and then by II.8.3 (Theorem 37) $\mathfrak{v} \cong so(6) \cong su(4)$ is a special unitary subalgebra. Hence $'\mathfrak{l}_0$ is simple and

$$'\mathfrak{l}_0/\mathfrak{t} \oplus \mathfrak{v} \oplus \mathfrak{a} \cong su(6)/u(5)$$

with the natural embedding.

In the case e_6

$$\mathfrak{a} \cong so(2), \qquad S_2 \in \mathrm{Int}_\mathfrak{g}(\mathfrak{a}), \qquad \mathfrak{v} \cong so(4) \cong su(2) \oplus su(2),$$

by II.8.2 (Theorem 36).

Thus we have

$$'\mathfrak{l}_0/\mathfrak{t} \oplus \mathfrak{a} \oplus \mathfrak{v},$$

where $\mathfrak{v}$ is a special subalgebra; this is possible only if

$$'\mathfrak{l}_0/\mathfrak{t} \oplus \mathfrak{a} \oplus \mathfrak{v} \cong su(3)/u(2) + su(3)/u(2)$$

with the natural embedding.

We note also that, by the construction, $'\mathfrak{l}_0 \oplus \mathfrak{w}^{(1)}$ is defined in $\mathfrak{l}^{(1)}$ uniquely. Therefore for f_4, e_6, e_7, e_8, respectively, $\mathfrak{l}^{(\alpha)}/'\mathfrak{l}_0 \oplus \mathfrak{w}^{(\alpha)}$ is

$$sp(3)/u(3), \qquad su(6)/su(3) \oplus su(3) \oplus u(1),$$
$$so(12)/u(6), \qquad e_7/e_6 \oplus so(2)$$

with the natural embeddings.

Thus we have obtained the theorem:

II.10.5. Theorem 47. *Under the assumptions and notations of II.10.2 (Theorem 46) for the Lie algebras of types* f_4, e_6, e_7, e_8, *respectively,* $\mathfrak{l}^{(\alpha)} / {'\mathfrak{l}_0} \oplus \mathfrak{w}^{(\alpha)}$ *is*

$$sp(3)/u(3), \qquad su(6)/su(3) \oplus su(3) \oplus u(1),$$

$$so(12)/u(6), \qquad e_7/e_6 \oplus so(2)$$

with the natural embeddings.

II.10.6. Definition 31. The hyper-involutive decompositions described by II.10.2 (Theorem 46), II.10.5 (Theorem 47) are called the basis hyper-involutive sums for the principal exceptional involutive automorphisms of types f_4, e_6, e_7, e_8.

II.10.7. Using the hyper-involutive decompositions and the results of II.1 one may write out the structure constants for the Lie algebras of types f_4, e_6, e_7, e_8 in the hyper-involutive base. These constants may be obtained analogously to the case of the Lie algebra g_2 (see II.6).

As a result, in such a way one may obtain the theorem of the existence and uniqueness for the Lie algebras of types f_4, e_6, e_7, e_8 as well as the theorem of the uniqueness of basis hyper-involutive sums for each of the algebras of types f_4, e_6, e_7, e_8.

We do not intend to do this, since the theorem of the existence and uniqueness of the Lie algebras of types f_4, e_6, e_7, e_8 is a well known result, which guarantees the validity of what has been said above.

The principal significance of the exploration discussed above, were it to be done, would consist in obtaining the canonical form of structure constants tensor for f_4, e_6, e_7, e_8.

Let us, finally, consider the construction of hyper-involutive decompositions with the principal unitary involutive automorphisms for the Lie algebras of types $so(n)$, $su(n)$, $sp(n)$.

II.10.8. For the algebra $\mathfrak{g} \cong su(n)$ the special unitary subalgebra is $\mathfrak{k} \cong so(6) \cong su(4)$. Therefore we might repeat (with small changes) the proof of II.10.2 (Theorem 46). However, the quickest is another way. Indeed, taking $su(3) \subset so(6) \cong su(4)$, as in II.10.2 (Theorem 46), we have simultaneously $su(3) \subset su(4) \subset su(n)$ with the natural embeddings.

Furthermore, we choose principal unitary involutive automorphisms $\widetilde{S}_1$, $\widetilde{S}_2$, $\widetilde{S}_3$ from $\mathrm{Int}_{su(n)}(su(3))$, therefore each of them generates the pair $su(3)/su(2) \oplus u(1)$ with the natural embeddings, moreover, $su(2) \subset su(3) \subset su(n)$.

It is convenient to represent $\widetilde{S}_1$, $\widetilde{S}_2$, $\widetilde{S}_3$ by diagonal matrices and $su(3) \subset su(n)$ by three-dimensional square matrices in the upper left corner (which can be easily obtained by re-numbering of rows and columns).

Thus

$$\overset{=}{\tilde{S}}_1 = \left\|\left\|\begin{array}{ccc|c} -1 & 0 & 0 & \\ 0 & -1 & 0 & \overset{=}{O} \\ 0 & 0 & 1 & \\ \hline & \overset{=}{O} & & \overset{=}{\mathrm{Id}} \end{array}\right\|\right\|, \qquad \overset{=}{\tilde{S}}_2 = \left\|\left\|\begin{array}{ccc|c} 1 & 0 & 0 & \\ 0 & -1 & 0 & \overset{=}{O} \\ 0 & 0 & -1 & \\ \hline & \overset{=}{O} & & \overset{=}{\mathrm{Id}} \end{array}\right\|\right\|,$$

$$\overset{=}{\tilde{S}}_3 = \left\|\left\|\begin{array}{ccc|c} -1 & 0 & 0 & \\ 0 & 1 & 0 & \overset{=}{O} \\ 0 & 0 & -1 & \\ \hline & \overset{=}{O} & & \overset{=}{\mathrm{Id}} \end{array}\right\|\right\|,$$

where $\overset{=}{O}$ are zero matrices and $\overset{=}{\mathrm{Id}}$ are the identity matrices,

$$\tilde{S}_\alpha : \overset{=}{\bar{a}} \mapsto \overset{=}{\tilde{S}}_\alpha \overset{=}{\bar{a}} \overset{=}{\tilde{S}}_\alpha, \quad \overset{=}{\bar{a}} \in su(n).$$

From this, in a straightforward way, we obtain the hyper-involutive decomposition

$$\mathfrak{g} = \tilde{l}_1 + \tilde{l}_2 + \tilde{l}_3, \qquad \mathfrak{g}/\tilde{l}_\alpha \cong su(n)/su(2) \oplus su(n-2) \oplus u(1),$$

$$\tilde{l}_\alpha/\tilde{l}_0 \cong su(n-2)/u(n-3) + su(2)/u(1) + u(1)/u(1).$$

Thus we have the theorem.

II.10.9. Theorem 48. *Let $\mathfrak{g}$ be a Lie algebra, $\mathfrak{g} \cong su(n)$, $n > 2$, and $\tilde{l}_1 = \mathfrak{p}^{(1)} \oplus \overset{\approx}{l}_1$ be its principal unitary unvolutive algebra of an involutive automorphism*

$$\tilde{S}_1 \in \mathrm{Int}_\mathfrak{g}(\mathfrak{p}^{(1)}) \cong SU(2),$$

that is,

$$\tilde{l}_1 \cong su(2) \oplus su(n-2) \oplus u(1).$$

Then there exists the hyper-involutive decomposition $\mathfrak{g} = \tilde{l}_1 + \tilde{l}_2 + \tilde{l}_3$ of the principal unitary involutive automorphisms $\tilde{S}_1$, $\tilde{S}_2$, $\tilde{S}_3$ such that

$$\tilde{l}_\alpha/\tilde{l}_0 \cong \overset{\approx}{l}_\alpha/\overset{\approx}{l}_0 + \mathfrak{p}^\alpha/\mathfrak{z}^{(\alpha)} + \mathfrak{w}^{(\alpha)}/\mathfrak{w}^{(\alpha)}$$

$$\cong su(n-2)/u(n-3) + su(2)/u(1) + u(1)/u(1)$$

with the natural embeddings.

Moreover, the conjugating automorphism p is the identity automorphism on $\overset{\approx}{l}_0$ and non-trivial on $\mathfrak{t} = \mathfrak{w}^{(\alpha)} \oplus \mathfrak{z}^{(\alpha)}$, $p\mathfrak{t} = \mathfrak{t}$, and

$$p\mathfrak{p}^{(1)} = \mathfrak{p}^{(2)}, \qquad p\mathfrak{p}^{(2)} = \mathfrak{p}^{(3)}, \qquad p\mathfrak{p}^{(3)} = \mathfrak{p}^{(1)}.$$

In addition, $p \in \mathrm{Int}_\mathfrak{g}(\mathfrak{b})$, where

$$\mathfrak{b} \subset \mathfrak{p}^{(1)} + \mathfrak{p}^{(2)} + \mathfrak{p}^{(3)} = \mathfrak{n} \cong su(3), \qquad \mathfrak{n}/\mathfrak{b} \cong su(3)/so(3),$$

with the natural embeddings.

II.10.10. For the Lie algebra $so(n)$ we might use the considerations of II.10.2 (Theorem 46) as well, since the special unitary subalgebra is $\mathfrak{k} \cong so(8)$.

But here another way is more effective.

Indeed, let us take

$$so(3) \subset su(3) \subset so(6) \subset so(8) \subset so(n)$$

with the natural embeddings. For certainty we may consider that $so(6)$ is represented by matrices in the upper left corner and the involutive automorphisms $\tilde{S}_\alpha \in \mathrm{Int}\,_\mathfrak{g}(su(3))$ are represented by diagonal matrices.

Thus we have

$$
\tilde{\tilde{S}}_1 =
\left\|
\begin{array}{cccccc|c}
-1 & 0 & 0 & 0 & 0 & 0 & \\
0 & -1 & 0 & 0 & 0 & 0 & \\
0 & 0 & 1 & 0 & 0 & 0 & \overline{\overline{O}} \\
0 & 0 & 0 & -1 & 0 & 0 & \\
0 & 0 & 0 & 0 & -1 & 0 & \\
0 & 0 & 0 & 0 & 0 & 1 & \\
\hline
& & \overline{\overline{O}} & & & & \overline{\overline{\mathrm{Id}}}
\end{array}
\right\| ,
$$

$$
\tilde{\tilde{S}}_2 =
\left\|
\begin{array}{cccccc|c}
1 & 0 & 0 & 0 & 0 & 0 & \\
0 & -1 & 0 & 0 & 0 & 0 & \\
0 & 0 & -1 & 0 & 0 & 0 & \overline{\overline{O}} \\
0 & 0 & 0 & 1 & 0 & 0 & \\
0 & 0 & 0 & 0 & -1 & 0 & \\
0 & 0 & 0 & 0 & 0 & -1 & \\
\hline
& & \overline{\overline{O}} & & & & \overline{\overline{\mathrm{Id}}}
\end{array}
\right\| ,
$$

$$
\tilde{\tilde{S}}_3 =
\left\|
\begin{array}{cccccc|c}
-1 & 0 & 0 & 0 & 0 & 0 & \\
0 & 1 & 0 & 0 & 0 & 0 & \\
0 & 0 & -1 & 0 & 0 & 0 & \overline{\overline{O}} \\
0 & 0 & 0 & -1 & 0 & 0 & \\
0 & 0 & 0 & 0 & 1 & 0 & \\
0 & 0 & 0 & 0 & 0 & -1 & \\
\hline
& & \overline{\overline{O}} & & & & \overline{\overline{\mathrm{Id}}}
\end{array}
\right\| ,
$$

where $\overline{\overline{O}}$ are zero matrices, and $\overline{\overline{\mathrm{Id}}}$ are the identity matrices,

$$\tilde{\tilde{S}}_\alpha : \bar{a} \mapsto \tilde{\tilde{S}}_\alpha\, \bar{a}\, \tilde{\tilde{S}}_\alpha, \qquad \bar{a} \in so(n) .$$

Hence in a straightforward way we arrive at the hyper-involutive decomposition:

$$\mathfrak{g} = \tilde{\mathfrak{l}}_1 + \tilde{\mathfrak{l}}_2 + \tilde{\mathfrak{l}}_3, \qquad \mathfrak{g}/\tilde{\mathfrak{l}}_\alpha \cong so(n)/so(4) \oplus so(n-4),$$

$$\tilde{\mathfrak{l}}_\alpha/\tilde{\mathfrak{l}}_0 \cong so(n-4)/(\,so(n-6) \oplus so(2)\,) + su(2)/so(2) + su(2)/so(2)$$

with the natural embeddings.

Thus we have proved the theorem:

II.10.11. Theorem 49. *Let $\mathfrak{g}$ be a Lie algebra, $\mathfrak{g} \cong so(n)$, $n > 5$, and let $\tilde{\mathfrak{l}}_1 = \mathfrak{p} \oplus \mathfrak{q} \oplus \tilde{\tilde{\mathfrak{l}}}_1$ be its principal di-unitary involutive algebra of an involutive automorphism $\widetilde{S}_1$,*

$$\widetilde{S}_1 \in \mathrm{Int}_{\mathfrak{g}}(\mathfrak{p}) \cong SU(2), \qquad \widetilde{S}_1 \in \mathrm{Int}_{\mathfrak{g}}(\mathfrak{q}) \cong SU(2),$$

that is, $\tilde{\mathfrak{l}}_1 \cong su(2) \oplus su(2) \oplus so(n-4)$.

Then there exists a hyper-involutive decomposition $\mathfrak{g} = \tilde{\mathfrak{l}}_1 + \tilde{\mathfrak{l}}_2 + \tilde{\mathfrak{l}}_3$ of principal di-unitary involutive automorphisms $\widetilde{S}_1, \widetilde{S}_2, \widetilde{S}_3$ such that

$$\tilde{\mathfrak{l}}_\alpha / \tilde{\mathfrak{l}}_0 = \tilde{\tilde{\mathfrak{l}}}_\alpha / \tilde{\tilde{\mathfrak{l}}}_0 + \mathfrak{p}^{(\alpha)} / \mathfrak{z}^\alpha + \mathfrak{q}^{(\alpha)} / \mathfrak{w}^{(\alpha)}$$

$$\cong so(n-4)/(\, so(n-6) \oplus so(2)\,) + su(2)/so(2) + su(2)/so(2)$$

with the natural embeddings.

The conjugating automorphism p is the identity automorphism on $\tilde{\tilde{\mathfrak{l}}}_0$, non-trivial on $\mathfrak{t} = \mathfrak{w}^{(\alpha)} + \mathfrak{z}^{(\alpha)}$, and $p\mathfrak{t} = \mathfrak{t}$.

Moreover,

$$p\mathfrak{p}^{(1)} = \mathfrak{p}^{(2)}, \qquad p\mathfrak{p}^{(2)} = \mathfrak{p}^{(3)}, \qquad p\mathfrak{p}^{(3)} = \mathfrak{p}^{(1)},$$

$$p\mathfrak{q}^{(1)} = \mathfrak{q}^{(2)}, \qquad p\mathfrak{q}^{(2)} = \mathfrak{q}^{(3)}, \qquad p\mathfrak{q}^{(3)} = \mathfrak{q}^{(1)}.$$

In addition, $p \in \mathrm{Int}_{\mathfrak{g}}(\mathfrak{b}) \cong SO(3)$, where

$$\mathfrak{b} \subset \mathfrak{p}^{(1)} + \mathfrak{p}^{(2)} + \mathfrak{p}^{(3)} = \mathfrak{n} \cong su(3), \qquad \mathfrak{n}/\mathfrak{b} \cong su(3)/so(3),$$

with the natural embeddings.

II.10.12. Definition 32. The hyper-involutive decompositions described by II.10.9 (Theorem 48), II.10.11 (Theorem 49) are called basis hyper-involutive for a principal central involutive automorphism of type U (or for the Lie algebra $su(n)$) and basis hyper-involutive for a principal involutive automorphism of type $U^{(2)}$ (or for the Lie algebra $so(n)$), respectively.

II.10.13. It is easily verified that for $sp(n)$, $n \neq 3$, there are no hyper-involutive decompositions analogous to the decompositions introduced above (that is, such that $\widetilde{S}_\alpha$ are principal involutive automorphisms of type U).

This may be achieved by means of the quaternionic model for $sp(n)$ and a representation of $\widetilde{S}_\alpha$ by diagonal matrices.

However, for $sp(3)$ such a decomposition exists (since in $sp(3)$ all special involutive automorphisms are principal) and is described in the next chapter.

CHAPTER II.11

SPECIAL HYPER-INVOLUTIVE SUMS

II.11.1. For the exceptional Lie algebras of types f and e one may construct hyper-involutive sums using only special unitary involutive algebras.

Let $\mathfrak{k}$ be a special unitary simple subalgebra of Lie algebra $\mathfrak{g}$ of an involutive automorphism S_1. Then, as we know, $\mathfrak{k} \cong so(m)$, $S_1 \in \text{Int}_\mathfrak{g}(\{t\zeta\})$, where $\{t\zeta\} \cong so(2)$ with the natural embedding into $\mathfrak{k} \cong so(m)$.

Since we are interested in the exceptional Lie algebras of types f and e we have $m > 8$. Therefore we may take in $\mathfrak{k} \cong so(m)$ a special unitary subalgebra $\mathfrak{q} \cong so(8)$ with the natural embedding into $\mathfrak{k}$.

Then evidently $S_1 \in \text{Int}_\mathfrak{g}(\mathfrak{q}) \cong \text{Spin}(8)$; moreover, there is also an involutive automorphism $S_2 \in \text{Int}_\mathfrak{g}(\mathfrak{q})$ such that $S_1 S_2 = S_2 S_1 = S_3$.

The restrictions of involutive automorphisms S_2 and S_3 to $\mathfrak{k}$ generate an involutive automorphism $\widetilde{S} = S_2|_\mathfrak{k} = S_3|_\mathfrak{k}$ with the involutive pair

$$\mathfrak{k}/\mathfrak{q} \oplus \tilde{\mathfrak{q}} \cong so(m)/so(8) \oplus so(m-8)$$

with the natural embedding.

Let us consider the involutive sum $\mathfrak{g} = \mathfrak{l}_1 + \mathfrak{l}_2 + \mathfrak{l}_3$ generated by the involutive automorphisms S_1, S_2, S_3. Then

$$\mathfrak{l}_1 = \mathfrak{k} \oplus \mathfrak{n}_1, \qquad \mathfrak{l}_0 = \mathfrak{q} \oplus \tilde{\mathfrak{q}} \oplus \mathfrak{n}_1,$$

and $\mathfrak{l}_2/\mathfrak{l}_0$, $\mathfrak{l}_3/\mathfrak{l}_0$ are the special unitary involutive pairs of the involutive automorphism S_1.

Furthermore, we have

$$\mathfrak{l}_2/\mathfrak{l}_0 = \mathfrak{l}_2^{(1)}/\mathfrak{q} \oplus \mathfrak{t} + \mathfrak{l}_2^{(2)}/\mathfrak{c},$$

where $\mathfrak{l}_2^{(1)}/\mathfrak{q} \oplus \mathfrak{t}$ is an elementary involutive pair.

But then $\mathfrak{l}_2^{(2)} \subset \mathfrak{l}_0$ (since $S_2 \in \text{Int}_\mathfrak{g}(\mathfrak{q})$, $S_3 \in \text{Int}_\mathfrak{g}(\mathfrak{q})$).

Therefore

$$\mathfrak{l}_2/\mathfrak{l}_0 = \mathfrak{l}_2^{(1)}/\mathfrak{q} \oplus \mathfrak{t} + \mathfrak{l}_2^{(2)}/\mathfrak{l}_2^{(2)},$$

where $\mathfrak{l}_2^{(1)}/\mathfrak{q} \oplus \mathfrak{t}$ is an elementary special unitary involutive pair with the special subalgebra $\mathfrak{q} \cong so(8)$.

Consequently by II.9.10 (Theorem 45)

$$\mathfrak{l}_2^{(1)}/\mathfrak{q} \oplus \mathfrak{t} \cong so(k)/so(8) \oplus so(k-8)$$

with the natural embedding.

(The case $\mathfrak{l}_2^{(1)} = \mathfrak{q}' \oplus \mathfrak{q}''$, where $\mathfrak{q}$ is a diagonal of the canonical involutive automorphism $\mathfrak{q}' \rightleftarrows \mathfrak{q}''$, is, of course, impossible, otherwise $S_\alpha \in \mathrm{Int}_\mathfrak{g}(\mathfrak{q})$, that is, the centre is non-trivial. But then S_α belongs to the centre in $\mathrm{Int}_\mathfrak{g}(\mathfrak{q}' \oplus \mathfrak{q}'')$, which is possible only if $\mathfrak{q}' \oplus \mathfrak{q}'' \subset \mathfrak{l}_2^{(1)} \subset \mathfrak{l}_0$. But then $\mathfrak{l}_2 \subset \mathfrak{l}_0$, which is impossible).

Since $S_2 \in \mathrm{Int}_\mathfrak{g}(\mathfrak{q}) \cong \mathrm{Spin}(8)$, it is evident that $S_2 \in \mathrm{Int}_\mathfrak{g}(\tilde{\tilde{\mathfrak{q}}})$, where $\tilde{\tilde{\mathfrak{q}}} \cong so(2)$, and $\tilde{\tilde{\mathfrak{q}}} \subset \mathfrak{q} \subset \mathfrak{l}_2^{(1)}$ is a chain isomorphic to

$$so(2) \subset so(8) \subset so(k)$$

with the natural embeddings.

Owing to the conjugacy of all subalgebras $so(2)$ in $so(k)$ by inner automorphisms we then have $S_2 \in \mathrm{Int}_\mathfrak{g}(\{t\zeta\})$, where $\{t\zeta\}$ is isomorphic to any $so(2)$ with the natural embedding into $\mathfrak{l}_2^{(1)} \cong so(k)$.

Therefore $\mathfrak{l}_2^{(1)}$ is a special unitary subalgebra of $\mathfrak{g}$.

Now, taking $\{t\zeta\} \subset \mathfrak{l}_2^{(1)} \dot{-} (\mathfrak{q} \oplus \mathfrak{t})$, $\{t\zeta\} \cong so(2)$, with the natural embedding into $\mathfrak{l}_2^{(1)}$, we see that $\mathfrak{g} = \mathfrak{l}_1 + \mathfrak{l}_3 + \mathfrak{l}_2$ is an iso-involutive sum and $\mathfrak{l}_3 = \varphi\,\mathfrak{l}_1$, where the conjugating automorphism $\varphi \in \mathrm{Int}_\mathfrak{g}(\{t\zeta\})$.

The arguments presented above may be repeated for the involutive pair $\mathfrak{l}_3/\mathfrak{l}_0$, where $\mathfrak{l}_2 = \psi\,\mathfrak{l}_1$, $\psi \in \mathrm{Int}_\mathfrak{g}(\mathfrak{l}_3)$.

As a result

$$\mathfrak{l}_\alpha/\mathfrak{l}_0 = \mathfrak{l}_\alpha^{(1)}/\mathfrak{q} \oplus \mathfrak{t}_\alpha + \mathfrak{l}_\alpha^{(2)}/\mathfrak{l}_\alpha^{(2)},$$

$$\mathfrak{l}_\alpha^{(1)}/\mathfrak{q} \oplus \mathfrak{t}_\alpha \cong so(k)/(\,so(8) \oplus so(k-8)\,).$$

Now, let us take

$$(\mathfrak{p}^{(1)} \oplus \mathfrak{p}^{(2)}) \oplus (\mathfrak{p}^{(3)} \oplus \mathfrak{p}^{(4)}) \cong so(4) \oplus so(4)$$

with the natural embedding into $\mathfrak{q} \cong so(8)$ and into $\mathfrak{l}_\alpha^{(1)} \cong so(k)$.

If $\tilde{S}_q \in \mathrm{Int}_\mathfrak{g}(\mathfrak{p}^{(a)}) \cong SU(2)$ ($a = 1,2,3,4$) then we have $\tilde{S}_1\tilde{S}_2 = \tilde{S}_3\tilde{S}_4 = S_1$, $\tilde{S}_1\tilde{S}_3 = \tilde{S}_2\tilde{S}_4 = S_2$, $\tilde{S}_1\tilde{S}_4 = \tilde{S}_2\tilde{S}_3 = S_3$.

The involutive automorphisms $\tilde{S}_q$ are principal (indeed, by construction $\tilde{S}_1$, $\tilde{S}_2$, S_1 generate an iso-involutive decomposition basis for $\tilde{S}_1$ and $\tilde{S}_2$; analogously, $\tilde{S}_3$, $\tilde{S}_4$, S_1 generate an iso-involutive decomposition basis for $\tilde{S}_3$ and $\tilde{S}_4$).

Then denoting the involutive algebras of the involutive automorphisms $\tilde{S}_a$ by $\tilde{\mathfrak{l}}_a$ we have $\tilde{\mathfrak{l}}_a = \mathfrak{p}^{(a)} \oplus \tilde{\tilde{\mathfrak{l}}}_a$, where $\tilde{\tilde{\mathfrak{l}}}_a$ is the maximal subalgebra of elements commuting with $\mathfrak{p}^{(a)}$, and $\mathfrak{p}^{(a)} \subset \tilde{\tilde{\mathfrak{l}}}_b$ for $a \neq b$.

We also note that

$$(\mathfrak{p}^{(1)} \oplus \mathfrak{p}^{(2)} \oplus \mathfrak{p}^{(3)} \oplus \mathfrak{p}^{(4)}) \subset \mathfrak{q} \subset \mathfrak{l}_\alpha^{(1)}$$

is isomorphic to

$$so(4) \oplus so(4) \subset so(8) \subset so(k)$$

with the natural embeddings.

Let $\tilde{\tilde{l}}_{\alpha b} = l_\alpha \cap \tilde{\tilde{l}}_b$, then evidently $\tilde{\tilde{l}}_b = \tilde{\tilde{l}}_{1b} + \tilde{\tilde{l}}_{2b} + \tilde{\tilde{l}}_{3b}$, where $\tilde{\tilde{l}}_{\alpha b}$ is the maximal subalgebra of elements of l_α which commute with $\mathfrak{p}^{(a)}$. In addition, evidently $\tilde{l}_{\alpha b} = l_\alpha \cap \tilde{l}_b = \tilde{\tilde{l}}_{\alpha b} \oplus \mathfrak{p}^{(a)}$ is a principal unitary involutive algebra in l_α of the involutive automorphism $\tilde{S}_p$.

Owing to what has been presented above we obtain

$$l_\alpha / \tilde{l}_{\alpha b} \oplus \mathfrak{p}^{(b)} = l_\alpha^{(1)} / (l_{\alpha b}^{(1)} \oplus \mathfrak{p}^{(b)}) + l_\alpha^{(2)} / l_\alpha^{(2)},$$

$$l_\alpha^{(1)} / (l_{\alpha b}^{(1)} \oplus \mathfrak{p}^{(b)}) \cong so(k) / (so(k-4) \oplus so(4))$$

with the natural embeddings.

Here

$$l_{\alpha b}^{(1)} = \tilde{l}_{\alpha b}^{(1)} \oplus \mathfrak{q}_{\alpha b} \cong so(k-4) \oplus su(2)$$

and the involutive pair

$$\tilde{l}_b / \tilde{l}_{\alpha b} = \tilde{l}_b / (l_{\alpha b}^{(1)} \oplus l_\alpha^{(2)}) = \tilde{l}_b / (\tilde{l}_{\alpha b}^{(1)} \oplus \mathfrak{q}_{\alpha b} \oplus l_\alpha^{(2)})$$

is the principal unitary involutive pair of the involutive automorphism S_α.

Moreover, $\tilde{l}_{\alpha b} \cong so(k-4)$ is the special unitary subalgebra of the involutive automorphism S_α, and $\mathfrak{q}_{\alpha b}$ coincides with $\mathfrak{p}^{(a)}$ for some $a \neq b$.

Let us also consider $\tilde{\tilde{l}}_{0b} = \tilde{\tilde{l}}_{\alpha b} \cap \tilde{\tilde{l}}_{\beta b}$ $(\alpha \neq \beta)$.

Obviously $\tilde{\tilde{l}}_{0b} = l_0 \cap \tilde{\tilde{l}}_b$ is a maximal subalgebra of elements from l_0 which commute with $\mathfrak{p}^{(b)} \subset \mathfrak{q}$, thus

$$\tilde{l}_{0b}^{(1)} = (\mathfrak{q} \cap \tilde{\tilde{l}}_b) \oplus (l_\alpha^{(2)} \oplus \mathfrak{t}_\alpha) = \bigcup_{a \neq b} \mathfrak{p}^{(a)} \oplus (l_\alpha^{(2)} \oplus \mathfrak{t}_\alpha).$$

And

$$\tilde{\tilde{l}}_{\alpha b} / \tilde{l}_{0b}^{(1)} = \tilde{l}_{\alpha b}^{(1)} / (\mathfrak{p}^{(q_1)} \oplus \mathfrak{p}^{(q_2)} \oplus \mathfrak{t}_\alpha) + \mathfrak{p}^{(q_3)} / \mathfrak{p}^{(q_3)} + l_\alpha^{(2)} / l_\alpha^{(2)},$$

where the involutive pair

$$\tilde{l}_{\alpha b}^{(1)} / (\mathfrak{p}^{(q_1)} \oplus \mathfrak{p}^{(q_2)} \oplus \mathfrak{t}_\alpha)$$

is principal di-unitary.

Thus $\tilde{\tilde{l}}_b = \tilde{\tilde{l}}_{1b} + \tilde{\tilde{l}}_{2b} + \tilde{\tilde{l}}_{3b}$ is an involutive sum, moreover,

$$\tilde{\tilde{l}}_{\alpha b} / \tilde{l}_{0b}^{(1)} \cong \mathfrak{a} / \mathfrak{c} + \mathfrak{q}_{\alpha b} / \mathfrak{q}_{\alpha b} + l_\alpha^{(2)} / l_\alpha^{(2)},$$

$$\mathfrak{a} / \mathfrak{c} \cong so(k-4) / so(4) \oplus so(k-8)$$

with the natural embeddings.

If $\mathfrak{g} \cong e_8$ then

$$\tilde{\mathfrak{l}}_1 \cong e_7 \oplus su(2), \qquad \tilde{\tilde{\mathfrak{l}}}_1/\tilde{\mathfrak{l}}_{\alpha 1} \cong e_7/so(12) \oplus su(2),$$

$$\tilde{\tilde{\mathfrak{l}}}_{\alpha 1}/\tilde{\tilde{\mathfrak{l}}}_{0 1} = \tilde{\mathfrak{l}}_{\alpha 1}^{(1)}/(\mathfrak{p}^{(2)} \oplus \mathfrak{p}^{(3)} \oplus \mathfrak{t}) + \mathfrak{p}^{(1)}/\mathfrak{p}^{(1)}$$

$$\cong so(12)/so(4) \oplus so(8) + su(2)/su(2),$$

with the natural embeddings.

Now we consider a diagonal $\mathfrak{d} = \Delta(\mathfrak{p}^{(2)} \oplus \mathfrak{p}^{(3)} \oplus \mathfrak{p}^{(1)})$ and the subalgebra $\mathfrak{f}$ of all elements of $\tilde{\tilde{\mathfrak{l}}}_1 \cong e_7$ which commute with $\mathfrak{d}$. It is evident that $\mathfrak{f} = \mathfrak{f}_1 + \mathfrak{f}_2 + \mathfrak{f}_3$, where $\mathfrak{f}_\alpha$ is the maximal subalgebra of elements from $\tilde{\tilde{\mathfrak{l}}}_{\alpha 1}$ which commute with $\mathfrak{d}$. Since $\mathfrak{p}^{(\alpha)}$ commutes with $\tilde{\tilde{\mathfrak{l}}}_{\alpha 1}$ we see that $\mathfrak{f}_1$ is a maximal subalgebra of elements from $\tilde{\tilde{\mathfrak{l}}}_{11}$ which commute with $\Delta(\mathfrak{p}^{(2)} \oplus \mathfrak{p}^{(2)})$ etc..

Let us note that $\Delta(\mathfrak{p}^{(2)} \oplus \mathfrak{p}^{(2)}) \cong so(3)$ with the natural embedding into $\tilde{\tilde{\mathfrak{l}}}_{11} \cong so(12)$, therefore $\mathfrak{f}_1 \cong so(9)$ with the natural embedding into $\tilde{\tilde{\mathfrak{l}}}_{11} \cong so(12)$. Analogously, $\mathfrak{f}_\alpha \cong so(9) \subset so(12) \cong \tilde{\tilde{\mathfrak{l}}}_{\alpha 1}$ with the natural embedding.

And from the construction, $\mathfrak{f}_0 = \mathfrak{f} \cap \tilde{\tilde{\mathfrak{l}}}_{01} = \mathfrak{t} \cong so(8)$ with the natural embedding into $\mathfrak{f}_\alpha \cong so(9) \subset so(12) \cong \tilde{\tilde{\mathfrak{l}}}_{\alpha 1}$.

According to the construction $\mathfrak{f}_\alpha$ is the special unitary subalgebra in $\mathfrak{f}$ of the involutive automorphism S_α.

Therefore

$$\mathfrak{f}/\mathfrak{f}_\alpha \cong \mathfrak{f}_4/so(9), \qquad \mathfrak{f}_\alpha/\mathfrak{f}_0 \cong so(9)/so(8),$$

where the latter is also a special unitary involutive pair.

Let us take

$$\mathfrak{h}_1 \oplus \mathfrak{h}_2 \oplus \mathfrak{h}_3 \oplus \mathfrak{h}_4 \cong so(4) \oplus so(4) \subset so(8)$$

$$\cong \mathfrak{t} \subset \mathfrak{f}_\alpha \cong so(9)$$

with the natural embedding, and also a diagonal

$$\Delta(\mathfrak{h}_1 \oplus \mathfrak{h}_2 \oplus \mathfrak{h}_3 \oplus \mathfrak{h}_4) = \mathfrak{a}$$

$$\cong \Delta(so(3) \oplus so(3)) \subset so(6) \subset so(8)$$

$$\cong \mathfrak{t} \subset \mathfrak{f}_\alpha \cong so(9)$$

with the natural embeddings.

Lastly, let us consider a maximal subalgebra $\mathfrak{e} \subset \mathfrak{f}$ of elements commuting with $\mathfrak{a}$. Evidently $\mathfrak{e} = \mathfrak{e}_1 + \mathfrak{e}_2 + \mathfrak{e}_3$, where $\mathfrak{e}_\alpha$ is a maximal subalgebra of elements of $\mathfrak{f}_\alpha$ which commute with $\mathfrak{a}$.

Then $e_0 = e_\alpha \cap e_\beta$ $(\alpha \neq \beta)$; moreover, $e_0 = e \cap f_0$ and it is a maximal subalgebra of elements from f_0 which commute with $\mathfrak{a}$.

By the construction we have

$$e_{(\alpha)} = \mathfrak{n}^{(\alpha)} \oplus \mathfrak{w}^{(\alpha)}, \qquad e_0 = \mathfrak{v}^{(\alpha)} \oplus \mathfrak{w}^{(\alpha)},$$

where $\mathfrak{n}^{(\alpha)} \cong so(3)$ with the natural embeddings into $f_\alpha \cong so(9)$. Thus

$$S_\alpha \in \mathrm{Int}_\mathfrak{g}(\mathfrak{n}^{(\alpha)}) \cong SU(2), \qquad \dim \mathfrak{w}^{(\alpha)} = \dim \mathfrak{v}^{(\alpha)} = 1.$$

Consequently $e/e_\alpha \cong su(3)/su(2) \oplus u(1)$. But then, according to II.3.10 (Theorem 6), II.3.14 (Theorem 10), we may take $\mathfrak{b} \subset e$ in such a way that $e/\mathfrak{b} \cong su(3)/so(3)$ with the natural embedding and so that $S_\alpha \in \mathrm{Int}_\mathfrak{g}(\mathfrak{b}) \cong SO(3)$ $(\alpha = 1, 2, 3)$ generate the written-above hyper-involutive decomposition for e, and consequently generate also the initial hyper-involutive decomposition for $\mathfrak{g}$.

The proof given above is suitable for $\mathfrak{g} \cong e_7$, $\mathfrak{g} \cong f_4$, as well, since we have passed step by step from e_8 to $e_7 \subset e_8$, and then to $f_4 \subset e_7$ performing the proof for f_4.

For the rest we should consider $\mathfrak{g} \cong e_6$.

Then

$$l_\alpha / l_0 = l_\alpha^{(1)}/\mathfrak{q} \oplus t_\alpha + l_\alpha^{(2)}/l_\alpha^{(2)}$$

$$\cong so(10)/so(8) \oplus so(2) + u(1)/u(1)$$

with the natural embedding.

Let us consider

$$so(4) \oplus so(4) \cong \mathfrak{h}_1 \oplus \mathfrak{h}_2 \oplus \mathfrak{h}_3 \oplus \mathfrak{h}_4 \subset \mathfrak{q} \cong so(8)$$

with the natural embedding, and also a diagonal

$$\mathfrak{a} = \Delta(\mathfrak{h}_1 \oplus \mathfrak{h}_2 \oplus \mathfrak{h}_3 \oplus \mathfrak{h}_4)$$

$$\cong \Delta\left(so(3) \oplus so(3) \right) \subset so(6) \subset so(8) \cong \mathfrak{q}$$

with the natural embedding.

Let e be the maximal subalgebra of elements from $\mathfrak{g}$ commuting with $\mathfrak{a}$. Evidently $e = e_1 + e_2 + e_3$, where e_α is the maximal subalgebra of elements of l_α commuting with $\mathfrak{a}$. $e_0 = e_\alpha \cap e_\beta$ $(\alpha \neq \beta)$ is then $e_0 = e \cap l_0$, which is the maximal subalgebra of elements of l_0 commuting with $\mathfrak{a}$.

Owing to the construction we have

$$e_\alpha = \mathfrak{n}^{(\alpha)} \oplus \mathfrak{m}^{(\alpha)} \oplus \mathfrak{w}^{(\alpha)} \oplus \mathfrak{v}^{(\alpha)},$$

where

$$\mathfrak{n}^{(\alpha)} \oplus \mathfrak{m}^{(\alpha)} \cong so(4), \qquad \dim \mathfrak{w}^{(\alpha)} = \dim \mathfrak{v}^{(\alpha)} = 1, \qquad \dim e_0 = 4.$$

Moreover,
$$\mathrm{Int}_{\mathfrak{g}}(\mathfrak{n}^{(\alpha)}) \cong \mathrm{Int}_{\mathfrak{g}}(\mathfrak{m}^{(\alpha)}) \cong SU(2).$$

However, $\mathrm{Int}_{\mathfrak{g}}(\mathfrak{n}^{(\alpha)} \oplus \mathfrak{m}^{(\alpha)}) \cong SU(2) \times SU(2)$ since $S_\alpha \in \mathrm{Int}_{\mathfrak{g}}(\Delta(\mathfrak{n}^{(\alpha)} \oplus \mathfrak{m}^{(\alpha)})) \cong SU(2)$.

Furthermore, the restriction $S_\alpha|_\mathfrak{e} \neq \mathrm{Id}$, thus

$$\mathrm{Id} \neq S_\alpha|_\mathfrak{e} \in \mathrm{Int}_{\mathfrak{g}}^\mathfrak{e}(\Delta(\mathfrak{n}^{(\alpha)} \oplus \mathfrak{m}^{(\alpha)})) = \mathrm{Int}_\mathfrak{e}(\Delta(\mathfrak{n}^{(\alpha)} \oplus \mathfrak{m}^{(\alpha)})),$$

whence either $S_\alpha|_\mathfrak{e} \in \mathrm{Int}_\mathfrak{e}(\mathfrak{n}^{(\alpha)})$ or $S_\alpha|_\mathfrak{e} \in \mathrm{Int}_\mathfrak{e}(\mathfrak{m}^{(\alpha)})$, or both are true.

Without loss of generality we may consider that $S_\alpha|_\mathfrak{e} \in \mathrm{Int}_\mathfrak{e}(\mathfrak{n}^{(\alpha)}) \cong SU(2)$ (otherwise we redenote $\mathfrak{n}^{(\alpha)}$ by $\mathfrak{m}^{(\alpha)}$ and vice versa).

If $S_\alpha|_\mathfrak{e} \notin \mathrm{Int}_\mathfrak{e}(\mathfrak{m}^{(\alpha)})$ (that is, $\mathrm{Int}_\mathfrak{e}(\mathfrak{m}^{(\alpha)}) \cong SO(3)$) then we consider a non-trivial unitary involutive automorphism $\widetilde{S}_\alpha \in \mathrm{Int}_{\mathfrak{g}}(\mathfrak{m}^{(\alpha)})$. This is a principal unitary involutive automorphism since $\mathfrak{m}^{(\alpha)} \cong su(2) \subset so(4)$ with the natural embeddings into the special subalgebra $\mathfrak{l}_\alpha^{(1)} \cong so(10)$.

In addition, by the assumption $\widetilde{S}_\alpha|_\mathfrak{e} = \mathrm{Id}$, and thus $\mathfrak{e}$ belongs to $\widetilde{\mathfrak{l}}_\alpha$ (being the involutive algebra of the involutive automorphism $\widetilde{S}_\alpha$).

But then, as we know, $\widetilde{\mathfrak{l}}_0 = \mathfrak{m}^{(\alpha)} \oplus \mathfrak{l}'_\alpha$, and consequently $\mathfrak{e} = \mathfrak{m}^{(\alpha)} \oplus \mathfrak{e}'_\alpha$.

If $\widetilde{S}_\alpha|_\mathfrak{e} = \mathrm{Id}$ for all $\alpha = 1, 2, 3$ then

$$\mathfrak{e} = \mathfrak{m}^{(1)} \oplus \mathfrak{m}^{(2)} \oplus \mathfrak{m}^{(3)} \oplus \mathfrak{e}'',$$

and evidently $\mathfrak{n}^{(\beta)} \subset \mathfrak{e}''$. Consequently $\mathfrak{n}^{(\alpha)}$ commutes with $\mathfrak{m}^{(\beta)}$.

But, since $S_\alpha \in \mathrm{Int}_{\mathfrak{g}}(\mathfrak{n}^{(\alpha)})$ we have

$$\mathfrak{m}^{(1)} \oplus \mathfrak{m}^{(2)} \oplus \mathfrak{m}^{(3)} \subset \mathfrak{e}_\alpha,$$

for any α, meaning that $\mathfrak{m}^{(1)} \oplus \mathfrak{m}^{(2)} \oplus \mathfrak{m}^{(3)} \subset \mathfrak{e}_0$. This contradicts the construction of $\mathfrak{e}_\alpha$ and $\mathfrak{e}_0$.

It remains to accept that for some $\alpha = \alpha_0$ we have

$$S_{\alpha_0}|_\mathfrak{e} \in \mathrm{Int}_\mathfrak{e}(\mathfrak{m}^{(\alpha_0)}), \qquad S_{\alpha_0}|_\mathfrak{e} \in \mathrm{Int}_\mathfrak{e}(\mathfrak{n}^{(\alpha_0)}), \qquad S_{\alpha_0}|_\mathfrak{e} \neq \mathrm{Id}.$$

Then
$$\mathfrak{e}/\mathfrak{e}_{\alpha_0} = \mathfrak{e}/\mathfrak{n}^{(\alpha_0)} \oplus \mathfrak{m}^{(\alpha_0)} \oplus \mathfrak{w}^{(\alpha_0)} \oplus \mathfrak{v}^{(\alpha_0)}$$

is a principal di-unitary involutive pair of an involutive automorphism $S_{\alpha_0}|_\mathfrak{e} \neq \mathrm{Id}$.

Now we can consider the decomposition of $\mathfrak{e}/\mathfrak{e}_{\alpha_0}$ into elementary involutive pairs. If we assume that

$$\mathfrak{e}/\mathfrak{e}_{\alpha_0} = \tilde{\mathfrak{e}}/(\mathfrak{n}^{(\alpha_0)} \oplus \mathfrak{m}^{(\alpha_0)} \oplus \mathfrak{e}_{\alpha_0}) + \widetilde{\tilde{\mathfrak{e}}}/\widetilde{\tilde{\mathfrak{e}}}_{\alpha_0},$$

where $\tilde{\mathfrak{e}}/(\mathfrak{n}^{(\alpha_0)} \oplus \mathfrak{m}^{(\alpha_0)} \oplus \mathfrak{e}_{\alpha_0})$ is elementary, then $\tilde{\mathfrak{e}}$ is simple and $S_{\alpha_0}|_{\tilde{\mathfrak{e}}} = \mathrm{Id}$, but then $S_{\alpha_0}|_\mathfrak{e} = \mathrm{Id}$, which is wrong.

Consequently we should assume that the decomposition into elementary involutive pairs has the form

$$e/e_{\alpha_0} = \tilde{e}/n^{(\alpha_0)} \oplus \tilde{\mathfrak{z}}^{(\alpha_0)} + \tilde{\tilde{e}}/m^{(\alpha_0)} \oplus \tilde{\tilde{\mathfrak{z}}}^{(\alpha_0)} + \tilde{\tilde{e}}/\tilde{e}.$$

However, $\tilde{\mathfrak{z}}^{(\alpha_0)}$ and $\tilde{\tilde{\mathfrak{z}}}^{(\alpha_0)}$ are non-trivial (otherwise we have involutive pairs of the form $\tilde{e}/n^{(\alpha_0)}$ or $\tilde{\tilde{e}}/m^{(\alpha_0)}$, that is, mono-unitary involutive pairs, where $\dim n^{(\alpha_0)} = \dim m^{(\alpha_0)} = 3$. But this is impossible according to I.8.2 (Theorem 31).

Thus the only possible case is

$$e/e_{\alpha_0} = \tilde{e}/n^{(\alpha_0)} \oplus \tilde{\mathfrak{z}}^{(\alpha_0)} + \tilde{\tilde{e}}/m^{(\alpha_0)} \oplus \tilde{\tilde{\mathfrak{z}}}^{(\alpha_0)},$$

where the involutive pairs in the right hand side are central unitary.

Consequently we have

$$\tilde{e}/n^{(\alpha_0)} \oplus \tilde{\mathfrak{z}}^{(\alpha_0)} \cong \tilde{\tilde{e}}/m^{(\alpha_0)} \oplus \tilde{\tilde{\mathfrak{z}}}^{(\alpha_0)} \cong su(3)/u(2)$$

with the natural embedding.

Thus

$$e = \tilde{e} \oplus \tilde{\tilde{e}} \cong su(3) \oplus su(3), \qquad S_\alpha(\tilde{e}) = \tilde{e}, \qquad S_\alpha(\tilde{\tilde{e}}) = \tilde{\tilde{e}}.$$

The involutive decomposition $e = e_1 + e_2 + e_3$ of the involutive automorphisms S_α ($\alpha = 1, 2, 3$) generates then the involutive decompositions

$$\tilde{e} = \tilde{e}_1 + \tilde{e}_2 + \tilde{e}_3, \quad \tilde{\tilde{e}} = \tilde{\tilde{e}}_1 + \tilde{\tilde{e}}_2 + \tilde{\tilde{e}}_3, \quad (\tilde{e}_\alpha = n^{(\alpha)}, \quad \tilde{\tilde{e}}_\alpha = m^{(\alpha)})$$

and simultaneously generates the involutive decomposition of the diagonal

$$\mathfrak{f} = \Delta(\tilde{e} \oplus \tilde{\tilde{e}}) \cong su(3),$$

$$\mathfrak{f} = \mathfrak{f}^{(1)} + \mathfrak{f}^{(2)} + \mathfrak{f}^{(3)} \quad (\mathfrak{f}^{(\alpha)} = \Delta(n^{(\alpha)} \oplus m^{(\alpha)}) \cong so(3))$$

with the natural embeddings into $\mathfrak{l}_{(\alpha)}^{(1)} \cong so(10)$.

Thus $S_\alpha \in \mathrm{Int}_\mathfrak{g}(\mathfrak{f}^{(\alpha)})$.

Consequently $\mathfrak{f} = \mathfrak{f}^{(1)} + \mathfrak{f}^{(2)} + \mathfrak{f}^{(3)}$ is a hyper-involutive decomposition for $\mathfrak{f} \cong su(3)$.

Let us now take $\mathfrak{b} \subset \mathfrak{f}$ in such a way that $S_\alpha \in \mathrm{Int}_\mathfrak{g}(\mathfrak{b})$ (that is, $\mathfrak{b} \cong so(3)$ with the natural embedding into $\mathfrak{f} \cong su(3)$, $\dim(\mathfrak{b} \cap \mathfrak{f}^{(\alpha)}) = 1$). The conjugating automorphism $\tilde{p} \in \mathrm{Int}_\mathfrak{f}(\mathfrak{b})$ of the hyper-involutive decomposition for $\mathfrak{f}$ generates the conjugating automorphism $p \in \mathrm{Int}_\mathfrak{g}(\mathfrak{b})$ of the involutive decomposition $\mathfrak{g} = \mathfrak{l}_1 + \mathfrak{l}_2 + \mathfrak{l}_3$ which, for this reason, is hyper-involutive.

Thus we have finished the construction of $\mathfrak{g} \cong e_6$.

Now we are going to formulate the results of the exploration presented above as a set of theorems.

II.11.2. Theorem 50. *Let* $\mathfrak{g} \cong f_4$, *and* $\mathfrak{l}_1$ *be its special non-principal unitary involutive algebra of an involutive automorphism* S_1.

Then there exists a hyper-involutive decomposition $\mathfrak{g} = \mathfrak{l}_1 + \mathfrak{l}_2 + \mathfrak{l}_3$ *such that*

$$\mathfrak{g}/\mathfrak{l}_\alpha \cong f_4/so(9), \qquad \mathfrak{l}_\alpha/\mathfrak{l}_0 \cong so(9)/so(8)$$

with the natural embeddings. (*Here* $\mathfrak{l}_0 = \mathfrak{l}_\alpha \cap \mathfrak{l}_\beta, \ \alpha \neq \beta$).

Moreover, the conjugating automorphism p *belongs to* $\text{Int}_\mathfrak{g}(\mathfrak{b}) \subset \text{Int}_\mathfrak{g}(\mathfrak{f})$, *where the involutive pair* $\mathfrak{f}/\mathfrak{b}$ *is isomorphic to* $su(3)/so(3)$ *with the natural embedding, and the subalgebra* $\mathfrak{f}$ *is the maximal subalgebra of elements of* $\mathfrak{g}$ *commuting with* $\Delta \subset \mathfrak{l}_0$, *where* Δ *is a diagonal in*

$$\mathfrak{r} \cong su(2) \oplus su(2) \oplus su(2) \oplus su(2)$$

$$\cong so(4) \oplus so(4) \subset so(8)$$

$$\cong \mathfrak{l}_0 \subset \mathfrak{l}_\alpha \cong so(9)$$

with the natural embeddings.

II.11.3. Theorem 51. *Let* $\mathfrak{g} \cong e_6$, *and* $\mathfrak{l}_1$ *be its special non-principal unitary involutive algebra of an involutive automorphism* S_1.

Then there exists a hyper-involutive decomposition $\mathfrak{g} = \mathfrak{l}_1 + \mathfrak{l}_2 + \mathfrak{l}_3$ *such that*

$$\mathfrak{g}/\mathfrak{l}_\alpha \cong e_6/so(10) \oplus u(1),$$

$$\mathfrak{l}_\alpha/\mathfrak{l}_0 \cong so(10)/so(8) \oplus so(2) + u(1)/u(1)$$

with the natural embeddings ($\mathfrak{l}_0 = \mathfrak{l}_\alpha \cap L_\beta, \ \alpha \neq \beta$).

Moreover, the conjugating automorphism p *belongs to* $\text{Int}_\mathfrak{g}(\mathfrak{b}) \subset \text{Int}_\mathfrak{g}(\mathfrak{f})$, *where the involutive pair is* $\mathfrak{f}/\mathfrak{b} \cong su(3)/so(3)$ *with the natural embedding, and the subalgebra* $\mathfrak{f}$ *is a diagonal of the canonical involutive automorphism* $\mathfrak{f}' \rightleftarrows \mathfrak{f}''$, *where* $\mathfrak{f}' \oplus \mathfrak{f}''$ *is the maximal subalgebra of elements from* $\mathfrak{g}$ *commuting with the subalgebra* $\Delta \subset \mathfrak{l}_0$, Δ *being a diagonal in*

$$\mathfrak{r} \cong su(2) \oplus su(2) \oplus su(2) \oplus su(2) \cong so(4) \oplus so(4) \subset so(8)$$

$$\cong \mathfrak{l}'_0 \subset \mathfrak{l}_0 \cong so(8) \oplus so(2) \subset so(10) \cong \mathfrak{l}'_\alpha \subset \mathfrak{l}_\alpha \cong so(10) \oplus u(1)$$

with the natural embedding (that is, Δ *is isomorphic to a diagonal in* $so(3) \oplus so(3)$ *with the natural embedding into* $so(8) \cong \mathfrak{l}'_0$).

II.11.4. Theorem 52. *Let* $\mathfrak{g} \cong e_7$, *and* $\mathfrak{l}_1$ *be its special unitary involutive algebra of an involutive automorphism* S_1. *Then there exists a hyper-involutive decomposition* $\mathfrak{g} = \mathfrak{l}_1 + \mathfrak{l}_2 + \mathfrak{l}_3$ *of involutive automorphisms* S_1, S_2, S_3 *such that*

$$\mathfrak{l}_\alpha = \tilde{\mathfrak{l}}_\alpha \oplus \mathfrak{p}^{(\alpha)} \cong so(12) \oplus su(2), \qquad \mathfrak{g}/\mathfrak{l}_\alpha \cong e_7/so(12) \oplus su(2),$$

$$\mathfrak{l}_\alpha/\mathfrak{l}_0 \cong so(12)/so(8) \oplus so(4) + su(2)/su(2)$$

with the natural embedding.

The subalgebra of all elements of $\mathfrak{g} \cong e_7$ *which commute with a diagonal in* $\mathfrak{p}^{(1)} + \mathfrak{p}^{(2)} + \mathfrak{p}^{(3)}$ *is* $\tilde{\mathfrak{g}} \cong f_4$, *and the involutive automorphisms* S_1, S_2, S_3 *induce in* $\tilde{\mathfrak{g}}$ *the hyper-involutive decomposition described by II.11.2 (Theorem 50).*

The hyper-involutive decompositions for $\mathfrak{g}$ *and* $\tilde{\mathfrak{g}}$ *have the same conjugating automorphism* p.

II.11.5. Theorem 53. *Let* $\mathfrak{g} \cong e_8$, *and* $\mathfrak{l}_1$ *be its special non-principal unitary involutive algebra of an involutive automorphism* S_1. *Then there exists the hyper-involutive decomposition* $\mathfrak{g} = \mathfrak{l}_1 + \mathfrak{l}_2 + \mathfrak{l}_3$ *of the involutive automorphisms* S_1, S_2, S_3 *such that*

$$\mathfrak{g}/\mathfrak{l}_\alpha \cong e_8/so(16), \qquad \mathfrak{l}_\alpha/\mathfrak{l}_0 = \mathfrak{l}_\alpha/\mathfrak{l}'_0 \oplus \mathfrak{l}''_0 \cong so(16)/so(8) \oplus so(8)$$

with the natural embeddings.

If

$$\mathfrak{p} \cong su(2) \subset so(4) \subset so(8) \cong \mathfrak{l}'_0$$

with the natural embeddings then the subalgebra of all elements from $\mathfrak{g} \cong e_8$ *commuting with* $\mathfrak{p}$ *is* $\tilde{\mathfrak{g}} \cong e_7$, *and the involutive automorphisms* S_1, S_2, S_3 *induce in* $\tilde{\mathfrak{g}}$ *the hyper-involutive decomposition described in II.11.4 (Theorem 52).*

The hyper-involutive decompositions for $\mathfrak{g}$ *and* $\tilde{\mathfrak{g}}$ *have the same conjugating automorphism* p.

II.11.6. Definition 33. The hyper-involutive decompositions described in II.11.2 (Theorem 50), II.11.3 (Theorem 51), II.11.4 (Theorem 52), II.11.5 (Theorem 53) are called special hyper-involutive sums for the Lie algebras of types f_4, e_6, e_7, e_8, respectively.

II.11.7. Remark. The special hyper-involutive sum for $\mathfrak{g} \cong e_7$ coincides with the involutive sum of II.8.3 (Theorem 37), and consequently II.11.4 (Theorem 52) contains the proof that this involutive sum is hyper-involutive.

II.11.8. Now we consider the situation for the Lie algebras of types $so(n)$, $su(n)$, $sp(n)$. Here the best way is to use their matrix realizations.

II.11.9. Let $\mathfrak{g} \cong so(n)$, and S_1, S_2, S_3 be its special non-principal involutive automorphisms. In this case the special non-principal unitary subalgebra has a form $so(8)$ with the natural embedding.

Therefore we may consider $\mathfrak{g} \cong so(n)$ as being realized by all skew-symmetric matrices, while the special unitary involutive automorphisms S_1, S_2, S_3 are represented by the matrices $\overline{\overline{S}}_1$, $\overline{\overline{S}}_2$, $\overline{\overline{S}}_3$ of the form

$$\overline{\overline{S}}_1 = \left\| \begin{array}{cc} -\overline{\overline{\mathrm{Id}}} & \overline{\overline{O}} \\ \overline{\overline{O}} & \overline{\overline{\mathrm{Id}}} \end{array} \right\| \begin{array}{l} \}8 \\ \}n-8 \end{array} , \qquad \overline{\overline{S}}_2 = \left\| \begin{array}{ccc} \overline{\overline{\mathrm{Id}}} & \overline{\overline{O}} & \overline{\overline{O}} \\ \overline{\overline{O}} & -\overline{\overline{\mathrm{Id}}} & \overline{\overline{O}} \\ \overline{\overline{O}} & \overline{\overline{O}} & \overline{\overline{\mathrm{Id}}} \end{array} \right\| \begin{array}{l} \}4 \\ \}8 \\ \}n-12 \end{array} ,$$

$$\underbrace{\qquad}_{8} \underbrace{\qquad}_{n-8} \qquad\qquad \underbrace{\quad}_{4} \underbrace{\quad}_{8} \underbrace{\quad}_{n-12}$$

$$\overline{\overline{S}}_3 = \left\| \begin{array}{cccc} -\overline{\overline{\mathrm{Id}}} & \overline{\overline{O}} & \overline{\overline{O}} & \overline{\overline{O}} \\ \overline{\overline{O}} & \overline{\overline{\mathrm{Id}}} & \overline{\overline{O}} & \overline{\overline{O}} \\ \overline{\overline{O}} & \overline{\overline{O}} & -\overline{\overline{\mathrm{Id}}} & \overline{\overline{O}} \\ \overline{\overline{O}} & \overline{\overline{O}} & \overline{\overline{O}} & \overline{\overline{\mathrm{Id}}} \end{array} \right\| \begin{array}{l} \}4 \\ \}4 \\ \}4 \\ \}n-12 \end{array} ,$$

$$\underbrace{\quad}_{4} \underbrace{\quad}_{4} \underbrace{\quad}_{4} \underbrace{\quad}_{n-12}$$

where $\overline{\overline{O}}$ are zero matrices and $\overline{\overline{\mathrm{Id}}}$ are the identity matrices.

Moreover,

$$S_\alpha: \bar{\bar{a}} \mapsto \bar{\bar{S}}_\alpha \, \bar{\bar{a}} \, \bar{\bar{S}}_\alpha, \quad \bar{\bar{a}} \in so(n).$$

In this way we see that this construction is possible only for $n \geq 12$. The involutive automorphisms S_1, S_2, S_3 generate the involutive decomposition

$$\mathfrak{g} = \mathfrak{l}_1 + \mathfrak{l}_2 + \mathfrak{l}_3 \quad (\mathfrak{l}_0 = \mathfrak{l}_\alpha \cap \mathfrak{l}_\beta, \ \alpha \neq \beta),$$

$$\mathfrak{g}/\mathfrak{l}_\alpha \cong so(n)/so(8) \oplus so(n-8),$$

$$\mathfrak{l}_\alpha/\mathfrak{l}_0 \cong \mathfrak{k}_\alpha/\mathfrak{m}_\alpha \oplus \mathfrak{n}_\alpha + \mathfrak{l}'_\alpha/\mathfrak{l}'_0 \oplus \mathfrak{t}_\alpha$$

$$\cong so(8)/so(4) \oplus so(4) + so(n-8)/so(n-12) \oplus so(4),$$

with the natural embeddings.

At the same time we have also

$$\mathfrak{k}_1 + \mathfrak{k}_2 + \mathfrak{k}_3 = \tilde{\mathfrak{k}} \cong so(12)$$

with the natural embeddings into $\mathfrak{g} \cong so(n)$ (note that $\tilde{\mathfrak{k}}$ may be characterized as the maximal subalgebra of elements in $\mathfrak{g}$ commuting with $\mathfrak{l}'_0 \cong so(n-12)$).

In addition we note that the involutive automorphisms S_1, S_2, S_3 generate, as well, in $\tilde{\mathfrak{k}}$ the involutive decomposition

$$\tilde{\mathfrak{k}} = \tilde{\mathfrak{k}}_1 + \tilde{\mathfrak{k}}_2 + \tilde{\mathfrak{k}}_3,$$

where $\tilde{\mathfrak{k}}_\alpha = \mathfrak{k}_\alpha \oplus \mathfrak{t}_\alpha$, and

$$\tilde{\mathfrak{k}}_0 = \tilde{\mathfrak{k}}_\alpha \underset{\alpha \neq \beta}{\cap} \tilde{\mathfrak{k}}_\beta = \mathfrak{m}_\alpha \oplus \mathfrak{n}_\alpha \oplus \mathfrak{t}_\alpha.$$

In the realization in our model we have:

$$\tilde{\mathfrak{k}}_\alpha = \mathfrak{k}_\alpha \oplus \mathfrak{t}_\alpha \cong so(8) \oplus so(4), \qquad \tilde{\mathfrak{k}}_0 \cong so(4) \oplus so(4) \oplus so(4),$$

$$\tilde{\mathfrak{k}}/\tilde{\mathfrak{k}}_\alpha \cong so(12)/so(8) \oplus so(4),$$

$$\tilde{\mathfrak{k}}_\alpha/\tilde{\mathfrak{k}}_0 \cong \mathfrak{k}_\alpha/\mathfrak{m}_\alpha \oplus \mathfrak{n}_\alpha + \mathfrak{t}_\alpha/\mathfrak{t}_\alpha \cong so(8)/so(4) \oplus so(4) + so(4)/so(4)$$

with the natural embeddings.

Let us take a diagonal Δ in $\mathfrak{m}_\alpha \oplus \mathfrak{n}_\alpha \oplus \mathfrak{t}_\alpha$, then

$$\Delta \cong \Delta\left(so(4) \oplus so(4) \oplus so(4)\right) \subset so(8) \oplus so(4) \subset so(12) \cong \tilde{\mathfrak{k}}$$

with the natural embedding.

We consider in $\tilde{\mathfrak{k}}$ the subalgebra $\mathfrak{b}$ of elements commuting with Δ. Since $\Delta \subset \tilde{\mathfrak{k}}_0$ we have $\mathfrak{b} = \mathfrak{b}^{(1)} + \mathfrak{b}^{(2)} + \mathfrak{b}^{(3)}$, where $\mathfrak{b}^{(\alpha)}$ is the maximal subalgebra of elements of $\tilde{\mathfrak{k}}_\alpha = \mathfrak{k}_\alpha \oplus \mathfrak{t}_\alpha$ commuting with Δ.

But then $\mathfrak{b}^{(\alpha)}$ is the maximal subalgebra of $\mathfrak{k}_\alpha$ which commutes with a diagonal $\mathfrak{d}$ of $\mathfrak{m}_\alpha \oplus \mathfrak{n}_\alpha$. However, $\mathfrak{d} \cong \Delta\,(\,so(4) \oplus so(4)\,)$ (that is, isomorphic to a diagonal in $so(4) \oplus so(4)$) with the natural embedding into $\mathfrak{k}_\alpha \cong so(8)$.

Therefore

$$\mathfrak{b}^{(\alpha)} \cong \Delta\,(\,so(2) \oplus so(2) \oplus so(2) \oplus so(2)\,)$$

with the natural embedding into $\mathfrak{k}_\alpha \cong so(8)$, and in addition

$$\mathfrak{b}^{(\alpha)} \subset \mathfrak{k}_\alpha \dot{-} (\mathfrak{m}_\alpha \oplus \mathfrak{n}_\alpha)\,.$$

Thus $\dim \mathfrak{b} = 3$, $\dim \mathfrak{b}^{(\alpha)} = 1$.

Finally, by construction $S_\alpha \in \mathrm{Int}_{\mathfrak{g}}(\mathfrak{b}^{(\alpha)})$ (since $\mathfrak{k}_\alpha \cong so(8)$ is a special unitary subalgebra in $so(12)$).

Now, taking an automorphism $p \in \mathrm{Int}_{\mathfrak{g}}(\mathfrak{b}) \cong SO(3)$ in such a way that $p\,\mathfrak{b}^{(1)} = \mathfrak{b}^{(2)}$, $p\,\mathfrak{b}^{(2)} = \mathfrak{b}^{(3)}$, $p\,\mathfrak{b}^{(3)} = \mathfrak{b}^{(1)}$, we see that p is the conjugating automorphism of the hyper-involutive decomposition $\mathfrak{g} = \mathfrak{l}_1 + \mathfrak{l}_2 + \mathfrak{l}_3$.

We have obtained the theorem:

II.11.10. Theorem 54. *Let* $\mathfrak{g} \cong so(n)$ *(* $n \geq 12$ *), and* $\mathfrak{l}_1 \cong so(8) \oplus so(n-8)$ *be its special unitary involutive algebra of an involutive automorphism* S_1.

Then there exists the hyper-involutive decomposition $\mathfrak{g} = \mathfrak{l}_1 + \mathfrak{l}_2 + \mathfrak{l}_3$ *of the involutive automorphisms* S_1, S_2, S_3 *such that*

$$\mathfrak{g}/\mathfrak{l}_\alpha \cong so(n)/so(8) \oplus so(n-8),$$

$$\mathfrak{l}_\alpha/\mathfrak{l}_0 = \mathfrak{k}_\alpha/\mathfrak{m}_\alpha \oplus \mathfrak{n}_\alpha = \mathfrak{l}'_\alpha/\mathfrak{l}'_0 \oplus \mathfrak{t}_\alpha$$

$$\cong so(8)/so(4) \oplus so(4) + so(n-8)/so(n-12) \oplus so(4)$$

with the natural embeddings.

Moreover,

$$S_\alpha \in \mathrm{Int}_{\mathfrak{g}}(\mathfrak{b}^{(\alpha)}) \subset \mathrm{Int}_{\mathfrak{g}}(\mathfrak{b}) \cong SO(3),$$

where $\mathfrak{b}^{(\alpha)}$ *is the maximal subalgebra of elements in* $\mathfrak{k}_\alpha \cong so(8)$ *commuting with a diagonal* $\mathfrak{d}$ *of*

$$\mathfrak{m}_\alpha \oplus \mathfrak{n}_\alpha \cong so(4) \oplus so(4)\,,$$

and $\dim \mathfrak{b}^{(\alpha)} = 1$.

II.11.11. We now consider the case $\mathfrak{g} \cong su(n)$. Then a special unitary non-principal subalgebra has the form $su(4)$ with the natural embedding into $su(n)$.

Let S_1, S_2, S_3 be special non-principal unitary involutive automorphisms such that

$$S_\alpha S_\beta = S_\gamma, \quad \alpha \neq \beta, \quad \beta \neq \gamma, \quad \gamma \neq \alpha.$$

Then we may consider that $\mathfrak{g} \cong su(n)$ is realized as the Lie algebra of all skew-Hermitian matrices and the involutive automorphisms S_1, S_2, S_3 are represented by the diagonal matrices $\bar{\bar{S}}_1, \bar{\bar{S}}_2, \bar{\bar{S}}_3$ from $\mathrm{Int}(su(n))$ of the form:

$$\bar{\bar{S}}_1 = \left\| \begin{array}{cc} -\bar{\bar{\mathrm{Id}}} & \bar{\bar{O}} \\ \bar{\bar{O}} & \bar{\bar{\mathrm{Id}}} \end{array} \right\| \begin{array}{l} \}4 \\ \}n-4 \end{array}, \qquad \bar{\bar{S}}_2 = \left\| \begin{array}{ccc} \bar{\bar{\mathrm{Id}}} & \bar{\bar{O}} & \bar{\bar{O}} \\ \bar{\bar{O}} & -\bar{\bar{\mathrm{Id}}} & \bar{\bar{O}} \\ \bar{\bar{O}} & \bar{\bar{O}} & \bar{\bar{\mathrm{Id}}} \end{array} \right\| \begin{array}{l} \}4 \\ \}2 \\ \}n-6 \end{array},$$

$$\underbrace{\phantom{-\bar{\bar{\mathrm{Id}}}}}_{4} \underbrace{\phantom{\bar{\bar{\mathrm{Id}}}}}_{n-4} \qquad\qquad \underbrace{\phantom{\bar{\bar{\mathrm{Id}}}}}_{2} \underbrace{\phantom{-\bar{\bar{\mathrm{Id}}}}}_{4} \underbrace{\phantom{\bar{\bar{\mathrm{Id}}}}}_{n-6}$$

$$\bar{\bar{S}}_3 = \left\| \begin{array}{cccc} -\bar{\bar{\mathrm{Id}}} & \bar{\bar{O}} & \bar{\bar{O}} & \bar{\bar{O}} \\ \bar{\bar{O}} & \bar{\bar{\mathrm{Id}}} & \bar{\bar{O}} & \bar{\bar{O}} \\ \bar{\bar{O}} & \bar{\bar{O}} & -\bar{\bar{\mathrm{Id}}} & \bar{\bar{O}} \\ \bar{\bar{O}} & \bar{\bar{O}} & \bar{\bar{O}} & \bar{\bar{\mathrm{Id}}} \end{array} \right\| \begin{array}{l} \}2 \\ \}2 \\ \}2 \\ \}n-6 \end{array},$$

$$\underbrace{\phantom{-\bar{\bar{\mathrm{Id}}}}}_{2} \underbrace{\phantom{\bar{\bar{\mathrm{Id}}}}}_{2} \underbrace{\phantom{-\bar{\bar{\mathrm{Id}}}}}_{2} \underbrace{\phantom{\bar{\bar{\mathrm{Id}}}}}_{n-6}$$

where $\bar{\bar{O}}$ are zero matrices, $\bar{\bar{\mathrm{Id}}}$ are the identity matrices. Moreover,

$$S_\alpha: \bar{a} \mapsto \bar{\bar{S}}_\alpha \, \bar{a} \, \bar{\bar{S}}_\alpha, \quad \bar{a} \in su(n).$$

Thus this construction is possible only for $n \geq 6$.

The involutive automorphisms S_1, S_2, S_3 generate the involutive decomposition $\mathfrak{g} = \mathfrak{l}_1 + \mathfrak{l}_2 = \mathfrak{l}_3$ ($\mathfrak{l}_0 = \mathfrak{l}_\alpha \cap \mathfrak{l}_\beta, \; \alpha \neq \beta$).

And we have

$$\mathfrak{g}/\mathfrak{l}_\alpha \cong su(n)/su(4) \oplus su(n-4) \oplus u(1),$$

$$\mathfrak{l}_\alpha/\mathfrak{l}_0 \cong \mathfrak{k}_\alpha/\mathfrak{m}_\alpha \oplus \mathfrak{n}_\alpha \oplus \mathfrak{z}_\alpha + \mathfrak{l}'_\alpha/\mathfrak{l}'_0 \oplus \mathfrak{t}_\alpha \oplus \mathfrak{w}^{(\alpha)} + \mathfrak{v}^{(\alpha)}/\mathfrak{v}^{(\alpha)}$$

$$\cong su(4)/su(2) \oplus su(2) \oplus u(1)$$

$$+ su(n-4)/su(n-6) \oplus su(2) \oplus u(1) + u(1)/u(1)$$

with the natural embeddings.

Moreover,

$$\mathfrak{k}_1 + \mathfrak{k}_2 + \mathfrak{k}_3 = \tilde{\mathfrak{k}} \cong su(6)$$

with the natural embedding into $\mathfrak{g} \cong su(n)$.

We now note that the involutive automorphisms S_1, S_2, S_3 generate in $\tilde{\mathfrak{k}}$, as well, an involutive decomposition $\tilde{\mathfrak{k}} = \tilde{\mathfrak{k}}_1 + \tilde{\mathfrak{k}}_2 + \tilde{\mathfrak{k}}_3$, where

$$\tilde{\mathfrak{k}}_\alpha = \mathfrak{k}_\alpha \oplus \mathfrak{t}_\alpha \oplus \tilde{\mathfrak{m}}_\alpha \cong su(4) \oplus su(2) \oplus u(1),$$

$$\tilde{\mathfrak{k}}_0 = \tilde{\mathfrak{k}}_\alpha \underset{\alpha \neq \beta}{\cap} \tilde{\mathfrak{k}}_\beta \cong \mathfrak{m}_\alpha \oplus \mathfrak{n}_\alpha \oplus \mathfrak{t}_\alpha \oplus \tilde{\mathfrak{w}}^{(\alpha)} \oplus \tilde{\tilde{\mathfrak{w}}}^{(\alpha)}$$

$$\cong su(2) \oplus su(2) \oplus su(2) \oplus u(1) \oplus u(1).$$

Taking the diagonal $\Delta = \Delta(\mathfrak{m}_\alpha \oplus \mathfrak{n}_\alpha \oplus \mathfrak{t}_\alpha)$, we have

$$\Delta \cong \Delta(\,su(2) \oplus su(2) \oplus su(2)\,) \subset su(4) \oplus su(2) \oplus u(1) \subset su(6) \cong \tilde{\mathfrak{k}}$$

with the natural embeddings.

Let us consider in $\tilde{\mathfrak{k}}$ the subalgebra $\tilde{\mathfrak{b}}$ of elements commuting with Δ.

Since $\Delta \subset \tilde{\mathfrak{k}}_0$ we have $\tilde{\mathfrak{b}} = \tilde{\mathfrak{b}}^{(1)} + \tilde{\mathfrak{b}}^{(2)} + \tilde{\mathfrak{b}}^{(3)}$, and $\tilde{\mathfrak{b}}^{(\alpha)}$ ($\alpha = 1, 2, 3$) is the maximal subalgebra of elements from $\tilde{\mathfrak{k}}_\alpha = \mathfrak{k}_\alpha \oplus \mathfrak{t}_\alpha \oplus \tilde{\mathfrak{w}}^{(\alpha)}$ commuting with Δ. But then $\tilde{\mathfrak{b}}^{(\alpha)} = \mathfrak{b}^{(\alpha)} \oplus \tilde{\mathfrak{w}}^{(\alpha)}$, where $\mathfrak{b}^{(\alpha)}$ is the maximal subalgebra of elements in $\mathfrak{k}_\alpha$ commuting with a diagonal $\mathfrak{d}$ of $\mathfrak{m}_\alpha \oplus \mathfrak{n}_\alpha$.

However, $\mathfrak{d} \cong \Delta(\,su(2) \oplus su(2)\,)$ (that is, isomorphic to a diagonal in $su(2) \oplus su(2)$) with the natural embedding into $\mathfrak{k}_\alpha \cong su(4)$.

Therefore $\mathfrak{b}^{(\alpha)} \cong \Delta(\,su(2) \oplus su(2)\,)$ with the natural embedding. (The easiest way to see this is to take into account that $su(4) \cong so(6) \cong \mathfrak{k}_\alpha$, then $\mathfrak{d} \cong so(3)$, $so(6) \cong \mathfrak{k}_\alpha$, with the natural embedding into $su(6) \cong \mathfrak{k}_\alpha$.)

Consequently $\mathfrak{b}^{(\alpha)} \cong so(3)$ with the natural embedding into $so(6) \cong \mathfrak{k}_\alpha$.

Thus $\mathfrak{b}^{(\alpha)} \cong \Delta(\,su(2) \oplus su(2)\,)$ with the natural embedding into $su(4) \cong \mathfrak{k}_\alpha$.

In addition,

$$\tilde{\mathfrak{b}}_0 = \tilde{\mathfrak{b}}^{(\alpha)} \cap \tilde{\mathfrak{k}}_0 = \tilde{\mathfrak{w}}^{(\alpha)} \oplus \tilde{\tilde{\mathfrak{w}}}^{(\alpha)},$$

whence we have the involutive decomposition

$$\tilde{\mathfrak{b}} = \tilde{\mathfrak{b}}^{(1)} + \tilde{\mathfrak{b}}^{(2)} + \tilde{\mathfrak{b}}^{(3)},$$

where

$$\tilde{\mathfrak{b}}^{(\alpha)}/\tilde{\mathfrak{b}}_0 \cong su(2)/u(1) + u(1)/u(1), \qquad \tilde{\mathfrak{b}}/\tilde{\mathfrak{b}}^{(\alpha)} \cong su(3)/u(2).$$

It is easy to see (using, for example, the model of skew-Hermitian matrices) that

$$\tilde{\mathfrak{b}} \cong \Delta(\,su(3) \oplus su(3)\,),$$

that is, $\tilde{\mathfrak{b}}$ is isomorphic to a diagonal in $su(3) \oplus su(3)$ with the natural embedding into $\tilde{\mathfrak{k}} \cong su(6)$.

Let us now choose $\mathfrak{h} \subset \tilde{\mathfrak{b}}$ in such a way that

$$\tilde{\mathfrak{b}}/\mathfrak{h} \cong su(3)/so(3),$$

with the natural embedding, so that $\{0\} \neq \mathfrak{h} \cap \tilde{\mathfrak{b}}^{(\alpha)} \subset (\tilde{\mathfrak{b}}^{(\alpha)} \dot{-} \tilde{\mathfrak{b}}_0)$.

Then, from the construction, $\mathfrak{h}^{(\alpha)} = \mathfrak{h} \cap \tilde{\mathfrak{b}}^{(\alpha)}$ is one-dimensional,

$$\mathfrak{h}^{(\alpha)} \cong \Delta(\,so(2) \oplus so(2)\,) \subset \Delta(\,su(2) \oplus su(2)\,) \subset su(4) \cong \mathfrak{k}_\alpha$$

with the natural embeddings.

But since $\mathfrak{k}_\alpha \cong su(4)$ is a special unitary subalgebra in $su(n)$ we have $S_\alpha \in \mathrm{Int}_\mathfrak{g}(\mathfrak{h}^{(\alpha)}) \subset \mathrm{Int}_\mathfrak{g}(\mathfrak{h}) \cong SO(3)$.

Taking now the conjugating automorphism $p \in \mathrm{Int}_\mathfrak{g}(\mathfrak{h})$ in such a way that

$$p^3 = \mathrm{Id}, \qquad p^{-1} S_1\, p = S_2, \qquad p^{-1} S_2\, p = S_3, \qquad p^{-1} S_3\, p = S_1$$

we see that the decomposition $\mathfrak{g} = \mathfrak{l}_1 + \mathfrak{l}_2 + \mathfrak{l}_3$ is hyper-involutive.

We have proved the theorem:

II.11.12. Theorem 55. *Let* $\mathfrak{g} \cong su(n)$ $(n \geq 6)$, *and*

$$\mathfrak{l}_1 \cong su(4) \oplus su(n-4) \oplus u(1)$$

be its special unitary involutive algebra of an involutive automorphism S_1.

Then there exists the hyper-involutive decomposition $\mathfrak{g} = \mathfrak{l}_1 + \mathfrak{l}_2 + \mathfrak{l}_3$ *of the involutive automorphisms* S_1, S_2, S_3 *such that*

$$\mathfrak{g}/\mathfrak{l}_\alpha \cong su(n)/su(4) \oplus su(n-4) \oplus u(1),$$

$$\mathfrak{l}_\alpha/\mathfrak{l}_0 = \mathfrak{k}_\alpha/\mathfrak{m}_\alpha \oplus \mathfrak{n}_\alpha \oplus \mathfrak{z}_\alpha + \mathfrak{l}'_\alpha/\mathfrak{l}'_0 \oplus \mathfrak{t}_\alpha \oplus \mathfrak{w}^{(\alpha)} + \mathfrak{v}^{(\alpha)}/\mathfrak{v}^{(\alpha)}$$

$$\cong su(4)/su(2) \oplus su(2) \oplus u(1)$$

$$+ su(n-4)/su(n-6) \oplus su(2) \oplus u(1) + u(1)/u(1)$$

$$(\mathfrak{l}_0 = \mathfrak{l}_\alpha \cap \mathfrak{l}_\beta, \ \alpha \neq \beta)$$

with the natural embeddings.

Moreover,

$$S_\alpha \in \text{Int}_\mathfrak{g}(\mathfrak{h}^{(\alpha)}) \subset \text{Int}_\mathfrak{g}(\mathfrak{h}) \cong SO(3), \qquad \dim \mathfrak{h}^{(\alpha)} = 1,$$

$$\mathfrak{h} \cong \Delta \left(so(3) \oplus so(3) \right) \subset \Delta \left(su(3) \oplus su(3) \right)$$

$$\cong \tilde{\mathfrak{b}} \subset \mathfrak{n}' \oplus \mathfrak{n}'' \cong su(3) \oplus su(3) \subset su(6) \cong \tilde{\mathfrak{k}} \subset \mathfrak{g} \cong su(n)$$

with the natural embeddings, where $\tilde{\mathfrak{b}}$ *is the maximal subalgebra of elements from* $\tilde{\mathfrak{k}}$ *commuting with the diagonal*

$$\Delta(\mathfrak{m}_\alpha \oplus \mathfrak{n}_\alpha \oplus \mathfrak{t}_\alpha) \cong \Delta \left(su(2) \oplus su(2) \oplus su(2) \right) \subset su(6) \cong \tilde{\mathfrak{k}}$$

with the natural embeddings.

II.11.13. Now we consider $\mathfrak{g} \cong sp(n)$. We regard $sp(n)$ as determined by quaternionic skew-Hermitian matrices of order n. Then a special unitary non-principal subalgebra has the form $sp(2)$ with the natural embeddings into $sp(n)$.

Let S_1, S_2, S_3 be special non-principal unitary involutive automorphisms, $S_\alpha S_\beta = S_\gamma$ $(\alpha \neq \beta, \ \beta \neq \gamma, \ \gamma \neq \alpha)$. Then we may consider that the involutive automorphisms S_1, S_2, S_3 are represented by the diagonal matrices $\tilde{S}_1, \tilde{S}_2, \tilde{S}_3$ from $\text{Int}(sp(n))$ of the form

$$\tilde{\tilde{S}}_1 = \left\| \begin{array}{cc|c} -1 & 0 & \overline{\overline{O}} \\ 0 & -1 & \\ \hline \underline{\underline{O}} & & \underline{\underline{\text{Id}}} \end{array} \right\|, \qquad \tilde{\tilde{S}}_2 = \left\| \begin{array}{c|cc|c} 1 & \overline{\overline{O}} & & \overline{\overline{O}} \\ \hline \underline{\underline{O}} & -1 & 0 & \overline{\overline{O}} \\ & 0 & -1 & \\ \hline \underline{\underline{O}} & \underline{\underline{O}} & & \underline{\underline{\text{Id}}} \end{array} \right\|,$$

$$\bar{\bar{S}}_3 = \left\| \begin{array}{ccc|c} -1 & 0 & 0 & \bar{\bar{O}} \\ \hline 0 & 1 & 0 & \bar{\bar{O}} \\ \hline 0 & 0 & -1 & \bar{\bar{O}} \\ \hline \bar{\bar{O}} & \bar{\bar{O}} & \bar{\bar{O}} & \bar{\bar{\mathrm{Id}}} \end{array} \right\| ,$$

where $\bar{\bar{O}}$ are zero matrices, and $\bar{\bar{\mathrm{Id}}}$ are the identity matrices. Moreover,

$$S_\alpha : \bar{\bar{a}} \mapsto \bar{\bar{S}}_\alpha \, \bar{\bar{a}} \, \bar{\bar{S}}_\alpha, \quad \bar{\bar{a}} \in sp(n).$$

Thus this construction is possible only for $n \geq 3$.

The involutive automorphisms S_1, S_2, S_3 generate the involutive decomposition

$$\mathfrak{g} = \mathfrak{l}_1 + \mathfrak{l}_2 + \mathfrak{l}_3 \quad (\mathfrak{l}_0 = \mathfrak{l}_\alpha \cap \mathfrak{l}_\beta, \ \alpha \neq \beta),$$

$$\mathfrak{g}/\mathfrak{l}_\alpha \cong sp(n)/sp(2) \oplus sp(n-2),$$

$$\mathfrak{l}_\alpha/\mathfrak{l}_0 = \mathfrak{k}_\alpha/\mathfrak{m}_\alpha \oplus \mathfrak{n}_\alpha + \mathfrak{l}'_\alpha/\mathfrak{l}'_0 \oplus \mathfrak{t}_\alpha$$

$$\cong sp(2)/sp(1) \oplus sp(1) + sp(n-2)/sp(n-3) \oplus sp(1)$$

with the natural embeddings.

Moreover,

$$\mathfrak{k}_1 + \mathfrak{k}_2 + \mathfrak{k}_3 = \tilde{\mathfrak{k}} \cong sp(3)$$

with the natural embeddings into $\mathfrak{g} \cong sp(n)$, and $\tilde{\mathfrak{k}} = \tilde{\mathfrak{k}}_1 + \tilde{\mathfrak{k}}_2 + \tilde{\mathfrak{k}}_3$, where $\tilde{\mathfrak{k}}_\alpha = \mathfrak{k}_\alpha \oplus \mathfrak{t}_\alpha$, is the involutive decomposition for $\tilde{\mathfrak{k}}$ of the involutive automorphisms S_1, S_2, S_3.

We also have

$$\tilde{\mathfrak{k}}_0 = \tilde{\mathfrak{k}}_\alpha \underset{\alpha \neq \beta}{\cap} \tilde{\mathfrak{k}}_\beta = \mathfrak{m}_\alpha \oplus \mathfrak{n}_\alpha \oplus \mathfrak{t}_\alpha \cong sp(1) \oplus sp(1) \oplus sp(1),$$

$$\tilde{\mathfrak{k}}_\alpha/\tilde{\mathfrak{k}}_0 \cong sp(2)/sp(1) \oplus sp(1) + sp(1)/sp(1)$$

with the natural embeddings into $\tilde{\mathfrak{k}} \cong sp(3)$.

Taking a diagonal $\Delta = \Delta(\mathfrak{m}_\alpha \oplus \mathfrak{n}_\alpha \oplus \mathfrak{t}_\alpha)$ we have

$$\Delta \cong \Delta\left(sp(1) \oplus sp(1) \oplus sp(1) \right) \subset sp(2) \oplus sp(1) \subset sp(3) \cong \tilde{\mathfrak{k}}$$

with the natural embeddings.

Let us consider in $\tilde{\mathfrak{k}}$ the subalgebra $\mathfrak{h}$ of all elements commuting with Δ.

Then $\mathfrak{h} = \mathfrak{h}_1 + \mathfrak{h}_2 + \mathfrak{h}_3$, where $\mathfrak{h}_\alpha$ is the maximal subalgebra of elements of $\tilde{\mathfrak{k}}_\alpha$ commuting with Δ. But then $\mathfrak{h}_\alpha$ is a subalgebra of all elements of $\mathfrak{k}_\alpha \cong sp(2)$ commuting with $\Delta(\mathfrak{m}_\alpha \oplus \mathfrak{n}_\alpha) \cong \Delta\left(sp(1) \oplus sp(1) \right)$.

Therefore $\mathfrak{h}_\alpha \cong so(2)$, $\dim \mathfrak{h}_\alpha = 1$.

Hence $\mathfrak{h} \cong so(3) \subset sp(3) \cong \tilde{\mathfrak{k}}$ with the natural embedding and

$$S_\alpha \in \mathrm{Int}_\mathfrak{g}(\mathfrak{h}_\alpha) \subset \mathrm{Int}_\mathfrak{g}(\mathfrak{h}) \cong SO(3).$$

Taking now the conjugating automorphism $p \in \mathrm{Int}_\mathfrak{g}(\mathfrak{h})$ in such a way that

$$p^3 = \mathrm{Id}, \qquad p^{-1} S_1 p = S_2, \qquad p^{-1} S_2 p = S_3, \qquad p^{-1} S_3 p = S_1,$$

we see that the decomposition $\mathfrak{g} = \mathfrak{l}_1 + \mathfrak{l}_2 + \mathfrak{l}_3$ is hyper-involutive.

Thus we have proved the theorem:

II.11.14. Theorem 56. *Let $\mathfrak{g} \cong sp(n)$ ($n \geq 3$), and $\mathfrak{l}_1 \cong sp(2) \oplus sp(n-2)$ be its special unitary involutive algebra of an involutive automorphism S_1.*

Then there exists the hyper-involutive decomposition $\mathfrak{g} = \mathfrak{l}_1 + \mathfrak{l}_2 + \mathfrak{l}_3$ of the involutive automorphisms S_1, S_2, S_3 such that

$$\mathfrak{g}/\mathfrak{l}_\alpha \cong sp(n)/sp(2) \oplus sp(n-2),$$

$$\mathfrak{l}_\alpha/\mathfrak{l}_0 = \mathfrak{k}_\alpha/\mathfrak{m}_\alpha \oplus \mathfrak{n}_\alpha + \mathfrak{l}'_\alpha/\mathfrak{l}'_0 \oplus \mathfrak{t}_\alpha$$

$$\cong sp(2)/sp(1) \oplus sp(1) + sp(n-2)/sp(n-3) \oplus sp(1),$$

$$\mathfrak{l}_0 = \mathfrak{l}_\alpha \cap \mathfrak{l}_\beta, \quad \alpha \neq \beta,$$

with the natural embeddings.

Moreover, $S_\alpha \in \mathrm{Int}_\mathfrak{g}(\mathfrak{h}_\alpha) \subset \mathrm{Int}_\mathfrak{g}(\mathfrak{h}) \cong SO(3)$, where $\mathfrak{h} \cong so(3) \subset sp(3) \subset sp(n)$ with the natural embeddings, and $\mathfrak{h}$ may be characterized as the maximal subalgebra of elements from $\tilde{\mathfrak{k}} = \mathfrak{k}_1 + \mathfrak{k}_2 + \mathfrak{k}_3$ commuting with the diagonal

$$\Delta(\mathfrak{m}_\alpha \oplus \mathfrak{n}_\alpha \oplus \mathfrak{t}_\alpha) \cong \Delta(sp(1) \oplus sp(1) \oplus sp(1)) \subset sp(3) \cong \tilde{\mathfrak{k}}$$

(with the natural embeddings).

II.11.15. Definition 34. The hyper-involutive decompositions described in II.11.10 (Theorem 54), II.11.12 (Theorem 55), II.11.14 (Theorem 56) are called the special hyper-involutive sums for the Lie algebras of type $so(n)$, $su(n)$, $sp(n)$, respectively.

II.11.16. Remark. It is easily seen that the method of proving II.11.10 (Theorem 54), II.11.12 (Theorem 55), II.11.14 (Theorem 56) permits us to find all hyper-involutive decompositions connected with inner involutive automorphisms (and not only with special) for the Lie algebras $so(n)$, $su(n)$, $sp(n)$.

PART THREE

Среди надрыва изломных Измов,
Столпотворения Чужих Миров,
Я славлю Солнце, пусть ярче брызнет
Снопами Света, Огнем Даров.

Не нам, горячим, слезиться Ленью,
Влачить тоскливо Упадка Тень.
Пусть все ликует, конец Сомненью,
Рассветом Гимнов пылает День.

Квантасмагор

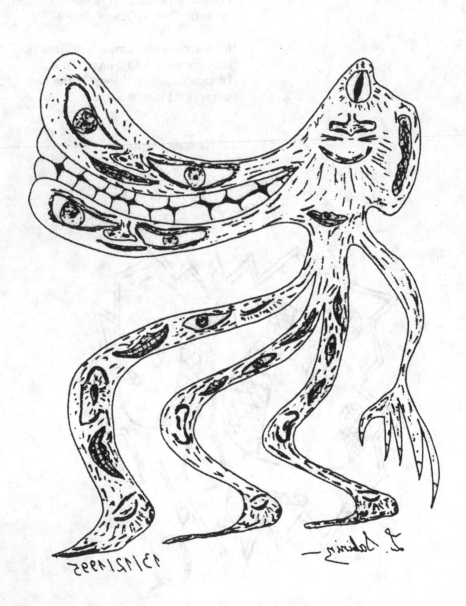

CHAPTER III.1

NOTATIONS, DEFINITIONS AND SOME PRELIMINARIES

In this Part we are going to consider some geometric applications of the theory presented above to symmetric and homogeneous spaces and to Lie groups.

First we consider the theory of mirrors of symmetric spaces. Along with one well known type of symmetric spaces (spaces of rank one) we introduce also two remarkable classes of symmetric spaces: principal and special U spaces.

Let G be a Lie group. It generates the corresponding Lie algebra which we denote $\mathfrak{g} = \ln G$ (as is well known from the theory of Lie groups—Lie algebras).

If H is a subgroup of a Lie group G then H generates the subalgebra $\mathfrak{h}$ of the Lie algebra $\mathfrak{g} = \ln G$.

We denote such an algebra by $\mathfrak{h} = \ln_G H$. (Thus $\ln_G G = \ln G$.)

In this case we also write

$$\ln(G, H) \overset{\text{def}}{=} (\ln G, \ln_G H),$$

and

$$\ln(G/H) \overset{\text{def}}{=} \ln G / \ln_G H.$$

III.1.1. Definition 35. Let G be a Lie group, S be its automorphism such that $S^2 = \text{Id}$, and let H be a maximal connected subgroup of G immobile under the action of S. Then we say that S is an involutive automorphism, H is an involutive (or characteristic) group of S, and G/H is an involutive pair of S.

We note that an involutive automorphism S of a Lie group G uniquely generates an involutive automorphism $\ln S$ of the Lie algebra $\ln G$; the converse is true (at least locally).

III.1.2. Definition 36. An involutive automorphism S of a Lie group G is said to be principal if $\ln S$ is a principal involutive automorphism of $\ln G$.

In this case the characteristic group H of the involutive automorphism S, the involutive pair G/H, and its corresponding symmetric space are said to be principal.

(Compare with I.1.10 Definition 1.)

III.1.3. Definition 37. A principal involutive automorphism S of a Lie group G is said to be principal orthogonal (of type O) if $\ln S$ is principal orthogonal (of type O) and principal unitary (of type U) if $\ln S$ is principal unitary (of type U).

Correspondingly we distinguish between orthogonal and unitary principal involutive groups, involutive pairs, and symmetric spaces.

(Compare with I.1.11 Definition 2.)

143

III.1.4. Definition 38. An involutive automorphism of a Lie group G is said to be central if $\ln S$ is central.

Correspondingly we define a central involutive group, involutive pair, symmetric space.

(Compare with I.1.12 Definition 3.)

III.1.5. Definition 39. A principal involutive automorphism S of a Lie group G is said to be principal di-unitary (of type $U^{(2)}$) if $\ln S$ is principal di-unitary (of type $U^{(2)}$).

Correspondingly we define a principal di-unitary group, involutive pair, symmetric space.

(Compare with I.1.13 Definition 4.)

III.1.6. Definition 40. An involutive automorphism S of a Lie group G is said to be mono-unitary (or of type $U^{(1)}$) if $\ln S$ is mono-unitary.

Correspondingly we define a mono-unitary involutive group, involutive pair, symmetric space.

(Compare with I.1.14 Definition 5.)

III.1.7. Definition 41. An involutive automorphism S of a Lie group G is said to be special if $\ln S$ is special.

Correspondingly we define a special involutive group, involutive pair, symmetric space.

(Compare with I.1.15 Definition 6.)

III.1.8. Definition 42. An involutive automorphism S of a Lie group G is said to be special orthogonal (of type O) if $\ln S$ is special orthogonal (of type O) and special unitary (of type U) if $\ln S$ is special unitary (of type U).

Respectively, we distinguish between a special orthogonal and special unitary involutive group, involutive pair, symmetric spaces.

(Compare with I.1.16 Definition 7.)

III.1.9. Definition 43. Let G be a simple compact Lie group, Q being its subgroup. We say that Q is a special unitary subgroup of G if $\ln_G Q$ is a special unitary subalgebra of $\ln G$.

(Compare with I.1.17 Definition 8.)

III.1.10. Definition 44. Let G/H be an involutive pair of an involutive automorphism S of a Lie group G. By the lower (upper) index of an involutive pair G/H, involutive automorphism S, and involutive group H we mean the lower (upper) index of the involutive automorphism $\ln S$.

If the lower and upper indices of an involutive pair coincide then we call it simply the index of the involutive automorphism, involutive group, involutive pair, or symmetric space, respectively.

(Compare with I.1.18 Definition 9.)

III.1.11. Definition 45. We say that a Lie group G is an involutive product of involutive groups $H_1, H_2, H_3 \subset G$, writing in that case

$$G = H_1 \boxtimes H_2 \boxtimes H_3,$$

if there exists the involutive decomposition

$$\ln G = \ln_G H_1 + \ln_G H_2 + \ln_G H_3$$

of the Lie algebras $\ln_G H_1, \ln_G H_2, \ln_G H_3 \subset \ln G$.

In this case we also say that there is the involutive decomposition of a Lie group G,

$$G = H_1 \boxtimes H_2 \boxtimes H_3,$$

into the involutive product of involutive groups $H_1, H_2, H_3 \subset G$.

(Compare with I.1.25 Definition 15.)

III.1.12. Definition 46. An involutive product (decomposition) of a Lie group G,

$$G = H_1 \boxtimes H_2 \boxtimes H_3,$$

is said to be iso-involutive if

$$\ln G = \ln_G H_1 + \ln_G H_2 + \ln_G H_3$$

is an iso-involutive sum.

(Compare with I.1.26 Definition 16.)

III.1.13. Definition 47. An involutive product $G = H_1 \boxtimes H_2 \boxtimes H_3$ is called hyper-involutive if the involutive decomposition

$$\ln G = \ln_G H_1 + \ln_G H_2 + \ln_G H_3$$

is hyper-involutive.

(Compare with I.1.27 Definition 17.)

III.1.14. Definition 48. Let G be a Lie group. By the curvature tensor of an involutive pair G/H (or symmetric space G/H) we mean the curvature tensor of an involutive pair $\ln G / \ln_G H$.

Correspondingly we speak of the curvature tensor of an iso-involutive group and of an involutive automorphism of a Lie group G.

(Compare with I.2.2 Definition 18.)

III.1.15. Definition 49. An iso-involutive decomposition of a Lie group G,

$$G = H_1 \boxtimes H_2 \boxtimes H_3,$$

is of the type 1 if

$$\ln G = \ln_G H_1 + \ln_G H_2 + \ln_G H_3$$

is an iso-involutive sum of type 1.

(Compare with I.3.6 Definition 19.)

III.1.16. Definition 50. A hyper-involutive decomposition of a Lie group G,

$$G = H_1 \boxtimes H_2 \boxtimes H_3,$$

is said to be simple if the hyper-involutive sum

$$\ln G = \ln_G H_1 + \ln_G H_2 + \ln_G H_3$$

is simple.

Otherwise we say that this hyper-involutive decomposition is general.

(Compare with II.1.7 Definition 25.)

III.1.17. Definition 51. Let G be a simple compact Lie group, H be its involutive group of a principal involutive automorphism S. We say that the involutive automorphism S, involutive group H, involutive pair G/H (or symmetric space G/H) are exceptional principal if $\ln S$ is an exceptional principal involutive automorphism.

Moreover, we say that S is of type:

1) F_4 if $\ln S$ is of type f_4;

2) E_6 if $\ln S$ is of type e_6;

3) E_7 if $\ln S$ is of type e_7;

4) E_8 if $\ln S$ is of type e_8.

(Compare with II.7.6 Definition 30.)

III.1.18. Definition 52. Let $M = G/H$ be a homogeneous reductive space [S. Kobayashi, K. Nomizu 63,69], and H be its stabilizer of a point $o \in M$. We say that a subsymmetry S_o is a geodesic subsymmetry if it is generated by an involutive automorphism S of a Lie group G in such a way that $\ln S \in \mathrm{Int}_{\ln G}\{t\zeta\}$, where $\zeta \in \mathfrak{m} = \ln G \dot{-} \ln_G H$ (meaning that $\mathfrak{m}$ is a reductive complement to $\ln_G H$, that is, $\ln G = \mathfrak{m} \dot{+} \ln_G H$, $[\mathfrak{m}\, \ln_G H] \subset \mathfrak{m}$).

Correspondingly we speak of a geodesic mirror [L.V. Sabinin, 58a,59a,59b].

III.1.19. Remark. We note that a subsymmetry S_x with respect to a point $x \in M$ is a (local or global) map of M onto itself generating an automorphism of G such that

$$S_x(x) = x, \qquad (S_x)^2 = \mathrm{Id}$$

[L.V. Sabinin, 58a, 59a, 59b].

The maximal connected set

$$K_x = \{y \in M \mid S_x(y) = y,\ S_x(x) = x\}$$

is called a *mirror of a subsymmetry S_x at the point $x \in M$*.

Evidently $x \in K_x$ [L.V. Sabinin, 58a,59a,59b].

III.1.20. Theorem 1. *Let G/H be a symmetric space of an involutive automorphism S ($H = \{h \in G \mid Sh = h\}$) with a simple compact Lie group G.*

Then in G/H there exist non-trivial (i.e., containing more than a one point) geodesic mirrors.

Proof. Let us take the compact group

$$\mathrm{Int}_{\mathfrak{g}}\left(\{\,t\zeta\,\}\right), \qquad \zeta \in \ln G \dot{-} \ln_G H,$$

which is possible by I.3.10 (Lemma 1), and let us take a non-trivial involutive element

$$A \in \mathrm{Int}_{\ln G}\left(\{t\zeta\}\right).$$

Then since $(\ln S)\,\zeta = -\zeta$ we have $(\ln S)\,A\,(\ln S)^{-1} = A^{-1} = A$, or

$$(\ln S)\,A = A\,(\ln S).$$

But A generates (at least locally) an involutive automorphism J of G ($\ln J = A$) such that $JS = SJ$. Therefore $J(H) = H$, that is, J is a subsymmetry of G/H with respect to a point $H = eH$ (e being the neutral element of G).

Since $A\zeta = \zeta$ we have $Ja = a$ if $\ln a = \tau\,\zeta \in \ln G \dot{-} \ln_G H$.

But then $a \notin H$, and consequently the mirror of the subsymmetry J is non-trivial.   ∎

III.1.21. Definition 53. We say that a mirror W of a homogeneous space G/H is special (unitary, orthogonal) if the corresponding subsymmetry generates in G a special (unitary, orthogonal) involutive automorphism.

Respectively, we speak of a special (unitary, orthogonal) subsymmetry.

III.1.22. Definition 54. We say that a mirror of a homogeneous space G/H is principal (unitary, orthogonal) if the corresponding subsymmetry generates in G a principal (unitary, orthogonal) involutive automorphism.

Respectively, we speak of a principal (unitary, orthogonal) subsymmetry.

III.1.23. Definition 55. A unitary special symmetric space G/H is said to be essentially unitary special if it is not unitary principal.

III.1.24. Definition 56. An essentially special but not principal symmetric space is called strictly special.

CHAPTER III.2

SYMMETRIC SPACES OF RANK 1

We are going to consider symmetric spaces G/H of rank 1 defined by an involutive automorphism S of a simple compact Lie group G, H being an isotropy subgroup.

The pair G/H generates the involutive pair $\ln G / \ln_G H$ of the involutive automorphism $\ln S$ of index 1.

We formulate first one evident result:

III. 2.1. Theorem 2. *A symmetric space G/H with a simple compact Lie group G is of rank 1 if and only if it has a geodesic mirror of rank 1.*

III. 2.2. Theorem 3. *Let $V = G/H$ be a symmetric space with a simple compact Lie group G, and W be its geodesic mirror of rank 1. Then*

$$W \cong SO(m+1)/SO(m)$$

with the natural embedding.

That is, W is of the constant curvature.

Proof. We consider the Lie algebra $\mathfrak{g} = \ln G$, the involutive algebra $\mathfrak{l}_1 = \ln_G H$ of an involutive automorphism S_1, and the involutive algebra $\mathfrak{l}_3$ of the involutive automorphism S_3 generated by the subsymmetry of the mirror W. Then we have an iso-involutive decomposition

$$\mathfrak{g} = \mathfrak{l}_1 + \mathfrak{l}_2 + \mathfrak{l}_3$$

of index 1.

Applying now I.6.1 (Theorem 19) and I.5.4 (Theorem 14) together with the decomposition of $\mathfrak{l}_3/\mathfrak{l}_0$, $\mathfrak{l}_0 = \mathfrak{l}_\alpha \underset{\alpha \neq \beta}{\cap} \mathfrak{l}_\beta$, into elementary involutive pairs we obtain the result. ∎

Now, III.2.1 (Theorem 2) and III.2.2 (Theorem 3) may be reformulated in the following way.

III.2.3. Theorem 4. *A symmetric space G/H with a simple compact Lie group G is of rank 1 if and only if G/H has a geodesic mirror of the constant curvature.*

III.2.4. Theorem 5. *Let G/H be a symmetric space with a simple compact Lie group G, and let its geodesic mirror W of index 1 be one-dimensional. Then*

$$G/H \cong SO(n+1)/SO(n)$$

with the natural embedding.

Proof. Passing to the corresponding Lie algebras and using I.6.3 (Theorem 21) we obtain the result.   ∎

III.2.5. Theorem 6. *Let G/H be a symmetric space with a simple compact Lie group G, and let its geodesic mirror W of rank 1 be two-dimensional. Then*

$$G/H \cong SU(n+1)/U(n)$$

with the natural embedding.

Proof. Passing to the corresponding Lie algebras and using I.6.4 (Theorem 22) we obtain the result.   ∎

III.2.6. Theorem 7. *Let G/H be a symmetric space with a simple compact Lie group G, and let its geodesic mirror W of rank 1 be four-dimensional. Then*

$$W \cong Sp(n+1)/Sp(n) \times Sp(1)$$

with the natural embedding.

Proof. Passing to the corresponding Lie algebras and using I.6.5. (Theorem 23) we obtain the result.   ∎

III.2.7. Theorem 8. *If G/H is a symmetric space of rank 1 with a simple compact Lie group G then G/H has no three-dimensional mirrors.*

Proof. Indeed, passing to the corresponding Lie algebras and applying I.6.10 (Theorem 28) we obtain the result.   ∎

III.2.8. Theorem 9. *Let G/H be a symmetric space of rank 1, and let G be a simple compact Lie group. If $H = H' \times H''$, $H' \neq \{e\}$, $H'' \neq \{e\}$, then either*

$$G/H \cong SU(n+1)/U(n)$$

or

$$G/H \cong Sp(n+1)/Sp(n) \times Sp(1)$$

with the natural embeddings.

Proof. Passing to the corresponding involutive pair $\ln(G/H)$ and applying I.8.1 (Theorem 30) we obtain the result.   ∎

CHAPTER III.3

PRINCIPAL SYMMETRIC SPACES

III.3.1. Theorem 10. *Let G/H be an irreducible symmetric space, G being a compact Lie group, $\dim H = 3$. Then $H \cong SO(3)$.*

Proof. Passing to the involutive pair $\mathfrak{g}/\mathfrak{l} = \ln G / \ln_G H$ and using I.8.2 (Theorem 31) we obtain the result. ∎

III.3.2. Theorem 11. *Let G be a simple compact Lie group, $\dim G > 3$. Then there exists a principal unitary symmetric space G/H.*

Proof. Passing to the Lie algebra $\ln G$ and using II.4.15 (Theorem 22) we obtain the result. ∎

III.3.3. Theorem 12. *If G/H is a principal orthogonal symmetric space with a simple compact Lie group G then either*

$$G/H \cong SO(n)/SO(n-3) \times SO(3)$$

or

$$G/H \cong SU(3)/SO(3)$$

with the natural embeddings.

Proof. Passing to the corresponding Lie algebras and using II.3.6 (Theorem 6), II.3.10 (Theorem 8), II.3.14 (Theorem 10) we obtain the result. ∎

III.3.4. Theorem 13. *Let $V = G/H$ be a principal orthogonal symmetric space with a compact simple Lie group G. Then for any point x of $V = G/H$ there exist three mirrors $V^{(1)}$, $V^{(2)}$, $V^{(3)}$ of the same dimension, passing through x, being generated by a discrete commutative group of subsymmetries $\{\mathrm{id}, S_x^{(1)}, S_x^{(2)}, S_x^{(3)}\}$, $S_x^{(i)} S_x^{(j)} = S_x^{(k)}$ ($i \neq j$, $j \neq k$, $k \neq i$).*

Moreover, there exists an inner automorphism $p \in H$ generated by an element of H such that $p^3 = \mathrm{id}$, $p V^{(1)} = V^{(2)}$, $p V^{(2)} = V^{(3)}$, $p V^{(3)} = V^{(1)}$.

Proof. Passing to the corresponding Lie algebras and using a hyper-involutive decomposition basis for an orthogonal involutive pair (see II.3.3 (Definition 26) and II.3.6 (Theorem 6), II.3.8 (Theorem 7), II.3.10 (Theorem 8), II.3.11 (Theorem 9), II.3.14 (Theorem 10) we obtain the result. ∎

III.3.5. Theorem 14. *If G/H is a principal orthogonal symmetric space with a simple compact Lie group G then G/H has a mirror W isomorphic either to a space of the constant curvature or to a direct product of spaces of the constant curvature and a one-dimensional space.*

Proof. Passing to the corresponding Lie algebras and constructing the hyper-involutive decomposition basis for a principal orthogonal involutive pair, we use then II.3.6 (Theorem 6), II.3.8 (Theorem 7), II.3.10 (Theorem 8), II.3.11 (Theorem 9), II.3.14 (Theorem 10).

This gives the result. ∎

III.3.6. Theorem 15. *Let G/H be a principal di-unitary symmetric space with a compact simple Lie group G.*
Then either
$$G/H \cong SO(n)/SO(n-4) \times SO(4)$$
or
$$G/H \cong G_2/SO(4)$$
with the natural embeddings.

Proof. Passing to the corresponding Lie algebras and using II.5.6 (Theorem 24), II.5.8 (Theorem 25), II.5.9 (Theorem 26), II.6.6 (Theorem 29) we obtain the result. ∎

III.3.7. Theorem 16. *Let G/H be a principal mono-unitary non-central symmetric space with a compact simple Lie group G. If H is semi-simple then $H = H' \times H''$, where $H' \cong SU(2)$, H'' is simple, and $\dim H'' > 3$.*

Proof. Passing to the corresponding Lie algebras we use II.7.3 (Theorem 33). This proves the theorem. ∎

III.3.8. Theorem 17. *Let G be a simple compact Lie group, G/H be a principal mono-unitary symmetric space of type F_4, then*
$$G/H \cong F_4/Sp(3) \times Sp(1)$$
with the natural embedding.

Proof. Passing to the corresponding Lie algebras we use II.8.1 (Theorem 35). Returning to the corresponding Lie groups we obtain the theorem. ∎

III.3.9. Theorem 18. *Let G be a simple compact Lie group, G/H be a principal mono-unitary symmetric space of type E_6, then*
$$G/H \cong E_6/SU(2) \times SU(6)$$
with the natural embedding.

Proof. We pass to the corresponding Lie algebras, make use of II.8.2 (Theorem 36), and return to the corresponding Lie groups. This proves the assertion of the theorem. ∎

III.3.10. Theorem 19. *Let G be a simple compact Lie group. If G/H is a principal mono-unitary symmetric space of type E_7 then*

$$G/H \cong E_7/SU(2) \times SO(12)$$

with the natural embedding.

Proof. Passing to the corresponding Lie algebras we make use of II.8.3 (Theorem 37).

Returning to the corresponding Lie groups we obtain the assertion of the theorem. ∎

III.3.11. Theorem 20. *Let G be a simple compact Lie group. If G/H is a principal mono-unitary symmetric space of type E_8 then*

$$G/H \cong E_8/SU(2) \times E_7$$

with the natural embedding.

Proof. Passing to the corresponding Lie algebras and using II.8.4 (Theorem 38) we obtain the result. ∎

III.3.12. Theorem 21. *Let G/H be a principal unitary symmetric space with a simple compact Lie group G, and let $G \ncong SO(k)$ $(k = 5, 6, 7, 8)$.*

Then G/H has a special geodesic mirror

$$W = \widetilde{G}/\widetilde{H} \cong SO(m)/SO(4) \times SO(m-4)$$

(with the natural embedding) such that $m = 5, 6, 8, 9, 10, 12, 16$.

Proof. Passing from G/H to the corresponding involutive pair $\mathfrak{g}/\mathfrak{l}_1 = \ln G/\ln_G H$ we construct an involutive decomposition $\mathfrak{g} = \mathfrak{l}_1 + \mathfrak{l}_2 + \mathfrak{l}_3$ basis for $\mathfrak{l}_1$ (see II.7.4 (Definition 29) and II.9.1 (Theorem 40)).

The special unitary involutive pair $\mathfrak{l}_3/\mathfrak{l}_0$ (see II.9.1 (Theorem 40)) generates in G/H a special geodesic mirror $W = \widetilde{G}/\widetilde{H} \cong SO(m)/SO(4) \times SO(m-4)$. Furthermore, the cases $m = 5, 6, 8, 9, 10, 12, 16$ follow from II.8.1 (Theorem 35), II.8.2 (Theorem 36), II.8.3 (Theorem 37), II.8.4 (Theorem 38), II.9.2 (Theorem 41), II.9.3 (Theorem 42), II.9.4 (Theorem 43). ∎

III.3.13. Theorem 22. *Under the assumptions and notations of III.3.12 (Theorem 21), if*

$$W \cong SO(5)/SO(4),$$

$$\cong SO(6)/SO(4) \times SO(2),$$

$$\cong SO(8)/SO(4) \times SO(4),$$

$$\cong SO(9)/SO(4) \times SO(5),$$

$$\cong SO(10)/SO(4) \times SO(6),$$

$$\cong SO(12)/SO(4) \times SO(8),$$

$$\cong SO(16)/SO(4) \times SO(12)$$

with the natural embeddings then correspondingly

$$G/H \cong Sp(n)/Sp(1) \times Sp(n-1), \quad n > 2;$$

$$\cong SU(n)/SU(2) \times U(n-2), \quad n > 4;$$

$$\cong SO(n)/SO(4) \times SO(n-4), \quad n > 8;$$

$$\cong F_4/SU(2) \times Sp(3);$$

$$\cong E_6/SU(2) \times SU(6);$$

$$\cong E_7/SU(2) \times SO(12);$$

$$\cong E_8/SU(2) \times E_7$$

with the natural embeddings.

(Here the groups are given up to local isomorphisms).

Proof. Passing to the corresponding Lie algebras and using II.9.1 (Theorem 40), II.8.1 (Theorem 35), II.8.2 (Theorem 36), II.8.3 (Theorem 37), II.8.4 (Theorem 38), II.9.2 (Theorem 41), II.9.3 (Theorem 42), II.9.4 (Theorem 43) we obtain the result. ∎

III.3.14. The maximal subgroup of a group G preserving the subset W invariant will be denoted by G^W

If a group G is compact, G/H is a symmetric space, and W is its mirror, then $W = \widetilde{G}^W/\widetilde{H}^W$ means the following: we take the maximal subgroup $G_1 \subset G$ which transforms W into itself. Then $G_1 = \widetilde{G} \times \widetilde{\widetilde{G}}$ (locally), where $\widetilde{G}$ is the maximal subgroup of G acting effectively on W, and $\widetilde{\widetilde{G}}$ is the subgroup of G acting on W in a trivial way. Furthermore we introduce $\widetilde{H} = H \cap \widetilde{G}$. After that we have the symmetric space $W = \widetilde{G}^W/\widetilde{H}^W$.

III.3.15. Theorem 23. *Let G/H be a symmetric space with a simple compact Lie group G, $H = H' \times H''$, and let $W = \widetilde{G}^W/(H' \times H'')^W$ be its mirror with a simple Lie group $\widetilde{G} \subset G$, $H' \subset H$ being isomorphic to $SU(2) \times SU(2)$ (locally).*

Then W is a special mirror and G/H is a principal unitary symmetric space.

Proof. We note first that the centre Z of H' consists of four involutive elements (that is $x^2 = e$ for $x \in Z$). Therefore the restriction of Z to $W = \widetilde{G}^W/(H' \times H'')^W$ consists of two elements. Indeed, Z^W (the restriction of Z to W) is the image of the natural morphism $\widetilde{G} \longrightarrow \widetilde{G}^W$, and the inverse image of the identity generates the centre $\widetilde{Z}$ of the group $\widetilde{G}$.

Let $G_1 = \widetilde{G} \times \widetilde{\widetilde{G}}$ be a group of all transformations of G which map W into itself. Evidently G_1 is the involutive group of the subsymmetry S generating the mirror W; consequently its centre Z_1 contains at most one non-trivial involutive element. Therefore $\widetilde{Z} \subset Z_1$ contains at most one nontrivial involutive element, as well.

If we assume that $\widetilde{Z}$ contains a single involutive element then $(H')^W$ is isomorphic to $SU(2) \times SU(2)$, that is, the involutive group $(H' \times H'')^W$ of a simple compact group $\widetilde{G}^W$ possesses the centre which consists of more than two elements, which is impossible.

Thus $\widetilde{Z}$ contains only one non-trivial involutive element q.

This means that either $(H')^W \cong SO(3) \times SU(2)$ or $(H')^W \cong SO(4)$.

Assuming the first possibility, we pass to the corresponding involutive algebras and make use of II.3.11 (Theorem 9).

As a result we see that

$$W = \widetilde{G}^W / (H' \times H'')^W \cong SO(7)/SO(3) \times SO(4),$$

where $\widetilde{G}$ is a simple compact special unitary subgroup of G.

However, this is impossible because of the classification of simple compact special unitary subalgebras (see II.9.10 (Theorem 45)).

Thus the only possibility is $(H')^W \cong SO(4)$, and the involutive element $q \in H'$ generates the mirror W.

But then W is a special mirror and

$$W = \widetilde{G}^W / (H' \times H'')^W \cong SO(m)/SO(4) \times SO(m-4), \quad m > 4,$$

by II.7.5 (Theorem 34) and II.5.6 (Theorem 24), II.5.8 (Theorem 25), II.5.9 (Theorem 26), II.5.10 (Theorem 27).

Applying arguments analogous to those considered above to the symmetric space M of involutive pair $H / H' \times H'' \times \widetilde{\widetilde{G}}$ and to the centre Z of $H' \cong SU(2) \times SU(2)$, we have either $(H')^M \cong SO(4)$ or $(H')^M \cong SO(3) \times SU(2)$.

But the first case is impossible (since then q commutes with elements of H, whence $H \subset \widetilde{G} \times \widetilde{\widetilde{G}}$, which is wrong, because H is the maximal subgroup of G).

Thus the only possibility is $(H')^M \cong SO(3) \times SU(2)$.

Passing from the groups to the corresponding Lie algebras we obtain that $\ln G / \ln_G H$ is a principal unitary involutive pair.

Consequently G/H is a principal unitary symmetric space.   ∎

III.3.16. Remark. In II.4.9 (Theorem 17), II.4.11 (Theorem 18), II.4.12 (Theorem 19), II.4.13 (Theorem 20) all the groups are indicated up to a local isomorphism.

CHAPTER III.4

ESSENTIALLY SPECIAL SYMMETRIC SPACES

III.4.1. Theorem 24. *If $V = G/H$ is an essentially special symmetric space with a simple compact Lie group G then for any point $x \in V = G/H$ there exist two principal mirrors W and W', and two related commutative principal subsymmetries, S and S', respectively, such that $W \cap W' = \{x\}$ (locally), W and W' are conjugated by an element from H, and*

$$\dim W + \dim W' = \dim G/H.$$

Proof. Passing to the corresponding Lie algebras $\ln G$ and $\mathfrak{l} = \ln_G H$ and selecting in $\mathfrak{l}$ a simple special subalgebra $\mathfrak{k} \subset \mathfrak{l}$, $\dim \mathfrak{k} > 3$, we construct by II.9.5 (Theorem 44) an iso-involutive decomposition

$$\mathfrak{g} = \mathfrak{l}_1 + \mathfrak{l}_2 + \mathfrak{l}_3$$

basis for $\mathfrak{l}_1$ and $\mathfrak{l}_2$.

By the same theorem $\mathfrak{l}_1$ and $\mathfrak{l}_2$ are principal unitary involutive algebras of involutive automorphisms S_1 and S_2, respectively.

The involutive automorphisms S_1 and S_2 of the Lie algebra $\mathfrak{g}$ generate in G/H the subsymmetries S and S' and the corresponding principal mirrors W, W' with the required properties. ∎

III.4.2. Theorem 25. *Let $V = G/H$ be a symmetric space generated by an involutive automorphism S with a simple compact Lie group G, and let S' and S'' be its principal subsymmetries at a point $x \in V$ with the mirrors W', W'', respectively, such that $S'S'' = S$ is a symmetry at the point x.*

If $\widetilde{G}^{W'}/\widetilde{H}^{W'}$ and $\widetilde{G}^{W''}/\widetilde{H}^{W''}$ are principal symmetric spaces then G/H is an essentially special symmetric space.

Proof. Passing to the corresponding Lie algebras we obtain the involutive sum generated by the subsymmetries S', S'', S

$$\mathfrak{g} = \mathfrak{l}' + \mathfrak{l}'' + \mathfrak{l} \qquad (\mathfrak{g} = \ln G, \ \mathfrak{l} = \ln_G H).$$

Since S' and S'' are principal subsymmetries their corresponding involutive algebras $\mathfrak{l}'$ and $\mathfrak{l}''$ are principal unitary algebras.

Moreover, from the conditions of the theorem it follows that

$$\mathfrak{l}' \cap \mathfrak{l}'' \cap \mathfrak{l} = \mathfrak{l}_0 = \mathfrak{p} \oplus \mathfrak{q} \oplus \tilde{\mathfrak{l}}_0,$$

where $\dim \mathfrak{p} = \dim \mathfrak{q} = 3$, and

$$\mathrm{Int}_{\mathfrak{g}}(\mathfrak{p}) \cong \mathrm{Int}_{\mathfrak{g}}(\mathfrak{q}) \cong SU(2).$$

Therefore the involutive sum constructed above is basis for the principal unitary involutive algebra $\mathfrak{l}'$, and by virtue of II.9.1 (Theorem 40) $\mathfrak{l}$ is a special involutive algebra.

Passing from the above Lie algebras to the corresponding Lie groups we obtain that G/H is an essentially special symmetric space.   ∎

III.4.3. Theorem 26. *If G/H is an essentially special symmetric space with a simple compact Lie group G and its principal unitary non-exceptional mirror $W = \widetilde{G}^W/\widetilde{H}^W$ is a principal unitary symmetric space then*

$$W \cong SO(n-4)/SO(4) \times SO(n-8), \quad n \geq 8;$$
$$\cong SU(n-2)/SU(2) \times U(n-4), \quad n > 4;$$
$$\cong Sp(n-1)/Sp(n-2) \times Sp(1), \quad n \geq 2,$$

and, respectively,

$$G/H \cong SO(n)/SO(8) \times SO(n-8),$$
$$\cong SU(n)/SU(4) \times U(n-4),$$
$$\cong Sp(n)/Sp(2) \times Sp(n-2).$$

(The groups are indicated up to local isomorphisms).

Proof. Passing to the corresponding Lie algebras and making use of II.9.2 (Theorem 41), II.9.3 (Theorem 42), II.9.4 (Theorem 43) we obtain the assertion of the theorem   ∎

III.4.4. Theorem 27. *If G/H is a special symmetric space with a simple compact Lie group G and its principal exceptional mirror $W = \widetilde{G}^W/\widetilde{H}^W$ is a principal unitary symmetric space then*

$$W \cong Sp(3)/Sp(2) \times Sp(1),$$
$$\cong SU(6)/S(U(4) \times U(2)),$$
$$\cong SO(12)/SO(8) \times SO(4),$$
$$\cong E_7/SO(12) \times SU(2)$$

and, respectively,

$$G/H \cong F_4/SO(9),$$
$$\cong E_6/SO(10) \times SO(2),$$
$$\cong E_7/SO(12) \times SU(2),$$
$$\cong E_8/SO(16).$$

(The groups are indicated up to local isomorphisms.)

Proof. Passing to the corresponding Lie algebras and using II.8.1 (Theorem 35), II.8.2 (Theorem 36), II.8.3 (Theorem 37), II.8.4 (Theorem 38) we obtain the assertion of the theorem. ∎

III.4.5. Theorem 28. *If G/H is an essentially special symmetric space with a simple compact Lie group G then*

$$G/H \cong SO(n)/SO(8) \times SO(n-8),$$
$$\cong SU(n)/S(U(4) \times U(n-4)),$$
$$\cong Sp(n)/Sp(2) \times Sp(n-2),$$
$$\cong F_4/SO(9),$$
$$\cong E_6/SO(10) \times SO(2),$$
$$\cong E_7/SO(12) \times SU(2),$$
$$\cong E_8/SO(16).$$

(The groups are indicated up to local isomorphisms.)

Proof. Passing to the corresponding Lie algebras and making use of II.9.10 (Theorem 45) we obtain the result. ∎

From this theorem there follows:

III.4.6. Theorem 29. *If G/H is a principal essentially special symmetric space with a simple compact Lie group G then*

$$G/H \cong SO(12)/SO(8) \times SO(4),$$
$$\cong SU(6)/S(U(4) \times U(2)),$$
$$\cong Sp(3)/Sp(2) \times Sp(1),$$
$$\cong E_7/SO(12) \times SU(2).$$

CHAPTER III.5

SOME THEOREMS ON SIMPLE COMPACT LIE GROUPS

III.5.1. Theorem 30. *If G is a simple compact Lie group, G/H is an involutive pair, $H \neq G$, and $\dim H = 3$ then $H \cong SO(3)$.*

Proof. We note first that evidently $\ln_G H \cong so(3)$.

Furthermore, $\mathrm{Int}_{\ln G}(\ln_G H) \cong SO(3)$, otherwise $\ln G / \ln_G H$ is a principal unitary involutive pair and $\dim H = 3$, which is impossible by I.8.2 (Theorem 31).

Consequently the involutive pair $\ln G / \ln_G H$ is a principal involutive pair of type O.

Since $H \neq G$ we obtain from II.3.6 (Theorem 6), II.3.10 (Theorem 8), II.3.14 (Theorem 10) that

$$\ln G / \ln_G H \cong su(3)/so(3)$$

with the natural embeddings, and there is an hyper-involutive decomposition basis for $\ln_G H$, which is described by II.3.6 (Theorem 6), that is,

$$\ln G = \mathfrak{l}_1 + \mathfrak{l}_2 + \mathfrak{l}_3,$$

$$\mathfrak{l}_0 = \mathfrak{l}_\alpha \cap \mathfrak{l}_\beta \quad (\alpha \neq \beta; \; \alpha, \beta = 1, 2, 3),$$

$$\mathfrak{l}_\alpha / \mathfrak{l}_0 \cong su(2)/so(2) + so(2)/so(2).$$

And $\mathfrak{l}_\alpha$ ($\alpha = 1, 2, 3$) is a principal unitary central involutive algebra, that is,

$$\mathfrak{l}_\alpha = \mathfrak{p}_\alpha \oplus \mathfrak{q}_\alpha \cong su(2) \oplus so(2)$$

and

$$\mathrm{Int}_{\ln G}(\mathfrak{p}_\alpha) \cong SU(2).$$

Finally,

$$\ln_G(H) \cap \mathfrak{p}_\alpha = \mathfrak{w}_\alpha \cong so(2).$$

We now consider a natural morphism

$$G \xrightarrow{\chi} \mathrm{Int}(\ln G).$$

Then

$$H \xrightarrow{\chi} \mathrm{Int}_{\ln G}(\ln_G H) \cong SO(3).$$

Assuming $H \cong SU(2)$ we have, on the one hand, that there is a strict morphism

$$\chi^{-1}\left(\mathrm{Int}_{\ln G}(\mathfrak{w}_\alpha)\right) \xrightarrow{\chi} \mathrm{Int}_{\ln G}(\mathfrak{w}_\alpha),$$

since $\chi^{-1}(\mathfrak{w}_\alpha) \subset H$; on the other hand, we have the isomorphism

$$\chi^{-1}\left(\mathrm{Int}_{\ln G}(\mathfrak{w}_\alpha)\right) \xrightarrow{\chi} \mathrm{Int}_{\ln G}(\mathfrak{w}_\alpha),$$

since

$$\chi^{-1}\left(\mathrm{Int}_{\ln G}(\mathfrak{w}_\alpha)\right) \subset \chi^{-1}\left(\mathrm{Int}_{\ln G}(\mathfrak{p}_\alpha)\right).$$

As a result the only possibility is $H \cong SO(3)$, which proves the theorem. ∎

III.5.2. Theorem 31. *If G is a simple compact connected Lie group and $\ln G \cong g_2$ then $G \cong \mathrm{Int}(g_2)$.*

Proof. We assume that G has a non-trivial centre and take an element $a_0 \neq e$ of this centre. We now consider a one-dimensional subgroup $a(t)$ passing through $a_0 = a(t_0)$ (as is well known, such a subgroup always exists for a connected compact Lie group) and a natural morphism $\chi: G \longrightarrow \mathrm{Int}(g_2)$.

The one-dimensional subgroup $a(t)$ belongs to some maximal torus T, but all maximal tori are conjugated in G by inner automorphisms.

Therefore we may assume that $\chi(T) \subset \mathrm{Int}_{g_2}(\mathfrak{p} \oplus \mathfrak{q})$, where $\mathfrak{p} \oplus \mathfrak{q} \cong so(4)$ is a principal di-unitary involutive algebra in g_2 of an involutive automorphism S (see II.6 and II.5.6 (Theorem 24). Owing to what has been said above we have the strict morphism

$$\chi^{-1}(\mathrm{Int}_{g_2}(\mathfrak{p} \oplus \mathfrak{q})) \xrightarrow{\chi} \mathrm{Int}_{g_2}(\mathfrak{p} \oplus \mathfrak{q}) \cong SO(4),$$

which is possible only if

$$\chi^{-1}(\mathrm{Int}_{g_2}(\mathfrak{p} \oplus \mathfrak{q})) \cong SU(2) \times SU(2).$$

If $\mathfrak{b}$ is a diagonal in $\mathfrak{p} \oplus \mathfrak{q}$ then owing to what has been considered above there is the strict morphism

$$SU(2) \cong \chi^{-1}(\mathrm{Int}_{g_2}(\mathfrak{b})) \xrightarrow{\chi} \mathrm{Int}_{g_2}(\mathfrak{b}) \cong SO(3).$$

Furthermore, we consider the involutive automorphism $S_1 = \mathrm{Int}_{g_2}(\zeta)$, where $\zeta \in \mathfrak{b}$, of a hyper-involutive decomposition basis for S (see II.5.6 (Theorem 24)) and the corresponding involutive algebra $\mathfrak{l}_1 = \mathfrak{n}_1 \oplus \mathfrak{m}_1 \cong so(4)$. Then evidently for $q = \chi^{-1}(S_1)$ we have: $q^4 = e$, $q^2 = a_0 \neq e$. By II.5.6 (Theorem 24) S_1 and S_2 are conjugated in $\mathrm{Int}(g_2)$, thus $\mathrm{Int}_{g_2}(\mathfrak{p} \oplus \mathfrak{q})$ and $\mathrm{Int}_{g_2}(\mathfrak{n}_1 \oplus \mathfrak{m}_1)$ are conjugated in $\mathrm{Int}(g_2)$. But then also $\chi^{-1}(\mathrm{Int}_{g_2}(\mathfrak{p} \oplus \mathfrak{q}))$ is conjugated with $\chi^{-1}(\mathrm{Int}_{g_2}(\mathfrak{n}_1 \oplus \mathfrak{m}_1))$ by inner automorphisms. Therefore

$$\chi^{-1}(\mathrm{Int}_{g_2}(\mathfrak{n}_1 \oplus \mathfrak{m}_1)) \cong SU(2) \times SU(2).$$

However, $\zeta \in \mathfrak{n}_1 \oplus \mathfrak{m}_1$, therefore

$$q \in Z \subset \chi^{-1}(\mathrm{Int}_{g_2}(\mathfrak{n}_1 \oplus \mathfrak{m}_1)) \cong SU(2) \times SU(2),$$

where Z is the centre of $\chi^{-1}(\mathrm{Int}_{g_2}(\mathfrak{n}_1 \oplus \mathfrak{m}_1))$. (Indeed, $\chi(q) = S_1$ is an element of the centre in $\mathrm{Int}_{g_2}(\mathfrak{n}_1 \oplus \mathfrak{m}_1)$, but the inverse image of the centre under the action of the morphism in our case is the set of elements of the centre.)

The latter is impossible since the centre in $SU(2) \times SU(2)$ consists of involutive elements and $q^2 = a_0 \neq e$. Thus we have a contradiction with the assumption that there exists an element $a_0 \neq e$ of the centre in G. ∎

III.5.3. Theorem 32. *If G is a simple compact connected Lie group and* $\ln G = f_4$ *then* $G \cong \text{Int}(f_4)$.

Proof. Let G have a non-trivial centre, and let $g \neq e$ be an element of this centre. Repeating the arguments given at the beginning of the previous theorem we obtain that g belongs to some maximal torus T in G.

Owing to the conjugacy of all maximal tori (by inner automorphisms of G) we may consider $\ln_G T \subset \mathfrak{l}$, where $\mathfrak{l}$ is a special unitary non-principal involutive algebra of f_4.

By II.8.1 (Theorem 35) we have $\mathfrak{l} \cong so(9)$ and then $\text{Int}_{f_4}(\mathfrak{l}) \cong \text{Spin}(9)$ (since it is known [L.S. Pontryagin 54,73], [C. Chevalley, 46] that the universal covering for $SO(9)$ has a two-element centre and the centre of $\text{Int}_{f_4}(\mathfrak{l})$ contains the non-trivial element S being an involutive automorphism of the involutive algebra $\mathfrak{l} \cong so(9)$).

We now consider a natural morphism

$$\chi \to G \longrightarrow \text{Int}(f_4).$$

By virtue of our assumption

$$\chi(g) = e, \qquad g \in \chi^{-1}\left(\text{Int}_{f_4}(\mathfrak{l})\right).$$

Therefore there exists the strict morphism $\chi^{-1}\left(\text{Int}_{f_4}(\mathfrak{l})\right) \longrightarrow \text{Int}_{f_4}(\mathfrak{l}) \cong \text{Spin}(9)$, which is impossible, since a morphism of a simple compact connected group onto its universal covering is an isomorphism. This proves our theorem. ∎

III.5.4. Remark. In an analogous way one might explore the centres of universal coverings for simple compact connected Lie groups when $\ln G = e_6$, $\ln G = e_7$, $\ln G = e_8$.

III.5.5. Let us now examine another problem, namely, the problem of describing of all inner involutive automorphisms of simple compact Lie algebras. For the exceptional simple compact algebras of type g_2, f_4, e_6, e_7, e_8 this problem is solved in a rather complicated way by the Root Method.

And in these cases the Mirror Geometry is very effective, as we shall see later. Indeed, because of the conjugacy of all maximal tori in a simple compact Lie algebra $\mathfrak{g}$ it is sufficient to find out inner automorphisms of $\mathfrak{g}$ in $\text{Int}_{\mathfrak{g}}(\mathfrak{l})$, where $\mathfrak{l}$ is a special unitary algebra of $\mathfrak{g}$.

However, in the case of the exceptional algebras

$$\mathfrak{l} \cong so(4), \quad so(9), \quad so(10) \oplus so(2), \quad so(12) \oplus so(3), \quad so(16),$$

as we already know, and thus we know $\text{Int}_{\mathfrak{g}}(\mathfrak{l})$.

Consequently the problem is reduced to finding involutive automorphisms in some covering groups for $\text{Int}(so(n))$, that is, it is reduced to finding involutive automorphisms in $SO(n)$, which is trivial.

We apply what has just been obtained to the algebras g_2 and f_4.

III.5.6. Theorem 33. *A non-trivial inner automorphism of a compact simple Lie algebra $\mathfrak{g}$ of type g_2 is unique, up to the conjugacy by inner automorphisms. That automorphism is a principal unitary involutive automorphism of $\mathfrak{g}$.*

Proof. Let $\mathfrak{g} \cong g_2$. Then its principal unitary involutive automorphism S has, as we know, the involutive algebra $\mathfrak{l} \cong su(2) \oplus su(2)$.

Then $S \in \mathrm{Int}_{\mathfrak{g}}(\mathfrak{l}) \cong SO(4)$. From the remarks preceding this theorem it is sufficient to find all inner involutive automorphisms of $\mathfrak{g}$ in $\mathrm{Int}_{\mathfrak{g}}(\mathfrak{l}) \cong SO(4)$. However, $SO(4) \cong \mathrm{Int}_{\mathfrak{g}}(\mathfrak{l})$ has only two, up to conjugacy, non-trivial involutive elements. Under the notations of II.5.6 (Theorem 24) those elements are S and S_1.

By II.5.6 (Theorem 24) these two elements are conjugated in $\mathrm{Int}(\mathfrak{g})$.

Thus in our case the unique non-trivial involutive automorphism of $\mathrm{Int}_{\mathfrak{g}}(\mathfrak{l})$ is S only (determining the principal unitary involutive algebra $\mathfrak{l}$). And the theorem is proved. ∎

III.5.7. Theorem 34. *Let $\mathfrak{g}$ be a simple compact Lie algebra of type f_4, and S be its non-trivial inner involutive automorphism. Then S is either principal unitary or special unitary non-principal.*

Proof. We consider a basis iso-involutive decomposition

$$\mathfrak{g} = \mathfrak{l}_1 + \mathfrak{l}_2 + \mathfrak{l}_3$$

for $\mathfrak{g} \cong f_4$, see II.8.1 (Theorem 35), where $\mathfrak{l}_1$ and $\mathfrak{l}_2$ are principal unitary involutive algebras, $\mathfrak{l}_1 \cong \mathfrak{l}_2 \cong su(2) \oplus sp(3)$, of the involutive automorphisms S_1 and S_2, and $\mathfrak{l}_3 \cong so(9)$ is a unitary special involutive algebra of the involutive automorphism S_3.

Note that in this case

$$\mathfrak{l}_0 = \mathfrak{l}_1 \cap \mathfrak{l}_2 \cap \mathfrak{l}_3 \cong so(4) \oplus so(5).$$

By the remarks of III.5.5 it is sufficient to find an involutive automorphism in $\mathrm{Int}_{\mathfrak{g}}(\mathfrak{l}_3) \cong \mathrm{Spin}(9)$. However, in $\mathrm{Int}_{\mathfrak{g}}(\mathfrak{l}_3) \cong \mathrm{Spin}(9)$ there are only three, up to conjugacy, involutive elements. Those elements are S_3, S_1, and one more involutive automorphism $\widetilde{S}$, generated by some subalgebra $\mathfrak{l}' \cong so(8)$ with the natural embedding into $so(9) \cong \mathfrak{l}_3$.

The involutive algebra $\widetilde{\mathfrak{l}}$ of the involutive automorphism $\widetilde{S}$ together with $\mathfrak{l}'$ generates the involutive pair $\widetilde{\mathfrak{l}}/\mathfrak{l}'$ of the involutive automorphism S_3, and $\mathfrak{l}' \cong so(8)$ is a special involutive algebra of the involutive automorphism S_3.

Therefore

$$\widetilde{\mathfrak{l}}/\mathfrak{l}' = \widetilde{\mathfrak{l}}'/\mathfrak{l}' + \widetilde{\mathfrak{l}}''/\{0\},$$

where $\widetilde{\mathfrak{l}}'$ is simple and compact and $\widetilde{\mathfrak{l}}'/\mathfrak{l}'$ is a special involutive pair of the involutive automorphism $\widetilde{S}$. Since $\mathfrak{l}' \cong so(8)$ by II.9.10 (Theorem 45) we have

$$\widetilde{\mathfrak{l}}'/\mathfrak{l}' \cong so(9)/so(8)$$

with the natural embedding.

However, $\mathfrak{l}_3/\mathfrak{l}'$ is also a special unitary involutive pair of the involutive automorphism $\widetilde{S}$, and $\widetilde{S} \in \mathrm{Int}_\mathfrak{g}(\mathfrak{l}') \cong \mathrm{Spin}(8)$. From this it follows that

$$\widetilde{S} \in \mathrm{Int}_\mathfrak{g}(\mathfrak{a}), \qquad \mathfrak{a} \cong so(3),$$

with the natural embeddings into $\widetilde{\mathfrak{l}}' \cong so(8)$ and into $\widetilde{\mathfrak{l}}' \cong so(9)$. Consequently $\widetilde{\mathfrak{l}}' \cong so(9)$ is a special unitary algebra in $\mathfrak{g}$.

But then from II.9.10 (Theorem 45) we have $\widetilde{\mathfrak{l}}'' = \{0\}$.

Thus $\widetilde{\mathfrak{l}} \cong so(9)$ is a special unitary involutive algebra of the involutive automorphism $\widetilde{S}$.

Analogous considerations can be given for $\widetilde{\widetilde{\mathfrak{l}}}$ which is the involutive algebra of the involutive automorphism $\widetilde{\widetilde{S}} = S_3\,\widetilde{S}$.

In this case

$$\mathfrak{g} = \mathfrak{l}_3 + \widetilde{\mathfrak{l}} + \widetilde{\widetilde{\mathfrak{l}}}, \qquad \mathfrak{l}_0 = \mathfrak{l}_3 \cap \widetilde{\mathfrak{l}} = \widetilde{\mathfrak{l}} \cap \widetilde{\widetilde{\mathfrak{l}}} = \mathfrak{l}_3 \cap \widetilde{\widetilde{\mathfrak{l}}} \cong so(8),$$

and it is easy to see that $\mathfrak{l}_3,\ \widetilde{\mathfrak{l}},\ \widetilde{\widetilde{\mathfrak{l}}}$ are pair-wise conjugated (indeed, taking $\{t\,\zeta\} \subset \mathfrak{l}_3 - \mathfrak{l}_0$ in such a way that $S_3 = \mathrm{Int}_\mathfrak{g}(t_0\,\zeta)$ we see that $\mathfrak{g} = \widetilde{\mathfrak{l}} + \widetilde{\widetilde{\mathfrak{l}}} + \mathfrak{l}_3$ is an iso-involutive sum with the conjugating automorphism $\varphi = \mathrm{Int}_\mathfrak{g}(\frac{1}{2}\,t_0\,\zeta)$, thus $\widetilde{\mathfrak{l}}$ and $\widetilde{\widetilde{\mathfrak{l}}}$ are conjugated in $\mathrm{Int}(\mathfrak{g})$).

Analogous arguments can be used for other pairs of involutive algebras, $\mathfrak{l}_3,\ \widetilde{\mathfrak{l}}$ and $\mathfrak{l}_3,\ \widetilde{\widetilde{\mathfrak{l}}}$.

But then $\widetilde{S},\ \widetilde{\widetilde{S}}$, and S_3 are also conjugated in $\mathrm{Int}(\mathfrak{g})$.

This shows that, except S_3 and S_1, there are no non-trivial inner involutive automorphisms of $\mathfrak{g} \cong f_4$ (up to the conjugacy in $\mathrm{Int}(\mathfrak{g})$).

This proves the theorem.   ∎

III.5.8. We can, performing analogous considerations and making use of basis involutive sums for e_6, e_7, e_8, find out their inner involutive automorphisms.

For example, the following result is true:

III.5.9. Theorem 35. *Let $\mathfrak{g}$ be a simple compact Lie algebra of type e_6 or e_8, S being a non-trivial inner involutive automorphism. Then S is either principal unitary or special unitary non-principal.*

CHAPTER III.6

TRI-SYMMETRIC AND HYPER-TRI-SYMMETRIC SPACES

III.6.1. Let G/H be a pair of Lie groups G and $H \subset G$. Then this pair uniquely defines a homogeneous space of left (or right) cosets of G by H with the motion group $\overset{*}{G} = G/Q$ and the stationary group $\overset{*}{H} = H/Q$, where Q is the maximal normal subgroup of G belonging to H. For this reason we denote by $\overset{*}{G}/\overset{*}{H}$, or $G\overset{*}{/}H$, the homogeneous space defined by a pair G/H.

III.6.2. Definition 57. We say that a homogeneous space G/H is an involutive product of homogeneous spaces $G_1\overset{*}{/}H_1$, $G_2\overset{*}{/}H_2$, and $G_3\overset{*}{/}H_3$ and write

$$G/H = G_1\overset{*}{/}H_1 \boxtimes G_2\overset{*}{/}H_2 \boxtimes G_3\overset{*}{/}H_3$$

if

$$G = G_1 \boxtimes G_2 \boxtimes G_3, \qquad H = H_1 \boxtimes H_2 \boxtimes H_3.$$

We say in this case that $\overset{*}{G}_\alpha/\overset{*}{H}_\alpha$ ($\alpha = 1, 2, 3$) are mirrors in G/H.

III.6.3. Definition 58. An involutive product

$$G/H = G_1\overset{*}{/}H_1 \boxtimes G_2\overset{*}{/}H_2 \boxtimes G_3\overset{*}{/}H_3$$

is called a tri-symmetric space [L.V. Sabinin, 61] with the mirrors $G_\alpha\overset{*}{/}H_\alpha$ ($\alpha = 1, 2, 3$) if

$$\mathfrak{k} = (\ln_G G_1) \cap (\ln_G G_2) \cap (\ln_G G_3) \subset \ln_G H;$$

moreover, it is said to be trivial if $\mathfrak{k} = \ln_G H$, semi-trivial if $\mathfrak{k} = \ln_G H_\alpha$ for some $\alpha = 1, 2, 3$, and non-trivial otherwise.

III.6.4. Remark. It is easy to see that in a tri-symmetric space

$$G/H = G_1\overset{*}{/}H_1 \boxtimes G_2\overset{*}{/}H_2 \boxtimes G_3\overset{*}{/}H_3$$

the mirrors $G_\alpha\overset{*}{/}H_\alpha$ ($\alpha = 1, 2, 3$) are symmetric spaces [L.V. Sabinin, 61].

163

III.6.5. Definition 59. A tri-symmetric space

$$G/H = G_1 \overset{*}{/} H_1 \boxtimes G_2 \overset{*}{/} H_2 \boxtimes G_3 \overset{*}{/} H_3$$

is called hyper-tri-symmetric if

$$G = G_1 \boxtimes G_2 \boxtimes G_3, \qquad H = H_1 \boxtimes H_2 \boxtimes H_3$$

are hyper-involutive products with the common conjugating automorphism p ($p^3 = \mathrm{id}$),

$$pG_1 = G_2, \qquad pG_2 = G_3, \qquad pG_3 = G_1,$$

$$pH_1 = H_2, \qquad pH_2 = H_3, \qquad pH_3 = H_1.$$

III.6.6. Remark. One may reformulate III.6.5 (Definition 59) as follows:

A homogeneous space $W = G/H$ is called a hyper-tri-symmetric space if for any point x of it there are three mirrors W', W'', W''', passing through x, which are generated by the commutative symmetries S', S'', S''', respectively, in such a way that for some neighbourhood of the point x

$$\dim(W' \cap W'' \cap W''') = 0,$$

$$\dim W' + \dim W'' + \dim W''' = \dim W,$$

and there exists an automorphism p of W such that $p^3 = \mathrm{id}$,

$$pW' = W'', \qquad pW'' = W''', \qquad pW''' = W'.$$

III.6.7. If a Lie group G is an involutive product $G = G_1 \boxtimes G_2 \boxtimes G_3$ then taking G/H, where $H = G_1 \cap G_2 \cap G_3$, we obtain a tri-symmetric space

$$G/H = G_1 \overset{*}{/} H \boxtimes G_2 \overset{*}{/} H \boxtimes G_3 \overset{*}{/} H.$$

It is easily seen that this example completely describes trivial tri-symmetric spaces [L.V. Sabinin, 61].

We are interested in non-trivial tri-symmetric spaces and, first of all, in hyper-tri-symmetric spaces with simple compact Lie groups of motions.

The trivial and semi-trivial tri-symmetric spaces G/H are not of interest since then either the stabilizer H is a non-maximal subgroup in G or G/H is a symmetric space.

We now present one evident result [L.V. Sabinin, 61].

III.6.8. Theorem 36. *Let G be a simple Lie group, and*

$$G/H = G_1 \overset{*}{/} H_1 \boxtimes G_2 \overset{*}{/} H_2 \boxtimes G_3 \overset{*}{/} H_3$$

be a tri-symmetric non-trivial space with a compact Lie group H.
   Then locally

$$G_\alpha = \widetilde{H}_\alpha \times \widetilde{G}_\alpha \quad (\alpha = 1, 2, 3),$$

where $\widetilde{G}_\alpha \neq \{e\}$, $\widetilde{H}_\alpha \subset H_\alpha$, and

$$G_\alpha \overset{*}{/} H_\alpha = \widetilde{G}_\alpha \overset{*}{/} (H_\alpha \cap \widetilde{G}_\alpha), \qquad H_\alpha \cap \widetilde{G}_\alpha \subset G_1 \cap G_2 \cap G_3.$$

From this theorem there follows:

III.6.9. Theorem 37. *If a tri-symmetric space*

$$G/H = G_1 \overset{*}{/} H_1 \boxtimes G_2 \overset{*}{/} H_2 \boxtimes G_3 \overset{*}{/} H_3$$

with a simple Lie group G and a compact Lie group H:
   a) *is non-trivial then G_1, G_2, G_3 can not be simple non-commutative Lie groups;*
   b) *has at least one simple and non-commutative Lie group G_α ($\alpha = 1, 2, 3$) then G/H is trivial or semi-trivial.*

This theorem may be slightly strengthened [L.V. Sabinin, 61]:

III.6.10. Theorem 38. *Let*

$$G/H = G_1 \overset{*}{/} H_1 \boxtimes G_2 \overset{*}{/} H_2 \boxtimes G_3 \overset{*}{/} H_3$$

be a tri-symmetric space with a compact Lie group H, and let a symmetric space $G_\alpha \overset{*}{/} G_0$, where $G_0 = G_1 \cap G_2 \cap G_3$, be irreducible for some $\alpha = 1, 2, 3$.
   Then G/H is semi-trivial or trivial.

Proof. This result follows immediately from III.6.8 (Theorem 36). ■

III.6.11. The forthcoming presentation concerns the classification of non--trivial tri-symmetric spaces with simple compact Lie groups of motion.

III.6.12. Remark. In order to avoid all ambiguities we note that

$$G \overset{*}{/} H = M \overset{*}{/} N \quad \text{means} \quad \overset{*}{G} = \overset{*}{M}, \quad \overset{*}{H} = \overset{*}{N}.$$

$$G \overset{*}{/} H = M/N \quad \text{means} \quad \overset{*}{G} = M, \quad \overset{*}{H} = N.$$

CHAPTER III.7

TRI-SYMMETRIC SPACES WITH
EXCEPTIONAL COMPACT GROUPS

III.7.1. We start with the exploration of the exceptional groups of types G_2, F_4, E_6, E_7, E_8.

Such a choice is intentional since in applications of the Root Method the exceptional Lie groups and algebras create the greatest difficulties.

The Mirror Geometry which has been developed in Part I and II gives us a simple and natural approach to problems related to the exceptional Lie groups and algebras. In what follows this will be seen.

III.7.2. Group of motions $G \cong G_2$. In this case there are no outer involutive automorphisms (see, for example, [S. Helgason 62, 78]) and a unique involutive automorphism of the Lie algebra $g_2 = \ln G_2$ (up to conjugacy in $\text{Int}(g_2)$) is the di-unitary principal involutive automorphism described in II.6, whence with the help of II.5.6 (Theorem 24) we obtain a unique (up to conjugacy in $\text{Int}(g_2)$) hyper-involutive decomposition

$$g_2 = \mathfrak{l}_1 + \mathfrak{l}_2 + \mathfrak{l}_3 \,,$$

where

$$\mathfrak{l}_\alpha \cong so(4), \qquad \mathfrak{l}_0 = \mathfrak{l}_\alpha \underset{\alpha \neq \beta}{\cap} \mathfrak{l}_\beta \cong so(2) \oplus so(2).$$

Taking into consideration III.6.8 (Theorem 36) and passing to the corresponding Lie algebras for the non-trivial tri-symmetric space

$$G_2/H = (Q_1 \overset{*}{/} H_1) \boxtimes (Q_2 \overset{*}{/} H_2) \boxtimes (Q_3 \overset{*}{/} H_3),$$

where $Q_1 \cong Q_2 \cong Q_3 \cong SO(4)$, we obtain $SO(4) \cong \widetilde{H}_\alpha \times \widetilde{G}_\alpha$ ($\alpha = 1, 2, 3$).

Consequently $so(4) \cong \ln{}_G \widetilde{H}_\alpha \oplus \ln{}_G \widetilde{G}_\alpha$, where $\ln{}_G \widetilde{H}_\alpha$ is a non-trivial ideal. Therefore $\ln{}_G \widetilde{H}_\alpha \cong su(2)$, and since $so(2) \oplus so(2) \cong \mathfrak{l}_0 \subset \ln{}_G \widetilde{H}_\alpha$ we obtain $\ln{}_G \widetilde{G}_\alpha = su(2)$.

Finally, we have

$$H_\alpha \cong SU(2) \times SO(2) \quad (\alpha = 1, 2, 3).$$

But $H = H_1 \boxtimes H_2 \boxtimes H_3$, where

$$\ln{}_G H = \ln{}_G H_1 + \ln{}_G H_2 + \ln{}_G H_3,$$

$$(\ln{}_G H_\alpha) \underset{\alpha \neq \beta}{\cap} (\ln{}_G H_\beta) = \mathfrak{l}_0 \cong so(2) \oplus so(2).$$

The latter is possible only if $\ln_G H \cong su(3)$. (Indeed, $\ln_G H/\ln_G H_\alpha$ is an involutive pair, where $\ln_G H_\alpha \cong su(2) \oplus so(2)$ is a unitary central involutive algebra which is elementary by I.8.2 (Theorem 31). Consequently $\ln_G H$ is simple. Lastly one should apply I.7.2 (Theorem 29).)

II.6.9 (Theorem 30) shows that the Lie algebra $su(3)$ is embedded into g_2 and acts irreducibly on $g_2/su(3)$.

Passing to the corresponding Lie groups G_2 and $SU(3)$ we have the theorem:

III.7.3. Theorem 39. *If G is a compact simple Lie group isomorphic to G_2 then there exists a unique non-trivial tri-symmetric space*

$$G/H = G_1 \overset{*}{/} H_1 \boxtimes G_2 \overset{*}{/} H_2 \boxtimes G_3 \overset{*}{/} H_3 \cong G_2/SU(3)$$

with the principal mirrors

$$G_\alpha \overset{*}{/} H_\alpha \cong SO(4) \overset{*}{/} SU(2) \times SO(2) = SU(2)/SO(2) \quad (\alpha = 1, 2, 3)$$

(with the natural embeddings).

This space is hyper-tri-symmetric and has an irreducible Lie group H.

III.7.4. Group of motions $G \cong F_4$.

In this case there are no outer involutive automorphism [S. Helgason 62, 78] and there exist only two essentially different involutive automorphisms in $f_4 = \ln F_4$, one of them is principal unitary, the other is non-principal special unitary (see III.5.7 (Theorem 34)).

However, in the last case the corresponding involutive algebra is $\mathfrak{l} \cong so(9)$. But by virtue of III.6.9 (Theorem 37) such an algebra can not generate a mirror in a non-trivial tri-symmetric space.

Thus we have to assume that in the tri-symmetric space

$$G/H = G_1 \overset{*}{/} H_1 \boxtimes G_2 \overset{*}{/} H_2 \boxtimes G_3 \overset{*}{/} H_3$$

G_α are principal involutive groups in $G \cong F_4$.

Therefore by II.8.1 (Theorem 35) $\ln_G G_\alpha \cong sp(3) \oplus su(2)$.

According to II.10.2 (Theorem 46) and II.10.5 (Theorem 47) we obtain a hyper-involutive decomposition for $\mathfrak{g} \cong f_4$, $\mathfrak{g} = \mathfrak{l}_1 + \mathfrak{l}_2 + \mathfrak{l}_3$, where

$$\mathfrak{l}_\alpha = \ln_G G_\alpha \cong sp(3) \oplus su(2),$$

$$\mathfrak{l}_\alpha/\mathfrak{l}_0 \cong sp(3)/u(3) + su(2)/so(2),$$

$$\mathfrak{l}_0 = \mathfrak{l}_\alpha \underset{\alpha \neq \beta}{\cap} \mathfrak{l}_\beta \quad (\alpha = 1, 2, 3).$$

It is easy to see that there is no other involutive decompositions for $\mathfrak{g} \cong f_4$ with the involutive algebras $\mathfrak{l}_\alpha \cong sp(3) \oplus su(2)$ (up to conjugacy in $\text{Int}(f_4)$, of course).

Therefore for the tri-symmetric spaces G/H, $G \cong F_4$, we have

$$G = G_1 \boxtimes G_2 \boxtimes G_3,$$

where

$$G_\alpha = G'_\alpha \times G''_\alpha \cong Sp(3) \times SU(2),$$

and

$$G_0 = G_\alpha \underset{\alpha \neq \beta}{\cap} G_\beta = \widetilde{G}_0 \times \widetilde{Z}' \times \widetilde{Z}''$$

$$\cong SU(3) \times U(1) \times U(1).$$

By III.6.8 (Theorem 36) either $H_\alpha \cong Sp(3) \times SO(2)$ or $H_\alpha \cong SU(2) \times U(3)$.

Let us note also that by II.10.2 (Theorem 46) and II.10.5 (Theorem 47)

$$\ln_G G''_1 + \ln_G G''_2 + \ln_G G''_3 = \mathfrak{n} \cong su(3)$$

is a subalgebra in $\mathfrak{g} \cong f_4 = \ln F_4$, and for this reason $G''_1 \boxtimes G''_2 \boxtimes G''_3 = N \cong SU(3)$ is a subgroup in $G \cong F_4$.

Suppose that $H_\alpha \cong Sp(3) \times SO(2)$ for all $\alpha = 1, 2, 3$.

Then evidently $H = G'_1 \boxtimes G'_2 \boxtimes G'_3$ and $\ln G = \ln_G H + \mathfrak{n}$, $\mathfrak{k} = (\ln_G H) \cap \mathfrak{n} = \mathfrak{z}' \oplus \mathfrak{z}'' \cong u(1) \oplus u(1)$, $\mathfrak{p} \perp \tilde{\mathfrak{n}}$, where $\ln_G H = \mathfrak{p} \dotplus \mathfrak{k}$, $\mathfrak{n} = \tilde{\mathfrak{n}} \dotplus \mathfrak{k}$.

Thus $[(\ln_G H)\tilde{\mathfrak{n}}] \subset \tilde{\mathfrak{n}}$. But then also $[(\ln_G H)\mathfrak{n}] \subset \mathfrak{n}$, and $\mathfrak{n}$ is an ideal of $\ln G$. The latter is impossible for a simple Lie algebra $\ln G$.

Consequently we are to assume that $H_\alpha \cong SU(2) \times U(3)$ at least for one α ($\alpha = 1, 2, 3$).

After re-numbering we may consider that $H_3 = G''_3 \times \widetilde{Z}' \times \widetilde{G}_0 \cong SU(2) \times U(3)$.

Passing from the involutive product to the corresponding involutive sum $\ln_G H = \ln_G H_1 + \ln_G H_2 + \ln_G H_3$ we also see that $\ln_G H/\ln_G H_3$ is a principal unitary involutive pair. Moreover, since $(\ln_G H_\alpha) \underset{\alpha \neq \beta}{\cap} (\ln_G H_\beta) = \ln_G G_0 \cong u(3) \oplus u(1)$ we see that $\ln_G H/\ln_G H_1$ is an involutive pair of index 1.

Therefore there are the following possibilities:
either

$$\ln_G H/\ln_G H_3 \cong su(3)/u(2) + su(3)/su(3)$$

or

$$\ln_G H/\ln_G H_3 \cong su(5)/su(2) \oplus su(3) \oplus u(1)$$

(owing to I.7.2 (Theorem 29)).

But then, correspondingly, either

$$\ln_G H_1 \cong \ln_G H_2 \cong su(2) \oplus u(3)$$

or

$$\ln_G H_1 \cong \ln_G H_2 \cong su(4)$$

(by virtue of I.6.4 (Theorem 22)).

However, the latter is impossible since either $H_1 \cong Sp(3) \times SO(2)$ or $H_1 \cong SU(2) \times U(3)$.

Thus we have

$$\ln_G H \cong su(3) \oplus su(3)$$

and

$$H \cong SU(3) \times SU(3).$$

We have proved the theorem:

III.7.5. Theorem 40. *If G is a compact simple Lie group isomorphic to F_4 then there exists the unique non-trivial tri-symmetric space*

$$G/H = G_1 \overset{*}{/} H_1 \boxtimes G_2 \overset{*}{/} H_2 \boxtimes G_3 \overset{*}{/} H_3 \cong F_4/SU(3) \times SU(3)$$

with the principal mirrors

$$G_\alpha \overset{*}{/} H_\alpha \cong SU(2) \times Sp(3) \overset{*}{/} SU(2) \times U(3) \cong Sp(3)/U(3)$$

(with the natural embeddings).

This space is hyper-tri-symmetric and has an irreducible group H.

III.7.6. Group of motions $G \cong E_6$.

In this case there are four essentially different involutive automorphisms of the Lie algebra e_6 [S. Helgason, 62,78] (up to conjugacy in $\text{Int}(e_6)$).

However, two of them possess the simple involutive algebras $sp(4)$ and f_4. For this reason, owing to III.6.9 (Theorem 37), such algebras can not generate mirrors in non-trivial tri-symmetric spaces.

Thus the only possibility is the involutive decomposition

$$\ln G = \ln_G G_1 + \ln_G G_2 + \ln_G G_3,$$

where the involutive algebras $\ln_G G_\alpha$ $(\alpha = 1, 2, 3)$ are either principal or non-principal special.

All such decompositions have been described in II.8.2 (Theorem 36), II.10.5 (Theorem 47), II.11.3 (Theorem 51).

However, the decompositions in II.8.2 (Theorem 36) and II.11.3 (Theorem 51) generates in the tri-symmetric space G/H the mirrors

$$SO(10) \times U(1) \overset{*}{/} SO(6) \times SO(4) \times U(1) \cong SO(10)/SO(6) \times SO(4)$$

and

$$SO(10) \times U(1) \overset{*}{/} SO(8) \times SO(2) \times U(1) \cong SO(10)/SO(8) \times SO(2),$$

respectively, which is impossible for the non-trivial tri-symmetric space G/H by virtue of III.6.10 (Theorem 38).

Thus the only possibility is

$$G/H = G_1 \overset{*}{/} H_1 \boxtimes G_2 \overset{*}{/} H_2 \boxtimes G_3 \overset{*}{/} H_3, \quad G \cong E_6,$$

G_α ($\alpha = 1, 2, 3$) are principal involutive groups in E_6, and there is the hyper-involutive decomposition described in II.10.5 (Theorem 47):

$$e_6 = \ln G = \ln_G G_1 + \ln_G G_2 + \ln_G G_3 = \mathfrak{l}_1 + \mathfrak{l}_2 + \mathfrak{l}_3,$$

where

$$\mathfrak{l}_\alpha \cong su(6) \oplus su(2),$$

$$\mathfrak{l}_\alpha/\mathfrak{l}_0 \cong su(6)/su(3) \oplus su(3) \oplus u(1) + su(2)/u(1),$$

$$\mathfrak{l}_0 = \mathfrak{l}_\alpha \underset{\alpha \neq \beta}{\cap} \mathfrak{l}_\beta \cong su(3) \oplus su(3) \oplus u(1) \oplus u(1) \quad (\alpha = 1, 2, 3).$$

There are no other decompositions with principal involutive algebras in the case of e_6 (of course, up to the conjugacy in $\mathrm{Aut}(e_6)$).

Further arguments are almost the same as in the case of F_4, up to details.

Thus we have the theorem:

III.7.7. Theorem 41. *If a compact simple Lie group G is isomorphic to E_6 then there exists the unique non-trivial symmetric space*

$$G/H = G_1 \overset{*}{/} H_1 \boxtimes G_2 \overset{*}{/} H_2 \boxtimes G_3 \overset{*}{/} H_3 \cong E_6/SU(3) \times SU(3) \times SU(3)$$

with the principle central mirrors

$$G_\alpha \overset{*}{/} H_\alpha \cong SU(2) \times SU(6) \overset{*}{/} SU(2) \times SU(3) \times SU(3) \times U(1)$$

$$\cong SU(6)/SU(3) \times SU(3) \times U(1)$$

(with the natural embeddings).

This space is hyper-tri-symmetric and has an irreducible group H.

III.7.8. Group of motions $\mathbf{G} \cong \mathbf{E_7}$. In this case we have only three essentially different involutive automorphisms of the Lie algebra e_7 [S. Helgason 62,78] with the involutive algebras $su(8)$, $so(12) \oplus su(2)$, $e_6 \oplus u(1)$. But we have to exclude the simple Lie algebra $su(8)$ from the consideration, because by III.6.9 (Theorem 37) such algebra can not generate a mirror in a non-trivial tri-symmetric space.

Suppose now that there exists the involutive decomposition for $\mathfrak{g} \cong e_7$, $\mathfrak{g} = \mathfrak{l}_1 + \mathfrak{l}_2 + \mathfrak{l}_3$, where $\mathfrak{l}_3 \cong e_6 \oplus u(1)$, $\mathfrak{l}_1 \cong so(12) \oplus su(2)$. Then there are two possibilities: either the centre of $\mathfrak{l}_3$, that is, $\mathfrak{z} \cong u(1)$, belongs to $\mathfrak{l}_0 = \mathfrak{l}_\alpha \underset{\alpha \neq \beta}{\cap} \mathfrak{l}_\beta$ or $\mathfrak{z} \subset \mathfrak{l}_3 \overset{\cdot}{-} \mathfrak{l}_0$.

We should exclude the first possibility because in this case for our tri-symmetric space G/H ($G \cong E_7$) the involutive algebra $\mathfrak{l}_3 = \ln_G G_3$ generates a mirror $G_3 \overset{*}{/} H_3$ with a simple compact group of motions $\overset{\cdot}{G}_3 \cong E_6$, which is impossible by virtue of III.6.10 (Theorem 38) for a non-trivial tri-symmetric space.

Thus the only case is $\mathfrak{z} \subset \mathfrak{l}_3 \overset{\cdot}{-} \mathfrak{l}_0$.

Then the involutive sum $\mathfrak{g} = \mathfrak{l}_1 + \mathfrak{l}_2 + \mathfrak{l}_3$, $\mathfrak{g} \cong e_7$, is iso-involutive with the conjugating automorphism $\varphi \in \mathrm{Int}_\mathfrak{g}(\mathfrak{z})$.

But then $\mathfrak{l}_2 \cong \mathfrak{l}_1 \cong so(12) \oplus su(2)$, which is impossible since it is easy to see that the composition of two commuting principal involutive automorphisms in the exceptional Lie algebras is special (or principal) involutive automorphism. For e_7 this means that $\mathfrak{l}_3 \cong so(12) \oplus su(2)$, which contradicts our assumption $\mathfrak{l}_3 \cong e_6 \oplus u(1)$.

As a result we have shown that for the non-trivial tri-symmetric space G/H $(G = E_7)$ either $\mathfrak{l}_\alpha = \ln_G G_\alpha \cong so(12) \oplus su(2)$ or $\mathfrak{l}_\alpha = \ln_G G_\alpha \cong e_6 \oplus u(1)$ $(\alpha = 1, 2, 3)$.

The case $\mathfrak{l}_\alpha = \ln_G G_\alpha \cong so(12) \oplus su(2)$ gives us two possible hyper-involutive decompositions for e_7 described in II.10.2 (Theorem 46), II.10.5 (Theorem 47), II.11.4 (Theorem 52).

According to II.11.4 (Theorem 52)

$$\mathfrak{l}_\alpha/\mathfrak{l}_0 \cong so(12)/so(8) \oplus so(4) + su(2)/su(2),$$

which means that the mirror

$$G_\alpha \overset{*}{/} H_\alpha \cong SO(12)/SO(8) \times SO(4) \quad (\alpha = 1, 2, 3)$$

has a simple Lie group of motions. But this case should be excluded from the consideration for non-trivial tri-symmetric spaces.

Thus in the case which is considered

$$\mathfrak{l}_\alpha = \ln_G G_\alpha \cong so(12) \oplus su(2)$$
$$\mathfrak{l}_\alpha/\mathfrak{l}_0 \cong so(12)/su(6) + su(2)/so(2).$$

Furthermore, by arguments analogous to the case of F_4 we obtain the non-trivial tri-symmetric space

$$E_7/H \cong E_7/SU(3) \times SU(6)$$

with the principal mirrors

$$G_\alpha \overset{*}{/} H_\alpha \cong SO(12)/SU(2) \overset{*}{/} U(6) \times SU(2) \cong SO(12)/U(6)$$

We are left with considering the case $\ln_G G_\alpha \cong e_6 \oplus u(1)$.

Taking into account all involutive algebras of e_6 (which we know) and calculating the corresponding dimensions (dim $\mathfrak{l}_0 = \dim e_7 - 3 \dim(\mathfrak{l}_\alpha - \mathfrak{l}_0)$, dim $e_7 = 133$, dim $\mathfrak{l}_\alpha = \dim e_6 + 1 = 79$), we conclude that if such a decomposition exists then

$$\mathfrak{l}_0 \cong f_4, \qquad \mathfrak{l}_\alpha/\mathfrak{l}_0 \cong e_6/f_4 + u(1)/\{0\}.$$

We note that in this case $\mathfrak{g} \cong e_7$ contains subalgebra $\mathfrak{l}_0 \oplus \mathfrak{m} \cong f_4 \oplus so(3)$ such that $\mathrm{Int}_\mathfrak{g}(\mathfrak{m}) \cong SO(3)$.

But this subalgebra has already been considered in II.11. It may be described by means of the special involutive decomposition from II.11.4 (Theorem 52).

Namely, there has been obtained

$$e_7 \cong \mathfrak{g} = \tilde{\mathfrak{l}}_1 + \tilde{\mathfrak{l}}_2 + \tilde{\mathfrak{l}}_3,$$

where

$$\mathfrak{g}/\tilde{\mathfrak{l}}_\alpha \cong e_7/so(12) \oplus su(2),$$

$$\tilde{\mathfrak{l}}_\alpha/\tilde{\mathfrak{l}}_0 \cong so(12)/so(8) \oplus so(4) + su(2)/su(2).$$

Thus

$$\tilde{\mathfrak{l}}_0 = \tilde{\mathfrak{l}}' \oplus \mathfrak{p}^{(1)} \oplus \mathfrak{p}^{(2)} \oplus \mathfrak{p}^{(3)} \cong so(8) \oplus su(2) \oplus su(2) \oplus su(2).$$

Taking a diagonal of $\mathfrak{p}^{(1)} \oplus \mathfrak{p}^{(2)} \oplus \mathfrak{p}^{(3)}$, $\mathfrak{m} = \Delta(\mathfrak{p}^{(1)} \oplus \mathfrak{p}^{(2)} \oplus \mathfrak{p}^{(3)}) \cong \Delta(su(2) \oplus su(2) \oplus su(2)) \cong so(3)$, we easily verify that the maximal subalgebra of e_7 commuting with $\mathfrak{m}$ is an algebra of type f_4 (first we should find the subalgebras of $\mathfrak{l}_\alpha \cong so(12)$ ($\alpha = 1, 2, 3$) commuting with $\mathfrak{m}$, this gives us $\mathfrak{n}_\alpha \cong so(9)$. Their involutive sums lead us to a subalgebra of type f_4). Moreover, it is easily seen that $\text{Int}_\mathfrak{g}(\mathfrak{m}) \cong SO(3)$ ($\mathfrak{g} \cong e_7$).

Using the hyper-involutive decomposition generated by commuting involutive automorphisms $S_\alpha \in \text{Int}_\mathfrak{g}(\mathfrak{m})$, $\mathfrak{g} \cong e_7$, $\alpha = 1, 2, 3$, the knowledge of all involutive subalgebras of e_7 [S. Helgason, 62,78], and computations of their dimensions we obtain that the hyper-involutive decomposition $\mathfrak{g} = \mathfrak{l}_1 + \mathfrak{l}_2 + \mathfrak{l}_3 \cong e_7$, where

$$\mathfrak{g}/\mathfrak{l}_\alpha \cong e_7/e_6 \oplus u(1),$$

$$\mathfrak{l}_\alpha/\mathfrak{l}_0 \cong e_6/f_4 + u(1)/\{0\},$$

indeed exists.

And by arguments analogous to the case of F_4 we arrive to the non-trivial tri-symmetric space

$$G/H \cong E_7/F_4 \times SO(3)$$

with the mirrors

$$G_\alpha \overset{*}{/} H_\alpha \cong E_6 \times U(1) \overset{*}{/} F_4 \times U(1) \cong E_6/F_4.$$

Thus we have proved the theorem:

III.7.9. Theorem 42. *If a compact simple Lie group G is isomorphic to E_7 then there exist only two non-trivial tri-symmetric spaces (up to isomorphism)*

$$G/H = G_1 \overset{*}{/} H_1 \boxtimes G_2 \overset{*}{/} H_2 \boxtimes G_3 \overset{*}{/} H_3,$$

namely:

1. $G/H \cong E_7/SU(3) \times SU(6)$ *with the principal mirrors*

$$G_\alpha \overset{*}{/} H_\alpha \cong SO(12) \times SU(2) \overset{*}{/} U(6) \times SU(2) \cong SO(12)/U(6);$$

2. $G/H \cong E_7/F_4 \times SO(3)$ *with the central mirrors*

$$G_\alpha \overset{*}{/} H_\alpha \cong E_6 \times U(1) \overset{*}{/} F_4 \times U(1) \cong E_6/F_4$$

(with the natural embeddings).

These spaces are hyper-tri-symmetric and have an irreducible Lie group H.

III.7.10. **Group of motions** $G \cong E_8$. In this case there are no outer involutive automorphisms [S. Helgason, 62,78] and there exist only two essentially different involutive automorphisms of $e_8 = \ln E_8$, one of them is a principal unitary with the involutive algebra $\mathfrak{l} \cong su(2) \oplus e_7$, and the other is non-principal special unitary with the simple involutive algebra $\mathfrak{l} \cong so(16)$.

We should exclude the second case from the consideration, since by III.6.9 (Theorem 37) a simple involutive algebra can not generate a mirror in a non-trivial tri-symmetric space.

Thus we should consider that in the tri-symmetric space $G/H = G_1 \overset{*}{/} H_1 \boxtimes G_2 \overset{*}{/} H_2 \boxtimes G_3 \overset{*}{/} H_3$, where $G \cong E_8$, G_α ($\alpha = 1, 2, 3$) are principal involutive groups. Therefore $\mathfrak{l}_\alpha = \ln_G G_\alpha \cong su(2) \oplus e_7$ and by virtue of II.10.2 (Theorem 46), II.10.5 (Theorem 47) we have the hyper-involutive decomposition

$$e_8 \cong \mathfrak{g} = \mathfrak{l}_1 + \mathfrak{l}_2 + \mathfrak{l}_3$$

with

$$\mathfrak{l}_\alpha / \mathfrak{l}_0 \cong e_7/e_6 \oplus u(1) + su(2)/so(2) \quad (\mathfrak{l}_0 = \mathfrak{l}_\alpha \underset{\alpha \neq \beta}{\cap} \mathfrak{l}_\beta).$$

It is easy to see that there are no other involutive decompositions of $\mathfrak{g} \cong e_8$ with the involutive algebras $\mathfrak{l}_\alpha \cong su(2) \oplus e_7$ (of course, up to the conjugacy in $\mathrm{Int}(e_8)$).

Furthermore, by arguments analogous to those which have been applied in the case F_4, we obtain the non-trivial tri-symmetric space $G/H \cong E_8/SU(3) \times E_6$ with the principal mirrors

$$G_\alpha \overset{*}{/} H_\alpha = E_7 \times SU(2) \overset{*}{/} E_6 \times U(1) \times SU(2) \cong E_7/E_6 \times U(1).$$

Thus the following is valid:

III.7.11. **Theorem 43.** *If a compact Lie group G is isomorphic to E_8 then there exists the unique non-trivial tri-symmetric space*

$$G/H = G_1 \overset{*}{/} H_1 \boxtimes G_2 \overset{*}{/} H_2 \boxtimes G_3 \overset{*}{/} H_3 \cong E_8/E_6 \times SU(3)$$

with the principal mirrors

$$G_\alpha \overset{*}{/} H_\alpha \cong E_7 \times SU(2) \overset{*}{/} E_6 \times U(1) \times SU(2) \cong E_7/E_6 \times U(1)$$

(with the natural embeddings).

This space is hyper-tri-symmetric and has an irreducible group H.

CHAPTER III.8

TRI-SYMMETRIC SPACES WITH
GROUPS OF MOTIONS SO(n), Sp(n), SU(n)

III.8.1. Group of motions $G \cong SO(n)$. We represent the corresponding algebra $\mathfrak{g} = \ln G \cong so(n)$ as the algebra of real skew-symmetric matrices $\overline{\overline{a}} \in so(n)$ with the standard operation of matrix commutator $\left[\overline{\overline{a}}\,\overline{\overline{c}}\right] = \overline{\overline{a}}\,\overline{\overline{c}} - \overline{\overline{c}}\,\overline{\overline{a}}$. Then involutive automorphisms are given by orthogonal matrices $\overline{\overline{S}} \in O(n)$ and act on $\mathfrak{g} = \ln G$ by the rule $\overline{\overline{a}} \mapsto S(\overline{\overline{a}}) = \overline{\overline{S}}\,\overline{\overline{a}}\,\overline{\overline{S}}^{-1}$, as is well known [E. Cartan, 49, 52].

Let S_1, S_2, S_3 be three pair-wise commuting involutive automorphisms. They are given by the matrices $\overline{\overline{S}}_1$, $\overline{\overline{S}}_2$, $\overline{\overline{S}}_3$, where $(\overline{\overline{S}}_\alpha)^2 = \pm\mathrm{Id}$, $\overline{\overline{S}}_\alpha\,\overline{\overline{S}}_\beta = \pm\overline{\overline{S}}_\beta\,\overline{\overline{S}}_\alpha$ ($\alpha \neq \beta$). Since we are interested in involutive automorphisms generating involutive decompositions we have $S_\alpha S_\beta = S_\gamma$ ($\alpha \neq \beta$, $\beta \neq \gamma$, $\gamma \neq \alpha$), and for the corresponding matrices $\overline{\overline{S}}_\alpha\,\overline{\overline{S}}_\beta = \pm\overline{\overline{S}}_\gamma$ ($\alpha \neq \beta$, $\beta \neq \gamma$, $\gamma \neq \alpha$).

Thus $\overline{\overline{S}}_1\,\overline{\overline{S}}_2 = \pm\overline{\overline{S}}_3$. But in the case of the sign minus in the right hand side we may assume $-\overline{\overline{S}}_3$ as a new $\overline{\overline{S}}_3$ (since $\overline{\overline{S}}_3$ and $-\overline{\overline{S}}_3$ generate the same involutive automorphism S_3).

Thus we assume $\overline{\overline{S}}_1\,\overline{\overline{S}}_2 = \overline{\overline{S}}_3$. And there are the following possibilities:

(1) 1. $(\overline{\overline{S}}_1)^2 = (\overline{\overline{S}}_2)^2 = (\overline{\overline{S}}_3)^2 = \overline{\overline{\mathrm{Id}}}$, then $\overline{\overline{S}}_1$, $\overline{\overline{S}}_2$, $\overline{\overline{S}}_3$ are pair-wise commutative;

(2) 2. $(\overline{\overline{S}}_1)^2 = (\overline{\overline{S}}_2)^2 = (\overline{\overline{S}}_3)^2 = -\overline{\overline{\mathrm{Id}}}$, then $\overline{\overline{S}}_1$, $\overline{\overline{S}}_2$, $\overline{\overline{S}}_3$ are pair-wise anticommutative;

(3) 3. $(\overline{\overline{S}}_1)^2 = (\overline{\overline{S}}_2)^2 = -\overline{\overline{\mathrm{Id}}}$, $(\overline{\overline{S}}_3)^2 = \overline{\overline{\mathrm{Id}}}$, then $\overline{\overline{S}}_1$, $\overline{\overline{S}}_2$, $\overline{\overline{S}}_3$ are pair-wise commutative;

(4) 4. $(\overline{\overline{S}}_1)^2 = (\overline{\overline{S}}_2)^2 = \overline{\overline{\mathrm{Id}}}$, $(\overline{\overline{S}}_3)^2 = -\overline{\overline{\mathrm{Id}}}$, then $\overline{\overline{S}}_1$, $\overline{\overline{S}}_2$, $\overline{\overline{S}}_3$ are pair-wise anticommutative.

The other possibilities are reduced to the listed above after re-numbering of the involutive automorphisms S_1, S_2, S_3.

III.8.2. We show first that the subcases 3 and 4 do not give us non-trivial tri-symmetric spaces.

Indeed, in the case 3 we have three pair-wise commutative orthogonal matrices which can be considered in a common canonical base. Then it is easy to see that

$n = 2m$,

$$so(2m) \cong \mathfrak{g} = \mathfrak{l}_1 + \mathfrak{l}_2 + \mathfrak{l}_3, \quad \mathfrak{l}_0 = \mathfrak{l}_\alpha \underset{\alpha \neq \beta}{\cap} \mathfrak{l}_\beta = u(k) \oplus u(m-k),$$

$$\mathfrak{l}_3/\mathfrak{l}_0 \cong so(2k)/u(k) + so(2m-2k)/u(m-k),$$

$$\mathfrak{l}_1/\mathfrak{l}_0 \cong \mathfrak{l}_2/\mathfrak{l}_0 \cong su(m)/su(k) \oplus su(m-k) \oplus u(1) + u(1)/u(1).$$

If we assume that there exists a non-trivial tri-symmetric space

$$G/H = G_1 \overset{*}{/} H_1 \boxtimes G_2 \overset{*}{/} H_2 \boxtimes G_3 \overset{*}{/} H_3,$$

generated by this involutive decomposition, then in this case the mirror

$$G_1 \overset{*}{/} H_1 = \overset{*}{G}_1/\overset{*}{G}_0 \cong SU(m)/S(U(k) \times U(m-k))$$

(with the natural embedding) has a simple group of motions $\overset{*}{G}_1 \cong SU(m)$, which is impossible by III.6.10 (Theorem 38).

III.8.3. Let us now consider the subcase 4. Then $\bar{\bar{S}}_1 \bar{\bar{S}}_2 = -\bar{\bar{S}}_2 \bar{\bar{S}}_1$, $(\bar{\bar{S}}_1)^2 = (\bar{\bar{S}}_2)^2 = \bar{\bar{\mathrm{Id}}}$. If $\bar{\bar{S}}_1 \bar{\zeta} = \lambda \bar{\zeta}$, then $\bar{\bar{S}}_1 (\bar{\bar{S}}_2 \bar{\zeta}) = -\lambda(\bar{\bar{S}}_2 \bar{\zeta})$, where $\lambda = \pm 1$. Therefore the maximal invariant subspaces of $\bar{\bar{S}}_1$ have the same dimension. Then transforming $\bar{\bar{S}}_1$ into the diagonal form and choosing a base invariant with respect to $\bar{\bar{S}}_2$ we obtain

$$\bar{\bar{S}}_1 = \left\| \begin{array}{c|c} \bar{\bar{\mathrm{Id}}} & \bar{\bar{O}} \\ \hline \bar{\bar{O}} & -\bar{\bar{\mathrm{Id}}} \end{array} \right\|, \quad \bar{\bar{S}}_2 = \left\| \begin{array}{c|c} \bar{\bar{O}} & \bar{\bar{\mathrm{Id}}} \\ \hline \bar{\bar{\mathrm{Id}}} & \bar{\bar{O}} \end{array} \right\|,$$

$$\underbrace{\qquad}_{m} \underbrace{\qquad}_{m} \qquad\qquad \underbrace{\qquad}_{m} \underbrace{\qquad}_{m}$$

$$\bar{\bar{S}}_3 = \bar{\bar{S}}_1 \bar{\bar{S}}_2 = \left\| \begin{array}{c|c} \bar{\bar{O}} & \bar{\bar{\mathrm{Id}}} \\ \hline -\bar{\bar{\mathrm{Id}}} & \bar{\bar{O}} \end{array} \right\|,$$

$$\underbrace{\qquad}_{m} \underbrace{\qquad}_{m}$$

where $\bar{\bar{O}}$ are zero matrices and $\bar{\bar{\mathrm{Id}}}$ are the identity matrices.

Matrices $\bar{\bar{a}} \in \mathfrak{l}_0 = \mathfrak{l}_\alpha \underset{\alpha \neq \beta}{\cap} \mathfrak{l}_\beta$ have evidently the form

$$\left\| \begin{array}{c|c} \bar{\bar{C}} & \bar{\bar{O}} \\ \hline \bar{\bar{O}} & \bar{\bar{C}} \end{array} \right\|, \quad \text{where } \bar{\bar{O}} \text{ is zero matrix, } \bar{\bar{C}} \text{ is a square matrix.}$$

Thus

$$\mathfrak{l}_0 \cong \Delta(so(m) \oplus so(m))$$

(diagonal in $so(m) \oplus so(m)$) and

$$\mathfrak{l}_1 \cong \mathfrak{l}_2 \cong so(m) \oplus so(m), \qquad \mathfrak{l}_3 \cong u(m).$$

Finally, we have

$$\mathfrak{l}_1/\mathfrak{l}_0 \cong \mathfrak{l}_2/\mathfrak{l}_0 \cong so(m) \oplus so(m)/\Delta(so(m) \oplus so(m)),$$

$$\mathfrak{l}_3/\mathfrak{l}_0 \cong su(m)/so(m) + u(1)/\{0\}.$$

If $m \neq 4$ then

$$G_1 \overset{*}{/} H_1 \cong SO(m) \times SO(m)/\Delta(SO(m) \times SO(m))$$

is irreducible, consequently by virtue of III.6.10 (Theorem 38) $G_1 \overset{*}{/} H_1$ can not generate a non-trivial tri-symmetric space.

If $m = 4$ we have

$$\mathfrak{g} \cong so(8), \qquad \mathfrak{l}_3/\mathfrak{l}_0 \cong su(4)/so(4),$$

$$\mathfrak{l}_1/\mathfrak{l}_0 \cong \mathfrak{l}_2/\mathfrak{l}_0 \cong so(4)/so(3) + so(4)/so(3).$$

And thus there exists the non-trivial tri-symmetric space

$$SO(8)/SO(3) \times SO(5) \cong G/H = G_1 \overset{*}{/} H_1 \boxtimes G_2 \overset{*}{/} H_2 \boxtimes G_3 \overset{*}{/} H_3,$$

where

$$so(3) \oplus so(5) \cong \ln_G H = \ln_G H_1 + \ln_G H_2 + \ln_G H_3,$$

$$\ln_G H_1 \cong \ln_G H_2 \cong so(3) \oplus so(4), \qquad \ln_G H_3 \cong u(1) \oplus so(3),$$

$$\ln_G H_0 = \ln_G (H_1 \cap H_2 \cap H_3) \cong so(3) \oplus so(3).$$

However, this space is $SO(8)/SO(3) \times SO(5)$ with the natural embedding, and thus is symmetric.

III.8.4. Let us now consider the subcase 1. Here $\bar{\bar{S}}_1$, $\bar{\bar{S}}_2$, $\bar{\bar{S}}_3$ are pair-wise commutative, therefore we can consider them with respect to the base in which $\bar{\bar{S}}_1$, $\bar{\bar{S}}_2$, $\bar{\bar{S}}_3$ have canonical forms simultaneously. Using the matrix model for $\bar{\bar{a}} \in so(n)$ we have

$$\bar{\bar{S}}_1 = \left\| \begin{array}{cccc} -\overline{\overline{\mathrm{Id}}} & \bar{\bar{O}} & \bar{\bar{O}} & \bar{\bar{O}} \\ \bar{\bar{O}} & -\overline{\overline{\mathrm{Id}}} & \bar{\bar{O}} & \bar{\bar{O}} \\ \bar{\bar{O}} & \bar{\bar{O}} & \overline{\overline{\mathrm{Id}}} & \bar{\bar{O}} \\ \bar{\bar{O}} & \bar{\bar{O}} & \bar{\bar{O}} & \overline{\overline{\mathrm{Id}}} \end{array} \right\| \begin{array}{l} \}p \\ \}q \\ \}r \\ \}k \end{array}, \qquad \bar{\bar{S}}_2 = \left\| \begin{array}{cccc} \overline{\overline{\mathrm{Id}}} & \bar{\bar{O}} & \bar{\bar{O}} & \bar{\bar{O}} \\ \bar{\bar{O}} & -\overline{\overline{\mathrm{Id}}} & \bar{\bar{O}} & \bar{\bar{O}} \\ \bar{\bar{O}} & \bar{\bar{O}} & -\overline{\overline{\mathrm{Id}}} & \bar{\bar{O}} \\ \bar{\bar{O}} & \bar{\bar{O}} & \bar{\bar{O}} & \overline{\overline{\mathrm{Id}}} \end{array} \right\| \begin{array}{l} \}p \\ \}q \\ \}r \\ \}k \end{array},$$
$$\underbrace{}_{p} \underbrace{}_{q} \underbrace{}_{r} \underbrace{}_{k} \qquad\qquad \underbrace{}_{p} \underbrace{}_{q} \underbrace{}_{r} \underbrace{}_{k}$$

$$\bar{\bar{S}}_3 = \left\| \begin{matrix} -\bar{\bar{\text{Id}}} & \bar{\bar{O}} & \bar{\bar{O}} & \bar{\bar{O}} \\ \bar{\bar{O}} & \bar{\bar{\text{Id}}} & \bar{\bar{O}} & \bar{\bar{O}} \\ \bar{\bar{O}} & \bar{\bar{O}} & -\bar{\bar{\text{Id}}} & \bar{\bar{O}} \\ \bar{\bar{O}} & \bar{\bar{O}} & \bar{\bar{O}} & \bar{\bar{\text{Id}}} \end{matrix} \right\| \begin{matrix} \}p \\ \}q \\ \}r \\ \}k \end{matrix} \ ,$$

$$\underbrace{\phantom{-\text{Id}}}_{p} \ \underbrace{\phantom{\text{Id}}}_{q} \ \underbrace{\phantom{-\text{Id}}}_{r} \ \underbrace{\phantom{\text{Id}}}_{k}$$

where $\bar{\bar{O}}$ mean zero matrices, $\bar{\bar{\text{Id}}}$ mean the identity matrices. Note that here some of p, q, r, k can be equal to zero or to one.

Thus

$$so(n) \cong \mathfrak{g} = \mathfrak{l}_1 + \mathfrak{l}_2 + \mathfrak{l}_3$$

$$\mathfrak{l}_0 = \mathfrak{l}_\alpha \underset{\alpha \neq \beta}{\cap} \mathfrak{l}_\beta \cong so(p) \oplus so(q) \oplus so(r) \oplus so(k),$$

$$\mathfrak{l}_1/\mathfrak{l}_0 \cong so(p+q)/so(p) \oplus so(q) + so(r+k)/so(r) \oplus so(k),$$

$$\mathfrak{l}_2/\mathfrak{l}_0 \cong so(q+r)/so(q) \oplus so(r) + so(p+k)/so(p) \oplus so(k),$$

$$\mathfrak{l}_3/\mathfrak{l}_0 \cong so(p+r)/so(p) \oplus so(r) + so(q+k)/so(q) \oplus so(k)$$

with the natural embeddings.

Therefore by virtue of III.6.8 (Theorem 36) in the decomposition $\ln_G H = \ln_G H_1 + \ln_G H_2 + \ln_G H_3$ we may take

$$\ln_G H_1 \cong so(p+q) \oplus so(r) \oplus so(k),$$

$$\ln_G H_2 \cong so(q+r) \oplus so(p) \oplus so(k)$$

(the other possibilities are analogous and can be obtained by substitutions of p, q, r, k).

Furthermore, we have

$$\ln_G H_3 = \left[\ln_G H_1 \ \ln_G H_2 \right] \cong so(q+r) \oplus so(p),$$

$$\ln_G H \cong so(p+q+r) \oplus so(k)$$

with the natural embeddings into $\mathfrak{g} = \ln G \cong so(p+q+r+k) = so(n)$. But then $G/H \cong SO(p+q+r+k)/SO(p+q+r) \times SO(k)$ is a symmetric space.

III.8.5. Remark. Let us also note that the case $p = q = r = 2$ gives some complication since there appear additional ideals in $\mathfrak{l}_0$.

However, if one looks for maximal subgroups $H \subset G$ then what has been said above is correct.

III.8.6. We now consider the subcase 2. Then $(\bar{\bar{S}}_1)^2 = (\bar{\bar{S}}_2)^2 = (\bar{\bar{S}}_3)^2 = -\bar{\bar{\text{Id}}}$ and $\bar{\bar{S}}_1$, $\bar{\bar{S}}_2$, $\bar{\bar{S}}_3$ are pair-wise anticommutative: $\bar{\bar{S}}_\alpha \bar{\bar{S}}_\beta + \bar{\bar{S}}_\beta \bar{\bar{S}}_\alpha = -2\delta_{\alpha\beta} \bar{\bar{\text{Id}}}$ $(\alpha, \beta, \gamma = 1, 2, 3)$.

Since $\bar{\bar{S}}_\alpha$ are orthogonal then evidently $\bar{\bar{S}}'_\alpha = -\bar{\bar{S}}_\alpha$ ($\bar{\bar{A}}'_\alpha$ stands for the transpose matrix of $\bar{\bar{A}}_\alpha$).

Such a system of matrices, as is well known, has four-dimensional invariant subspaces. Thus this case is possible only if $\mathfrak{g} \cong so(4m)$.

By the choice of appropriate base in these four dimensional invariant subspaces this system of matrices can be transformed to the common canonical form

$$\delta_1 = \begin{Vmatrix} 0 & -1 & 0 & 0 \\ 1 & 0 & 0 & 0 \\ 0 & 0 & 0 & 1 \\ 0 & 0 & -1 & 0 \end{Vmatrix}, \qquad \delta_2 = \begin{Vmatrix} 0 & 0 & 1 & 0 \\ 0 & 0 & 0 & 1 \\ -1 & 0 & 0 & 0 \\ 0 & -1 & 0 & 0 \end{Vmatrix},$$

$$\delta_3 = \begin{Vmatrix} 0 & 0 & 0 & -1 \\ 0 & 0 & 1 & 0 \\ 0 & -1 & 0 & 0 \\ 1 & 0 & 0 & 0 \end{Vmatrix}.$$

Thus

$$\bar{\bar{S}}_1 = \begin{Vmatrix} \delta_1 & \bar{\bar{O}} & \cdots & \bar{\bar{O}} \\ 0 & \delta_1 & \cdots & \bar{\bar{O}} \\ \bar{\bar{O}} & \cdots & \delta_1 & \bar{\bar{O}} \\ \bar{\bar{O}} & \bar{\bar{O}} & \cdots & \delta_1 \end{Vmatrix}, \qquad \bar{\bar{S}}_2 = \begin{Vmatrix} \delta_2 & \bar{\bar{O}} & \cdots & \bar{\bar{O}} \\ 0 & \delta_2 & \cdots & \bar{\bar{O}} \\ \bar{\bar{O}} & \cdots & \delta_2 & \bar{\bar{O}} \\ \bar{\bar{O}} & \bar{\bar{O}} & \cdots & \delta_2 \end{Vmatrix},$$

$$\bar{\bar{S}}_3 = \left. \begin{Vmatrix} \delta_3 & \bar{\bar{O}} & \cdots & \bar{\bar{O}} \\ \bar{\bar{O}} & \delta_3 & \cdots & \bar{\bar{O}} \\ \bar{\bar{O}} & \cdots & \delta_3 & \bar{\bar{O}} \\ \bar{\bar{O}} & \bar{\bar{O}} & \cdots & \delta_3 \end{Vmatrix} \right\} m.$$

$$\underbrace{\qquad\qquad\qquad\qquad}_{m}$$

Now, since $S_\alpha(\bar{a}) = \bar{\bar{S}}_\alpha \bar{a} \bar{\bar{S}}_\alpha^{-1}$, $\bar{\bar{S}}_\alpha \in so(4m) \cong \mathfrak{g}$ we find $\mathfrak{l}_\alpha$ as a maximal subspace of $\bar{a} \in \mathfrak{g}$ such that $\bar{\bar{S}}_\alpha \bar{a} \bar{\bar{S}}_\alpha^{-1} = \bar{a}$, that is, $\bar{a}\,\bar{\bar{S}}_\alpha = \bar{\bar{S}}_\alpha\,\bar{a}$. Therefore $\mathfrak{l}_\alpha \cong u(2m)$ (obviously $\bar{\bar{S}}_\alpha$ generates in $\mathfrak{l}_\alpha$ a one-dimensional centre). Furthermore, we have $\bar{\bar{S}}_\alpha \bar{\bar{S}}_\beta \bar{\bar{S}}_\alpha^{-1} = -\bar{\bar{S}}_\beta$ ($\alpha \neq \beta$), that is, $\bar{\bar{S}}_\alpha \notin \mathfrak{l}_0 = \mathfrak{l}_\alpha \underset{\alpha \neq \beta}{\cap} \mathfrak{l}_\beta$, $\bar{\bar{S}}_\alpha \in \mathfrak{l}_\alpha \dot{-} \mathfrak{l}_0$, and $\mathfrak{g} = \mathfrak{l}_1 + \mathfrak{l}_2 + \mathfrak{l}_3$ is a hyper-involutive sum.

The subalgebra $\mathfrak{l}_0$ can be found with the usage of these concrete forms of matrices $\bar{\bar{S}}_1, \bar{\bar{S}}_2, \bar{\bar{S}}_3$ as the maximal subalgebra in $so(4m)$ which commutes with $\bar{\bar{S}}_1, \bar{\bar{S}}_2, \bar{\bar{S}}_3$.

This gives us $\mathfrak{l}_0 \cong sp(m)$, as is well known.

One could also use the consideration of dimensions:

$$\dim \mathfrak{g} = \dim so(4m) = 2m\,(4m-1) = 3\,\dim(\mathfrak{l}_1 \dot{-} \mathfrak{l}_0) + \dim \mathfrak{l}_0,$$
$$\dim \mathfrak{l}_1 = \dim u(2m) = (2m)^2 = \dim(\mathfrak{l}_1 \dot{-} \mathfrak{l}_0) + \dim \mathfrak{l}_0.$$

Whence $\dim(\mathfrak{l}_1 \dot{-} \mathfrak{l}_0) = m\,(2m-1)$, $\dim \mathfrak{l}_0 = m\,(2m+1)$.

Therefore $\mathfrak{l}_\alpha / \mathfrak{l}_0 \cong su(2m)/sp(m) + u(1)/\{0\}$.

Using, finally, III.6.8 (Theorem 36) we obtain the tri-symmetric space

$$SO(4m)/SO(3) \times Sp(m) \cong G/H = G_1 \overset{*}{/} H_1 \boxtimes G_2 \overset{*}{/} H_2 \boxtimes G_3 \overset{*}{/} H_3,$$

where

$$G_\alpha \overset{*}{/} H_\alpha \cong U(2m) \overset{*}{/} Sp(m) \times U(1) \cong SU(2m)/Sp(m).$$

Thus we have proved the theorem:

III.8.7. Theorem 44. *If a compact simple Lie group G is isomorphic to $SO(n)$ then all non-trivial non-symmetric tri-symmetric spaces with the group of motions G and the maximal Lie subgroup H have the form*

$$G/H = G_1 \overset{*}{/} H_1 \boxtimes G_2 \overset{*}{/} H_2 \boxtimes G_3 \overset{*}{/} H_3 \cong SO(4m)/SO(3) \times Sp(m) \quad (m > 2)$$

with the central mirrors

$$G_\alpha \overset{*}{/} H_\alpha \cong U(2m) \overset{*}{/} Sp(m) \times U(1) \cong SU(2m)/Sp(m)$$

(with the natural embeddings).

These spaces are hyper-tri-symmetric and have an irreducible group H.

III.8.8. Group of motions $\mathbf{G} \cong \mathbf{Sp(n)}$. We consider the group $Sp(n)$ as the group of all quaternionic matrices preserving the standard inner product in $\mathbb{H}^n$ ($\mathbb{H}$ stands for the skew-field of quaternions). See [C. Chevalley, 46]. Then the Lie algebra $sp(n)$ is the algebra of all skew-symmetric quaternionic $n \times n$ matrices.

Thus $\bar{a} \in sp(n)$ means that $(\bar{a})' = -\bar{\bar{a}}$ (here $\bar{a} \mapsto \bar{\bar{a}}$ is the operation of the quaternionic conjugation).

As is well known, all involutive automorphisms of $sp(n)$ have the form

$$\bar{\bar{a}} \mapsto S(\bar{\bar{a}}) = \bar{\bar{S}}\,\bar{\bar{a}}\,\bar{\bar{S}}^{-1}, \qquad (\bar{\bar{S}})^2 = \pm \mathrm{Id}, \qquad \bar{\bar{S}} \in Sp(n).$$

In the following consideration the arguments are similar to the case $G \cong SO(n)$, thus for the matrices $\bar{\bar{S}}_1$, $\bar{\bar{S}}_2$, $\bar{\bar{S}}_3$ of the involutive automorphisms S_1, S_2, S_3 of the involutive decomposition $sp(n) \cong \mathfrak{g} = \mathfrak{l}_1 + \mathfrak{l}_2 + \mathfrak{l}_3$ we have the following possibilities:

1. $(\bar{\bar{S}}_\alpha)^2 = \bar{\bar{\mathrm{Id}}}$, $\bar{\bar{S}}_1$, $\bar{\bar{S}}_2$, $\bar{\bar{S}}_3$ are pair-wise commutative, $\bar{\bar{S}}_1 \bar{\bar{S}}_2 = \bar{\bar{S}}_3$;
2. $(\bar{\bar{S}}_\alpha)^2 = -\bar{\bar{\mathrm{Id}}}$, $\bar{\bar{S}}_1$, $\bar{\bar{S}}_2$, $\bar{\bar{S}}_3$ are pair-wise anticommutative, $\bar{\bar{S}}_1 \bar{\bar{S}}_2 = \bar{\bar{S}}_3$;
3. $(\bar{\bar{S}}_1)^2 = (\bar{\bar{S}}_2)^2 = -\bar{\bar{\mathrm{Id}}}$, $(\bar{\bar{S}}_3)^2 = \bar{\bar{\mathrm{Id}}}$, $\bar{\bar{S}}_1$, $\bar{\bar{S}}_2$, $\bar{\bar{S}}_3$ are pair-wise commutative, $\bar{\bar{S}}_1 \bar{\bar{S}}_2 = \bar{\bar{S}}_3$;
4. $(\bar{\bar{S}}_1)^2 = (\bar{\bar{S}}_2)^2 = \bar{\bar{\mathrm{Id}}}$, $(\bar{\bar{S}}_3)^2 = -\bar{\bar{\mathrm{Id}}}$, $\bar{\bar{S}}_1$, $\bar{\bar{S}}_2$, $\bar{\bar{S}}_3$ are pair-wise anticommutative, $\bar{\bar{S}}_1 \bar{\bar{S}}_2 = \bar{\bar{S}}_3$.

The other possibilities are reduced to those listed above after some re-numbering of the involutive automorphisms S_1, S_2, S_3.

III.8.9. The subcases 3 and 4 do not generate non-trivial tri-symmetric spaces. Indeed, in the case 3 we have

$$sp(n) \cong \mathfrak{g} = \mathfrak{l}_1 + \mathfrak{l}_2 + \mathfrak{l}_3,$$

$$\mathfrak{l}_\alpha \underset{\alpha \neq \beta}{\cap} \mathfrak{l}_\beta = \mathfrak{l}_0 \cong su(p) \oplus su(n-p) \oplus u(1) \oplus u(1),$$

$$\mathfrak{l}_1/\mathfrak{l}_0 \cong \mathfrak{l}_2/\mathfrak{l}_0 \cong su(n)/su(p) \oplus su(n-p) \oplus u(1) + u(1)/u(1),$$

$$\mathfrak{l}_3/\mathfrak{l}_0 \cong sp(p)/u(p) + sp(n-p)/u(n-p).$$

If there were to exist a non-trivial tri-symmetric space generated by this involutive decomposition then the symmetric space

$$G_1 \overset{*}{/} H_1 = G_1 \overset{*}{/} G_0 \cong SU(n)/S(U(p) \times U(n-p))$$

(with the natural embedding) should be irreducible, which is impossible because of III.6.10 (Theorem 38).

III.8.10. Now we consider the subcase 4. Here, as well as in the case $G \cong SO(n)$, $n = 2m$,

$$\bar{\bar{S}}_1 \bar{\bar{S}}_3 \bar{\bar{S}}_1^{-1} = -\bar{\bar{S}}_3 \qquad \bar{\bar{S}}_2 \bar{\bar{S}}_3 \bar{\bar{S}}_2^{-1} = -\bar{\bar{S}}_3, \qquad \bar{\bar{S}}_3 \in sp(2m).$$

Thus for the involutive decomposition $sp(2m) \cong \mathfrak{g} = \mathfrak{l}_1 + \mathfrak{l}_2 + \mathfrak{l}_3$ we have $\bar{\bar{S}}_3 \notin \mathfrak{l}_0$, $\bar{\bar{S}}_3 \in \mathfrak{l}_3 \dot{-} \mathfrak{l}_0$, and $\bar{\bar{S}}_3$ generates a one-dimensional centre in $\mathfrak{l}_3 \cong u(2m)$. Consequently this sum is iso-involutive. Furthermore, obviously

$$\mathfrak{l}_1 \cong \mathfrak{l}_2 \cong sp(m) \oplus sp(m),$$

$$\mathfrak{l}_1/\mathfrak{l}_0 \cong \mathfrak{l}_2/\mathfrak{l}_0 \cong sp(m) \oplus sp(m)/\Delta(sp(m) \oplus sp(m)),$$

$$\mathfrak{l}_3/\mathfrak{l}_0 \cong su(2m)/sp(m) + u(1)/\{0\}.$$

But then in the tri-symmetric space

$$G/H = G_1 \overset{*}{/} H_1 \boxtimes G_2 \overset{*}{/} H_2 \boxtimes G_3 \overset{*}{/} H_3$$

we would have the irreducible symmetric mirror

$$G_1 \overset{*}{/} H_1 = \overset{*}{G}_1/\overset{*}{G}_0 = Sp(m) \times Sp(m)/\Delta(Sp(m) \times Sp(m)),$$

which is impossible for non-trivial tri-symmetric space by virtue of III.6.10 (Theorem 38).

III.8.11. We now consider the subcase 1. By the method analogous to the method given for the case of $SO(n)$ we conclude that all non-trivial tri-symmetric spaces appearing here are symmetric.

III.8.12. Lastly, we consider the subcase 2.

Here we have

$$\bar{\bar{S}}_\alpha \bar{\bar{S}}_\beta + \bar{\bar{S}}_\beta \bar{\bar{S}}_\alpha = -2\delta_{\alpha\beta}\,\mathrm{Id}\,, \qquad \bar{\bar{S}}_\alpha \in sp(n),$$

and we can put $\bar{\bar{S}}_\alpha = i_\alpha\,\mathrm{Id}$ (where i_1, i_2, i_3 are i, j, k, respectively, that is, the standard basis elements in the algebra of quaternions $\mathbb{H}$). Up to automorphisms of $sp(n)$ there are no other possibilities to choose $\bar{\bar{S}}_\alpha$ ($\alpha = 1, 2, 3$). Therefore we have

$$sp(n) \cong \mathfrak{g} = \mathfrak{l}_1 + \mathfrak{l}_2 + \mathfrak{l}_3,$$

$$\mathfrak{l}_0 = \mathfrak{l}_\alpha \underset{\alpha\neq\beta}{\cap} \mathfrak{l}_\beta \cong so(n)\,,$$

$$\mathfrak{l}_\alpha/\mathfrak{l}_0 \cong su(n)/so(n) + u(1)/\{0\}$$

with the natural embeddings.

Evidently here the decomposition $\mathfrak{g} = \mathfrak{l}_1 + \mathfrak{l}_2 + \mathfrak{l}_3$ is hyper-involutive. Using III.6.8 (Theorem 36) we see that for the non-trivial tri-symmetric space G/H, $G \cong Sp(n)$, we have to take

$$\ln_G H = \ln_G H_1 + \ln_G H_2 + \ln_G H_3 \cong so(n) \oplus so(3),$$

where $\ln_G G_\alpha \cong u(n)$.

Thus we have:

III.8.13. Theorem 45. *If a compact simple Lie group G is isomorphic to $Sp(n)$ ($n > 2$) then any non-trivial tri-symmetric non-symmetric space with the group of motions G has the form:*

$$G/H = G_1 \overset{*}{/} H_1 \boxtimes G_2 \overset{*}{/} H_2 \boxtimes G_3 \overset{*}{/} H_3 \cong Sp(n)/SO(3) \times SO(n)$$

with the central mirrors

$$G_\alpha \overset{*}{/} H_\alpha \cong U(n) \overset{*}{/} SO(n) \times U(1) \cong SU(n)/SO(n)$$

(with the natural embeddings).

This space is tri-symmetric and has an irreducible isotropy group H.

III.8.14. Group of motions $G \cong SU(n)$ $(n > 1)$. We represent the Lie algebra $\mathfrak{g} = \ln G \cong su(n)$ by means of all skew-Hermitian complex matrices $\bar{\bar{a}} \in su(n)$, $(\bar{\bar{a}})' = -\bar{\bar{a}}$, with the standard operation $[\bar{\bar{a}}\,\bar{\bar{c}}] = \bar{\bar{a}}\,\bar{\bar{c}} - \bar{\bar{c}}\,\bar{\bar{a}}$.

There are only two essentially different outer involutive automorphisms with the involutive algebras $so(n)$ and $sp(m)$, in this case $n = 2m$ [C. Chevalley, 46].

These involutive algebras are simple non-one-dimensional (except $n = 2, 4$), and consequently by III.6.10 (Theorem 38) can not generate mirrors in non-trivial tri-symmetric spaces.

The cases $n = 2, 4$ give us $su(2) \cong so(3)$ and $su(4) \cong so(6)$, which have already been explored in III.8.1.

Thus the involutive automorphisms S_α of the involutive decomposition

$$su(n) \cong \ln G = \mathfrak{l}_1 + \mathfrak{l}_2 + \mathfrak{l}_3$$

can be only inner automorphisms.

The involutive automorphism S_α acts as $S_\alpha(\bar{\bar{a}}) = \bar{\bar{S}}_\alpha\,\bar{\bar{a}}\,\bar{\bar{S}}_\alpha^{-1}$, where $\bar{\bar{S}}_\alpha \in SU(n)$.

Evidently we have $(\bar{\bar{S}}_\alpha)^2 \in Z$, $(\bar{\bar{S}}_\alpha \bar{\bar{S}}_\beta)(\bar{\bar{S}}_\beta \bar{\bar{S}}_\alpha)^{-1} \in Z$, where Z is the centre of $SU(n)$. Furthermore, $S_1 S_2 = S_3$, whence $\bar{\bar{S}}_3^{-1} \bar{\bar{S}}_1 \bar{\bar{S}}_2 \in Z$. The elements of Z have the form

$$\lambda_k = e^{i\,2\pi k/n}\,\bar{\bar{\mathrm{Id}}} \quad (k = 0, \ldots, n-1).$$

Thus

$$(\bar{\bar{S}}_\alpha)^2 = \mu(\alpha)\,\bar{\bar{\mathrm{Id}}}, \qquad \bar{\bar{S}}_\alpha \bar{\bar{S}}_\beta = \nu(\alpha, \beta)\,\bar{\bar{S}}_\beta \bar{\bar{S}}_\alpha,$$

$$|\mu(\alpha)| = |\nu(\alpha, \beta)| = 1 \quad (\alpha \neq \beta).$$

Taking $(1/\sqrt{\mu(\alpha)})\,\bar{\bar{S}}_\alpha$ instead of $\bar{\bar{S}}_\alpha$, we evidently obtain unitary matrices generating the same involutive automorphisms S_α $(\alpha = 1, 2, 3)$. Thus we may assume that

$$\bar{\bar{S}}_\alpha \in U(n), \qquad (\bar{\bar{S}}_\alpha)^2 = \bar{\bar{\mathrm{Id}}}, \qquad \bar{\bar{S}}_\alpha \bar{\bar{S}}_\beta = \chi(\alpha\,\beta)\,\bar{\bar{S}}_\beta \bar{\bar{S}}_\alpha \,(\alpha \neq \beta).$$

Obviously $\chi(\alpha, \beta)\chi(\beta, \alpha) = 1$. In addition, we have

$$\bar{\bar{S}}_\alpha(\bar{\bar{S}}_\alpha \bar{\bar{S}}_\beta)\bar{\bar{S}}_\alpha = \chi(\alpha, \beta)\,\bar{\bar{S}}_\alpha(\bar{\bar{S}}_\beta \bar{\bar{S}}_\alpha)\bar{\bar{S}}_\alpha$$

And, using $\bar{\bar{S}}_\beta \bar{\bar{S}}_\alpha = \chi(\alpha, \beta)\bar{\bar{S}}_\alpha \bar{\bar{S}}_\beta$, we arrive at $(\chi(\alpha, \beta))^2 = 1$, or $\chi(\alpha, \beta) = \pm 1$.

We now consider all possible subcases:

1. $\bar{\bar{S}}_1 \bar{\bar{S}}_2 = \bar{\bar{S}}_2 \bar{\bar{S}}_1$, $\bar{\bar{S}}_2 \bar{\bar{S}}_3 = \bar{\bar{S}}_3 \bar{\bar{S}}_2$, $\bar{\bar{S}}_3 \bar{\bar{S}}_1 = \bar{\bar{S}}_1 \bar{\bar{S}}_3$.

Since $\bar{\bar{S}}_\alpha \in U(n)$ and $(\bar{\bar{S}}_\alpha)^2 = \bar{\bar{\mathrm{Id}}}$ we have $((\bar{\bar{S}}_\alpha)')^* = \bar{\bar{S}}_\alpha$, and the system of commutative unitary matrices can be transformed to diagonal forms simultaneously with ± 1 along the principal diagonal by means of unitary transformation of base.

Then for $su(n) \cong \mathfrak{g} = \mathfrak{l}_1 + \mathfrak{l}_2 + \mathfrak{l}_3$ we have

$$\mathfrak{l}_1 \cong s(u(p+q) \oplus u(r+k)), \qquad \mathfrak{l}_2 \cong s(u(q+r) \oplus u(p+k)),$$

$$\mathfrak{l}_3 \cong s(u(p+r) \oplus u(q+k)),$$

$$\mathfrak{l}_0 \cong \mathfrak{l}_\alpha \underset{\alpha \neq \beta}{\cap} \mathfrak{l}_\beta \cong s(u(p) \oplus u(q) \oplus u(r) \oplus u(k)),$$

$$\mathfrak{l}_1 / \mathfrak{l}_0 \cong su(p+q)/s(u(p) \oplus u(q)) + su(r+k)/s(u(r) \oplus u(k)) + u(1)/u(1),$$

$$\mathfrak{l}_2 / \mathfrak{l}_0 \cong su(q+r)/s(u(q) \oplus u(r)) + su(p+k)/s(u(p) \oplus u(k)) + u(1)/u(1),$$

$$\mathfrak{l}_3 / \mathfrak{l}_0 \cong su(p+r)/s(u(p) \oplus u(r)) + su(q+k)/s(u(q) \oplus u(k)) + u(1)/u(1).$$

Furthermore, by arguments analogous to the case $G \cong SO(n)$, we conclude that all tri-symmetric spaces generated by such an involutive decomposition are symmetric.

Since $\bar{\bar{S}}_1 \bar{\bar{S}}_2 = \tau \bar{\bar{S}}_3$ ($\tau = \exp 2\pi i k/n$), then if $\bar{\bar{S}}_1 \bar{\bar{S}}_2 = \bar{\bar{S}}_2 \bar{\bar{S}}_1$ we have $\bar{\bar{S}}_2 = \tau \bar{\bar{S}}_1 \bar{\bar{S}}_3$ and $\bar{\bar{S}}_2 = \tau \bar{\bar{S}}_3 \bar{\bar{S}}_1$, that is, $\bar{\bar{S}}_1 \bar{\bar{S}}_3 = \bar{\bar{S}}_3 \bar{\bar{S}}_1$.

Analogously $\bar{\bar{S}}_2 \bar{\bar{S}}_3 = \bar{\bar{S}}_3 \bar{\bar{S}}_2$.

Therefore we have to consider the only case in which all $\bar{\bar{S}}_\alpha$ ($\alpha = 1, 2, 3$) are pair-wise anticommutative.

Thus

2. $\bar{\bar{S}}_\alpha \bar{\bar{S}}_\beta + \bar{\bar{S}}_\beta \bar{\bar{S}}_\alpha = 2 \delta_{\alpha\beta} \mathrm{Id}$, $\bar{\bar{S}}_\alpha \in U(n)$.

If $\bar{\bar{S}}_1 \bar{\bar{S}}_2 = \tau \bar{\bar{S}}_3$, then $(\bar{S}'_2)^* (\bar{S}'_1)^* = \overset{*}{\tau}(\bar{S}'_3)^*$ and $\bar{\bar{S}}_2 \bar{\bar{S}}_1 = \overset{*}{\tau} \bar{\bar{S}}_3$.

Thus $\overset{*}{\tau} = -\tau$, that is, $\bar{\bar{S}}_1 \bar{\bar{S}}_2 = \sigma i \bar{\bar{S}}_3$ (σ is real). However, $\sigma i \bar{\bar{S}}_3$ is also unitary, whence $\sigma = \pm 1$. And we obtain $\bar{\bar{S}}_1 \bar{\bar{S}}_2 = \sigma i \bar{\bar{S}}_3$.

But then $\bar{\bar{S}}_3 \bar{\bar{S}}_1 = \pm \sigma i \bar{\bar{S}}_2$, $\bar{\bar{S}}_2 \bar{\bar{S}}_3 = \pm \sigma i \bar{\bar{S}}_1$ ($\sigma = +1$ or $\sigma = -1$).

Replacing $\bar{\bar{S}}_3$ by $(-\sigma) i \bar{\bar{S}}_3$, $\bar{\bar{S}}_1$ by $(\sigma i) \bar{\bar{S}}_1$, $\bar{\bar{S}}_2$ by $(\sigma i) \bar{\bar{S}}_2$, we do not change the automorphisms S_α ($\alpha = 1, 2, 3$).

Thus we can consider

$$\bar{\bar{S}}_\alpha \bar{\bar{S}}_\beta + \bar{\bar{S}}_\beta \bar{\bar{S}}_\alpha = -2 \delta_{\alpha\beta} \mathrm{Id}, \qquad \bar{\bar{S}}_\alpha \in U(n),$$

$$\bar{\bar{S}}_1 \bar{\bar{S}}_2 = \bar{\bar{S}}_3, \qquad \bar{\bar{S}}_3 \bar{\bar{S}}_1 = \bar{\bar{S}}_2, \qquad \bar{\bar{S}}_2 \bar{\bar{S}}_3 = \bar{\bar{S}}_1.$$

The anticommutativity of $\bar{\bar{S}}_1$ and $\bar{\bar{S}}_2$ imply $n = 2m$, therefore we may choose a base in such a way that

$$\bar{\bar{S}}_1 = \left\| \begin{array}{c|c} i\bar{\bar{\mathrm{Id}}} & \bar{\bar{O}} \\ \hline \bar{\bar{O}} & -i\bar{\bar{\mathrm{Id}}} \end{array} \right\|, \qquad \bar{\bar{S}}_2 = \left\| \begin{array}{c|c} \bar{\bar{O}} & \bar{\bar{\mathrm{Id}}} \\ \hline -\bar{\bar{\mathrm{Id}}} & \bar{\bar{O}} \end{array} \right\|,$$

$$\underbrace{\qquad}_{m} \underbrace{\qquad}_{m} \qquad \underbrace{\qquad}_{m} \underbrace{\qquad}_{m}$$

$$\bar{\bar{S}}_3 = \left\| \begin{array}{c|c} \bar{\bar{O}} & i\bar{\bar{\mathrm{Id}}} \\ \hline i\bar{\bar{\mathrm{Id}}} & \bar{\bar{O}} \end{array} \right\|,$$

$$\underbrace{}_{m} \underbrace{}_{m}$$

where $\bar{\bar{O}}$ are zero matrices and $\bar{\bar{\mathrm{Id}}}$ are the identity matrices.

Therefore for the involutive decomposition $su(2m) \cong \mathfrak{g} = \mathfrak{l}_1 + \mathfrak{l}_2 + \mathfrak{l}_3$ we obtain

$$\mathfrak{l}_\alpha \cong su(m) \oplus su(m) \oplus u(1),$$

$$\mathfrak{l}_0 = \mathfrak{l}_\alpha \underset{\alpha \neq \beta}{\cap} \mathfrak{l}_\beta \cong \Delta(su(m) \oplus su(m)) \cong su(m)$$

(here Δ means a diagonal of a canonical involutive automorphism),

$$\mathfrak{l}_\alpha / \mathfrak{l}_0 \cong su(m) \oplus su(m) / \Delta(su(m) \oplus su(m)) + u(1)/\{0\}$$

(with the natural embeddings).

The union of centers of $\mathfrak{l}_1$, $\mathfrak{l}_2$, $\mathfrak{l}_3$ generates a three-dimensional maximal subalgebra of elements of $\mathfrak{g}$ commuting with $\mathfrak{l}_0$, which is isomorphic to $so(3)$.

Thus in the non-trivial tri-symmetric space G/H, $G \cong SU(2m)$, we have by III.6.8 (Theorem 36)

$$\ln_G H \cong su(m) \oplus so(3).$$

And we obtain the theorem:

III.8.15. Theorem 46. *If G is a simple compact Lie group isomorphic to $SU(n)$ then any non-trivial tri-symmetric non-symmetric space with the group of motions G has the form:*

$$G/H = G_1 \overset{*}{/} H_1 \boxtimes G_2 \overset{*}{/} H_2 \boxtimes G_3 \overset{*}{/} H_3$$

$$\cong SU(2m)/SO(3) \times SU(m) \quad (m > 2)$$

with the central mirrors

$$G_\alpha \overset{*}{/} H_\alpha \cong S(U(m) \times U(m)) \overset{*}{/} \Delta(SU(m) \times SU(m)) \times U(1)$$

$$\cong SU(m) \times SU(m)/\Delta(SU(m) \times SU(m))$$

(with the natural embeddings).

This space is hyper-tri-symmetric and has an irreducible isotropy group H.

PART FOUR

Смените латы на заплаты
И откажитесь от зарплаты.
Порвите путы шалопуты
И устремитесь в высь Лапуты.

Квантасмагор

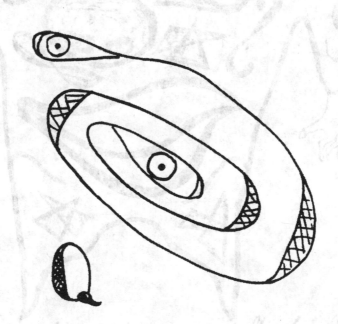

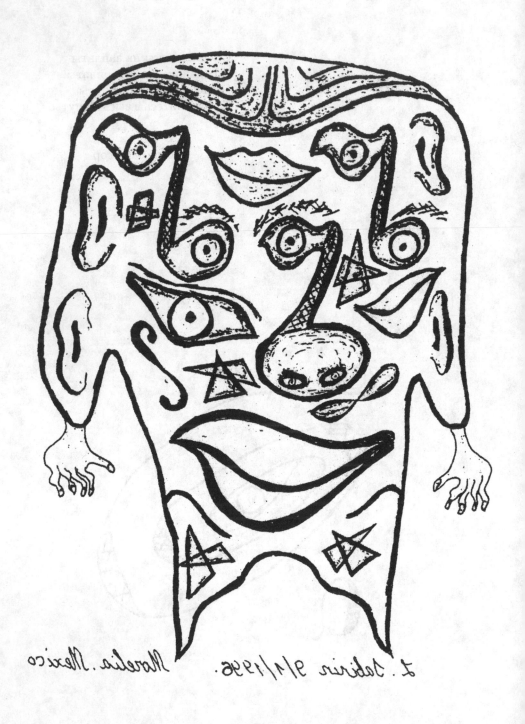

CHAPTER IV.1

SUBSYMMETRIC RIEMANNIAN HOMOGENEOUS SPACES

IV.1.1. We consider a homogeneous C^∞-smooth manifold $M = G/H$ with a Lie group G and its closed subgroup H. This means, in particular, that H does not contain a non-trivial normal subgroup of G.

The elements of M are left cosets of the group G by the subgroup H, that is, those of the form gH, $g \in G$.

The group G acts on G/H in the canonical way by left translations,

$$L_g(aH) \stackrel{\text{def}}{=} (ga)H.$$

This action, as is known, is faithful and transitive. Since G and H are Lie groups this action is C^∞-smooth, see [S. Kobayasi, K. Nomizu, 63,69].

In the following we consider homogeneous spaces G/H together with such a smooth canonical action of a Lie group G.

A homogeneous space can be defined in a different way.

IV.1.2. Definition 60. Let M be a C^∞-smooth manifold, and G be a Lie group. We say that there is defined a representation, or action T, of a Lie group G on a manifold M if a map

$$G \ni g \stackrel{T}{\mapsto} T_g \in \text{Diff } M$$

with the properties

$$T_{g_1} \circ T_{g_2} = T_{g_1 g_2}, \qquad (T_g)^{-1} = T_{g^{-1}}, \qquad T_e = \text{id}_M$$

is given.

Here e means the identity element of G.

Note that this means that T is a morphism of a Lie group G into the group of diffeomorphisms of a manifold M.

We consider only C^∞-smooth actions, that is, such actions that

$$T: (g, x) \in G \times M \mapsto T_g x \in M$$

is a C^∞-smooth map.

IV.1.3. Definition 61. An action (representation) is called faithful (or effective) if

$$T_g = \text{id}_M \implies g = e.$$

An action is called transitive if for any $x, y \in M$ there is $g \in G$ such that $T_g x = y$.

IV.1.4. Definition 62. A manifold M is called a homogeneous space if a faithful transitive smooth action T of a Lie group G is defined on M.

In this case we write (M, T, G).

IV.1.5. Definition 63. Let (M, T, G) be a smooth homogeneous space. A closed subgroup

$$G_x = \{g \in G \mid T_g x = x, \ x \in M\}$$

of a group G is called a stationary subgroup (stabilizer, or isotropy group) of a point $x \in M$.

IV.1.6. In a homogeneous space all stationary subgroups are conjugated (and consequently isomorphic), since if $T_g x = y$ then $g\, G_x g^{-1} = G_y$.

If we now consider G/G_x with a canonical action by left translations (see IV.1.1) then we obtain an action isomorphic (equivalent) to the action T.

Indeed, let

$$\varphi \colon g\, G_x \in G/G_x \mapsto T_g x \in M.$$

The map φ is correctly defined, since if $g\, G_x = \tilde{g}\, G_x$ then $\tilde{g} = gh$, $h \in G_x$, and

$$T_{\tilde{g}} x = T_{gh} x = (T_g \circ T_h) x = T_g(T_h x) = T_g x.$$

Since G acts on M transitively φ is surjective:

$$\forall y \in M, \ \exists g \in G \colon T_g x = y.$$

But also φ is injective, since if $\varphi(g\, G_x) = \varphi(\tilde{g}\, G_x)$ then $T_g x = T_{\tilde{g}} x$, or $T_g T_{\tilde{g}^{-1}} x = x$, or $g^{-1}\tilde{g} \in G_x$, which means that $\tilde{g} \in g\, G_x$, whence $\tilde{g}\, G_x = g\, G_x$.

Thus $\varphi \colon G/G_x \to M$ is a bijection (moreover, a diffeomorphism).

Finally, we have

$$\varphi(L_a(g\, G_x)) = \varphi((ag)G_x) = T_{ag} x = T_a(T_g x) = T_a \varphi(g\, G_x),$$

or

$$\varphi \circ L_a \circ \varphi^{-1} = T_a,$$

which shows that we have an isomorphism (moreover, smooth isomorphism) of actions.

IV.1.7. Definition 64. A homogeneous space (M, G, T) is called a Riemannian homogeneous space if there is given a Riemannian metric g on M (that is, a tensor field which is twice covariant, symmetric, and positive-definite at any point) and T_a $(a \in G)$ are isometries of this metric (that is, for any smooth curve $\gamma \colon [0, 1] \to M$ its length $\ell(\gamma)$ coincides with the length $\ell(T_a \circ \gamma)$ of the transformed curve).

We use in this case the notation (M, G, T, g).

IV.1.8. It is well known [S. Kobayasi, K. Nomizu, 63,69] that a homogeneous space is Riemannian if and only if its stationary group at some (and consequently at any) point is compact.

Therefore any Riemannian homogeneous space may be considered as G/H, where G is a Lie group and H is its closed compact subgroup.

IV.1.9.   Definition 65. A diffeomorphism $s\colon M \to M$ of a homogeneous space (M, G, T) is called a mirror subsymmetry (reflection) if:

1. there exists $x \in M$ such that $sx = x$;
2. $s \circ s = \mathrm{id}_M$;
3. s is an automorphism of the action T, that is, $s \circ T_g \circ s^{-1} = T_{\sigma(g)}$, where $\sigma\colon G \to G$ is a smooth map.

We often write s_x instead of s, explicitly indicating a point $x \in M$ which is immobile under the action of s_x.

IV.1.10.   Remark. The map $\sigma\colon G \to G$ of IV.1.9 (Definition 65) is an involutive automorphism of the Lie group G. Indeed,

$$s \circ T_{g\tilde{g}} \circ s^{-1} = (s \circ T_g \circ s^{-1}) \circ (s \circ T_{\tilde{g}} \circ s^{-1})$$

implies

$$T_{\sigma(g\,\tilde{g})} = T_{\sigma(g)\sigma(\tilde{g})}$$

whence, by the faithfulness of the action,

$$\sigma(g\,\tilde{g}) = \sigma(g)\sigma(\tilde{g}),$$

thus σ is an endomorphism of G.

Furthermore, $T_g = s \circ (s \circ T_g \circ s^{-1}) \circ s^{-1}$ gives us $T_g = T_{\sigma(\sigma g)}$, or $g = (\sigma \circ \sigma)g$, or $\sigma \circ \sigma = \mathrm{id}_G$.

Thus σ is an involutive automorphism of G.

Moreover, since $s_x\, x = x$ we obtain, for $g \in G_x$,

$$x = (s \circ T_g \circ s^{-1})x = T_{\sigma(g)}x,$$

that is, $\sigma G_x = G_x$.

Thus σ is an involutive automorphism of G transforming G_x into itself.

IV.1.11.   Definition 66. The set $\mathfrak{k}$ of all immobile points of a mirror subsymmetry s_x is called a mirror. This set is a submanifold of M, and may be not connected.

Considering $G^+ = \{g \in G \mid \sigma g = g\}$ we have a subgroup of G which is called a mirror subgroup of a mirror subsymmetry s_x. (See Definition 65.)

IV.1.12. In the natural way any mirror $\mathfrak{k}$ can be equipped with an action (may be not faithful) of the group G^+. Indeed, if $y \in \mathfrak{k}$, $g \in G^+$ then

$$s_x(T_g y) = T_{\sigma(g)}(s_x y) = T_g y,$$

so that

$$y \in \mathfrak{k}, \ g \in G^+ \implies T_g y \in \mathfrak{k}.$$

IV.1.13. Remark. It is verified that the connected component of the identity, $(G^+)^\circ$, in G acts transitively on the connected component of a point $x \in \mathfrak{k}$ of the corresponding mirror $\mathfrak{k}$.

IV.1.14. Let us now clarify how one can restore a mirror subsymmetry s_x by $\sigma \colon G \to G$. We have a homogeneous space G/H ($H = G_x$) and an involutive automorphism σ such that $\sigma H = H$.

We consider $\tilde{\sigma} \colon gH \mapsto (\sigma g)H$. Then

$$\tilde{\sigma}(L_a(gH)) = \tilde{\sigma}((ag)H) = (\sigma(ag))H = [(\sigma a)(\sigma g))]H = L_{\sigma a}(\tilde{\sigma}(gH)).$$

Thus

$$\tilde{\sigma} \circ L_a \circ \tilde{\sigma}^{-1} = L_{\sigma a}.$$

Since $\varphi \circ L_a \circ \varphi^{-1} = T_a$ (see IV.1.6) then

$$(\varphi \circ \tilde{\sigma} \circ \varphi^{-1}) \circ T_a \circ (\varphi \circ \tilde{\sigma} \circ \varphi^{-1})^{-1} = T_{\sigma a}$$

and

$$s_x = \varphi \circ \tilde{\sigma} \circ \varphi^{-1},$$

where

$$\varphi(g\,G_x) = T_g x, \qquad \tilde{\sigma}(g\,G_x) = (\sigma g)\,G_x.$$

Thus the exploration of homogeneous spaces with mirror subsymmetries is reduced to the exploration of pairs (G, H) of Lie groups, where H is a closed subgroup of G, with an involutive automorphism

$$\sigma \colon G \to G, \qquad \sigma H = H.$$

We say in this case that a *mirror subsymmetric triplet* (G, H, σ) is given.

IV.1.15. Definition 67. A homogeneous space (M, G, T) with a mirror subsymmetry s_x is called a mirror subsymmetric homogeneous space (or a homogeneous space with a mirror).

The dimension of the component of connectedness of a point x of the mirror $\mathfrak{k}$ generated by a mirror subsymmetry s_x is called the order of the mirror $\mathfrak{k}$ and of the mirror subsymmetry s_x.

IV.1.16. Remark. Mirror subsymmetries can be constructed at any point $y \in M$ if one takes $g(y) \in G$ in such a way that $T_{g(y)}\, x = y$ (this is possible owing to the transitivity of the action T).

Then

$$s_y = T_{g(y)} \circ s_x \circ T_{g(y)}^{-1}.$$

IV.1.17.   Definition 68. A homogeneous mirror subsymmetric space (M, G, T) with a Riemannian metric g such that s_x is its isometry is called a homogeneous Riemannian mirror subsymmetric space and is denoted (M, G, T, s_x, g).

IV.1.18. Remark. Thus in the above case T_a $(a \in G)$ and s_x are isometries.

The following results are true [L.V. Sabinin, 58a,b]:

IV.1.19. Theorem 1. *Any homogeneous Riemannian space with a non-trivial stationary group G_x has an isometric mirror subsymmetry s_x (i.e., it is a mirror subsymmetric space).*

IV.1.20. Theorem 2. *A mirror of a mirror subsymmetry of a Riemannian space is a totally geodesic submanifold.*

IV.1.21. Remark. A homogeneous mirror subsymmetric space M may have different subsymmetries s_x.

The minimum of orders of different mirror subsymmetries at a point $x \in M$ is called a *mirror rank* of a mirror subsymmetric homogeneous space.

IV.1.22. The smallest possible order of a mirror is zero. In this case the point x is an isolated immobile point of a mirror subsymmetry s_x and is called simply a mirror symmetry (or Cartan symmetry).

If the mirror rank of a homogeneous mirror subsymmetric space is zero then such a space is said to be symmetric.

The theory of symmetric spaces goes back to P.A. Shirokov [P.A. Shirokov, 25], E. Cartan [E. Cartan, 49,52], and has been well developed (see, for example, [S. Helgason, 62,78]).

IV.1.23. If M is a symmetric homogeneous space then, taking $g \in G^+$, we have

$$\sigma(g) = g, \qquad (s_x \circ T_g \circ s_x^{-1})(x) = T_{\sigma(g)}x = T_g x.$$

That is,

$$s_x(T_g x) = T_g x.$$

If g is taken from a sufficiently small neighbourhood Ω_e of the identity e in the group G^+ then $T_g x$ belongs to an arbitrarily taken neighbourhood of x. But $x \in M$ is the isolated immobile point of s_x. Therefore $T_g x = x$.

According to the Lie group theory [S. Helgason, 62,78], [L.S. Pontryagin, 73], as is well known, any neighbourhood of the identity of a Lie group generates its connected component.

Thus $T_g x = x$ for any $g \in (G^+)^\circ$ and we obtain $(G^+)^\circ \subset G_x$.

It may be shown that $G_x \subset G^+$ (see [S. Kobayashi, K. Nomizu, 69], vol. II, Chapter 9,10).

CHAPTER IV.2

SUBSYMMETRIC HOMOGENEOUS
SPACES AND LIE ALGEBRAS

IV.2.1. As is well known, a Lie group is uniquely determined locally by its Lie algebra. Moreover, any connected simply connected global Lie group is uniquely determined by its Lie algebra. All other Lie groups may be obtained from simply connected Lie groups after an appropriate factorization by a discrete central normal subgroup [S. Helgason, 62,78].

And the exploration of Lie algebras as linear (bilinear) structures is much simpler than the exploration of a non-linear object such as a Lie group. For this reason it is natural to pass from a homogeneous space G/H to a pair (doublet) of Lie algebras $(\mathfrak{g}, \mathfrak{h})$, where $\mathfrak{g}$ is the Lie algebra of the group G and $\mathfrak{h}$ is the subalgebra of $\mathfrak{g}$ corresponding to H ($\mathfrak{g} = \ln G$, $\mathfrak{h} = \ln_G H$.)

Since by the definition of a homogeneous space (see IV.1) a subgroup H does not contain a non-trivial normal subgroup and to any normal subgroup of G there corresponds an ideal of the Lie algebra $\mathfrak{g}$ we have a pair (doublet) $(\mathfrak{g}, \mathfrak{h})$ such that $\mathfrak{h}$ does not contain non-trivial ideals of $\mathfrak{g}$. We say that such a doublet is *exact*, or *effective*.

The Lie algebra $\mathfrak{g}$ of a group G is identified in the usual way with the tangent space $T_e(G)$ to the group G at the identity $e \in G$. Then the subalgebra $\mathfrak{h}$ is a tangent space $T_e(H)$ canonically embedded into $T_e(G)$ as a subspace.

An automorphism φ of a Lie group G generates, as is well known [S. Helgason, 62,78], [L.S. Pontryagin, 73], an automorphism of the corresponding Lie algebra, being the tangent map $\varphi_{*,e} \colon \mathfrak{g} \to \mathfrak{g}$ at a point e. If $\varphi^2 = \mathrm{id}$ then $\varphi_{*,e}^2 = \mathrm{id}$.

A mirror subsymmetric triplet (G, H, s) of a group G with a subgroup H and a mirror symmetry s generates a subsymmetric triplet $(\mathfrak{g}, \mathfrak{h}, s_{*,e})$, where $\mathfrak{g} = \ln G$ $\mathfrak{h} = \ln_G H$; we note that $s_{*,e}\mathfrak{h} = \mathfrak{h}$ since $sH = H$.

Conversely, having given a mirror subsymmetric triplet $(\mathfrak{g}, \mathfrak{h}, \sigma)$ of a Lie algebra $\mathfrak{g}$ we can reconstruct, by means of the exponential map, the corresponding mirror subsymmetric triplet (G, H, s), where

$$G = \exp \mathfrak{g}, \qquad H = \exp \mathfrak{h}, \qquad s = \exp \circ \sigma \circ \exp^{-1} .$$

Sometimes we write this briefly as $\exp(\mathfrak{g}, \mathfrak{h}, \sigma) = (G, H, s)$.

As a result the problem of the local classification of mirror subsymmetric homogeneous spaces is reduced to the problem of classification of mirror subsymmetric exact triplets of Lie algebras.

192

IV.2.2. Let (G_1, H_1, s_1) and (G_2, H_2, s_2) be two mirror subsymmetric triplets of Lie groups. Then we can introduce their *direct product*, which is a mirror subsymmetric triplet, by the rule:

$$(G_1, H_1, s_1) \times (G_2, H_2, s_2) = (G_1 \times G_2, H_1 \times H_2, s_1 \times s_2).$$

Here $G_1 \times G_2$ and $H_1 \times H_2$ are the direct products of groups and

$$(s_1 \times s_2)(g_1, g_2) \overset{\text{def}}{=} (s_1 g_1, s_2 g_2); \qquad g_1 \in G_1, \quad g_2 \in G_2.$$

A mirror subsymmetric triplet is called *decomposable* (*reducible*) if it is diffeomorphic to a direct product of two non-trivial mirror subsymmetric triplets. Otherwise we say that it is *non-decomposable* (*irreducible*).

Under a diffeomorphism of triplets (G, H, s) and $(\widetilde{G}, \widetilde{H}, \widetilde{s})$ we mean a diffeomorphism of Lie groups $\varphi \colon G \to \widetilde{G}$ such that

$$\varphi H = \widetilde{H}, \qquad \varphi \circ s \circ \varphi^{-1} = \widetilde{s}.$$

Having given some triplet we can prolong the process of its decomposition up to the direct product of non-decomposable triplets.

Analogously one may introduce the notion of a *direct product* for two mirror subsymmetric triplets of Lie algebras:

$$(\mathfrak{g}_1, \mathfrak{h}_1, \sigma_1) \times (\mathfrak{g}_2, \mathfrak{h}_2, \sigma_2) \overset{\text{def}}{=} (\mathfrak{g}_1 \times \mathfrak{g}_2, \mathfrak{h}_1 \times \mathfrak{h}_2, \sigma_1 \times \sigma_2).$$

Sometimes one says in this case that this is a (exterior) direct sum of mirror subsymmetric triplets and writes

$$(\mathfrak{g}_1, \mathfrak{h}_1, \sigma_1) \oplus (\mathfrak{g}_2, \mathfrak{h}_2, \sigma_2) \overset{\text{def}}{=} (\mathfrak{g}_1 \oplus \mathfrak{g}_2, \mathfrak{h}_1 \oplus \mathfrak{h}_2, \sigma_1 \oplus \sigma_2).$$

A mirror subsymmetric triplet of a Lie algebra is *decomposable* (*reducible*) if it is isomorphic to a direct product of two non-trivial mirror subsymmetric triplets, otherwise one says that it is *non-decomposable* (*irreducible*).

Note that under an isomorphism of triplets $(\mathfrak{g}, \mathfrak{h}, \sigma)$ and $(\widetilde{\mathfrak{g}}, \widetilde{\mathfrak{h}}, \widetilde{\sigma})$ we understand an isomorphism $\chi \colon \mathfrak{g} \to \widetilde{\mathfrak{g}}$ such that $\chi \mathfrak{h} = \widetilde{\mathfrak{h}}$, $\chi \circ \sigma \circ \chi^{-1} = \widetilde{\sigma}$.

Having given some triplet of a Lie algebra we can decompose it up to the direct product of several irreducible triplets.

IV.2.3. Remark. A decomposition of an exact triplet gives us a direct product of exact irreducible triplets.

IV.2.4. The classification of mirror subsymmetric triplets is reduced evidently to the classification of irreducible triplets.

IV.2.5. Remark. There exists an obvious relation between the irreducibility of triplets of Lie groups, (G, H, s), and the corresponding triplets of Lie algebras, $(\mathfrak{g}, \mathfrak{h}, \sigma)$. They are simultaneously either reducible or irreducible.

This is the direct consequence of the canonical functorial relation between Lie groups and Lie algebras.

IV.2.6. A Lie group G acts onto itself by the *adjoint action* (*representation*):

$$(\operatorname{Ad} g): G \to G, \qquad (\operatorname{Ad} g)p \stackrel{\text{def}}{=} g\,p\,g^{-1} \quad (g, p \in G).$$

This action generates the *linear adjoint action* (representation) of G on its Lie algebra $\mathfrak{g}$:

$$\operatorname{ad} g \stackrel{\text{def}}{=} (\operatorname{Ad} g)_{*,e} \quad (g \in G).$$

Here $\operatorname{Ad} g$ is an (inner) automorphism of the group G and $\operatorname{ad} g$ is an automorphism of the corresponding Lie algebra $\mathfrak{g}$.

Let $\mathfrak{g}$ be a Lie algebra and

$$(\operatorname{ad} x)\, y \stackrel{\text{def}}{=} [x\,y] \quad (x, y \in \mathfrak{g}).$$

The map

$$x \mapsto \operatorname{ad} x \quad (x \in \mathfrak{g})$$

then defines the *linear adjoint action* (*representation*) of a Lie algebra $\mathfrak{g}$.

By the Jacobi identity evidently

$$\operatorname{ad}[x\,y] = \operatorname{ad} x \circ \operatorname{ad} y - \operatorname{ad} y \circ \operatorname{ad} x.$$

More generally, let G be a Lie group, $\mathfrak{g}$ be its Lie algebra, and V be a vector space. We say that the *linear action* (*representation*) *of a Lie group G on a vector space V is given* if there is defined a morphism of groups:

$$T: G \to \operatorname{Aut}(V).$$

We say that the linear action (representation) of a Lie algebra $\mathfrak{g}$ on a vector space V is given if there is defined

$$\ell: \mathfrak{g} \to \operatorname{End}(V)$$

such that

$$\ell[x\,y] = \ell(x) \circ \ell(y) - \ell(y) \circ \ell(x) \quad (x, y \in \mathfrak{g}).$$

IV.2.7. Remark. Sometimes, for the sake of brevity, we say 'linear representation V', meaning, of course, that V is given together with some action.

A linear action (representation) of a Lie group G (respectively of a Lie algebra $\mathfrak{g}$) on V is called *completely reducible* (or *semi-simple*) if with respect to an action of $T(g)$ (respectively, of $\ell(g)$) any invariant subspace $W \subset V$ has an invariant complement $U \subset V$, $W \dotplus U = V$.

A linear action (representation) on a vector space V is called *irreducible* (*simple*) if the only invariant subspaces are V and $\{0\}$ (trivial).

A completely irreducible finite-dimensional linear action (representation) can be decomposed into a direct sum of irreducible representations, which means that

$$V = V_1 \dotplus \cdots \dotplus V_m$$

and $T(g)V_\alpha \subset V_\alpha$ ($\alpha = 1, \ldots, n$) in the case of a Lie group (and $\ell(\xi)V_\alpha \subset V_\alpha$ in the case of a Lie algebra).

Then $T_i(g) = T(g)|_{V_i}$ in the Lie group case (and $\ell_i(\xi) = \ell(\xi)|_{V_i}$ in the Lie algebra case) defines a linear action (representation).

IV.2.8. A Lie group G is said to be *simple* if $\dim G \neq 1$ and all its normal Lie subgroups are trivial (only G and e). G is called *semi-simple* if it is isomorphic to a direct product of simple non-one-dimensional Lie groups [S. Helgason, 62,78], [L.S. Pontryagin, 73].

There are the other, equivalent, definitions of the semi-simplicity.

A Lie algebra $\mathfrak{g}$ is said to be *simple* if $\dim \mathfrak{g} \neq 1$ and its ideals are trivial (only $\mathfrak{g}$ and $\{0\}$). $\mathfrak{g}$ is called *semi-simple* if it is isomorphic to a direct sum of non-one-dimensional simple subalgebras.

There are also the different, equivalent, definitions of the semi-simplicity. See [S. Helgason, 62,78].

A simple (respectively, semi-simple) Lie group has a simple (respectively, semi-simple) Lie algebra. And vice versa.

We note the well known result that *any linear representation of a semi-simple Lie group (Lie algebra) is completely reducible* [S. Helgason, 62,78], [L.S. Pontryagin, 73]

A Lie group is said to be *compact* if it is compact with respect to its topology.

A (real) Lie algebra $\mathfrak{g}$ is *compact* if there exists on $\mathfrak{g}$ a positive-definite symmetric (Hermitian, in the case of the field $\mathbb{C}$) ad $\mathfrak{g}$-invariant form $\langle x, y \rangle$, that is,

$$\langle (\mathrm{ad}z)x, y \rangle + \langle x, (\mathrm{ad}z)y \rangle = 0 \quad (\forall\, x, y, z \in \mathfrak{g}).$$

The Lie algebra of a compact Lie group is compact.

As is well known [S. Helgason, 62,78], [L.S. Pontryagin, 73], *any linear representation of a compact Lie group or Lie algebra is completely reducible.*

Moreover, for any linear representation ρ of a compact Lie group G (or of a compact Lie algebra) $\rho\colon G \to \mathrm{Aut}(V)$ (or $\mathfrak{g} \to \mathrm{End}(V)$) on a vector space V, there exists a positive-definite symmetric bilinear (Hermitian, in the case of the field $\mathbb{C}$) form $\varphi(x, y)$ which is ρ-invariant, that is,

$$\varphi(\rho(g)x, \rho(g)y) = \varphi(x, y), \quad (g \in G); \quad x, y \in V$$

(respectively, $\varphi(\rho(\xi)x, y) + \varphi(x, \rho(\xi)y) = 0, \ \xi \in \mathfrak{g}; \ x, y \in V$).

This result can be proved by means of an invariant integration [S. Helgason, 62,78]; if $\alpha(x, y)$ is a positive-definite symmetric (Hermitian) form on a vector space V and ρ is a representation of a compact Lie group G then

$$\varphi(x, y) = \int_G \alpha(\rho(g)x, \rho(g)y)\, d\mu_g$$

is a ρ-invariant positive-definite form.

Here μ_g is an invariant measure on the Lie group G.

The integral above has a sense owing to the compactness of G.

A compact Lie group G is locally isomorphic to the direct product of a torus (that is a connected commutative compact Lie group) with a semi-simple compact Lie group.

IV.2.9. Having given a Lie group or (its) Lie algebra $\mathfrak{g}$ one can introduce on $\mathfrak{g}$ $(= T_e(G))$ the *Killing form* (or the *Cartan metric*)

$$B(\xi, \eta) = -\operatorname{trace}(\operatorname{ad}\xi \circ \operatorname{ad}\eta) \quad (\xi, \eta \in \mathfrak{g}).$$

(Note that in some books the Killing form is defined as $-B(\xi, \eta)$.)
It is easily verified that B is $\operatorname{Ad} G$-invariant and $\operatorname{ad}\mathfrak{g}$-invariant, that is,

$$B((\operatorname{ad} g)\xi, (\operatorname{ad} g)\eta) = B(\xi, \eta),$$

$$B((\operatorname{ad}\zeta)\xi, \eta) + B(\xi, (\operatorname{ad}\zeta)\eta) = 0,$$

for any $g \in G$; $\xi, \eta, \zeta \in \mathfrak{g}$.

A Lie group G (or a Lie algebra $\mathfrak{g}$) is semi-simple if and only if its Cartan metric (Killing form) B is non-degenerate $(B(\mathfrak{g}, \zeta) = \{0\} \implies \zeta = 0)$.

It is known that the Cartan metric of a compact Lie group (or Lie algebra) is positive semi-definite (that is, $B(\zeta, \zeta) \geq 0$).

Lastly, a Lie group (Lie algebra) is semi-simple and compact if and only if its Cartan metric is positive-definite.

IV.2.10. A doublet of Lie algebras $(\mathfrak{g}, \mathfrak{h})$ $(\mathfrak{h} \in \mathfrak{g})$ is called *reductive* if $\mathfrak{g} = \mathfrak{m} \dotplus \mathfrak{h}$ (direct sum of vector spaces) and

$$(\operatorname{ad}\mathfrak{h})\mathfrak{m} \subset \mathfrak{m}.$$

A homogeneous space G/H is called *reductive* if its corresponding doublet $(\mathfrak{g}, \mathfrak{h})$ is reductive.

IV.2.11. Definition 69. A mirror subsymmetric triplet $(\mathfrak{g}, \mathfrak{h}, \sigma)$ is said to be reductive if

$$\mathfrak{g} = \mathfrak{m} \dotplus \mathfrak{h}, \qquad (\operatorname{ad}\mathfrak{h})\mathfrak{m} \subset \mathfrak{m}, \qquad \sigma\mathfrak{m} \subset \mathfrak{m}.$$

Respectively, a homogeneous mirror subsymmetric space (G, H, s) is said to be reductive if its corresponding triplet of Lie algebras, $(\mathfrak{g}, \mathfrak{h}, \sigma)$ (here $\sigma = s_{*,e}$), is reductive.

If $\mathfrak{h}$ is compact then the doublet $(\mathfrak{g}, \mathfrak{h})$ of Lie algebras is reductive.
Indeed, let us consider the linear representation

$$\operatorname{ad}\zeta : \mathfrak{g} \to \mathfrak{g}, \quad \zeta \in \mathfrak{g},$$

and, furthermore,

$$\rho(\zeta) = \operatorname{ad}\zeta, \quad \zeta \in \mathfrak{h}.$$

This is a linear representation of the Lie algebra $\mathfrak{h}$ on $\mathfrak{g}$.

By the properties of linear representations of compact Lie algebras [S. Helgason, 62, 78] there exists a positive-definite ρ-invariant form $\alpha(x, y)$ on $\mathfrak{g}$.

Evidently $\rho(\mathfrak{h})\mathfrak{h} \subset \mathfrak{h}$. But then the orthogonal complement $\mathfrak{m} = \mathfrak{h}^{\perp \alpha}$ of the subalgebra $\mathfrak{h}$ to $\mathfrak{g}$ is ρ-invariant.

Thus

$$\mathfrak{g} = \mathfrak{m} + \mathfrak{h}, \qquad (\operatorname{ad}\mathfrak{h})\mathfrak{m} \subset \mathfrak{m}.$$

If $\mathfrak{h}$ is compact then a mirror subsymmetric triplet of Lie algebras $(\mathfrak{g}, \mathfrak{h}, \sigma)$ is reductive. Indeed, as above, we take a positive-definite $\operatorname{ad}\mathfrak{h}$-invariant form $\alpha(x, y)$ on $\mathfrak{g}$.

Furthermore, we consider the form $\gamma(x, y) = \alpha(x, y) + \alpha(\sigma x, \sigma y)$. This form is evidently σ-invariant, $\operatorname{ad}\mathfrak{h}$-invariant, and positive-definite.

Introducing the orthogonal complement $\mathfrak{m} = \mathfrak{h}^{\perp \gamma}$ of $\mathfrak{h}$ in $\mathfrak{g}$ respective to γ and taking into account the invariance of $\mathfrak{h}$ under the action of $\operatorname{ad}\mathfrak{h}$ and σ we obtain

$$(\operatorname{ad}\mathfrak{h})\mathfrak{m} \subset \mathfrak{m}, \qquad \sigma\mathfrak{m} \subset \mathfrak{m}.$$

Hence it follows that *a homogeneous mirror subsymmetric Riemannian space (G, H, s) is reductive.*

IV.2.12. Remark. Using the complete reductivity of linear representations of semi-simple Lie groups and Lie algebras one can show in the similar way that if $\mathfrak{h}$ is semi-simple then the doublet of Lie algebras $(\mathfrak{g}, \mathfrak{h})$ is reductive [S. Helgason, 62, 78].

IV.2.13. We now show that if a mirror subsymmeric triplet $(\mathfrak{g}, \mathfrak{h}, \sigma)$ is exact and $\mathfrak{h}$ is a compact subalgebra then the Cartan metric of the Lie algebra $\mathfrak{g}$, $B(x, y)$, is positive-definite on $\mathfrak{h}$.

Indeed, let us take a positive-definite symmetric $\operatorname{ad}\mathfrak{h}$-invariant form $\gamma(x, y)$ on $\mathfrak{g}$. By the $\operatorname{ad}\mathfrak{h}$-invariance of γ we have

$$(\operatorname{ad}x)^* = -(\operatorname{ad}x), \qquad x \in \mathfrak{h}.$$

(Here A^* means the dual endomorphism of A with respect to γ.)

As is well known, AA^* is a self-dual endomorphism (as well as A^*A). In our case obviously $AA^* = A^*A$, thus $A = \operatorname{ad}x$ is a normal endomorphism.

Therefore AA^* is non-negative, that is, $\gamma(AA^*x, x) \geq 0$ for $x \neq 0$. For this reason all proper values of AA^* are non-negative and therefore

$$B(x, x)|_{\mathfrak{h}} = -\operatorname{trace}(\operatorname{ad}x)^2 = \operatorname{trace}(AA^*) \geq 0,$$

$$\operatorname{trace}(AA^*) = 0 \iff A = 0.$$

Then

$$B(x, x)|_{\mathfrak{h}} = -\operatorname{trace}(\operatorname{ad}x)^2 = \operatorname{trace}(AA^*) \geq 0 \quad (x \in \mathfrak{h}),$$

$$0 = B(x, x)|_{\mathfrak{h}} = \operatorname{trace}(AA^*) = 0 \iff \operatorname{ad}x = A = 0.$$

Now we consider a vector subspace $\mathfrak{k} = \{x \in \mathfrak{h} \mid \operatorname{ad}x = 0\}$ of $\mathfrak{g}$. Then

$$[\mathfrak{k}\,\mathfrak{g}] = (\operatorname{ad}\mathfrak{k})\mathfrak{g} = \{0\},$$

that is, $\mathfrak{k}$ is an ideal of $\mathfrak{g}$, $\mathfrak{k} \subset \mathfrak{h}$.

Owing to the exactness of the Lie triplet $(\mathfrak{g}, \mathfrak{h}, \sigma)$ we have $\mathfrak{k} = \{0\}$.

We have proved that $B|_{\mathfrak{h}}$ is positive-definite.

CHAPTER IV.3

MIRROR SUBSYMMETRIC LIE
TRIPLETS OF RIEMANNIAN TYPE

IV.3.1. We consider an exact Lie triplet $(\mathfrak{g}, \mathfrak{h}, \sigma)$, where $\mathfrak{g}$ is a Lie algebra, $\mathfrak{h}$ its compact subalgebra, and σ is an involutive automorphism of $\mathfrak{g}$ such that $\sigma\mathfrak{h} = \mathfrak{h}$.

Such a triplet is called *a mirror subsymmetric triplet of Riemannian type (Lie-Riemannian triplet).*

By what has been considered above (see [S. Kobayashi, K. Nomizu, 63,69]) such a triplet $(\mathfrak{g}, \mathfrak{h}, \sigma)$ is canonically reductive. Indeed, taking $\mathfrak{m} = \mathfrak{h}^{\perp B}$ (that is, the subspace orthogonal to $\mathfrak{h}$ in $\mathfrak{g}$ with respect to the Cartan metric B) we have $\mathfrak{g} = \mathfrak{m} \dotplus \mathfrak{h}$ since the Cartan metric B is positive-definite on $\mathfrak{h}$.

Because $(\operatorname{ad}\mathfrak{h})\,\mathfrak{h} \subset \mathfrak{h}$ and the Cartan metric is $\operatorname{ad}\mathfrak{h}$-invariant we have $(\operatorname{ad}\mathfrak{h})\,\mathfrak{m} \subset \mathfrak{m}$.

Finally, the invariance of the Cartan metric and of the subalgebra $\mathfrak{h}$ under the action of σ gives us $\sigma\mathfrak{m} \subset \mathfrak{m}$.

IV.3.2. In what follows we consider canonically reductive Lie triplets $(\mathfrak{g}, \mathfrak{h}, \sigma)$, being exact, if it is not said otherwise.

Sometimes it is convenient to consider a Lie triplet together with $\mathfrak{m} = \mathfrak{h}^{\perp B}$ and to say that *a Lie mirror subsymmetric quadruplet of Riemannian type* $(\mathfrak{g}, \mathfrak{h}, \mathfrak{m}, \sigma)$ *is given.*

In this case we should remember that

$$\mathfrak{m} = \mathfrak{h}^{\perp B}, \qquad \mathfrak{g} = \mathfrak{m} \dotplus \mathfrak{h}, \qquad (\operatorname{ad}\mathfrak{h})\,\mathfrak{m} \subset \mathfrak{m}, \qquad \sigma\mathfrak{m} \subset \mathfrak{m},$$
$$(\sigma\mathfrak{h}) \subset \mathfrak{h}, \qquad \sigma \in \operatorname{Aut}\mathfrak{g}, \qquad \sigma^2 = \operatorname{Id} \quad (\sigma \neq \operatorname{Id}). \tag{IV.1}$$

The decomposition

$$x = (1/2)(\operatorname{Id} + \sigma)\,x + (1/2)(\operatorname{Id} - \sigma)\,x \quad (x \in \mathfrak{g}) \tag{IV.2}$$

shows that

$$\mathfrak{g} = \mathfrak{g}^+ \dotplus \mathfrak{g}^-, \qquad \mathfrak{g}^+ = (1/2)(\operatorname{Id} + \sigma)\,\mathfrak{g}, \qquad \mathfrak{g}^- = (1/2)(\operatorname{Id} - \sigma)\,\mathfrak{g}, \tag{IV.3}$$

where

$$\mathfrak{g}^+ = \{x \in \mathfrak{g} \mid \sigma x = x\}, \qquad \mathfrak{g}^- = \{x \in \mathfrak{g} \mid \sigma x = -x\}. \tag{IV.4}$$

Evidently $\mathfrak{g}^+$ is a subalgebra of the Lie algebra $\mathfrak{g}$.

Since $\sigma\mathfrak{m} \subset \mathfrak{m}$, $(\sigma\mathfrak{h}) \subset \mathfrak{h}$, it is easily seen that

$$\left.\begin{aligned}
\mathfrak{g}^+ &= \mathfrak{m}^+ + \mathfrak{h}^+, \qquad \mathfrak{g}^- = \mathfrak{m}^- + \mathfrak{h}^-, \\
\mathfrak{m}^+ &= \{x \in \mathfrak{m} \mid \sigma x = x\} = [(1/2)(\mathrm{Id} + \sigma)]\,\mathfrak{m}, \\
\mathfrak{m}^- &= \{x \in \mathfrak{m} \mid \sigma x = -x\} = [(1/2)(\mathrm{Id} - \sigma)]\,\mathfrak{m}, \\
\mathfrak{h}^+ &= \{x \in \mathfrak{h} \mid \sigma x = x\} = [(1/2)(\mathrm{Id} + \sigma)]\,\mathfrak{h}, \\
\mathfrak{h}^- &= \{x \in \mathfrak{h} \mid \sigma x = -x\} = [(1/2)(\mathrm{Id} - \sigma)]\,\mathfrak{h}.
\end{aligned}\right\} \tag{IV.5}$$

Using the definitions of $\mathfrak{m}^+$, $\mathfrak{m}^-$, $\mathfrak{h}^+$, $\mathfrak{h}^-$, the reductivity of the pair $(\mathfrak{g}, \mathfrak{h})$, and that σ is an involutive automorphism we have

$$\begin{aligned}
[\mathfrak{m}^+\, \mathfrak{m}^+] &\subset \mathfrak{m}^+ + \mathfrak{h}^+ = \mathfrak{g}^+, \qquad & [\mathfrak{h}^+\, \mathfrak{h}^+] &\subset \mathfrak{h}^+, \\
[\mathfrak{m}^+\, \mathfrak{m}^-] &\subset \mathfrak{m}^- + \mathfrak{h}^- = \mathfrak{g}^-, \qquad & [\mathfrak{m}^+\, \mathfrak{h}^-] &\subset \mathfrak{m}^-, \\
[\mathfrak{m}^-\, \mathfrak{h}^+] &\subset \mathfrak{m}^-, \qquad [\mathfrak{h}^-\, \mathfrak{h}^-] \subset \mathfrak{h}^+, \qquad & [\mathfrak{h}^+\, \mathfrak{h}^-] &\subset \mathfrak{h}^-, \\
[\mathfrak{m}^-\, \mathfrak{m}^-] &\subset \mathfrak{m}^+ + \mathfrak{h}^+ = \mathfrak{g}^+, \qquad & [\mathfrak{m}^-\, \mathfrak{h}^-] \subset \mathfrak{m}^+, \quad [\mathfrak{m}^+\, \mathfrak{h}^+] &\subset \mathfrak{m}^+.
\end{aligned} \tag{IV.6}$$

Let us note that $\mathfrak{m}^- \neq \{0\}$. (Otherwise $\mathfrak{m}^+ = \mathfrak{m}$ and then $[\mathfrak{m}\,\mathfrak{h}^-] = \{0\}$. In addition $\mathfrak{h}^- \neq \{0\}$, otherwise $\sigma = \mathrm{Id}$, which is impossible.)

Furthermore, the subalgebra generated by $\mathfrak{h}^-$, that is,

$$\mathfrak{k} = \mathfrak{h}^- + \langle[\mathfrak{h}^-\,\mathfrak{h}^-]\rangle = \{\eta + \sum_i [\xi_i\,\zeta_i] \mid \eta, \xi_i, \zeta_j \in \mathfrak{h}^-\},$$

satisfies the conditions

$$[\mathfrak{h}\,\mathfrak{k}] \subset \mathfrak{k}, \qquad [\mathfrak{m}\,\mathfrak{k}] = \{0\}.$$

Consequently $\mathfrak{k}$ is a non-trivial ideal in $\mathfrak{g}$ and $\mathfrak{k} \subset \mathfrak{h}$, which is impossible since the pair $(\mathfrak{g}, \mathfrak{h})$ is exact.

Moreover, if $\mathfrak{m}^+ = \{0\}$ then $\mathfrak{h}^- = \{0\}$ and the pair $(\mathfrak{g}, \mathfrak{h})$ is symmetric. Indeed, if $\mathfrak{m}^+ = \{0\}$, $\mathfrak{h}^- \neq \{0\}$ then $\mathfrak{m}^- = \mathfrak{m}$ and, by (IV.3.6), we have $[\mathfrak{m}\,\mathfrak{h}^-] = \{0\}$. Hence by the Jacobi identity it follows that $\mathfrak{k} = \mathfrak{h}^- + \langle[\mathfrak{h}^-\,\mathfrak{h}^-]\rangle$ is an ideal of $\mathfrak{g}$, $\mathfrak{k} \subset \mathfrak{h}$, which contradicts the exactness of the pair $(\mathfrak{g}, \mathfrak{h})$.

IV.3.3. $\mathfrak{g}^+ = \{\eta \in \mathfrak{g} \mid \sigma\eta = \eta\}$ is called a *mirror of the Lie algebra* $\mathfrak{g}$, or of the pair $(\mathfrak{g}, \mathfrak{h})$. One says also that $\mathfrak{g}^+$ is an involutive algebra of the Lie algebra $\mathfrak{g}$ (see Part I).

We say that $\mathfrak{m}^+$ is a *mirror of the mirror subsymmetric reductive triplet* $(\mathfrak{g}, \mathfrak{h}, \sigma)$, or *quadruplet* $(\mathfrak{g}, \mathfrak{h}, \mathfrak{m}, \sigma)$.

IV.3.4. Remark. Passing from a Lie algebra to the corresponding Lie group by means of the exponential map we obtain $\exp \mathfrak{m}^+$ which generates a mirror in the homogeneous space $\exp \mathfrak{g}/\exp \mathfrak{h}$.

This explains the accepted terminology.

IV.3.5. The dimension of a mirror, $\dim \mathfrak{m}^+ = N$, is called a *mirror order* (*index*) of a *mirror subsymmetric reductive triplet* $(\mathfrak{g}, \mathfrak{h}, \sigma)$, or *quadruplet* $(\mathfrak{g}, \mathfrak{h}, \mathfrak{m}, \sigma)$, and is denoted m.ord $(\mathfrak{g}, \mathfrak{h}, \sigma)$ (ord $(\mathfrak{g}, \mathfrak{h}, \mathfrak{m}, \sigma)$, respectively).

In this case we say that σ is a mirror subsymmetry of the mirror order N.

Any reductive pair $(\mathfrak{g}, \mathfrak{h})$ has, generally speaking, subsymmetries of different mirror orders. The minimal possible mirror order of mirror subsymmetries of the given reductive pair $(\mathfrak{g}, \mathfrak{h})$ is called the *mirror class* of this pair.

Analogously one can define the *mirror order* and the *mirror class* of a reductive triplet (G, H, σ) and of a homogeneous space G/H (G and H being Lie groups).

IV.3.6. Remark. The mirror order and the mirror class of a reductive triplet for the Lie groups case coincide with the mirror order and the mirror class of the corresponding reductive triplet of Lie algebra and reductive pair, respectively.

IV.3.7. Remark. Mirror subsymmetric Lie–Riemannian triplets of the mirror order zero are simply involutive triplets, that is, $(\mathfrak{g}, \mathfrak{h}, \sigma)$, where $\sigma\eta = \eta \iff \eta \in \mathfrak{h}$. They correspond to Riemannian symmetric spaces G/H.

Symmetric Riemannian spaces (and the corresponding Lie–Riemannian triplets of Lie algebras) have been classified by E. Cartan (see, for example, [S. Helgason, 62, 78]).

Mirror subsymmetric Riemannian spaces of the class one (and the corresponding triplets of Lie algebras) have been explored as well [L.V. Sabinin, 58b, 59b].

Thus now we are going to explore the case of the mirror order (or of the mirror class) two.

IV.3.8. Now, we introduce the canonical base of a Lie algebra $\mathfrak{g}$ generated by a mirror subsymmetric triplet of Riemannian type $(\mathfrak{g}, \mathfrak{h}, \sigma)$. Such a triplet is canonically reductive with $\mathfrak{m} = \mathfrak{h}^{\perp_B}$.

In fact, we consider a quadruplet $(\mathfrak{g}, \mathfrak{h}, \mathfrak{m}, \sigma)$. Since this quadruplet is of Riemannian type there exists a σ-invariant and ad η-invariant positive-definite bilinear form $g(\xi, \eta)$ on $\mathfrak{m}$ [S. Helgason, 62,78]. Choosing one such form we use it furthermore. Evidently $\mathfrak{m}^+$ and $\mathfrak{m}^-$ are orthogonal with respect to this form, as well as with respect to the Cartan metric B. In addition, $\mathfrak{h}^+$ and $\mathfrak{h}^-$ are orthogonal to $\mathfrak{m}^+$ and $\mathfrak{m}^-$ with respect to B.

Finally, $\mathfrak{h}^+$ is orthogonal to $\mathfrak{h}^-$ with respect to B.

Now we take bases in $\mathfrak{m}^+$ and $\mathfrak{m}^-$ which are orthonormal with respect to B and orthogonal with respect to g.

Furthermore, we take bases in $\mathfrak{h}^+$ and $\mathfrak{h}^-$ which are orthogonal with respect to B (this is possible since for an exact triplet of Riemannian type the Cartan metric is positive-definite on $\mathfrak{h}$ [S. Helgason, 62,78]).

Thus we have

$$\left.\begin{array}{ll} X_a \ (a = 1, 2, \ldots, m) & \text{– orthogonal base in } \mathfrak{m}^+, \\[2mm] X_i \ (i = m+1, \ldots, n) & \text{– orthogonal base in } \mathfrak{m}^-, \\[2mm] Y_{\hat{\alpha}} \ (\hat{\alpha} = n+1, \ldots, s) & \text{– orthogonal base in } \mathfrak{h}^-, \\[2mm] Y_\alpha \ (\alpha = s+1, \ldots, r) & \text{– orthogonal base in } \mathfrak{h}^+. \end{array}\right\} \quad \text{(IV.7)}$$

The union of the bases constructed above constitutes an orthogonal base in $\mathfrak{g}$ which is called a *canonical* base.

Let us note that in this case $\{X_J\}_{J=1,2,\ldots,n}$ is an orthogonal base on $\mathfrak{m}$ and $\{Y_A\}_{A=n+1,\ldots,r}$ is an orthogonal base on $\mathfrak{h}$.

IV.3.9. Remark. A canonical base is taken with some arbitrariness since the choice of orthogonal bases on $\mathfrak{m}^+$, $\mathfrak{m}^-$, $\mathfrak{h}^+$, $\mathfrak{h}^-$ is not unique.

IV.3.10. We now write the structure equations of a Lie algebra $\mathfrak{g}$ in a canonical base. Owing to the action of σ they have the form

$$
\begin{aligned}
[X_i \, X_j] = S_{ij}^a X_a - b_{ij}^\alpha Y_\alpha, \qquad & [X_a \, X_b] = S_{ab}^c X_c - b_{ab}^\alpha Y_\alpha, \\
[X_i \, X_a] = S_{ia}^j X_j - b_{ia}^{\hat\lambda} Y_{\hat\lambda}, \qquad & [X_i \, Y_{\hat\lambda}] = a_{i\hat\lambda}^c X_c, \\
[X_a \, Y_{\hat\alpha}] = a_{a\hat\alpha}^j X_j, \qquad & [X_i \, Y_\alpha] = a_{i\alpha}^j X_j, \\
[X_b \, Y_\alpha] = a_{b\alpha}^c X_c, \qquad & [Y_{\hat\alpha} \, Y_{\hat\beta}] = c_{\hat\alpha\hat\beta}^\gamma Y_\gamma, \\
[Y_\alpha \, Y_\beta] = c_{\alpha\beta}^\gamma Y_\gamma, \qquad & [Y_{\hat\alpha} \, Y_\beta] = c_{\hat\alpha\beta}^{\hat\gamma} Y_{\hat\gamma}; \\
a, b, c = 1, 2, \ldots, m; \quad & i, j = m+1, \ldots, n; \\
\hat\alpha, \hat\beta = n+1, \ldots, s; \quad & \alpha, \beta = s+1, \ldots, r.
\end{aligned}
\right\} \tag{IV.8}
$$

By the choice of a canonical base we have

$$
\begin{aligned}
g(X_a, X_b) = g_{ab} = \delta_{ab}, \qquad & g(X_i, X_j) = g_{ij} = \delta_{ij}, \\
g(X_a, X_i) = g_{ai} = 0, \qquad & B(X_a, X_b) = B_{ab} = \lambda(a)\delta_{ab}, \\
B(X_i, X_j) = \lambda(i)\delta_{ij}, \qquad & B(X_a, X_i) = B_{ai} = 0, \\
B(Y_\alpha, Y_\beta) = B_{\alpha\beta} = \delta_{\alpha\beta}, \qquad & B(Y_{\hat\alpha}, Y_{\hat\beta}) = B_{\hat\alpha\hat\beta} = \delta_{\hat\alpha\hat\beta}, \\
B(Y_\alpha, Y_{\hat\beta}) = B_{\alpha\hat\beta} = 0, \qquad & B(X_a, Y_\alpha) = B_{a\alpha} = 0, \\
B(X_a, Y_{\hat\alpha}) = B_{a\hat\alpha} = 0, \qquad & B(X_i, Y_\alpha) = B_{i\alpha} = 0, \\
B(X_i, Y_{\hat\alpha}) = B_{i\hat\alpha} = 0. &
\end{aligned}
\right\} \tag{IV.9}
$$

In a brief presentation this is

$$
\begin{aligned}
g(X_I, X_J) = g_{IJ} = \delta_{IJ}, \qquad & B(X_I, X_J) = B_{IJ} = \lambda(I)\delta_{IJ}, \\
B(Y_A, Y_D) = B_{AD} = \delta_{AD}, \qquad & B(Y_A, X_J) = B_{AJ}.
\end{aligned}
\right\} \tag{IV.10}
$$

In the formulas (IV.9), (IV.10) the σ-invariance of the forms B and g has been used.

Taking into account the ad $\mathfrak{g}$-invariance of the form B, the ad $\mathfrak{h}$-invariance of the form g, and (IV.2) we obtain

$$\lambda(I)S_{JK}^I + \lambda(J)S_{IK}^J = 0, \tag{IV.11}$$

$$\lambda(I)a_{JK}^I + \lambda(J)a_{IK}^J = 0, \tag{IV.12}$$

$$-\lambda(I)a_{JA}^I + b_{JI}^A = 0, \tag{IV.13}$$

$$a_{KA}^I + a_{IA}^K = 0, \tag{IV.14}$$

$$c_{BD}^A + c_{AD}^B = 0. \tag{IV.15}$$

IV.3.11. Let us write out the Jacobi identities for the structure (IV.8) in a canonical base.

The relations

$$[X_a\,[X_b\,X_c]] + [X_c\,[X_a\,X_b]] + [X_b\,[X_c\,X_a]] = 0$$

give us

$$S_{bc}^d\,S_{ad}^e + S_{ab}^d\,S_{cd}^e + S_{ca}^d\,S_{bd}^e - a_{a\alpha}^e\,b_{bc}^\alpha - a_{b\alpha}^e\,b_{ca}^\alpha - a_{c\alpha}^e\,b_{ab}^\alpha = 0, \tag{IV.16}$$

$$b_{ad}^\beta\,S_{bc}^d + b_{cd}^\beta\,S_{ab}^d + b_{bd}^\beta\,S_{ca}^d = 0. \tag{IV.17}$$

The relations
$$[[X_i\,X_j]\,X_k] + [[X_k\,X_i]\,X_j] + [[X_j\,X_k]\,X_i] = 0$$
give us

$$S_{ij}^a\,S_{ak}^p + S_{ki}^a\,S_{aj}^p + S_{jk}^a\,S_{ai}^p + b_{ij}^\alpha\,a_{k\alpha}^p + b_{ki}^\alpha\,a_{j\alpha}^p + b_{jk}^\alpha\,a_{i\alpha}^p = 0, \tag{IV.18}$$

$$S_{ij}^a\,b_{ka}^{\hat{\lambda}} + S_{ki}^a\,b_{ja}^{\hat{\lambda}} + S_{jk}^a\,b_{ia}^{\hat{\lambda}} = 0. \tag{IV.19}$$

The relations

$$[[X_a\,X_b]\,X_k] + [[X_k\,X_a]\,X_b] + [[X_b\,X_k]\,X_a] = 0$$

give us

$$S_{ab}^c\,S_{ck}^j + S_{ka}^p\,S_{pb}^j - S_{kb}^p\,S_{pa}^j + b_{ab}^\alpha\,a_{k\alpha}^i + b_{ka}^{\hat{\mu}}\,a_{b\hat{\mu}}^j - b_{kb}^{\hat{\mu}}\,a_{a\hat{\mu}}^j = 0, \tag{IV.20}$$

$$S_{ab}^c\,b_{kc}^{\hat{\lambda}} + S_{ka}^p\,b_{bp}^{\hat{\lambda}} - S_{kb}^i\,b_{ia}^{\hat{\lambda}} = 0. \tag{IV.21}$$

The relations
$$[[X_i\,X_j]\,X_c] + [[X_c\,X_i]\,X_j] + [[X_j\,X_c]\,X_i] = 0$$

give us

$$S_{ij}^a S_{ac}^d + S_{ci}^p S_{pj}^d - S_{cj}^p S_{pi}^d + b_{ij}^\alpha a_{c\alpha}^d + b_{ci}^{\hat\lambda} a_{j\hat\lambda}^d - b_{cj}^{\hat\lambda} a_{i\hat\lambda}^d = 0, \tag{IV.22}$$

$$S_{ij}^a b_{ac}^\alpha + S_{ci}^p b_{pj}^\alpha - S_{cj}^p b_{pi}^\alpha = 0. \tag{IV.23}$$

The relations

$$[[X_a X_b] Y_\alpha] + [[Y_\alpha X_a] X_b] + [[X_b Y_\alpha] X_a] = 0$$

give us

$$S_{ab}^d a_{d\alpha}^e - a_{a\alpha}^d S_{db}^e + a_{b\alpha}^d S_{da}^e = 0, \tag{IV.24}$$

$$b_{ab}^\lambda c_{\alpha\lambda}^\gamma + a_{a\alpha}^d b_{db}^\gamma - a_{b\alpha}^d b_{da}^\gamma = 0. \tag{IV.25}$$

The relations

$$[[X_i X_j] Y_\alpha] + [[Y_\alpha X_i] X_j] + [[X_j Y_\alpha] X_i] = 0$$

give us

$$S_{ij}^d a_{d\alpha}^a - a_{i\alpha}^k S_{kj}^a + a_{j\alpha}^k S_{ki}^a = 0, \tag{IV.26}$$

$$b_{ij}^\lambda c_{\alpha\lambda}^\gamma + a_{i\alpha}^k b_{kj}^\gamma - a_{j\alpha}^k b_{ki}^\gamma = 0. \tag{IV.27}$$

The relations

$$[[X_i X_a] Y_\alpha] + [[Y_\alpha X_i] X_a] + [[X_a Y_\alpha] X_i] = 0$$

give us

$$S_{ia}^j a_{j\alpha}^p - a_{i\alpha}^q S_{qa}^p + a_{a\alpha}^c S_{ci}^p = 0, \tag{IV.28}$$

$$b_{ai}^{\hat\lambda} c_{\hat\lambda\alpha}^{\hat\mu} + b_{qa}^{\hat\mu} a_{i\alpha}^q + b_{ic}^{\hat\mu} a_{a\alpha}^c = 0. \tag{IV.29}$$

The relations

$$[[X_a X_b] Y_{\hat\alpha}] + [[Y_{\hat\alpha} X_a] X_b] + [[X_b Y_{\hat\alpha}] X_a] = 0$$

give us

$$S_{ab}^c a_{c\hat\alpha}^p - a_{a\hat\alpha}^q S_{qb}^p + a_{b\hat\alpha}^q S_{qa}^p = 0, \tag{IV.30}$$

$$b_{ab}^\lambda c_{\hat\alpha\lambda}^{\hat\mu} + a_{a\hat\alpha}^q b_{qb}^{\hat\mu} - a_{b\hat\alpha}^q b_{qa}^{\hat\mu} = 0. \tag{IV.31}$$

The relations

$$[[X_i X_j] Y_{\hat\alpha}] + [[Y_{\hat\alpha} X_i] X_j] + [[X_j Y_{\hat\alpha}] X_i] = 0$$

give us

$$S_{ij}^d \, a_{d\hat{\alpha}}^p + a_{j\hat{\alpha}}^d \, S_{di}^p - a_{i\hat{\alpha}}^d \, S_{dj}^p = 0, \tag{IV.32}$$

$$b_{ij}^\lambda \, c_{\hat{\alpha}\lambda}^{\hat{\mu}} + a_{j\hat{\alpha}}^d \, b_{id}^{\hat{\mu}} - a_{i\hat{\alpha}}^d \, b_{jd}^{\hat{\mu}} = 0. \tag{IV.33}$$

The relations

$$[[X_i \, X_a] \, Y_{\hat{\alpha}}] + [[Y_{\hat{\alpha}} \, X_i] \, X_a] + [[X_a \, Y_{\hat{\alpha}}] \, X_i] = 0$$

give us

$$S_{ia}^p \, a_{p\hat{\alpha}}^d + a_{i\hat{\alpha}}^c \, S_{ac}^d + a_{a\hat{\alpha}}^p \, S_{pi}^d = 0, \tag{IV.34}$$

$$b_{ia}^{\hat{\lambda}} \, c_{\hat{\alpha}\hat{\lambda}}^\beta + a_{i\hat{\alpha}}^c \, b_{ca}^\beta - a_{a\hat{\alpha}}^p \, b_{pi}^\beta = 0. \tag{IV.35}$$

It remains to consider the relations connected with the symmetric pair $(\mathfrak{h}^-, \mathfrak{h}^+)$, that is, with Y_α, $Y_{\hat{\alpha}}$ $(\alpha = s+1, \ldots, r; \, \hat{\alpha} = n+1, \ldots, s)$.

The relations

$$[[Y_{\hat{\alpha}} \, Y_{\hat{\beta}}] \, Y_{\hat{\gamma}}] + [[Y_{\hat{\gamma}} \, Y_{\hat{\alpha}}] \, Y_{\hat{\beta}}] + [[Y_{\hat{\beta}} \, Y_{\hat{\gamma}}] \, Y_{\hat{\alpha}}] = 0$$

give us

$$c_{\hat{\alpha}\hat{\beta}}^\lambda \, c_{\hat{\gamma}\lambda}^{\hat{\mu}} + c_{\hat{\gamma}\hat{\alpha}}^\lambda \, c_{\hat{\beta}\lambda}^{\hat{\mu}} + c_{\hat{\beta}\hat{\gamma}}^\nu \, c_{\hat{\alpha}\lambda}^{\hat{\mu}} = 0. \tag{IV.36}$$

The relations

$$[[Y_\alpha \, Y_\beta] \, Y_\gamma] + [[Y_\gamma \, Y_\alpha] \, Y_\beta] + [[Y_\beta \, Y_\gamma] \, Y_\alpha] = 0$$

give us

$$c_{\alpha\beta}^\lambda \, c_{\gamma\lambda}^\mu + c_{\gamma\alpha}^\lambda \, c_{\beta\lambda}^\mu + c_{\beta\gamma}^\lambda \, c_{\alpha\lambda}^\mu = 0. \tag{IV.37}$$

The relations

$$[[Y_{\hat{\alpha}} \, Y_{\hat{\beta}}] \, Y_\gamma] + [[Y_\gamma \, Y_{\hat{\alpha}}] \, Y_{\hat{\beta}}] + [[Y_{\hat{\beta}} \, Y_\gamma] \, Y_{\hat{\alpha}}] = 0$$

give us

$$c_{\hat{\alpha}\hat{\beta}}^\lambda \, c_{\lambda\gamma}^\mu + c_{\gamma\hat{\alpha}}^{\hat{\nu}} \, c_{\hat{\nu}\hat{\beta}}^\mu + c_{\hat{\beta}\gamma}^{\hat{\nu}} \, c_{\hat{\nu}\hat{\alpha}}^\mu = 0. \tag{IV.38}$$

The relations

$$[[Y_{\hat{\alpha}} \, Y_\beta] \, Y_\gamma] + [[Y_\gamma \, Y_{\hat{\alpha}}] \, Y_\beta] + [[Y_\beta \, Y_\gamma] \, Y_{\hat{\alpha}}] = 0$$

give us

$$c_{\hat{\alpha}\beta}^{\hat{\nu}} \, c_{\hat{\nu}\gamma}^{\hat{\mu}} + c_{\gamma\hat{\alpha}}^{\hat{\nu}} \, c_{\hat{\nu}\beta}^{\hat{\mu}} + c_{\beta\gamma}^\lambda \, c_{\lambda\hat{\alpha}}^{\hat{\mu}} = 0. \tag{IV.39}$$

The relations

$$[[Y_\alpha \, Y_\beta] \, X_a] + [[X_a \, Y_\alpha] \, Y_\beta] + [[Y_\beta \, X_a] \, Y_\alpha] = 0$$

give us

$$-c_{\alpha\beta}^{\gamma}\, a_{a\gamma}^{c} + a_{a\alpha}^{d}\, a_{d\beta}^{c} - a_{a\beta}^{d}\, a_{d\alpha}^{c} = 0. \tag{IV.40}$$

The relations

$$[[Y_{\alpha}\, Y_{\beta}]\, X_{i}] + [[X_{i}\, Y_{\alpha}]\, Y_{\beta}] + [[Y_{\beta}\, X_{i}]\, Y_{\alpha}] = 0$$

give us

$$-c_{\alpha\beta}^{\gamma}\, a_{i\gamma}^{p} + a_{i\alpha}^{q}\, a_{q\beta}^{p} - a_{i\beta}^{q}\, a_{q\alpha}^{p} = 0. \tag{IV.41}$$

The relations

$$[[Y_{\hat{\alpha}}\, Y_{\hat{\beta}}]\, X_{a}] + [[X_{a}\, Y_{\hat{\alpha}}]\, Y_{\hat{\beta}}] + [[Y_{\hat{\beta}}\, X_{a}]\, Y_{\hat{\alpha}}] = 0$$

give us

$$-c_{\hat{\alpha}\hat{\beta}}^{\gamma}\, a_{a\gamma}^{c} + a_{a\hat{\alpha}}^{q}\, a_{q\hat{\beta}}^{c} - a_{a\hat{\beta}}^{q}\, a_{q\hat{\alpha}}^{c} = 0. \tag{IV.42}$$

The relations

$$[[Y_{\hat{\alpha}}\, Y_{\beta}]\, X_{a}] + [[X_{a}\, Y_{\hat{\alpha}}]\, Y_{\beta}] + [[Y_{\beta}\, X_{a}]\, Y_{\hat{\alpha}}] = 0$$

give us

$$-c_{\hat{\alpha}\beta}^{\hat{\gamma}}\, a_{a\hat{\gamma}}^{p} + a_{a\hat{\alpha}}^{q}\, a_{q\beta}^{p} - a_{a\beta}^{c}\, a_{c\hat{\alpha}}^{p} = 0. \tag{IV.43}$$

The relations

$$[[Y_{\hat{\alpha}}\, Y_{\hat{\beta}}]\, X_{i}] + [[X_{i}\, Y_{\hat{\alpha}}]\, Y_{\hat{\beta}}] + [[Y_{\hat{\beta}}\, X_{i}]\, Y_{\hat{\alpha}}] = 0$$

give us

$$-c_{\hat{\alpha}\hat{\beta}}^{\gamma}\, a_{i\gamma}^{q} + a_{i\hat{\alpha}}^{c}\, a_{c\hat{\beta}}^{q} - a_{i\hat{\beta}}^{c}\, a_{c\hat{\alpha}}^{q} = 0. \tag{IV.44}$$

The relations

$$[[Y_{\hat{\alpha}}\, Y_{\beta}]\, X_{i}] + [[X_{i}\, Y_{\hat{\alpha}}]\, Y_{\beta}] + [[Y_{\beta}\, X_{i}]\, Y_{\hat{\alpha}}] = 0$$

give us

$$-c_{\hat{\alpha}\beta}^{\hat{\gamma}}\, a_{i\hat{\gamma}}^{c} + a_{i\hat{\alpha}}^{d}\, a_{d\beta}^{c} - a_{i\beta}^{q}\, a_{q\hat{\alpha}}^{c} = 0. \tag{IV.45}$$

The relations (IV.16)–(IV.45) present all possible Jacobi identities.

There are also the relations concerning the orthogonality of the base with respect to the Cartan metric

$$B_{ij} = B(X_{i}, X_{j}) = -\text{trace}(\text{ad}X_{i}\,\text{ad}X_{j})$$

$$= S_{ik}^{a}\, S_{ja}^{k} - S_{ia}^{p}\, S_{jp}^{a} - a_{i\alpha}^{p}\, b_{pj}^{\alpha} - a_{j\alpha}^{p}\, b_{pi}^{\alpha} \tag{IV.46}$$

$$- a_{i\hat{\alpha}}^{d}\, b_{dj}^{\hat{\alpha}} - a_{j\hat{\alpha}}^{d}\, b_{di}^{\hat{\alpha}} = \lambda(i)\delta_{ij}.$$

$$B_{ab} = B(X_a, X_b) = - S^c_{ad} S^d_{bc} - S^k_{ap} S^p_{bk} - a^d_{ba} b^\alpha_{da}$$

$$- a^d_{a\alpha} b^\alpha_{db} - a^k_{a\hat\alpha} b^{\hat\alpha}_{kb} - a^k_{b\hat\alpha} b^{\hat\alpha}_{ka} = \lambda(a)\delta_{ab}. \qquad \text{(IV.47)}$$

$$B_{\alpha\beta} = B(Y_\alpha, Y_\beta)$$

$$= - c^\gamma_{\alpha\delta} c^\delta_{\beta\gamma} - c^{\hat\gamma}_{\hat\delta\alpha} c^{\hat\delta}_{\hat\gamma\beta} - a^a_{b\alpha} a^b_{a\beta} - a^q_{\alpha p} a^p_{q\beta} = \delta_{\alpha\beta}. \qquad \text{(IV.48)}$$

$$B_{\hat\alpha\hat\beta} = B(Y_{\hat\alpha}, Y_{\hat\beta})$$

$$= - a^k_{a\hat\alpha} a^a_{k\hat\beta} - a^b_{p\hat\alpha} a^p_{b\hat\beta} - c^{\hat\mu}_{\hat\alpha\lambda} c^\lambda_{\hat\beta\hat\mu} - c^\mu_{\hat\alpha\hat\lambda} c^{\hat\lambda}_{\hat\beta\mu} = \delta_{\hat\alpha\hat\beta}. \qquad \text{(IV.49)}$$

The equalities

$$B_{i\alpha} = 0, \qquad B_{ia} = 0, \qquad B_{i\hat\alpha} = 0,$$

$$B_{a\alpha} = 0, \qquad B_{a\hat\alpha} = 0, \qquad B_{\beta\hat\alpha} = 0$$

are satisfied automatically since σ is an involutive automorphism preserving the Cartan metric.

IV.3.12. Now, using the equalities written above we obtain some information about the structure of a two-dimensional mirror.

Let $(\mathfrak{g}, \mathfrak{h}, \mathfrak{m}, \sigma)$ be a mirror subsymmetric quadruplet of Riemannian type and of the mirror order two, that is, $\dim \mathfrak{g}^+ - \dim \mathfrak{h}^+ = \dim \mathfrak{m}^+ = 2$.

There are two possibilities:

I. $[\mathfrak{h}^+ \mathfrak{m}^+] = \{0\}$;

II. $[\mathfrak{h}^+ \mathfrak{m}^+] \neq \{0\}$.

Owing to the invariance of the Cartan metric we have

$$B((\operatorname{ad} X)Y, Z) + B(Y, (\operatorname{ad} X)Z) = 0,$$

and if $X, Z \in \mathfrak{m}^+$, $Y \in \mathfrak{h}^+$ we obtain in the case I $(\operatorname{ad} X)Y = [X Y] = 0$. Thus $B(Y, [X Z]) = 0$, or $[\mathfrak{m}^+\mathfrak{m}^+] \perp_B \mathfrak{h}^+$, implying $[\mathfrak{m}^+\mathfrak{m}^+] \subset \mathfrak{m}^+$.

As a result in the case I

$$[\mathfrak{h}^+ \mathfrak{m}^+] = \{0\} \implies [\mathfrak{m}^+\mathfrak{m}^+] \subset \mathfrak{m}^+,$$

which means that $\mathfrak{h}^+$ and $\mathfrak{m}^+$ are ideals of $\mathfrak{g}^+$. Then $\mathfrak{g}^+ = \mathfrak{m}^+ \dotplus \mathfrak{h}^+$ is a direct sum of ideals, $\mathfrak{g}^+ = \mathfrak{m}^+ \oplus \mathfrak{h}^+$. Since $\mathfrak{m}^+$ is a two-dimensional subalgebra then either $[\mathfrak{m}^+ \mathfrak{m}^+] = \{0\}$, and $\mathfrak{m}^+$ is abelian, or $[\mathfrak{m}^+ \mathfrak{m}^+] = \{tX\}_{t\in\mathbb{R}}$ $(X \neq 0, X \in \mathfrak{m}^+)$.

In the latter case $[\mathfrak{m}^+ X] \subset [\mathfrak{m}^+ \mathfrak{m}^+] = \mathbb{R}X$ and $\mathbb{R}X$ is a one-dimensional ideal in $\mathfrak{m}^+$. In this case the restriction of the Cartan metric of the algebra $\mathfrak{g}$ to $\mathfrak{m}^+$, that is, $B|_{\mathfrak{m}^+}$, is a degenerate form and $B(X, X) = 0$. Indeed, by the ad-invariance of the Cartan metric we have for some $Z \in \mathfrak{m}^+$ such that $[Z X] = \lambda X$ $(\lambda \neq 0)$,

$$B((\operatorname{ad} Z)X, X) + B(X, (\operatorname{ad} Z)X) = 0,$$

or $2B([Z\,X], X) = 0$. But $[Z\,X] = \lambda X$ $(\lambda \neq 0)$, whence $2\lambda B(X, X) = 0$, that is, $B(X, X) = 0$.

Another possibility is $[\mathfrak{m}^+ \mathfrak{h}^+] \neq \{0\}$.

Let us consider $[\mathfrak{m}^+ \mathfrak{m}^+]$, since $\dim \mathfrak{m}^+ = 2$ we have $[\mathfrak{m}^+ \mathfrak{m}^+] = \mathbb{R}Z$, where either $Z \neq 0$ or $Z = 0$.

Then from the Jacobi identities we have $[\mathfrak{h}^+ \mathfrak{m}^+] \subset \mathfrak{m}^+$, and consequently

$$[\mathfrak{h}^+ [\mathfrak{m}^+ \mathfrak{m}^+]] \subset [\mathfrak{m}^+ [\mathfrak{h}^+ \mathfrak{m}^+]] + [\mathfrak{m}^+ [\mathfrak{m}^+ \mathfrak{h}^+]] \subset [\mathfrak{m}^+ \mathfrak{m}^+].$$

Thus $[\mathfrak{h}^+ Z] \subset \mathbb{R}Z$. Decomposing Z along $\mathfrak{m}^+$ and $\mathfrak{h}^+$, $Z = Z_{\mathfrak{m}+} + Z_{\mathfrak{h}+}$, we obtain

$$[\mathfrak{h}^+ Z_{\mathfrak{m}+}] \subset \mathbb{R}Z_{\mathfrak{m}+}, \qquad [\mathfrak{h}^+ Z_{\mathfrak{h}+}] \subset \mathbb{R}Z_{\mathfrak{h}+}.$$

If we assume that $Z_{\mathfrak{m}+} \neq 0$ (thus $Z \neq 0$ as well) then taking $Z'_{\mathfrak{m}+} \perp_{\mathfrak{m}+} Z_{\mathfrak{m}+}$ with respect to the invariant metric g we obtain $[\mathfrak{h}^+ Z'_{\mathfrak{m}+}] \subset \mathbb{R}Z'_{\mathfrak{m}+}$.

Indeed,

$$g((\operatorname{ad} Y)Z_{\mathfrak{m}+}, Z'_{\mathfrak{m}+}) + g(Z_{\mathfrak{m}+}, (\operatorname{ad} Y)Z'_{\mathfrak{m}+}) = 0 \quad (\forall Y \in \mathfrak{h}^+)$$

implies

$$g(\lambda Z_{\mathfrak{m}+}, Z'_{\mathfrak{m}+}) + g(Z_{\mathfrak{m}+}, [Y\,Z'_{\mathfrak{m}+}]) = 0.$$

Furthermore, we have

$$g(Z_{\mathfrak{m}+}, [Y\,Z'_{\mathfrak{m}+}]) = 0 \implies [Y\,Z'_{\mathfrak{m}+}] \perp_{\mathfrak{m}+} Z_{\mathfrak{m}+},$$

or

$$[Y\,Z'_{\mathfrak{m}+}] = \mu Z'_{\mathfrak{m}+}.$$

Moreover, using again the $\operatorname{ad} \mathfrak{h}$-invariance of the form g we have

$$g((\operatorname{ad} Y)Z_{\mathfrak{m}+}, Z_{\mathfrak{m}+}) + g(Z_{\mathfrak{m}+}, (\operatorname{ad} Y)Z_{\mathfrak{m}+}) = 0 \quad (\forall Y \in \mathfrak{h}^+),$$

which implies

$$2\lambda\, g(Z_{\mathfrak{m}+}, Z_{\mathfrak{m}+}) = 0,$$

whence $\lambda = 0$. (Here $[Y\,Z_{\mathfrak{m}+}] = \lambda Z_{\mathfrak{m}+}$.)

And analogously

$$g((\operatorname{ad} Y)Z'_{\mathfrak{m}+}, Z'_{\mathfrak{m}+}) + g(Z'_{\mathfrak{m}+}, (\operatorname{ad} Y)Z'_{\mathfrak{m}+}) = 0 \quad (\forall Y \in \mathfrak{h}^+)$$

implies

$$2\mu\, g(Z'_{\mathfrak{m}+}, Z'_{\mathfrak{m}+}) = 0,$$

whence $\mu = 0$. (Here $[Y\,Z'_{\mathfrak{m}+}] = \mu Z'_{\mathfrak{m}+}$.)

In such a way

$$[Y\,Z_{\mathfrak{m}+}] = [Y\,Z'_{\mathfrak{m}+}] = 0 \quad (\forall Y \in \mathfrak{h}^+).$$

But then
$$[Y\,X] = 0 \quad (\forall Y \in \mathfrak{h}^+, \ \forall X \in \mathfrak{m}^+)$$

since $Z_{\mathfrak{m}+}$, $Z'_{\mathfrak{m}+}$ constitute a base in $\mathfrak{m}^+$.

Thus $[\mathfrak{h}^+\mathfrak{m}^+] = \{0\}$, contradicting the condition $([\mathfrak{h}^+\mathfrak{m}^+] \neq \{0\})$.

Consequently $Z_{\mathfrak{m}+} = 0$. But then $[\mathfrak{m}^+\mathfrak{m}^+] = \mathbb{R}Z$, where, according to what has been obtained before, $Z \in \mathfrak{h}^+$, thus $[\mathfrak{m}^+\mathfrak{m}^+] \subset \mathfrak{h}^+$, showing that the triplet $(\mathfrak{g}^+, \mathfrak{h}^+, \mathfrak{m}^+)$ is symmetric.

Moreover, $[\mathfrak{h}^+Z] = 0$, since taking an orthogonal base X_1, X_2 on $\mathfrak{m}^+$ we have

$$g((\mathrm{ad}\,Y)X_a, X_a) + g(X_a, (\mathrm{ad}\,Y)X_a) = 0 \quad (\forall Y \in \mathfrak{h}^+),$$

or $g((\mathrm{ad}\,Y)X_a, X_a) = 0$, or $(\mathrm{ad}\,Y)X_a \perp_{\mathfrak{m}^+} X_a$ $(\forall Y \in \mathfrak{h}^+)$.

This implies
$$(\mathrm{ad}\,Y)X_1 = \lambda X_2, \qquad (\mathrm{ad}\,Y)X_2 = \mu X_1.$$

But
$$g((\mathrm{ad}\,Y)X_1, X_2) + g(X_1, (\mathrm{ad}\,Y)X_2) = 0,$$

which gives
$$\lambda\,g(X_2, X_2) + \mu\,g(X_1, X_1) = 0,$$

or $\lambda + \mu = 0$.

Thus now $(\mathrm{ad}\,Y)X_1 = \lambda X_2$, $(\mathrm{ad}\,Y)X_2 = -\lambda X_1$. But then by the Jacobi identity

$$[Y[X_1\,X_2]] = -[X_2[Y\,X_1]] - [X_1[X_2\,Y]] = -\lambda[X_2\,X_2] - \lambda[X_1\,X_1] = 0.$$

Therefore $[\mathfrak{h}^+[\mathfrak{m}^+\mathfrak{m}^+]] = \{0\}$, in particular, $[\mathfrak{h}^+\,Z] = \{0\}$, which means $\mathbb{R}Z = [\mathfrak{m}^+\mathfrak{m}^+] = \mathfrak{z}$ is a one-dimensional ideal of $\mathfrak{h}^+$ if $Z \neq 0$, or which is the same $[\mathfrak{m}^+\mathfrak{m}^+] \neq \{0\}$. Otherwise

$$B((\mathrm{ad}\,X_1)Z,\ X_2) + B(Z,\ (\mathrm{ad}\,X_1)X_2) = 0$$

implies
$$B(Z, [X_1\,X_2]) = 0, \quad \text{or} \quad B(Z, Z) = 0, \quad \text{or} \quad Z = 0,$$

which contradicts that $B|_{\mathfrak{h}^+}$ is a positive-definite form.

The orthogonal to $\mathfrak{z}$ complement $\tilde{\mathfrak{h}}^+$, with respect to the positive-definite $B|_{\mathfrak{h}^+}$, is an ideal in $\mathfrak{h}^+$ such that $\mathfrak{h}^+ = \tilde{\mathfrak{h}}^+ \oplus \mathfrak{z}$.

If $Z = 0$ then, taking

$$\tilde{\mathfrak{h}}^+ = \{\xi \in \mathfrak{h}^+ \mid [\xi\mathfrak{m}^+] = \{0\}\},$$

we obtain an ideal of $\mathfrak{h}^+$ such that $\dim \mathfrak{h}^+ = \dim \tilde{\mathfrak{h}}^+ + 1$.

Taking the orthogonal complement to $\tilde{\mathfrak{h}}^+$ in $\mathfrak{h}^+$ with respect to the positive-definite $B_{\mathfrak{h}^+}$ we obtain a one-dimensional ideal $\mathfrak{z}$ of $\mathfrak{h}^+$ such that $\mathfrak{h}^+ = \tilde{\mathfrak{h}}^+ \oplus \mathfrak{z}$.

Evidently $[\mathfrak{z}\,\mathfrak{m}^+] \neq \{0\}$ (otherwise $[\mathfrak{h}^+\mathfrak{m}^+] \neq \{0\}$, which is impossible).

Finally, we have obtained the decomposition

$$\mathfrak{g}^+ = \tilde{\mathfrak{h}}^+ \oplus (\mathfrak{z} + \mathfrak{m}^+), \qquad \mathfrak{h}^+ = \tilde{\mathfrak{h}}^+ \oplus \mathfrak{z},$$

where $\mathfrak{z} \dotplus \mathfrak{m}^+$ is a three-dimensional ideal of $\mathfrak{g}^+$, $\dim \mathfrak{z} = 1$, $\dim \mathfrak{m}^+ = 2$, and either $[\mathfrak{m}^+\mathfrak{m}^+] = \mathfrak{z}$, $[\mathfrak{z}\,\mathfrak{m}^+] \neq \{0\}$, or $[\mathfrak{m}^+\mathfrak{m}^+] = \{0\}$, $[\mathfrak{z}\,\mathfrak{m}^+] \neq \{0\}$.

Owing to the ad-invariance of the form B we then obtain $B|_{\mathfrak{m}^+} = c\,g|_{\mathfrak{m}^+}$. Indeed, if X_1, X_2 is an orthogonal base of $\mathfrak{m}^+$, with respect to B and g, which is normed with respect to g, and $Y \in \mathfrak{z}$ then

$$1 = g(X_1, X_1) = g(X_2, X_2), \qquad B(X_1, X_1) = \lambda,$$

$$B(X_2, X_2) = \mu, \qquad g(X_1, X_2) = B(X_1, X_2) = 0.$$

In addition,

$$B([Y\,X_2],\, X_1) + B(X_2,\, [Y\,X_1]) = 0$$

implies

$$B(\alpha X_1, X_1) + B(X_2, (-\alpha)X_2) = 0 \quad (\alpha \neq 0)$$

and then

$$\alpha\{B(X_1, X_1) - B(X_2, X_2)\} = 0,$$

or

$$B(X_1, X_1) = B(X_2, X_2).$$

This means that $\lambda = \mu$. But then

$$B(X_1, X_1) = c\,g(X_1, X_1), \qquad B(X_2, X_2) = c\,g(X_2, X_2),$$

which implies

$$B(X, X') = c\,g(X, X') \quad (\forall X, X' \in \mathfrak{m}^+).$$

Finally, we have

$$B|_{\mathfrak{m}^+} = c\,g|_{\mathfrak{m}^+}.$$

In the first case ($Y \in \mathfrak{z} \neq \{0\}$) we have for the orthogonal base X_1, X_2 on $\mathfrak{m}^+$

$$B([X_1\,Y], X_2) + B(Y, [X_1\,X_2]) = 0,$$

which gives

$$B(\sigma X_2, X_2) + B(Y, \tau Y) = 0 \quad (\tau \neq 0, \sigma \neq 0),$$

or

$$c = B(X_2, X_2) = -\left(\frac{\tau}{\sigma}\right) B(Y, Y) \neq 0.$$

In the second case because $[X_1\,X_2] = 0$ the same equality gives us

$$B([X_1\,Y], X_2) = 0, \qquad \text{or} \quad \sigma B(X_2, X_2) = 0 \quad (\sigma \neq 0).$$

Thus

$$c = B(X_2, X_2) = 0.$$

We present the results of this exploration in the following theorem:

IV.3.13.   Theorem 3. Let $(\mathfrak{g}, \mathfrak{h}, \mathfrak{m}, \sigma)$ be an exact mirror subsymmetric quadruplet of Riemannian type and of order two. Then for the triplet $(\mathfrak{g}^+, \mathfrak{h}^+, \mathfrak{m}^+)$ we have $\dim \mathfrak{m}^+ = 2$ and there are the following possibilities:

I. $\mathfrak{g}^+ = \mathfrak{h}^+ \oplus \mathfrak{m}^+$ (direct sum of ideals), moreover

    a) either $[\mathfrak{m}^+\mathfrak{m}^+] = \mathfrak{n} \neq \{0\}$ is a one-dimensional ideal in $\mathfrak{m}^+$ and then $B|_{\mathfrak{n}} = 0$ (solvable case)

    b) or $[\mathfrak{m}^+\mathfrak{m}^+] = \{0\}$, that is, $\mathfrak{m}^+$ is an abelian ideal of $\mathfrak{g}^+$ (abelian case).

II. $\mathfrak{g}^+ = \tilde{\mathfrak{h}}^+ \oplus (\mathfrak{z} \dotplus \mathfrak{m}^+)$ (direct sum of ideals of $\mathfrak{g}^+$), $\mathfrak{h}^+ = \tilde{\mathfrak{h}}^+ \oplus \mathfrak{z}$ (direct sum of ideals of $\mathfrak{h}^+$), $\dim \mathfrak{z} = 1$, moreover

    a) either $[\mathfrak{m}^+\mathfrak{m}^+] = \mathfrak{z}$, $\quad B|_{\mathfrak{m}^+} = c\, g|_{\mathfrak{m}^+} \quad$ $(c > 0$, elliptic case),

    b) or $[\mathfrak{m}^+\mathfrak{m}^+] = \mathfrak{z}$, $\quad B|_{\mathfrak{m}^+} = c\, g|_{\mathfrak{m}^+} \quad$ $(c < 0$, hyperbolic case),

    c) or $[\mathfrak{m}^+\mathfrak{m}^+] = \{0\}$, $\quad B|_{\mathfrak{m}^+} = c\, g|_{\mathfrak{m}^+} \quad$ $(c = 0$, parabolic case).

Thus there exist possibly five types of different two-dimensional mirrors (and exact mirror subsymmetric quadruplets) of order two.

IV.3.14.   Remark. The classification of two-dimensional mirrors presented above could be obtained from the tensor relations (IV.16)–(IV.49) as well.

IV.3.15. It is worthwhile writing out the foregoing results by means of structure constants. See (IV.8)–(IV.10). In this language we obtain

$$S^a_{bc} = S_{bc}\, S^a, \qquad a^c_{ba} = a^c_b \tau_\alpha, \qquad b^\alpha_{cd} = \tau^\alpha\, b_{cd},$$
$$c^\alpha_{\beta\gamma} \tau^\gamma = 0, \qquad \tau_\alpha = g_{\alpha\beta}\tau^\beta, \tag{IV.50}$$

where the following subcases can occur:

    1. $\tau^\alpha = 0$ (then $\tau_\beta = 0$ as well), not all S^a are zero.
       $S_{12} = -S_{21} = 1$, $\quad S_{11} = S_{22} = 0$.
       In this case $B_{ab}S^b = 0$ (solvable case);
    2. $\tau^\alpha = 0$ (then $\tau_\beta = 0$ as well), $S^a = 0$ (abelian case);
    3. Not all τ^α are zero (then not all τ_β are zero as well),
       $S^a = 0$, $B_{ab} = c\, g_{ab}$, $c > 0$ (elliptic case);
    4. Not all τ^α are zero (then not all τ_β are zero as well),
       $S^a = 0$, $B_{ab} = c\, g_{ab}$, $c < 0$ (hyperbolic case);
    5. Not all τ^α are zero (then not all τ_β are zero as well),
       $S^a = 0$, $B_{ab} = 0$ (parabolic case).

IV.3.16.   Remark. Substituting from (IV.50) into the Jacobi identities which have been received in the canonical base (see IV.3.11, (IV.8), (IV.16)–(IV.45)) and solving the equations obtained in that way one can, in principle, have the classification of mirror subsymmetric quadruplets of Riemannian type with two-dimensional mirrors.

In this way the classification of mirror subsymmetric quadruplets of Riemannian type with one-dimensional mirrors has been realized [L.V. Sabinin, 58a, 58b].

The case of a two-dimensional mirror is more complicated and it is useful to develop some geometric approach making the solution easier. This is done in what follows.

CHAPTER IV.4

MOBILE MIRRORS. ISO-INVOLUTIVE DECOMPOSITIONS

IV.4.1. Let $(\mathfrak{g}, \mathfrak{h}, \sigma)$ be a mirror subsymmetric triplet. We say that it is of *mobile type* (i.e., has a mobile mirror) if $[\mathfrak{h}^- \, \mathfrak{g}^+]$ does not belong to $\mathfrak{g}^+$. Otherwise (that is, $[\mathfrak{h}^- \, \mathfrak{g}^+] \subset \mathfrak{g}^+$) we say that it is of *immobile type* (i.e., has an immobile mirror).

Since $[\mathfrak{h}^- \, \mathfrak{g}^+] \subset \mathfrak{g}^-$ (see IV.3.6) we may also say that the triplet (mirror) is mobile if $[\mathfrak{h}^- \, \mathfrak{g}^+] \neq \{0\}$ and immobile if $[\mathfrak{h}^- \, \mathfrak{g}^+] = \{0\}$.

IV.4.2. Now we consider an exact mirror subsymmetric quadruplet of Riemannian type $(\mathfrak{g}, \mathfrak{h}, \mathfrak{m}, \sigma)$.

If it is of immobile type then $\mathfrak{h}^- = \{0\}$, otherwise $\mathfrak{h}^-$ generates a non-trivial ideal of $\mathfrak{g}$ in $\mathfrak{h}$ (see [S. Helgason, 62, 78]), which is impossible since $(\mathfrak{g}, \mathfrak{h})$ is an exact pair.

Thus in this case the mirror is immobile if and only if $\mathfrak{h}^- = \{0\}$.

If the mirror is mobile then $[\zeta \, \mathfrak{m}^+] \neq \{0\}$ for any $0 \neq \zeta \in \mathfrak{h}^-$, otherwise ζ commuting with $\mathfrak{m}^+$ again generates a non-trivial ideal of $\mathfrak{g}$ belonging to $\mathfrak{h}$, which is impossible since $(\mathfrak{g}, \mathfrak{h})$ is an exact pair.

IV.4.3. If the mirror is mobile then we can transform it in an iso-involutive way which we describe now [L.V. Sabinin, 65, 68].

Let (G, H, s) be a mirror subsymmetric triplet of Riemannian type (with a Lie group G and its subgroup H) connected with a homogeneous space $G/H = M$, and let $(\mathfrak{g}, \mathfrak{h}, \mathfrak{m}, \sigma)$ be a corresponding quadruplet of Lie algebras with a mobile mirror. Taking $0 \neq \zeta \in \mathfrak{h}^-$ and its corresponding one-dimensional subgroup $\{g(t)\}_{t \in \mathbb{R}} \subset H$ with a canonical parameter t we have evidently $g(t) = \exp \zeta t$, therefore

$$s \circ T_{g(t)} = T_{g(-t)} \circ s = T_{[g(t)]^{-1}} \circ s.$$

Indeed,

$$s \circ T_{g(t)} = s \circ T_{\exp(\zeta t)} = T_{\exp(\sigma \zeta t)} \circ s = T_{\exp(-\zeta t)} \circ s = T_{[g(t)]^{-1}} \circ s.$$

Taking the closure $T = \overline{\{g(t)\}}_{t \in \mathbb{R}}$ of $g(t)$ we obtain, since H (stationary group of a point) is closed,

$$T \subset H, \qquad s \circ T_g \circ s^{-1} = T_{g^{-1}}$$

for any $g \in T$.

Finally, it is obvious that T is a connected commutative group.

Thus T is a connected commutative closed subgroup of $H \subset G$, that is, a torus, with the property

$$s \circ T_g \circ s^{-1} = T_{g^{-1}} \quad (\forall g \in T).$$

Any torus is a product of one-dimensional tori, $T = T_1 \times \cdots \times T_m$, where $T_\alpha \cong SO(2)$.

Now we consider one such torus, for example T_1.

As a result we have proved the theorem:

IV.4.4. Theorem 4. *If the mirror of a mirror subsymmetric triplet of Riemannian type (G, H, s) is mobile then there exists a one-dimensional subgroup $T = \{h(t)\}_{t \in \mathbb{R}} \cong SO(2)$ (that is, a closed one-dimensional subgroup), $T \subset H$, such that*

$$s \circ T_{h(t)} \circ s^{-1} = T_{[h(t)]^{-1}}.$$

IV.4.5. Now let t_0 be the smallest positive number such that $T_{h(t_0)} = \mathrm{id}$ (such number exists because $\{h(t)\}_{t \in \mathbb{R}} \cong SO(2)$).

We introduce $\varphi = T_{h(t_0/4)}$ and $s_3 = T_{h(t_0/2)}$. Then $\varphi^2 = s_3$, $(s_3)^2 = \mathrm{id}$.

Let us also introduce $s_1 = s$, $s_2 = \varphi \circ s_1 \circ \varphi^{-1}$.

In this way we have obtained an iso-involutive discrete group of isometries of a Riemannian space $M = G/H$ preserving the point $o\,(= H)$ immobile.

We denote this group I.I.Gr$(s_1, s_2, s_3; \varphi)$. It consists of the elements s_1, s_2, s_3; φ, φ^{-1}, id satisfying the relations

$$(s_i)^2 = \mathrm{id}, \qquad s_i \circ s_j = s_k \quad (i \neq j,\ j \neq k,\ k \neq i),$$
$$\varphi^2 = s_3, \qquad \varphi \circ s_1 \circ \varphi^{-1} = s_2, \qquad s_1 \circ \varphi \circ s_1^{-1} = \varphi^{-1}. \tag{IV.51}$$

From these there easily follow other relations, for example,

$$\varphi \circ s_2 \circ \varphi^{-1} = s_1, \qquad \varphi \circ s_1 \circ \varphi = s_1 \qquad s_2 \circ \varphi \circ s_2^{-1} = \varphi^{-1}. \tag{IV.52}$$

Thus in the case of a mobile mirror we have proved the theorem:

IV.4.6. Theorem 5. *Any mirror subsymmetry s_1 of a point $z \in M$ of a mirror subsymmetric homogeneous space $M = G/H$ (H being compact) can be included into the iso-involutive discrete group I.I.Gr$(s_1, s_2, s_3;\ \varphi)$ with the mirror subsymmetries s_1, s_2, s_3 of the point $z \in M$.*

IV.4.7. The mirrors $K_z^{(1)}$, $K_z^{(2)}$, $K_z^{(3)}$, corresponding to mirror subsymmetries s_1, s_2, s_3, have the common intersection $K_z^{(0)} = K_z^{(i)} \underset{i \neq j}{\cap} K_z^{(j)} = K_z^{(1)} \cap K_z^{(2)} \cap K_z^{(3)}$

and are pairwise orthogonal at $z \in M$ out of its common part $K_z^{(0)}$. See [L.V. Sabinin 65, 68, 70, 72].

More generally, we can introduce the following:

IV.4.8. Definition 70. If three mirror subsymmetries $\underset{x}{s_1}, \underset{x}{s_2}, \underset{x}{s_3}$ (with the immobile point $x \in M$) of a homogeneous Riemannian space $M = G/H$ (H being the stabilizer of $x \in M$) with the properties

$$(\underset{x}{s_i})^2 = \text{id}, \qquad \underset{x}{s_i} \circ \underset{x}{s_j} = \underset{x}{s_k} \quad (i \neq j, \ j \neq k, \ k \neq i) \qquad \text{(IV.53)}$$

are given then the group generated by these subsymmetries is called an involutive group and is denoted $\text{I.Gr}(\underset{x}{s_1}, \underset{x}{s_2}, \underset{x}{s_3})$.

For the homogeneous mirror subsymmetric space $M = G/H$ with an involutive group of automorphisms $\text{I.Gr}(\underset{x}{s_1}, \underset{x}{s_2}, \underset{x}{s_3})$ we use the notation $(G, H, \text{I.Gr}(\underset{x}{s_1}, \underset{x}{s_2}, \underset{x}{s_3}))$, or $(G, H, \underset{x}{s_1}, \underset{x}{s_2}, \underset{x}{s_3})$, for short.

Correspondingly, in the iso-involutive case we write $(G, H, \text{I.I.Gr}(\underset{x}{s_1}, \underset{x}{s_2}, \underset{x}{s_3}, \varphi))$, or more briefly $(G, H, \underset{x}{s_1}, \underset{x}{s_2}, \underset{x}{s_3}, \varphi)$.

IV.4.9. Definition 71. A homogeneous space M/G is called tri-symmetric if it has an involutive group $\text{I.Gr}(\underset{x}{s_1}, \underset{x}{s_2}, \underset{x}{s_3})$ such that in some neighbourhood of $x \in M$ the intersection of the corresponding mirrors consists of a one point,

$$K_x^{(0)} = K_x^{(1)} \cap K_x^{(2)} \cap K_x^{(3)} = \{x\}.$$

IV.4.10. The constructions introduced above for a homogeneous space G/H can be repeated for the doublet of the corresponding Lie algebras $(\mathfrak{g}, \mathfrak{h})$.

Indeed, an involutive group $\text{I.Gr}(\underset{x}{s_1}, \underset{x}{s_2}, \underset{x}{s_3})$ of a space G/H generates the involutive group of automorphisms of the Lie algebra $\mathfrak{g}$, $\text{I.Gr}(\sigma_1, \sigma_2, \sigma_3)$, and since $\underset{x}{s_i} H = H$ we have $\sigma_i \mathfrak{h} = \mathfrak{h}$.

Thus we have obtained a mirror triplet $(\mathfrak{g}, \mathfrak{h}, \text{I.Gr}(\sigma_1, \sigma_2, \sigma_3)$ (or more briefly $(\mathfrak{g}, \mathfrak{h}, \sigma_1, \sigma_2, \sigma_3)$ with the properties

$$\sigma_i \mathfrak{h} = \mathfrak{h}, \qquad \sigma^2 = \text{id}, \qquad \sigma_i \sigma_j = \sigma_k \quad (i \neq j, \ j \neq k, \ k \neq i).$$

IV.4.11. Definition 72. A mirror triplet $(\mathfrak{g}, \mathfrak{h}, \text{I.Gr}(\sigma_1, \sigma_2, \sigma_3))$ is said to be tri-symmetric if

$$\mathfrak{g}_1^+ \cap \mathfrak{g}_2^+ = \mathfrak{g}_2^+ \cap \mathfrak{g}_3^+ = \mathfrak{g}_3^+ \cap \mathfrak{g}_1^+ = \mathfrak{g}_1^+ \cap \mathfrak{g}_2^+ \cap \mathfrak{g}_3^+ \subset \mathfrak{h}$$

(here $\mathfrak{g}_i^+ = \{x \in \mathfrak{g} \mid \sigma_i x = x\}$).

IV.4.12. In what follows, we are interested in *reductive* triplets $(\mathfrak{g}, \mathfrak{h}, \text{I.Gr}(\sigma_1, \sigma_2, \sigma_3))$, $\mathfrak{g}$ and $\mathfrak{h}$ being Lie algebras, that is, we consider such triplets that $\mathfrak{g} = \mathfrak{m} + \mathfrak{h}$, $[\mathfrak{m} \, \mathfrak{h}] \subset \mathfrak{m}$, $\sigma_i \mathfrak{m}_i = \mathfrak{m}_i$ ($i = 1, 2, 3$).

The latter inclusion is valid for a reductive pair $(\mathfrak{g}, \mathfrak{h})$ if we introduce another $\mathfrak{m}$.

Indeed, if $\mathfrak{g} = \mathfrak{m} + \mathfrak{h}$, $\sigma_1 \mathfrak{h} = \mathfrak{h}$, where σ_1 is an involutive automorphism, then we can consider $\mathfrak{m} = \mathfrak{n} + (\mathfrak{m} \cap \sigma_1 \mathfrak{m})$ and furthermore $\tilde{\mathfrak{m}} = \mathfrak{a} + \mathfrak{c}$, where

$$\mathfrak{a} = \{x + \sigma_1 x \mid x \in \mathfrak{n}\}, \qquad \mathfrak{c} = (\mathfrak{m} \cap \sigma_1 \mathfrak{m}).$$

Evidently $\mathfrak{g} = \tilde{\mathfrak{m}} \dotplus \mathfrak{h}$, $[\tilde{\mathfrak{m}}\,\mathfrak{h}] \subset \mathfrak{m}$.

In addition,

$$\mathfrak{a} = \{x + \sigma_1 x \mid x \in \mathfrak{m}\} \subset \mathfrak{a} \dotplus \mathfrak{c} = \tilde{\mathfrak{m}},$$

which is easily seen. Thus $\tilde{\mathfrak{m}} = \tilde{\mathfrak{a}} \dotplus \mathfrak{c}$.

But $\sigma_1 \tilde{\mathfrak{a}} = \tilde{\mathfrak{a}}$, consequently $\sigma_1 \tilde{\mathfrak{m}} = \tilde{\mathfrak{m}}$.

Now, starting from $\tilde{\mathfrak{m}}$, we take

$$\tilde{\tilde{\mathfrak{m}}} = \mathfrak{t} \dotplus \mathfrak{d}, \qquad \mathfrak{t} = \{x + \sigma_2 x \mid x \in \tilde{\mathfrak{m}}\}, \qquad \mathfrak{d} = (\tilde{\mathfrak{m}} \cap \sigma_2 \tilde{\mathfrak{m}}).$$

Then, as before,

$$\mathfrak{g} = \tilde{\tilde{\mathfrak{m}}} \dotplus \mathfrak{h}, \qquad [\tilde{\tilde{\mathfrak{m}}}\,\mathfrak{h}] \subset \tilde{\tilde{\mathfrak{m}}}, \qquad \sigma_2 \tilde{\tilde{\mathfrak{m}}} = \tilde{\tilde{\mathfrak{m}}}.$$

Moreover, since $\sigma_1 \sigma_2 = \sigma_2 \sigma_1 = \sigma_3$ and $\sigma_1 \tilde{\mathfrak{m}} = \tilde{\mathfrak{m}}$ we have

$$\sigma_1 \mathfrak{t} = \{\sigma_1 x + \sigma_2(\sigma_1 x) \mid x \in \tilde{\mathfrak{m}}\} = \{y + \sigma_2 y \mid y \in \tilde{\mathfrak{m}}\} = \mathfrak{t},$$

$$\sigma_1 \mathfrak{d} = (\sigma_1 \tilde{\mathfrak{m}} \cap \sigma_1 \sigma_2 \tilde{\mathfrak{m}}) = (\sigma_1 \tilde{\mathfrak{m}} \cap \sigma_2 \sigma_1 \tilde{\mathfrak{m}}) = (\tilde{\mathfrak{m}} \cap \sigma_2 \tilde{\mathfrak{m}}) = \mathfrak{d}.$$

Thus $\sigma_1 \tilde{\tilde{\mathfrak{m}}} = \tilde{\tilde{\mathfrak{m}}}$.

As a result

$$\mathfrak{g} = \tilde{\tilde{\mathfrak{m}}} \dotplus \mathfrak{h}, \qquad [\tilde{\tilde{\mathfrak{m}}}\,\mathfrak{h}] \subset \tilde{\tilde{\mathfrak{m}}},$$

$$\sigma_1 \tilde{\tilde{\mathfrak{m}}} = \tilde{\tilde{\mathfrak{m}}}, \qquad \sigma_2 \tilde{\tilde{\mathfrak{m}}} = \tilde{\tilde{\mathfrak{m}}}, \qquad \sigma_3 \tilde{\tilde{\mathfrak{m}}} = \sigma_1 \sigma_2 \tilde{\tilde{\mathfrak{m}}} = \tilde{\tilde{\mathfrak{m}}}.$$

From what has been obtained above we always should take $\mathfrak{g} = \mathfrak{m} \dotplus \mathfrak{h}$, $[\mathfrak{m}\,\mathfrak{h}] \subset \mathfrak{m}$, in such a way that $\sigma_\alpha \mathfrak{m} = \mathfrak{m}$ $(\alpha = 1, 2, 3)$.

In this case we simply say that a reductive quadruplet $(\mathfrak{g}, \mathfrak{h}, \mathfrak{m}, \sigma_1, \sigma_2, \sigma_3)$ is given.

IV.4.13. Remark. In the case when the Cartan metric B is non-degenerate on $\mathfrak{h}$ (for example, if $\mathfrak{h}$ is semi-simple or compact, see [S. Helgason 62, 78]) we can simply take a unique $\mathfrak{m} \perp_B \mathfrak{h}$. Then

$$\mathfrak{g} = \mathfrak{m} \dotplus \mathfrak{h}, \qquad [\mathfrak{m}\,\mathfrak{h}] \subset \mathfrak{m}, \qquad \sigma_\alpha \mathfrak{m} = \mathfrak{m} \quad (\alpha = 1, 2, 3),$$

because of the invariance of the Cartan metric with respect to $\operatorname{ad} \zeta$ and σ_α ($\alpha = 1, 2, 3$).

Briefly, we frequently say a σ_α-invariant doublet $(\mathfrak{g}, \mathfrak{h})$, or a σ_α-invariant triplet $(\mathfrak{g}, \mathfrak{h}, \mathfrak{m})$, instead of $(\mathfrak{g}, \mathfrak{h}, \sigma_1, \sigma_2, \sigma_3)$, or $(\mathfrak{g}, \mathfrak{h}, \mathfrak{m}, \sigma_1, \sigma_2, \sigma_3)$, respectively.

CHAPTER IV.5

HOMOGENEOUS RIEMANNIAN SPACES
WITH TWO-DIMENSIONAL MIRRORS

IV.5.1. In the case of a simple compact Lie group, taking into account IV.3.13 (Theorem 1), the positive-definiteness of a bilinear metric form g, and the positive-definiteness of the Cartan metric for a compact simple Lie group, we see that the only possible cases are the abelian case I b) or the elliptic case II a).

Let us show that the abelian case can not occur.

Owing to the simplicity of the Lie algebra $\mathfrak{g}$ we have an exact involutive pair $(\mathfrak{g}, \mathfrak{g}^+)$, where $\mathfrak{g}^+ = \mathfrak{m}^+ \dotplus \mathfrak{h}^+$ and $\mathfrak{m}^+$ is a two-dimensional abelian ideal of $\mathfrak{g}^+$ ($\mathfrak{g}^+ = \mathfrak{m}^+ \oplus_{\mathfrak{g}^+} \mathfrak{h}^+$). This means that $\mathfrak{m}^+$ belongs to the centre of $\mathfrak{g}^+$, $\dim \mathfrak{m}^+ = 2$.

There exists $0 \neq \zeta \in \mathfrak{m}^+$ such that $\mathrm{ad}|_{\mathfrak{g}^-} \zeta$ is a degenerate non-zero endomorphism. Indeed, $\mathrm{ad}\,\eta|_{\mathfrak{g}^-} \neq 0$ for any $\eta \neq 0$, $\eta \in \mathfrak{m}^+$ (otherwise, $\mathrm{ad}\,\eta|_{\mathfrak{g}} = 0$, because $\mathrm{ad}\,\eta|_{\mathfrak{g}^+} = 0$, implying the existence of a non-trivial centre of $\mathfrak{g}$, which is impossible).

Now let $\xi, \eta \in \mathfrak{m}^+$ be linearly independent (that is, $\{\xi, \eta\}$ is a base on $\mathfrak{m}^+$).

If $\mathrm{ad}\,\xi|_{\mathfrak{g}^-}$ (or $\mathrm{ad}\,\eta|_{\mathfrak{g}^-}$) is degenerate then we define $\zeta = \xi$ (or $\zeta = \eta$).

If $(\mathrm{ad}\,\xi|_{\mathfrak{g}^-})$ and $(\mathrm{ad}\,\eta|_{\mathfrak{g}^-})$ are non-degenerate then

$$\mathrm{ad}(\xi - \lambda\eta)|_{\mathfrak{g}^-} = \mathrm{ad}\,\xi|_{\mathfrak{g}^-} - \lambda\,\mathrm{ad}\,\eta|_{\mathfrak{g}^-}$$

is degenerate for some λ. Indeed, its degeneracy is equivalent to the degeneracy of

$$(\mathrm{ad}\,\xi|_{\mathfrak{g}^-})((\mathrm{ad}\,\eta|_{\mathfrak{g}^-})^{-1}) - \lambda\,\mathrm{Id},$$

which means

$$\det((\mathrm{ad}\,\xi|_{\mathfrak{g}^-})(\mathrm{ad}\,\eta|_{\mathfrak{g}^-})^{-1} - \lambda\,\mathrm{Id}) = 0.$$

Consequently we should take $\lambda = \lambda_0 \in \mathbb{R}$ which is a proper value of the endomorphism $F = (\mathrm{ad}\,\xi|_{\mathfrak{g}^-})(\mathrm{ad}\,\eta|_{\mathfrak{g}^-})^{-1}$.

Such $\lambda \in \mathbb{R}$ exists since F is self-adjoint with respect to the positive-definite Cartan–Killing metric form B. Indeed, owing to the ad-invariance of B

$$
\begin{aligned}
B((\mathrm{ad}\,\xi|_{\mathfrak{g}^-})(\mathrm{ad}\,\eta|_{\mathfrak{g}^-})^{-1}x, y) &= -B((\mathrm{ad}\,\eta|_{\mathfrak{g}^-})^{-1}x, (\mathrm{ad}\,\xi|_{\mathfrak{g}^-})y) \\
&= B(x, (\mathrm{ad}\,\eta|_{\mathfrak{g}^-})^{-1}(\mathrm{ad}\,\xi|_{\mathfrak{g}^-})y) \\
&= B(x, (\mathrm{ad}\,\xi|_{\mathfrak{g}^-})(\mathrm{ad}\,\eta|_{\mathfrak{g}^-})^{-1}y).
\end{aligned}
$$

(We have used that the inverse of a self-adjoint endomorphism (if exists) is self-adjoint, and also the commutativity, $[\xi\,\eta] = 0$, implying $\mathrm{ad}\,\xi\,\mathrm{ad}\,\eta = \mathrm{ad}\,\eta\,\mathrm{ad}\,\xi$.)

Thus taking $\zeta = \xi - \lambda_0 \eta$ we see that $\mathrm{ad}\,\zeta|_{\mathfrak{g}^-} \neq 0$ ($\zeta \in \mathfrak{m}^+$) and is degenerate. Furthermore, for any $\tau \in \mathfrak{g}^+$, $\mathrm{ad}\,\tau|_{\mathfrak{g}^-}$ and $\mathrm{ad}\,\zeta|_{\mathfrak{g}^-}$ commute, whence

$$\mathfrak{g}^0 = \{x \in \mathfrak{g}^- \mid [\zeta\,x] = 0\}$$

is invariant under the action of $\mathrm{ad}\,\mathfrak{g}^+$ (owing to the construction it is evident that $\mathfrak{g}^- \supsetneq \mathfrak{g}^0 \supsetneq \{0\}$).

Let us introduce $\tilde{\mathfrak{g}}$ which is the orthogonal complement of $\mathfrak{g}^0$ in $\mathfrak{g}^-$ with respect to the Cartan metric B. Then $\tilde{\mathfrak{g}}$ is $\mathrm{ad}\,\mathfrak{g}^+$-invariant because $\mathfrak{g}^0$ and B are $\mathrm{ad}\,\mathfrak{g}^+$-invariant. Thus

$$\mathfrak{g}^0 \dotplus \tilde{\mathfrak{g}} = \mathfrak{g}^-, \qquad [\mathfrak{g}^0\,\mathfrak{g}^+] \subset \mathfrak{g}^0, \qquad [\tilde{\mathfrak{g}}\,\mathfrak{g}] \subset \tilde{\mathfrak{g}}.$$

Now we consider $[\mathfrak{g}^0\,\tilde{\mathfrak{g}}] = \mathfrak{p}$.

Evidently $\mathfrak{p} \subset \mathfrak{g}^+$, since $\mathfrak{g}^0$ and $\tilde{\mathfrak{g}}$ are subspaces of $\mathfrak{g}^-$, and $(\mathfrak{g}, \mathfrak{g}^+)$ is an involutive pair.

Furthermore, for $z \in \mathfrak{g}^+$, $a \in \mathfrak{g}^0$, $q \in \tilde{\mathfrak{g}}$, owing to the ad-invariance of the Cartan metric B we have

$$B([q\,z],\,a) + B(z,\,[q\,a]) = 0.$$

That is, $[q\,a] \in \mathfrak{p} \subset \mathfrak{g}^+$ is orthogonal to $\mathfrak{g}^+$, hence $[q\,a] = 0$.

Thus $[\mathfrak{g}^0\,\tilde{\mathfrak{g}}] = \{0\}$. Now taking $\mathfrak{g}^0 \dotplus \langle [\mathfrak{g}^0\,\mathfrak{g}^0] \rangle = \mathfrak{q}$ we see that $\mathfrak{q}$ is an ideal of $\mathfrak{g}$, since, because of what has been said above and Jacobi identities, we have

$$[\mathfrak{q}\,\tilde{\mathfrak{g}}] = \{0\} \subset \mathfrak{q},$$

$$[\mathfrak{q}\,\mathfrak{g}^+] \subset \mathfrak{q}, \qquad [\mathfrak{q}\,\mathfrak{g}^0] \subset \mathfrak{g}^0 \subset \mathfrak{q}.$$

Obviously $\mathfrak{q} \subsetneq \mathfrak{g}$ since $\tilde{\mathfrak{g}} \not\subset \mathfrak{q}$, $\tilde{\mathfrak{g}} \neq \{0\}$, and $\mathfrak{q} \neq \{0\}$ since $\mathfrak{g}^0 \neq \{0\}$.

Thus $\mathfrak{q}$ is a proper ideal of a simple algebra $\mathfrak{g}$, which is impossible.

We have obtained:

IV.5.2. Theorem 6. *A two-dimensional mirror of a Riemannian homogeneous space G/H with a simple compact Lie group G is of elliptic type.*

IV.5.3. Remark. In the consideration presented above we actually have used the results of E. Cartan (See [E. Cartan, 49, 52], [S. Helgason, 62, 78]) about the reducibility of the group of rotations of a symmetric Riemannian space.

In particular, we have repeated in the compact case the following result of [E. Cartan, 49]:

IV.5.4. Theorem 7. *The centre of the group of rotations of a symmetric Riemannian space G/H with a simple Lie group G is at most one-dimensional.*

Or, analogously, in the Lie algebras setting:

IV.5.5. Theorem 8. *Let $(\mathfrak{g}, \mathfrak{h})$ be an involutive pair of Lie algebras, where $\mathfrak{g}$ is simple and $\mathfrak{h}$ is its compact subalgebra. Then the centre of $\mathfrak{h}$ is at most one-dimensional.*

IV.5.6. In the case of a homogeneous Riemannian space with a simple compact group of motions and with two-dimensional mirrors there exist two types of mirrors: *unitary* and *orthogonal*. Indeed, we know by IV.5.2 (Theorem 6) that in this case a two-dimensional mirror is of elliptic type. This means that for the corresponding doublet $(\mathfrak{g}, \mathfrak{h})$ we have

$$\mathfrak{g}^+ = (\mathfrak{m}^+ \dotplus \mathfrak{z}) \oplus_{\mathfrak{g}^+} \tilde{\mathfrak{h}},$$

where

$$\mathfrak{z} \oplus_{\mathfrak{h}^+} \tilde{\mathfrak{h}} = \mathfrak{h}^+,$$

and $\mathfrak{z}$ is the one-dimensional centre of $\mathfrak{h}^+$, moreover, $\mathfrak{m}^+ \dotplus \mathfrak{z} = \mathfrak{b}$ is a three-dimensional simple compact subalgebra.

Thus

$$\mathfrak{b} \cong so(3) \cong su(2) \cong sp(1).$$

Let us now consider $e^{\operatorname{ad}\mathfrak{b}} = B$. This is a simple compact connected three-dimensional Lie group which is a subgroup of Aut $\mathfrak{g}$. Such a group is isomorphic either to $SO(3)$ or to $SU(2)$. In the first case we say that a mirror (and a corresponding involutive automorphism) is of *orthogonal type* (*of type O*), in the second case we say that a mirror is of *unitary type* (*of type U*).

IV.5.7. Remark. Instead of $e^{\operatorname{ad}\mathfrak{b}} = B \subset \text{Aut}\,\mathfrak{g}$ we might have used $\exp\mathfrak{b}$, a subgroup of the Lie group $G = \exp\mathfrak{g}$, which again is isomorphic to either $SO(3)$ or $SU(2)$.

IV.5.8. Remark. We note that $SO(3) \cong SU(2)/Z_2$, where $Z_2 = \{\text{Id}, -\text{Id}\}$ is the discrete centre of $SU(2)$.

Here $(-\text{Id})$ is the unique non-trivial involutive element of $SU(2)$. Moreover, any one-dimensional subgroup of $SU(2)$ passes through this element.

IV.5.9. By IV.5.6 the involutive automorphism s, related to a two-dimensional mirror of unitary type of a pair $(\mathfrak{g}, \mathfrak{h})$ with a simple compact Lie algebra $\mathfrak{g}$, coincides with the involutive central element $\sigma \in e^{\operatorname{ad}\mathfrak{b}} \cong SU(2)$ (we recall that $\mathfrak{b} = (\mathfrak{m}^+ \dotplus \mathfrak{z})$, $\mathfrak{g}^+ = \mathfrak{b} \oplus_{\mathfrak{g}^+} \tilde{\mathfrak{h}}$, $\mathfrak{h}^+ = \tilde{\mathfrak{h}} \oplus_{\mathfrak{h}^+} \mathfrak{z}$).

Indeed, σ is an involutive automorphism of a Lie algebra $\mathfrak{g}$ preserving any $\zeta \in \mathfrak{g}^+$ (since, owing to IV.5.7, σ belongs to any one-dimensional subgroup $\{e^{\operatorname{ad} t\zeta}\}_{t\in\mathbb{R}}$ of $e^{\operatorname{ad}\mathfrak{b}} \cong SU(2)$ we obtain $\sigma = e^{\tau\operatorname{ad}\eta}$ for any $\eta \in \mathfrak{b}, \tau \in \mathbb{R}$). And since $\mathfrak{g}^+ \ni \zeta = \zeta_1 + \zeta_2$ ($\zeta_1 \in \mathfrak{b}$, $\zeta_2 \in \tilde{\mathfrak{h}}$) and $(\operatorname{ad}\zeta_1)\zeta_1 = 0$, $(\operatorname{ad}\zeta_1)\zeta_2 = 0$ we have

$$\sigma\zeta = \sigma\zeta_1 + \sigma\zeta_2 = (e^{\tau\operatorname{ad}\zeta_1})\zeta_1 + (e^{\tau\operatorname{ad}\zeta_1})\zeta_2$$

$$= \left(\zeta_1 + \frac{\tau}{1!}(\operatorname{ad}\zeta_1)\zeta_1 + \dots\right) + \left(\zeta_2 + \frac{\tau}{1!}(\operatorname{ad}\zeta_1)\zeta_2 + \dots\right) = \zeta_1 + \zeta_2 = \zeta.$$

If we assume that the subalgebra $\mathfrak{i}$ of immobile elements of the involutive automorphism σ is larger than $\mathfrak{g}^+$ then this contradicts the maximality of the involutive algebra of a simple Lie algebra (see [E. Cartan, 49, 52], [S. Helgason, 62, 78]). The proof is reduced to the theorem of E. Cartan on the irreducibility of the $\mathrm{ad}\,\mathfrak{g}^+$-action on $\mathfrak{g}$ in the case of a simple compact Lie algebra $\mathfrak{g} = \mathfrak{g}^+ + \mathfrak{g}^-$. Indeed, $\mathfrak{i} = \mathfrak{g}^+ + \mathfrak{l}$, where $\mathfrak{l} \subset \mathfrak{g}^-$, and then $[\mathfrak{g}^+\,\mathfrak{l}] \subset \mathfrak{l}$ since $[\mathfrak{g}^+\,\mathfrak{i}] \subset \mathfrak{i}$ and $\mathfrak{g}^+ \subset \mathfrak{i}$.

In addition, $\mathfrak{k} = \mathfrak{i} + \langle[\mathfrak{i}\,\mathfrak{i}]\rangle$ is an ideal of $\mathfrak{g}$, which is possible only if $\mathfrak{k} = \{0\}$ or $\mathfrak{k} = \mathfrak{g}$. The first case implies $\mathfrak{l} = \{0\}$, contradicting the assumptions. The second case implies $\mathfrak{l} = \mathfrak{g}^-$, $\langle[\mathfrak{l}\,\mathfrak{l}]\rangle = \mathfrak{g}^+$, which is impossible because then $\mathfrak{i} = \mathfrak{g}^+ + \mathfrak{l} = \mathfrak{g}^+ + \mathfrak{g}^- = \mathfrak{g}$, $\sigma = \mathrm{id}$.

IV.5.10. Remark. The involutive subalgebras of the form $\mathfrak{h} = \mathfrak{b} \oplus_\mathfrak{h} \tilde{\mathfrak{h}}$, where $\mathfrak{b}$ is a three-dimensional simple compact subalgebra, have been considered and studied in [L.V. Sabinin 69,70]. See on this matter also Part I and II of this monograph.

Such involutive subalgebras and their corresponding involutive automorphisms and involutive pairs are called principal (see Parts I and II of this monograph).

Moreover, we have to distinguish between *principal orthogonal (of type O)* and *principal unitary (of type U)* cases when $e^{\mathrm{ad}\,\mathfrak{b}} \cong SO(3)$ or $e^{\mathrm{ad}\,\mathfrak{b}} \cong SU(2)$, respectively.

IV.5.11. We now consider a homogeneous Riemannian space $M = G/H$ with a mobile two-dimensional mirror of unitary type and a simple compact Lie group G. Then an 'iso-involutive rotation' of such a mirror gives us a new mirror whose intersection with the initial mirror is trivial (that is, consists locally of one point), and consequently G/H is a tri-symmetric space.

We prove this assertion in terms of Lie algebras.

IV.5.12. Theorem 9. *Let $\mathfrak{g}$ be a simple compact Lie algebra, $(\mathfrak{g}, \mathfrak{h})$ be a doublet of Lie algebras with a mobile mirror $\mathfrak{m}^+$ $(\mathfrak{g}^+ = \mathfrak{m}^+ + \mathfrak{h}^+)$, $\dim \mathfrak{m}^+ = 2$, generated by a principal unitary involutive automorphism σ.*

Then there exists the iso-involutive group,

$$\sigma_1 = \sigma, \quad \sigma_2 = \varphi \circ \sigma_1 \circ \varphi^{-1}, \quad \sigma_3 = \varphi^2, \quad \varphi, \quad \varphi^{-1}, \quad \mathrm{id},$$

such that $\mathfrak{g}$ and $\mathfrak{h}$ are invariant under its action, and there exists the corresponding iso-involutive decomposition

$$\mathfrak{g} = \mathfrak{g}_1^+ + \mathfrak{g}_2^+ + \mathfrak{g}_3^+ \quad (\mathfrak{g}_i^+ = \{x \in \mathfrak{g} \mid \sigma_i\,x = x\})$$

which is tri-symmetric, that is,

$$\mathfrak{g}_1^+ \cap \mathfrak{g}_2^+ = \mathfrak{g}_2^+ \cap \mathfrak{g}_3^+ = \mathfrak{g}_3^+ \cap \mathfrak{g}_1^+ = \mathfrak{g}^0 \subset \mathfrak{h}.$$

Proof. We note that $\mathfrak{g} = \mathfrak{m} + \mathfrak{h}$ (reductive decomposition) and

$$\mathfrak{m} = \mathfrak{m}_1^+ + \mathfrak{m}_2^+ + \mathfrak{m}_3^+, \qquad \mathfrak{g}_i^+ = \mathfrak{m}_i^+ + \mathfrak{h}_i^+ \quad (i = 1, 2, 3).$$

The condition of the tri-symmetry means

$$\mathfrak{m}_1^+ \cap \mathfrak{m}_2^+ = \mathfrak{m}_2^+ \cap \mathfrak{m}_3^+ = \mathfrak{m}_3^+ \cap \mathfrak{m}_1^+ = \{0\}.$$

Suppose that $\mathfrak{m}_1^+ \cap \mathfrak{m}_2^+ \neq \{0\}$. Then there exists at least a one-dimensional intersection $\mathfrak{m}^0 = \mathfrak{m}_1^+ \cap \mathfrak{m}_2^+$. However, $\mathfrak{g}_i^+ = \mathfrak{m}_i^+ + \mathfrak{h}_i^+ = (\mathfrak{m}_i^+ + \mathfrak{z}_i) \oplus \tilde{\mathfrak{h}}_i^+$, where $\mathfrak{m}_i^+ + \mathfrak{z}_i = \mathfrak{b}_i$ is a three-dimensional simple compact ideal of $\mathfrak{g}_i^+$. Therefore $\mathfrak{m}^0 = \mathfrak{b}_1 \cap \mathfrak{b}_2$ under our assumptions. Since σ_1 and σ_2 are principal unitary involutive automorphisms (owing to the conjugacy of σ_1 and σ_2 by φ in $\mathrm{Aut}\,\mathfrak{g}$) then $e^{\mathrm{ad}\,\mathfrak{b}_1} \cong e^{\mathrm{ad}\,\mathfrak{b}_2} \cong SU(2)$. Taking $\zeta \in \mathfrak{m}^0$ we obtain $\sigma_1 = e^{\tau\,\mathrm{ad}\,\zeta}$. But then $\sigma_1 \in e^{\mathrm{ad}\,\mathfrak{b}_2}$, $\sigma_2 \in e^{\mathrm{ad}\,\mathfrak{b}_2}$. Since $e^{\mathrm{ad}\,\mathfrak{b}_2}$ contains the only non-trivial automorphism σ_2 (because in its isomorphic image $SU(2)$ this is true) we have $\sigma_1 = \sigma_2$. The latter is impossible by the construction of an iso-involutive group (see I.1.22 (Definition 1.3)). ∎

The proof of the analogous result in the group language follows immediately.

IV.5.13. Theorem 10. *Let $\mathfrak{g}$ be a simple compact Lie algebra non-isomorphic to $so(n)$ or $su(3)$, and $(\mathfrak{g}, \mathfrak{h})$ be a doublet having a two-dimensional mobile mirror of an involutive automorphism σ.*

Then there exists the iso-involutive group,

$$\sigma_1 = \sigma, \quad \sigma_2 = \varphi \circ \sigma_1 \circ \varphi^{-1}, \quad \sigma_3 = \varphi^2, \quad \varphi, \quad \varphi^{-1}, \quad \mathrm{id},$$

and the corresponding iso-involutive decomposition $\mathfrak{g} = \mathfrak{g}_1^+ + \mathfrak{g}_2^+ + \mathfrak{g}_3^+$ such that the pair $(\mathfrak{g}, \mathfrak{h})$ is invariant under the action of this group and is tri-symmetric.

Proof. All simple compact Lie algebras over $\mathbb{R}$ are known, as well as their involutive automorphisms, see [S. Helgason, 62, 78].

Then we can select all principal involutive automorphisms. The quite remarkable result is that if $\mathfrak{g}$ is non-isomorphic to $so(n)$ or $su(3)$, that is,

$$\mathfrak{g} \cong su(n) \ (n \geq 4), \quad sp(n) \ (n \geq 2), \quad g_2, \quad f_4, \quad e_6, \quad e_7, \quad e_8$$

then all its principal involutive automorphisms are unitary [L.V. Sabinin, 69, 70a, 70b].

Now, by virtue of IV.5.12 (Theorem 9) we obtain the desired result. ∎

IV.5.14. Remark. Of course, $so(n)$, along with the principal involutive algebra $so(3) \oplus so(n-3)$ (with the natural embedding) of type O, has also the principal unitary involutive algebra

$$su(2) \oplus su(2) \oplus so(n-4) = so(4) \oplus so(n-4)$$

(with the natural embedding).

Analogously $su(3)$ has the principal orthogonal involutive algebra $so(3)$ and the principal unitary algebra

$$s(u(2) \oplus u(1))$$

(with the natural embeddings).

Therefore the Theorem 10 (see IV.5.13) can be extended up to $so(n)$ with the involutive algebra $so(4) \oplus so(n-4)$ and up to $su(3)$ with the involutive algebra $s(u(2) \oplus u(1))$.

However, in what follows we do not use this result.

CHAPTER IV.6

HOMOGENEOUS RIEMANNIAN SPACES WITH GROUPS
SO(n), SU(3) AND TWO-DIMENSIONAL MIRRORS

IV.6.1. We are going to classify all homogeneous Riemannian spaces with simple compact Lie groups of motions. In particular, we will use for this purpose the well known classification of pairs $(\mathfrak{g}, \mathfrak{h})$ with a simple compact Lie algebra $\mathfrak{g}$ having a tri-symmetric decomposition, see Part III of this monograph and [L.V. Sabinin, 72].

Therefore we now consider the case when a pair $(\mathfrak{g}, \mathfrak{h})$ with a two-dimensional mobile mirror, perhaps, have no tri-symmetric decomposition. As before, we consider a simple compact Lie algebra $\mathfrak{g}$. According to the results of IV.5.3 such case is possible only if

$$\mathfrak{g} \cong so(n), \qquad \mathfrak{g}^+ = \mathfrak{b} \oplus \tilde{\mathfrak{h}} \cong so(3) \oplus so(n-3)$$

or

$$\mathfrak{g} \cong su(3), \qquad \mathfrak{g}^+ = \mathfrak{b} \oplus \tilde{\mathfrak{h}} \cong so(3) \oplus \{0\}$$

(with the natural embeddings).

(Of course, for the algebras $so(n)$, $su(3)$ the tri-symmetric cases are possible, in principle.)

Let us explore first $(\mathfrak{g}, \mathfrak{g}^+) = (so(n), so(3) \oplus so(n-3))$.

Since $\mathfrak{h}^+ = \mathfrak{z} \oplus \tilde{\mathfrak{h}} \cong so(2) \oplus so(n-3)$ the subalgebra $\mathfrak{h}$ has the central involutive algebra $\mathfrak{h}^+$ (with the one-dimensional centre $\mathfrak{z} \cong so(2)$). The compact involutive pair $(\mathfrak{h}, \mathfrak{h}^+)$ possesses a decomposition into elementary involutive pairs,

$$\mathfrak{h} / \mathfrak{h}^+ = \sum_{i=1}^{N} (\mathfrak{h}_i / \mathfrak{h}_i^+),$$

which means that

$$\mathfrak{h} = \oplus_{i=1}^{N} \mathfrak{h}_i, \qquad \mathfrak{h}^+ = \oplus_{i=1}^{N} \mathfrak{h}_i^+$$

are direct sums of ideals and a further decomposition is impossible, that is, either $\mathfrak{h}_i$ is simple (or one-dimensional) or $\mathfrak{h}_i$ is isomorphic to the direct product of a simple Lie algebra with itself, $\mathfrak{h}^+$ being a diagonal of such a product. (See I.1.2.)

The case $n = 4$ is impossible since then $\mathfrak{g} \cong so(4)$ which is not simple.

The case $n = 3$ is also impossible since then there are no two-dimensional mirrors.

Let us now exclude the case $n > 5$.

If $n > 5$ then $so(n-3)$ is either simple or semi-simple (when $n = 7$, $so(4) = su(2) \oplus su(2)$).

Therefore we have the decomposition

$$\mathfrak{h}/\mathfrak{h}^+ = (\mathfrak{h}_1, \mathfrak{z} \oplus \tilde{\mathfrak{h}}_1) + (\mathfrak{h}_2, \tilde{\mathfrak{h}}_2)$$

(here $\mathfrak{z} \cong so(2)$, $\tilde{\mathfrak{h}} = \tilde{\mathfrak{h}}_1 \oplus \tilde{\mathfrak{h}}_2 \cong so(n-3)$, $\mathfrak{h}_1$ is either simple or $\mathfrak{h}_1 = \mathfrak{z}$ and then $\tilde{\mathfrak{h}}_1 = \{0\}$).

If $\tilde{\mathfrak{h}}_1 \neq \{0\}$ then the subalgebra $\mathfrak{i} + \tilde{\mathfrak{h}}_1$, where $\mathfrak{i}$ consists of all vectors commuting with $\tilde{\mathfrak{h}}_1$, is larger then $\mathfrak{g}^+$ (since besides $\mathfrak{g}^+$ it contains vectors from $\mathfrak{h}_2$).

Since an involutive algebra $\mathfrak{g}^+$ of a simple compact Lie algebra $\mathfrak{g}$ is a maximal subalgebra we have $\mathfrak{i} + \tilde{\mathfrak{h}}_1 = \mathfrak{g}^+$. Thus $\mathfrak{h}_2 = \tilde{\mathfrak{h}}_2 \subset \mathfrak{h}^+$.

Furthermore, if $\mathfrak{h}_2 \neq \{0\}$ then the subalgebra $\mathfrak{i} + \mathfrak{h}_2$, where $\mathfrak{i}$ consists of all vectors commuting with $\mathfrak{h}_2$, is larger than $\mathfrak{g}^+$ (since it contains also vectors of $\mathfrak{h}_1$). But then $\mathfrak{h}_1 = \mathfrak{z} \oplus \tilde{\mathfrak{h}}_1$ and by the above $\mathfrak{h}_1 = \mathfrak{z} \subset \mathfrak{h}^+$.

Thus we have obtained that $\tilde{\mathfrak{h}}_1 \neq \{0\}$, $\mathfrak{h}_2 \neq \{0\}$ imply $\mathfrak{h} = \mathfrak{h}_1 \oplus \mathfrak{h}_2 \subset \mathfrak{h}^+$, and consequently the mirror is immobile. This contradicts the conditions.

As a result, if $\tilde{\mathfrak{h}}_1 \neq \{0\}$ then $\mathfrak{h}_2 = \tilde{\mathfrak{h}}_2 = \{0\}$.

And

$$\mathfrak{h}/\mathfrak{h}^+ = (\mathfrak{h}, \mathfrak{z} \oplus \tilde{\mathfrak{h}}),$$

where $\mathfrak{h}$ is simple compact, and $\mathfrak{z} \oplus \tilde{\mathfrak{h}} \cong so(2) \oplus so(n-3)$ with the natural embedding into $\mathfrak{g} \cong so(n)$.

Owing to the well known classification of E. Cartan of involutive automorphisms and involutive algebras of simple compact Lie algebras (see, for example, [S. Helgason, 62,78]) the pair $(\mathfrak{h}, \mathfrak{z} \oplus \tilde{\mathfrak{h}})$ is none but $(so(n-1), so(2) \oplus so(n-3))$ with the natural embedding and

$$(\mathfrak{g}, \mathfrak{h}) \cong (so(n), so(n-1)).$$

We should now verify that the above pair is also a pair with the natural embedding, and consequently is involutive.

Because of the decomposition $\mathfrak{g} = \mathfrak{m} + \mathfrak{h}$ we have

$$\dim \mathfrak{m} = \dim \mathfrak{g} - \dim \mathfrak{h} = \frac{1}{2}n(n-1) - \frac{1}{2}(n-1)(n-2) = n-1.$$

In addition, $\operatorname{ad}\mathfrak{h}$ acts exactly on the $(n-1)$-dimensional subspace $\mathfrak{m}$ by means of skew-symmetric endomorphisms and

$$\dim \operatorname{ad}\mathfrak{h} = \dim \mathfrak{h} = \frac{1}{2}(n-1)(n-2).$$

The latter means that $\operatorname{ad}\mathfrak{h}$ is the algebra of all skew-symmetric automorphisms. Whence it is easily seen that $(\mathfrak{g}, \mathfrak{h}) \cong (so(n), so(n-1))$ is an involutive pair with the natural embedding. The simplest way to verify this is to pass to Lie groups from Lie algebras. This gives us a homogeneous Riemannian space

$$G/H \cong SO(n)/SO(n-1).$$

It is of maximal mobility, that is,

$$\frac{1}{2}(\dim G/H)(\dim G/H - 1) = \dim H.$$

But then it is, as is well known, a space of the constant curvature, that is, in particular, a symmetric space (see [S. Kobayashi, K. Nomizu 63, 69]).

IV.6.2. Remark. Of course, the decomposition

$$(\mathfrak{h}, \mathfrak{z} \oplus \tilde{\mathfrak{h}}) \cong (so(n-1), so(2) \oplus so(n-3))$$

(with the natural embedding) can be obtained without the classification of E. Cartan.

IV.6.3. Another possibility (again $n > 5$) is $\tilde{\mathfrak{h}}_1 = \{0\}$. But then

$$\mathfrak{h}/\mathfrak{h}^+ = (\mathfrak{h}_1, \mathfrak{z}) + (\mathfrak{h}_2, \tilde{\mathfrak{h}}_2), \qquad so(n-3) \cong \tilde{\mathfrak{h}}_2 \neq \{0\}.$$

Considering as before the subalgebra $\mathfrak{i} + \tilde{\mathfrak{h}}_2$, $\tilde{\mathfrak{h}}_2 \neq \{0\}$, where $\mathfrak{i}$ is the subalgebra of all vectors commuting with $\tilde{\mathfrak{h}}_2$, we obtain $\mathfrak{g}^+ \subset \mathfrak{i} + \tilde{\mathfrak{h}}_2$.

Moreover, $\mathfrak{h}_1 \subset \mathfrak{i} + \tilde{\mathfrak{h}}_2$. Since the involutive algebra $\mathfrak{g}^+$ is a maximal subalgebra of $\mathfrak{g}$ we have $\mathfrak{g}^+ = \mathfrak{i} + \tilde{\mathfrak{h}}_2$, whence $\mathfrak{h}_1 \subset \mathfrak{h}^+$, that is, $\mathfrak{h}_1 = \mathfrak{z}$.

In this case

$$\mathfrak{h}/\mathfrak{h}^+ = (\mathfrak{z}, \mathfrak{z}) + (\mathfrak{h}_2, \tilde{\mathfrak{h}}_2), \qquad \mathfrak{z} \cong so(2), \qquad \tilde{\mathfrak{h}}_2 \cong so(n-3).$$

Now let us note that $\mathfrak{h}^+$ is a subalgebra of the Lie algebra $\mathfrak{k}$ of all elements commuting with $\mathfrak{z} \cong so(2)$.

Evidently $\mathfrak{k} = \mathfrak{z} \oplus \tilde{\mathfrak{k}} \cong so(2) \oplus so(n-2)$ and $\tilde{\mathfrak{h}}_2 \subset \tilde{\mathfrak{k}}$.

Thus

$$(\tilde{\mathfrak{k}}, \mathfrak{h}_2) \cong (so(n-2), so(n-3))$$

is an involutive pair with the natural embedding and $\tilde{\mathfrak{k}} \supset \mathfrak{h}_2 \supsetneqq \tilde{\mathfrak{h}}_2$. Owing to the maximality of the involutive algebra $\tilde{\mathfrak{h}}_2$ in $\tilde{\mathfrak{k}}$ we have $\mathfrak{h}_2 = \tilde{\mathfrak{k}}$.

Therefore in this case

$$(\mathfrak{g}, \mathfrak{h}) \cong (so(n), so(2) \oplus so(n-2))$$

(with the natural embedding) which is an involutive pair.

IV.6.4. Let us consider, finally, the case $n = 5$. Here $\mathfrak{h}^+ \cong so(2) \oplus so(n-3)$ turns into $\mathfrak{h}^+ = \mathfrak{z} \oplus \tilde{\mathfrak{z}} \cong so(2) \oplus so(2)$. The decomposition for $(\mathfrak{h}, \mathfrak{h}^+)$ now takes the form

$$(\mathfrak{h}, \mathfrak{h}^+) = (\mathfrak{h}_1, \mathfrak{z}_1) + (\mathfrak{h}_2, \mathfrak{z}_2) + (\mathfrak{h}_3, \{0\}),$$

where $\mathfrak{h}_1$ (respectively $\mathfrak{h}_2$) is either simple or coincides with $\mathfrak{z}_1$ (respectively, with $\mathfrak{z}_2$), $\mathfrak{h}_3$ is commutative, $\mathfrak{z}_1$ and $\mathfrak{z}_2$ are one-dimensional and, generally speaking, do not coincide with $\mathfrak{z}$ and $\tilde{\mathfrak{z}}$, correspondingly.

Let us show first that $\mathfrak{h}_3 = \{0\}$. Indeed, any $x \in \tilde{\mathfrak{z}}$ has the form $x = x_1 + x_2$, where $x_1 \in \mathfrak{z}_1$, $x_2 \in \mathfrak{z}_2$, and therefore commutes with any element of $\mathfrak{h}_3$. Consequently the algebra $\mathfrak{k}$ of all elements commuting with $\tilde{\mathfrak{z}}$ contains $\mathfrak{g}^+$ and $\mathfrak{h}_3$. But the involutive algebra $\mathfrak{g}^+$ is maximal in $\mathfrak{g}$, hence $\mathfrak{k} = \mathfrak{g}^+$, $\mathfrak{h}_3 = \{0\}$.

Thus

$$(\mathfrak{h}, \mathfrak{h}^+) = (\mathfrak{h}_1, \mathfrak{z}_1) + (\mathfrak{h}_2, \mathfrak{z}_2).$$

If $\mathfrak{h}_1$ and $\mathfrak{h}_2$ are simple then

$$(\mathfrak{h}_1, \mathfrak{z}_1) \cong (\mathfrak{h}_2, \mathfrak{z}_2) \cong (so(3), so(2))$$

with the natural embedding. We have used the evident property that a simple compact Lie algebra with a one-dimensional involutive algebra is $so(3)$ with $so(2)$ (with the natural embedding up to isomorphism). Thus in our case

$$(\mathfrak{h}, \mathfrak{h}^+) \cong (so(4), so(2) \oplus so(2)),$$

$$(\mathfrak{g}, \mathfrak{h}) \cong (so(5), so(4)).$$

Owing to the arguments given above we obtain again the case of maximal mobility (see the considerations before IV.6.2). Therefore this is an involutive pair (with the natural embedding).

We recall that $so(5) \cong sp(2)$, consequently this pair can also be represented as $(sp(2), sp(1) \oplus sp(1))$.

The next possibility is $\mathfrak{h}_1 = \mathfrak{z}_1$, $\mathfrak{h}_2$ is simple (the case $\mathfrak{h}_2 = \mathfrak{z}_2$, $\mathfrak{h}_1$ is simple follows from that after re-numbering).

In the matrix realization of $\mathfrak{g} \cong so(5)$ by the set of all 5×5 skew-symmetric matrices we may put that the basis element from $\mathfrak{z} \cong so(2)$ has the form

$$\bar{\bar{a}}_1 = \left\| \begin{array}{cc|c} 0 & -1 & \bar{\bar{O}} \\ 1 & 0 & \\ \hline \bar{\bar{O}} & & \bar{\bar{O}} \end{array} \right\|$$

and the basis element from $\bar{\mathfrak{z}} \cong so(2)$ has the form

$$\bar{\bar{a}}_2 = \left\| \begin{array}{c|cc|c} \bar{\bar{O}} & \bar{\bar{O}} & & \bar{\bar{O}} \\ \hline \bar{\bar{O}} & 0 & -1 & \bar{\bar{O}} \\ & 1 & 0 & \\ \hline \bar{\bar{O}} & \bar{\bar{O}} & & \bar{\bar{O}} \end{array} \right\|$$

Some non-zero element of $\mathfrak{z}_1$ then has the form $\bar{\bar{a}} = \lambda \bar{\bar{a}}_1 + \mu \bar{\bar{a}}_2$ ($\lambda, \mu \neq 0$), that is,

$$\left\| \begin{array}{cc|cc|c} 0 & -\lambda & \bar{\bar{O}} & & \bar{\bar{O}} \\ \lambda & 0 & & & \\ \hline \bar{\bar{O}} & & 0 & -\mu & \bar{\bar{O}} \\ & & \mu & 0 & \\ \hline \bar{\bar{O}} & & \bar{\bar{O}} & & 0 \end{array} \right\|.$$

Now we consider the subalgebra $\mathfrak{k}$ of all elements $\bar{\bar{q}}$ commuting with $\bar{\bar{a}}$ (and consequently with $\mathfrak{z}_1$ as well).

If $\lambda^2 \neq \mu^2$, $\lambda \neq 0$, $\mu \neq 0$, then

$$\bar{\bar{q}} = \left\| \begin{array}{ccc} \begin{matrix} 0 & -\alpha \\ \alpha & 0 \end{matrix} & \bar{\bar{O}} & \bar{\bar{O}} \\ \bar{\bar{O}} & \begin{matrix} 0 & -\beta \\ \beta & 0 \end{matrix} & \bar{\bar{O}} \\ \bar{\bar{O}} & \bar{\bar{O}} & 0 \end{array} \right\|$$

with arbitrarily taken $\alpha, \beta \in \mathbb{R}$. That is, $\mathfrak{k} = \mathfrak{z} \oplus \tilde{\mathfrak{z}}$. We should exclude this case from the consideration, since according to the condition the at least three-dimensional subalgebra $\mathfrak{h}_2 \cong so(3)$ commutes with $\mathfrak{z}_1$.

Thus we have only the following possibilities:

1. $\lambda \neq 0$, $\mu = 0$, then, of course, $\lambda^2 \neq \mu^2$, that is, $\mathfrak{z}_1 = \mathfrak{z}$. In this case any element of $\mathfrak{k}$ has the form:

$$\left\| \begin{array}{cc} \begin{matrix} 0 & -\alpha \\ \alpha & 0 \end{matrix} & \bar{\bar{O}} \\ \bar{\bar{O}} & * \end{array} \right\| \qquad (\forall \alpha \in \mathbb{R}),$$

that is, $\mathfrak{k} \cong so(2) \oplus so(3)$. But also $\mathfrak{h} = \mathfrak{z}_1 \oplus \mathfrak{h}_2 \cong so(2) \oplus so(3)$. Since $\mathfrak{h} \subset \mathfrak{k}$ and $\dim \mathfrak{h} = \dim \mathfrak{k}$ we have $\mathfrak{k} = \mathfrak{h}$. Moreover, here $\mathfrak{z}_1 = \mathfrak{z}$, $\mathfrak{z}_2 = \tilde{\mathfrak{z}}$.

2. $\lambda = 0$, $\mu \neq 0$, that is, $\mathfrak{z}_1 = \tilde{\mathfrak{z}}$. Any element $\bar{\bar{q}}$ of $\mathfrak{k}$ has the form:

$$\left\| \begin{array}{ccc} * & & * \\ & \begin{matrix} 0 & -\beta \\ \beta & 0 \end{matrix} & \\ * & & * \end{array} \right\|.$$

Again, $\mathfrak{k} \cong so(2) \oplus so(3)$. But also $\mathfrak{h} = \mathfrak{z}_1 \oplus \mathfrak{h}_2 \cong so(2) \oplus so(3)$. Since $\mathfrak{h} \subset \mathfrak{k}$, $\dim \mathfrak{h} = \dim \mathfrak{k}$, we have $\mathfrak{k} = \mathfrak{h}$. Moreover, here $\mathfrak{z}_1 = \tilde{\mathfrak{z}}$, $\mathfrak{z}_2 = \mathfrak{z}$, which is impossible since then the subalgebra $\mathfrak{i}$ of elements commuting with $\mathfrak{z}_1 = \tilde{\mathfrak{z}}$ contains $\mathfrak{g}^+$, but $\mathfrak{h}$ is not a subset of $\mathfrak{g}^+$ in contradiction with the maximality of the involutive algebra $\mathfrak{g}^+$.

3. $\lambda^2 = \mu^2$, $\lambda \neq 0$, $\mu \neq 0$, that is, $\lambda = \pm\mu \neq 0$.

Without loss of generality we can set $\lambda = \pm\mu = 1$. The basis element of $\mathfrak{z}_1$ in this case is

$$\bar{\bar{a}} = \bar{\bar{a}}_1 + \bar{\bar{a}}_2 = \left\| \begin{array}{c|c|c} \begin{matrix} 0 & -1 \\ 1 & 0 \end{matrix} & \bar{\bar{O}} & \bar{\bar{O}} \\ \hline \bar{\bar{O}} & \begin{matrix} 0 & \mp 1 \\ \pm 1 & 0 \end{matrix} & \bar{\bar{O}} \\ \hline \bar{\bar{O}} & \bar{\bar{O}} & 0 \end{array} \right\|.$$

For the subalgebra $\mathfrak{k}$ of all elements $\bar{\bar{q}}$ commuting with $\mathfrak{z}_1$ we have

$$\bar{\bar{q}} = \left\| \begin{array}{c|c|c} \begin{matrix} 0 & -\alpha \\ \alpha & 0 \end{matrix} & \begin{matrix} \gamma & \delta \\ \mp\delta & \pm\gamma \end{matrix} & \bar{\bar{O}} \\ \hline \begin{matrix} -\gamma & \pm\delta \\ -\delta & \mp\gamma \end{matrix} & \begin{matrix} 0 & -\beta \\ \beta & 0 \end{matrix} & \bar{\bar{O}} \\ \hline \bar{\bar{O}} & \bar{\bar{O}} & 0 \end{array} \right\|.$$

This means that $\mathfrak{k} \cong so(2) \oplus su(2)$ because of the decomposition

$$\left\| \begin{array}{c|c|c} \begin{matrix} 0 & -\alpha \\ \alpha & 0 \end{matrix} & \begin{matrix} \gamma & \delta \\ \mp\delta & \pm\gamma \end{matrix} & \bar{\bar{O}} \\ \hline \begin{matrix} -\gamma & \pm\delta \\ -\delta & \mp\gamma \end{matrix} & \begin{matrix} 0 & -\beta \\ \beta & 0 \end{matrix} & \bar{\bar{O}} \\ \hline \bar{\bar{O}} & \bar{\bar{O}} & 0 \end{array} \right\| = \left\| \begin{array}{c|c|c} \begin{matrix} 0 & -\frac{\alpha\pm\beta}{2} \\ \frac{\alpha\pm\beta}{2} & 0 \end{matrix} & \bar{\bar{O}} & \bar{\bar{O}} \\ \hline \bar{\bar{O}} & \begin{matrix} 0 & -\frac{\beta\pm\alpha}{2} \\ \frac{\beta\pm\alpha}{2} & 0 \end{matrix} & \bar{\bar{O}} \\ \hline \bar{\bar{O}} & \bar{\bar{O}} & 0 \end{array} \right\|$$

$$+ \left\| \begin{array}{c|c|c} \begin{matrix} 0 & -\frac{\alpha\mp\beta}{2} \\ \frac{\alpha\mp\beta}{2} & 0 \end{matrix} & \begin{matrix} \gamma & \delta \\ \mp\delta & \pm\gamma \end{matrix} & \bar{\bar{O}} \\ \hline \begin{matrix} -\gamma & \pm\delta \\ -\delta & \mp\gamma \end{matrix} & \begin{matrix} 0 & -\frac{\beta\pm\alpha}{2} \\ \frac{\beta\pm\alpha}{2} & 0 \end{matrix} & \bar{\bar{O}} \\ \hline \bar{\bar{O}} & \bar{\bar{O}} & 0 \end{array} \right\|.$$

But $\mathfrak{h} \cong so(2) \oplus so(3)$ as well, therefore $\dim \mathfrak{k} = \dim \mathfrak{h}$, $\mathfrak{h} \subset \mathfrak{k}$, whence $\mathfrak{k} = \mathfrak{h}$. Moreover, two subcases are possible, as is indicated by the sign ($\pm$) in the above formulas.

Now we summarize the previous considerations in the following theorem.

IV.6.5. Theorem 11. *Let* $(\mathfrak{g}, \mathfrak{h})$ *be a pair with a simple compact Lie algebra* $\mathfrak{g} \cong so(n)$ *and with a two-dimensional mobile mirror* $(\mathfrak{g}^+, \mathfrak{h}^+)$ *of orthogonal type* $(\mathfrak{g}^+ = \mathfrak{m}^+ + \mathfrak{h}^+, \dim \mathfrak{m}^+ = 2, \mathfrak{h} \supsetneq \mathfrak{h}^+)$.

Then the following cases are possible:

1). $(\mathfrak{g}, \mathfrak{h}) \cong (so(n), so(n-1))$ *is an involutive pair,*

$(\mathfrak{g}^+, \mathfrak{h}^+) \cong (so(3) \oplus so(n-3), so(2) \oplus so(n-3)),$

$(\mathfrak{h}, \mathfrak{h}^+) \cong (so(n-1), so(2) \oplus so(n-3))$

(with the natural embeddings).

In this case the mirror $(\mathfrak{g}^+, \mathfrak{h}^+)$ can be moved either in a tri-symmetric or in a non-tri-symmetric way.

2). $(\mathfrak{g}, \mathfrak{h}) \cong (so(n), so(2) \oplus so(n-2))$ *is an involutive pair,*

$(\mathfrak{g}^+, \mathfrak{h}^+) \cong (so(3) \oplus so(n-3), so(2) \oplus so(n-3)),$

$(\mathfrak{h}, \mathfrak{h}^+) \cong (so(2) \oplus so(n-2), so(2) \oplus so(n-3))$

(with the natural embeddings).

In this case the mirror $(\mathfrak{g}^+, \mathfrak{h}^+)$ can be moved in a non-tri-symmetric way only.

3). $(\mathfrak{g}, \mathfrak{h}) \cong (so(5), u(1) \oplus su(2)) \cong (sp(2), u(1) \oplus su(2)),$

$(\mathfrak{g}^+, \mathfrak{h}^+) \cong (so(3) \oplus so(2), so(2) \oplus so(2)),$

$(\mathfrak{h}, \mathfrak{h}^+) \cong (u(1) \oplus su(2), u(1) \oplus u(1))$

(with the natural embeddings).

Here $(\mathfrak{g}, \mathfrak{h})$ is not an involutive pair with the non-maximal subalgebra $\mathfrak{h} \subset \mathfrak{g}$. In this case the mirror $(\mathfrak{g}^+, \mathfrak{h}^+)$ can be moved only in a tri-symmetric way.

IV.6.6. Remark. In the case 3) the enlargement of $\mathfrak{h}^+ \cong so(2) \oplus so(2)$ up to the maximal involutive algebra $\mathfrak{h}' \cong so(4)$ gives us the case 1) with $n = 5$. This exceptional case exists because $so(4)$ is not simple.

IV.6.7. Now we consider the pair $(\mathfrak{g}, \mathfrak{h})$, $\mathfrak{g} \cong su(3)$, $\mathfrak{g}^+ \cong so(3)$ (with the natural embeddings), that is $\mathfrak{g} \cong su(3)$ with a two-dimensional mobile orthogonal mirror. Here $(\mathfrak{g}^+, \mathfrak{h}^+) \cong (so(3), so(2))$ with the natural embedding, $(\mathfrak{h}, \mathfrak{h}^+)$ has the form $(\mathfrak{h}, \mathfrak{z})$, where $\mathfrak{z} = \mathfrak{h}^+ \cong so(2)$ is one-dimensional.

Furthermore, we have the compact pair

$$(\mathfrak{h}, \mathfrak{z}) = (\mathfrak{h}_1, \mathfrak{z}) + (\mathfrak{h}_2, \{0\}),$$

where $\mathfrak{h}_2$ is commutative and either $\mathfrak{h}_1$ is simple three-dimensional or $\mathfrak{h}_1 = \mathfrak{z}$ and $\mathfrak{h}_2 \neq \{0\}$.

In any case, then $\mathfrak{h}_2 = \tilde{\mathfrak{z}}$ is at most one-dimensional since the rank of $su(3)$ is two. Thus $\mathfrak{z} \oplus \mathfrak{h}_2$ is at most a two-dimensional commutative subalgebra. Therefore,

$$(\mathfrak{h}, \mathfrak{h}^+) = (\mathfrak{h}_1, \mathfrak{z}) + (\tilde{\mathfrak{z}}, \{0\}),$$

$\mathfrak{z} = \mathfrak{h}^+ \cong so(2)$, $\tilde{\mathfrak{z}} \cong so(2)$ or $\tilde{\mathfrak{z}} = \{0\}$, where $\mathfrak{h}_1 \cong so(3) \cong su(2)$ or $\mathfrak{h}_1 = \mathfrak{z} \cong so(2)$.

Consequently the following cases are possible:

1) $(\mathfrak{h}, \mathfrak{h}^+) = (\mathfrak{h}_1, \mathfrak{z}) + (\tilde{\mathfrak{z}}, \{0\})$, $\mathfrak{h}_1 \cong so(3)$, $\mathfrak{z} = \tilde{\mathfrak{z}} \cong so(2)$.

Here $\mathfrak{h}_1$ is a simple three-dimensional subalgebra commuting with a one-dimensional subalgebra of $\mathfrak{g} \cong su(3)$. This is possible only if $\mathfrak{h}_1 \cong su(2)$, $\tilde{\mathfrak{z}} \cong u(1)$ with the natural embeddings into $su(3)$ (being verified in the matrix model of $su(3)$ in a straightforward way). Then $\mathfrak{h} = (\mathfrak{h}_1 \oplus \mathfrak{z}) \cong su(2) \oplus u(1)$ is an involutive algebra in $\mathfrak{g} \cong su(3)$.

This case is tri-symmetric (moreover, two-symmetric).

2) $(\mathfrak{h}, \mathfrak{h}^+) = (\mathfrak{h}, \mathfrak{z})$, $\mathfrak{h} \cong so(3)$, $\mathfrak{z} \cong so(2)$.

Let σ be an involutive automorphism with the involutive algebra $\mathfrak{g}^+ \cong so(3)$ (this automorphism is outer), and let σ_1 be an involutive automorphism from $e^{\operatorname{ad}\mathfrak{z}} \cong SO(2)$. Then $\mathfrak{h}$ and $\mathfrak{g}^+$ are invariant under the actions of σ and σ_1.

Moreover, $\sigma\sigma_1 = \sigma_1\sigma = \sigma_2$ is again an involutive automorphism. Evidently vectors of $\mathfrak{z} = \mathfrak{h}^+$ are immobile under the action of σ_1.

There are only two possibilities: either σ acts on $\mathfrak{h}$ without immobile non-zero vectors or σ acts on $\mathfrak{h}$ as the identity map. (Otherwise we have a two-dimensional commutative involutive subalgebra of the three-dimensional compact algebra $\mathfrak{h}$, which is impossible).

If the first possibility is valid then $\sigma_2 = \sigma\sigma_1$ preserves all elements of $\mathfrak{h}$. We have shown that there exists an involutive automorphism (either inner σ_1 or outer σ_2) commuting with σ such that its involutive algebra $\mathfrak{k}$ contains $\mathfrak{h}$. But all involutive automorphisms and involutive algebras of $su(3)$ are well known (see, for example, [S. Helgason, 62,78]).

Therefore either $\mathfrak{k} \cong su(3)$ or $\mathfrak{k} \cong su(2) \oplus u(1)$ with the natural embeddings into $\mathfrak{g} \cong su(3)$. In our case $\mathfrak{h}$ is three-dimensional and belongs to $\mathfrak{k}$, consequently either $\mathfrak{h} \cong so(3)$ or $\mathfrak{h} \cong su(2)$, with the natural embeddings into $\mathfrak{g} \cong su(3)$.

In the first case we obtain the involutive pair $(\mathfrak{g}, \mathfrak{h}) \cong (su(3), so(3))$, in the second case we obtain $(\mathfrak{g}, \mathfrak{h}) \cong (su(3), su(2))$ (which is non-involutive but can be enlarged up to involutive after changing $su(2)$ for $su(2) \oplus u(1)$, which gives the subcase 1)).

Both possibilities are, indeed, realized (as involutive algebras of the involutive automorphisms σ_1 and $\sigma_2 = \sigma\sigma_1$). Both these possibilities are tri-symmetric.

3). $(\mathfrak{h}, \mathfrak{h}^+) = (\mathfrak{z}, \mathfrak{z}) + (\tilde{\mathfrak{z}}, \{0\})$, $\mathfrak{z}$, $\tilde{\mathfrak{z}}$ are commutative one-dimensional.

Here $\mathfrak{h} = (\mathfrak{z} \oplus \tilde{\mathfrak{z}}) \cong so(2) \oplus u(1)$ is a maximal two-dimensional subalgebra of $su(3)$. This case can be obtained from the preceding case by the restriction of $\mathfrak{h} \cong su(2) \oplus u(1)$ up to $so(2) \oplus u(1)$ (with the natural embeddings).

This case is tri-symmetric.

We summarize the results obtained above in the following theorem.

IV.6.8. Theorem 12. *Let $(\mathfrak{g}, \mathfrak{h})$, $\mathfrak{g} \cong su(3)$, be a pair with a two-dimensional mobile mirror $(\mathfrak{g}^+, \mathfrak{h}^+)$ $(\mathfrak{g}^+ = \mathfrak{m}^+ + \mathfrak{h}^+$, $\dim \mathfrak{m}^+ = 2$, $\mathfrak{h} \neq \mathfrak{h}^+)$ of orthogonal type. Then there exist only the following possibilities:*

1. $(\mathfrak{g}, \mathfrak{h}) \cong (su(3), su(2) \oplus u(1))$ *is an involutive pair,*

$(\mathfrak{g}^+, \mathfrak{h}^+) \cong (so(3), so(2))$,

$(\mathfrak{h}, \mathfrak{h}^+) \cong (su(2) \oplus u(1), so(2) \oplus u(1))$,

with the natural embeddings.

2. $(\mathfrak{g}, \mathfrak{h}) \cong (su(3), so(3))$ *is an involutive pair,*

$(\mathfrak{g}^+, \mathfrak{h}^+) \cong (so(3), so(2))$,

$(\mathfrak{h}, \mathfrak{h}^+) \cong (so(3), so(2))$,

with the natural embeddings.

3. $(\mathfrak{g}, \mathfrak{h}) \cong (su(3), su(2))$ is a non-involutive pair,

$(\mathfrak{g}^+, \mathfrak{h}^+) \cong (so(3), so(2))$,

$(\mathfrak{h}, \mathfrak{h}^+) \cong (su(2), so(2))$,

with the natural embeddings.

This is a tri-symmetric case.

4. $(\mathfrak{g}, \mathfrak{h}) \cong (su(3), so(2) \oplus u(1))$ is a non-involutive pair,

$(\mathfrak{g}^+, \mathfrak{h}^+) \cong (so(3), so(2))$,

$(\mathfrak{h}, \mathfrak{h}^+) \cong (su(2) \oplus u(1), so(2))$,

with the natural embeddings.

This is a tri-symmetric case.

IV.6.9. The methods presented in previous chapters can be applied as well to the case of unitary two-dimensional mirrors.

Let us consider the case $(\mathfrak{g}, \mathfrak{h})$ with $\mathfrak{g} \cong so(n)$ and with a mobile two-dimensional mirror $(\mathfrak{g}^+, \mathfrak{m}^+)$ of unitary type. Then (see [L.V. Sabinin 69, 70b], [S. Helgason, 62, 78])

$$\mathfrak{g}^+ \cong so(4) \oplus so(n-4) \cong su(2) \oplus su(2) \oplus so(n-4),$$

and evidently

$$\mathfrak{h}^+ \cong u(1) \oplus su(2) \oplus so(n-4).$$

As before, we explore the involutive pair $(\mathfrak{h}, \mathfrak{h}^+)$. First, let $n \neq 6$.

Then there exists the decomposition

$$(\mathfrak{h}, \mathfrak{h}^+) = (\mathfrak{h}_1, \tilde{\mathfrak{h}}_1 \oplus \mathfrak{z}) + (\mathfrak{h}_2, \tilde{\mathfrak{h}}_2), \qquad \mathfrak{z} \cong u(1),$$

where $\mathfrak{h}_1$ is either simple or $\mathfrak{h}_1 = \mathfrak{z}$, $\tilde{\mathfrak{h}}_1 = \{0\}$.

If $\tilde{\mathfrak{h}}_1 \neq \{0\}$ then $\mathfrak{h}_2 = \tilde{\mathfrak{h}}_2 \subset \mathfrak{h}^+$ (otherwise $\tilde{\mathfrak{h}}_1 + \mathfrak{i}$, where $\mathfrak{i}$ is the subalgebra of all vectors commuting with $\tilde{\mathfrak{h}}_1$, is larger then the involutive algebra $\mathfrak{g}^+$, which is impossible in a simple compact Lie algebra $\mathfrak{g}$). But then $\mathfrak{h}_2 = \tilde{\mathfrak{h}}_2 = \{0\}$ (otherwise $\tilde{\mathfrak{h}}_2 + \mathfrak{j}$, where $\mathfrak{j}$ is the subalgebra of all vectors commuting with $\tilde{\mathfrak{h}}_2$, and evidently containing the involutive algebra $\mathfrak{g}^+$, should coincide with $\mathfrak{g}^+$ owing to the maximality of $\mathfrak{g}^+$. But then $\mathfrak{h}_1 \subset \mathfrak{h}^+$, that is, $\mathfrak{h}_1 = \mathfrak{z} \oplus \tilde{\mathfrak{h}}_1$, implying $\tilde{\mathfrak{h}}_1 = \{0\}$, which is impossible in this case).

Furthermore, if $\tilde{\mathfrak{h}}_1 = \{0\}$ then for $\tilde{\mathfrak{h}}_2 \neq \{0\}$ we obtain $\mathfrak{h}_1 = \mathfrak{z}$ (again using the coincidence of $\mathfrak{g}^+$ with $\tilde{\mathfrak{h}}_2 + \mathfrak{k}$, where $\mathfrak{k}$ is the subalgebra of all elements commuting with $\tilde{\mathfrak{h}}_2$). And if $\tilde{\mathfrak{h}}_2 = \{0\}$ we have

$$(\mathfrak{h}_1, \mathfrak{z} + \tilde{\mathfrak{h}}_1) \cong (so(3), so(2)), \qquad (\mathfrak{h}_2, \tilde{\mathfrak{h}}_2) = (\mathfrak{h}_2, \{0\}).$$

But since in the unitary case the involutive automorphism σ of the involutive algebra $\mathfrak{g}^+$ belongs to $e^{\mathrm{ad}\,\mathfrak{z}}$ ($\mathfrak{z}$ being commutative with $\mathfrak{h}_2$) we have $\mathfrak{h}_2 \subset \mathfrak{g}^+$, and consequently $\mathfrak{h}_2 = \{0\}$.

As a result for $n \neq 6$ we have the following possibilities:

I. $(\mathfrak{h}, \mathfrak{h}^+)$ is an involutive pair, where $\mathfrak{h}$ is simple and

$$\mathfrak{h}^+ \cong u(1) \oplus su(2) \oplus so(n-4).$$

Moreover, $(\mathfrak{h}, \mathfrak{h}^+)$ is an exact principal unitary involutive pair $(\sigma \in e^{\mathrm{ad}b} \cong SU(2))$.

The E. Cartan's list of involutive pairs and involutive automorphisms of simple compact Lie algebras (see [S. Helgason, 62,78] or works [L.V. Sabinin, 69,70] on principal involutive automorphisms of simple compact Lie algebras) shows that this is possible only if either $n = 5$, and then

$$(\mathfrak{h}, \mathfrak{h}^+) \cong (su(3), u(1) \oplus su(2)),$$

or $n = 7$, and then

$$(\mathfrak{h}, \mathfrak{h}^+) \cong (su(4), u(1) \oplus su(2) \oplus su(2)),$$

or $n = 10$, and then (since $so(6) \cong su(4)$)

$$(\mathfrak{h}, \mathfrak{h}^+) \cong (su(6), u(1) \oplus su(2) \oplus su(4)),$$

with the natural embeddings.

In the first case we have

$$(\mathfrak{g}, \mathfrak{h}) \cong (so(5), su(3)).$$

Taking into account that $\mathfrak{g} = \mathfrak{m} + \mathfrak{h}$ we have $\dim \mathfrak{m} = 2$, $\dim \mathfrak{h} = 8$, which is impossible since $\mathrm{ad}\,\mathfrak{h}$ acts on $\mathfrak{m}$ exactly by skew-symmetric endomorphisms, that is,

$$\dim \mathfrak{h} = \dim \mathrm{ad}\,\mathfrak{h} \leq \frac{1}{2}(\dim \mathfrak{m})(\dim \mathfrak{m} - 1),$$

or $8 \leq 1$, giving a contradiction.

In the second case we have

$$(\mathfrak{g}, \mathfrak{h}) \cong (so(7), su(4)), \quad \dim \mathfrak{h} = 15, \quad \dim \mathfrak{m} = 6.$$

Thus

$$\dim \mathfrak{h} = \dim \mathrm{ad}\,\mathfrak{h} = \frac{1}{2}(\dim \mathfrak{m})(\dim \mathfrak{m} - 1),$$

that is, the pair

$$(\mathfrak{g}, \mathfrak{h}) \cong (so(7), su(4)) \cong (so(7), so(6))$$

is of the maximal mobility and then (see [S. Kobayashi, K. Nomizu, 63,69]) $\mathfrak{h} \cong su(4) \cong so(6)$ is embedded into $\mathfrak{g} \cong so(7)$ in the natural way, which means that $(\mathfrak{g}, \mathfrak{h})$ is an involutive pair. But more thoroughful consideration shows that such case is impossible. Indeed, then $\mathfrak{h}^+ = \mathfrak{z} \oplus \mathfrak{b}_1 \oplus \mathfrak{b}_2 \cong u(1) \oplus su(2) \oplus so(3))$. Moreover, $e^{\mathrm{ad}b_1} \cong SU(2)$, $e^{\mathrm{ad}b_2} \cong SO(3)$. Therefore in the restriction to $\mathfrak{h}$ we have $e^{\mathrm{ad}b_1} \cong SU(2)$, $e^{\mathrm{ad}b_2} \cong SO(3)$.

But in $\mathfrak{h} \cong su(4)$ we have $e^{\operatorname{ad} b_2} \cong SU(2)$ for $\mathfrak{h}^+$. This shows the impossibility of this case.

Finally, we should consider $n = 10$, that is,

$$(\mathfrak{g}, \mathfrak{h}) \cong (so(10), su(6)),$$
$$(\mathfrak{g}^+, \mathfrak{h}^+) \cong (so(4) \oplus so(6), u(1) \oplus su(2) \oplus so(6)).$$

In this case $\dim \mathfrak{h} = 35$ and the considerations of dimensions do not give any essential information. However, if σ_1 is the involutive automorphism of the involutive algebra $\mathfrak{g}^+$ then it is principal unitary. Consequently, using the iso-involutive rotation of the two-dimensional mirror, we obtain the tri-symmetric decomposition

$$\mathfrak{m} = \mathfrak{m}_1^+ + \mathfrak{m}_2^+ + \mathfrak{m}_3^+,$$

where $\dim \mathfrak{m} = 10$, $\dim \mathfrak{m}_1^+ = \dim \mathfrak{m}_2^+ = 2$, and as a result $\dim \mathfrak{m}_3^+ = 6$.

Simultaneously two other involutive automorphisms, σ_2 and $\sigma_3 = \sigma_1\sigma_2 = \sigma_2\sigma_1$, appear. Since

$$(\mathfrak{h}, \mathfrak{h}^+) \cong (su(6), u(1) \oplus su(2) \oplus su(4))$$

with the natural embeddings, we consider $\mathfrak{h} = \mathfrak{h}^+ + \mathfrak{h}^-$ and take $\varphi \in e^{\operatorname{ad}\zeta}$ $(\zeta \in \mathfrak{h}^-)$ in such a way that $\sigma_3 = \varphi^2$ has the algebra of immobile elements $\mathfrak{h}_3^+ \cong su(5) \oplus u(1)$ (with the natural embedding into $su(6)$), which is possible as is verified in the matrix model. Then we have

$$\mathfrak{g}_3^+ = \mathfrak{m}_3^+ + \mathfrak{h}_3^+, \qquad \mathfrak{h}_3^+ = \mathfrak{z} \oplus \mathfrak{h}' \cong u(1) \oplus su(5),$$

and $\operatorname{ad} \mathfrak{h}'$ acts on $\mathfrak{m}_3^+$ by skew-symmetric endomorphisms. We have $\dim \mathfrak{m}_3^+ = 6$, $\dim \mathfrak{h}' = 24$, but the maximal dimension of the algebra of all skew-symmetric endomorphisms is $\frac{1}{2}(6 \times 5) = 15$. Since $\mathfrak{h}' \cong su(5)$ is simple, the above is possible only if $\operatorname{ad} \mathfrak{h}'$ acts on $\mathfrak{m}_3^+$ in the trivial way.

But then either

$$(\mathfrak{g}_3^+, \mathfrak{h}_3^+) \cong (\tilde{\mathfrak{g}}, \mathfrak{z}) + (\mathfrak{h}', \mathfrak{h}') + (\tilde{\tilde{\mathfrak{g}}}, \{0\}),$$

where $\tilde{\mathfrak{g}}$ is three-dimensional simple, $\tilde{\tilde{\mathfrak{g}}}$ is abelian four-dimensional, or $\mathfrak{h}'$ and $\mathfrak{z}$, and consequently $\mathfrak{h}_3^+$, act on $\mathfrak{m}_3^+$ in the trivial way. In any of the above cases the involutive algebra $\mathfrak{g}_3^+$ has either a four-dimensional or a six-dimensional commutative ideal, which is impossible for a simple compact Lie algebra $\mathfrak{g}$ (see I.1).

Thus this case is also impossible.

2. Another possibility is $(n \neq 6)$

$$(\mathfrak{h}, \mathfrak{h}^+) = (\mathfrak{z}, \mathfrak{z}) + (\mathfrak{h}_2, \tilde{\mathfrak{h}}_2),$$

where $\mathfrak{z} \cong u(1)$, $\tilde{\mathfrak{h}}_2 \cong su(2) \oplus so(n-4)$.

We note that here $\mathfrak{h}_2$ is simple, otherwise

$$(\mathfrak{h}_2, \tilde{\mathfrak{h}}_2) = (\mathfrak{h}_2', \tilde{\mathfrak{h}}_2') + (\mathfrak{h}_2'', \tilde{\mathfrak{h}}_2''),$$

where $\mathfrak{h}_2' \cong su(2)$, $\mathfrak{h}_2'' \cong so(n-4)$.

Again, as in the previous considerations, $\tilde{\mathfrak{h}}_2' + \mathfrak{i}$ (or $\tilde{\mathfrak{h}}_2'' + \mathfrak{i}$), where $\mathfrak{i}$ is the subalgebra of all elements commuting with $\tilde{\mathfrak{h}}_2'$ (respectively, with $\tilde{\mathfrak{h}}_2''$), is larger than $\mathfrak{g}^+$, which is impossible in the case of a mobile mirror.

Note that $\mathfrak{h}_2$ commutes with $\mathfrak{z} \cong u(1)$.

The maximal subalgebra of all elements commuting with $\mathfrak{z} \cong u(1)$ (which is isomorphic to a diagonal in $so(2) \oplus so(2)$) is, however,

$$\mathfrak{z} \oplus \tilde{\mathfrak{h}}_2 \cong u(1) \oplus su(2) \oplus so(n-4).$$

Consequently in this case $\mathfrak{h} = \mathfrak{h}^+$, which contradicts the mobility of the mirror.

3. We consider lastly $\mathfrak{g} \cong so(6) \cong su(4)$.

Here

$$(\mathfrak{g}^+, \mathfrak{h}^+) \cong (so(4) \oplus so(2), u(1) \oplus su(2) \oplus so(2)),$$

$$(\mathfrak{h}, \mathfrak{h}^+) = (\mathfrak{h}, \mathfrak{b} \oplus \mathfrak{z} \oplus \tilde{\mathfrak{z}}),$$

$$\mathfrak{b} \oplus \mathfrak{z} \oplus \tilde{\mathfrak{z}} \cong su(2) \oplus u(1) \oplus so(2).$$

The decomposition which we are interested in is:
either

1. $(\mathfrak{h}, \mathfrak{h}^+) = (\mathfrak{h}_1, \mathfrak{b} \oplus \mathfrak{z}_1) + (\mathfrak{h}_2, \mathfrak{z}_2)$, where $\mathfrak{h}_1$ is simple, $\mathfrak{b}$ is three-dimensional simple compact, $\mathfrak{z}_1$ and $\mathfrak{z}_2$ are one-dimensional (but, generally speaking, different from $\mathfrak{z}$ and $\tilde{\mathfrak{z}}$), $\mathfrak{h}_2$ is simple and three-dimensional or $\mathfrak{h}_2$ is commutative;
   or

2. $(\mathfrak{h}, \mathfrak{h}^+) = (\mathfrak{h}_1, \mathfrak{z}_1) + (\mathfrak{h}_2, \mathfrak{z}_2) + (\mathfrak{h}_3, \mathfrak{b})$, where $\mathfrak{h}_1$ is simple three-dimensional, $\mathfrak{h}_2$ is commutative, $\mathfrak{h}_3$ is semi-simple, $\mathfrak{b}$ is three-dimensional, and $\mathfrak{z}_1$, $\mathfrak{z}_2$ are one-dimensional (not coinciding, in general, with $\mathfrak{z}$, $\tilde{\mathfrak{z}}$).

In the first case taking into account the list of involutive pairs (E. Cartan) and that $\mathfrak{b} \oplus \mathfrak{z}_1$ is a unitary principal involutive algebra of $\mathfrak{h}_1$ we obtain

$$(\mathfrak{h}_1, \mathfrak{b} \oplus \mathfrak{z}_1) \cong (su(3), su(2) \oplus u(1)).$$

Furthermore, $\mathfrak{h}_2 = \mathfrak{z}_2$, otherwise the subalgebra $\mathfrak{b} \oplus \mathfrak{i}$, where $\mathfrak{i}$ consists of all elements commuting with $\mathfrak{b}$, is larger then $\mathfrak{g}^+$ (which is impossible since $\mathfrak{g}^+$ is an involutive algebra). The basis element of $\mathfrak{z}_2$ in the matrix representation has the form

$$\left\| \begin{array}{c|c|c} \begin{matrix} 0 & \lambda \\ -\lambda & 0 \end{matrix} & \overline{\overline{O}} & \overline{\overline{O}} \\ \hline \overline{\overline{O}} & \begin{matrix} 0 & \mu \\ -\mu & 0 \end{matrix} & \overline{\overline{O}} \\ \hline \overline{\overline{O}} & \overline{\overline{O}} & \begin{matrix} 0 & \nu \\ -\nu & 0 \end{matrix} \end{array} \right\|,$$

and $\mathfrak{h}_1 \cong su(3)$ commutes with it.

But, as the matrix model shows, this is possible only if

$$\left\|\begin{array}{c|c|c} \begin{matrix} 0 & 1 \\ -1 & 0 \end{matrix} & \overline{\overline{O}} & \overline{\overline{O}} \\ \hline \overline{\overline{O}} & \begin{matrix} 0 & 1 \\ -1 & 0 \end{matrix} & \overline{\overline{O}} \\ \hline \overline{\overline{O}} & \overline{\overline{O}} & \begin{matrix} 0 & 1 \\ -1 & 0 \end{matrix} \end{array}\right\|$$

(after re-numbering of rows and columns).

Then $\mathfrak{h} = \mathfrak{h}_1 \oplus \mathfrak{z}_2 \cong u(3)$ with the natural embedding into $so(6) \cong su(4)$. This gives us the involutive pair

$$(\mathfrak{g}, \mathfrak{h}) \cong (so(6), u(3)) \cong (su(4), u(3))$$

with the natural embeddings.

Here

$$(\mathfrak{g}^+, \mathfrak{h}^+) \cong ((so(4) \oplus so(2), u(1) \oplus su(2) \oplus so(2)),$$

$$(\mathfrak{h}, \mathfrak{h}^+) \cong (u(3), u(2) \oplus u(1)) \cong (su(3), u(2)) + (u(1), u(1))$$

with the natural embeddings.

The second possibility, that is,

$$(\mathfrak{h}, \mathfrak{h}^+) = (\mathfrak{h}_1, \mathfrak{z}_1) + (\mathfrak{h}_2, \mathfrak{z}_2) + (\mathfrak{h}_3, \mathfrak{b})$$

is not valid if $\mathfrak{h}_3 \neq \mathfrak{b}$.

Indeed, here $(\mathfrak{h}_3, \mathfrak{b})$ is an involutive pair, therefore, as is easily seen, dim $\mathfrak{h}_3 = 6$, dim $\mathfrak{m} = 3$, that is, $(\mathfrak{h}_3, \mathfrak{b}) \cong (so(4), so(3))$.

But $\sigma_1 \in e^{\mathrm{ad}\mathfrak{b}}$ since σ_1 is di-unitary, consequently the restriction of σ_1 on $\mathfrak{h}_3$ is a unitary principal involutive automorphism, which is wrong for $(so(4), so(3))$. This contradiction shows that in our case $\mathfrak{h}_3 = \mathfrak{b}$.

Furthermore, we have $\mathfrak{h}_2 = \mathfrak{z}_2$, $\mathfrak{h}_1 = \mathfrak{z}_1$ (otherwise the subalgebra $\mathfrak{b} \oplus i$, where i consists of all elements commuting with $\mathfrak{b}$, is larger than the involutive algebra $\mathfrak{g}^+$, which is impossible). But then we obtain an immobile mirror, which is wrong.

Thus we have the theorem:

IV.6.10. Theorem 13. *Let* $\mathfrak{g} \cong so(n)$, $n > 4$, *and a pair* $(\mathfrak{g}, \mathfrak{h})$ *has a two-dimensional mobile mirror of unitary type. Then this pair is involutive and*

$$(\mathfrak{g}, \mathfrak{h}) \cong (so(6), u(3)) \cong (su(4), u(3)).$$

Furthermore,

$$(\mathfrak{g}^+, \mathfrak{h}^+) \cong (so(4) \oplus so(2), u(1) \oplus su(2) \oplus su(2)),$$

$$(\mathfrak{h}, \mathfrak{h}^+) \cong (u(3), u(2) \oplus u(1)) \cong (su(3), u(2)) + (u(1), u(1))$$

with the natural embeddings.
This case is tri-symmetric.

CHAPTER IV.7

HOMOGENEOUS RIEMANNIAN SPACES WITH
SIMPLE COMPACT LIE GROUPS OF MOTIONS
$G \ncong SO(n), SU(3)$ AND TWO-DIMENSIONAL MIRRORS

IV.7.1. In this case the possible Lie algebras $\mathfrak{g}$ are:

$$su(n) \quad (n \geq 4), \qquad sp(n) \quad (n \geq 2), \qquad g_2, \quad f_4, \quad e_6, \quad e_7, \quad e_8.$$

We consider first the case of a mobile two-dimensional mirror. By IV.5.12 (Theorem 10) in our case the pair $(\mathfrak{g}, \mathfrak{h})$ gives the iso-involutive decomposition

$$\mathfrak{g} = \mathfrak{g}_1^+ + \mathfrak{g}_2^+ + \mathfrak{g}_3^+,$$

with respect to which $(\mathfrak{g}, \mathfrak{h})$ is invariant and tri-symmetric, and $\dim \mathfrak{m}_1^+ = \dim \mathfrak{m}_2^+$.

We consider separately two subcases: the case of the non-symmetric pair $(\mathfrak{g}, \mathfrak{h})$ and the case of the symmetric pair $(\mathfrak{g}, \mathfrak{h})$.

1. $(\mathfrak{g}, \mathfrak{h})$ is non-symmetric. In the language of the corresponding homogeneous spaces we have a tri-symmetric non-symmetric homogeneous space G/H with a simple compact Lie group of motions G which is isomorphic to $SU(n)$ $(n \geq 4)$, $Sp(n)$ $(n \geq 2)$, G_2, F_4, E_6, E_7, E_8. All such spaces are known (see Part III or [L.V. Sabinin, 70, 72]).

We write out the corresponding table in the language of Lie algebras:

$\mathfrak{g}/\mathfrak{h}$	mirror	$(\mathfrak{g}_\alpha^+, \mathfrak{h}_\alpha^+)$
$g_2/su(3)$	with the principal mirror	$su(2)/so(2)$
$f_4/su(3) \oplus su(3)$	with the principal mirror	$sp(3)/u(3)$
$e_6/su(3) \oplus su(3) \oplus su(3)$	with the principal mirror	$su(6)/s(u(3) \oplus u(3))$
$e_7/f_4 \oplus so(3)$	with the central mirror	e_6/f_4
$e_7/su(6) \oplus su(3)$	with the principal mirror	$so(12)/u(6)$
$e_8/e_6 \oplus su(3)$	with the principal mirror	$e_7/e_6 \oplus u(1)$
$so(4m)/sp(m) \oplus so(3)$	with the central mirror	$su(2m)/sp(m)$
$su(2m)/su(m) \oplus so(3)$	with the central mirror	$su(m) \oplus su(m)/su(m)$
$sp(n)/so(n) \oplus so(3)$	with the central mirror	$su(n)/so(n)$

We note that from the first point of view the maximality of a subgroup H in G (of a subalgebra $\mathfrak{h}$ in $\mathfrak{g}$) is required in the cited works. But in fact, if $G \not\cong SO(n)$ then the proofs in the above cited works do not require the maximality of H in G ($\mathfrak{h}$ in $\mathfrak{g}$) as is easily seen.

We are interested in the cases with two-dimensional mirrors, $\dim \mathfrak{m}_1^+ = 2$, which can be separated out from the above table by a simple calculation of mirror dimensions, $\dim \mathfrak{g}_1^+ - \dim \mathfrak{h}_1^+ = \dim \mathfrak{m}_1^+ = 2$. As a result the only possibility is $(\mathfrak{g}, \mathfrak{h}) \cong (g_2, su(3))$.

Thus we have:

IV.7.2. Theorem 14. *Let $\mathfrak{g}$ be a simple compact Lie algebra non-isomorphic to $so(n)$ or $su(3)$, and let $(\mathfrak{g}, \mathfrak{h})$ be a non-symmetric pair with a mobile two-dimensional mirror. Then*

$$(\mathfrak{g}, \mathfrak{h}) \cong g_2/su(3),$$

$$(\mathfrak{g}^+, \mathfrak{h}^+) \cong (su(2) \oplus su(2), so(2) \oplus su(2))$$

$$\cong (su(2), so(2)) + (su(2), su(2)),$$

$$(\mathfrak{h}, \mathfrak{h}^+) \cong (su(3), u(2))$$

with the natural embeddings.

In this case the mirror is of unitary type and the case is tri-symmetric.

IV.7.3. Remark. Note that in the above case we have a hyper-tri-symmetric case, that is, all three mirrors are isomorphic and there exists an automorphism p ($p^3 = \mathrm{Id}$) such that

$$p\,\mathfrak{g}_1^+ = \mathfrak{g}_2^+, \qquad p\,\mathfrak{g}_2^+ = \mathfrak{g}_3^+, \qquad p\,\mathfrak{g}_3^+ = \mathfrak{g}_1^+, \qquad p\,\mathfrak{h} = \mathfrak{h}$$

(see [L.V. Sabinin, 70,72]).

2. Now, let the pair $(\mathfrak{g}, \mathfrak{h})$ be symmetric. Since the simple compact Lie algebra $\mathfrak{g}$ is non-isomorphic to $so(n)$ or $su(3)$ we see that the principal involutive algebra of the mirror $\mathfrak{g}_1^+$ is of unitary type (see Part I, II or [L.V. Sabinin, 70,72]).

That is,

$$\mathfrak{g}_1^+ = \mathfrak{m}_1^+ + \mathfrak{h}_1^+ = \mathfrak{b}_1 \oplus \tilde{\mathfrak{h}}_1,$$

$$\mathfrak{b}_1 = \mathfrak{m}_1^+ + \mathfrak{z}_1, \qquad \mathfrak{h}_1^+ = \mathfrak{z}_1 \oplus \tilde{\mathfrak{h}}_1,$$

where $\dim \mathfrak{z}_1 = 1$, $\dim \mathfrak{b}_1 = 3$, and the involutive automorphism σ_1 of the mirror $\mathfrak{g}_1^+$ belongs to $\exp(\mathrm{ad}\,\mathfrak{b}_1) \cong SU(2)$ (see IV.5).

Taking now any $\zeta \in \mathfrak{m}_1^+$ we have

$$\sigma_1 \in \exp(\mathrm{ad}(\mathbb{R}\,\zeta)).$$

Therefore if $[\eta\,\zeta] = 0$ ($\eta \in \mathfrak{m}$) then $\sigma_1\eta = \eta$. But then $\eta \in \mathfrak{g}^+ \cap \mathfrak{m} = \mathfrak{m}_1^+$, whence $\eta = \lambda\,\zeta$.

This means that the involutive pair $(\mathfrak{g}, \mathfrak{h})$ is of rank 1. All involutive pairs of rank 1 with a simple compact Lie algebra $\mathfrak{g}$ non-isomorphic to $so(n)$ or $su(n)$ are known (see, for example, [S. Helgason 62, 78]).

Those are:

$$(su(n), u(n-1)), \quad n > 4; \quad (sp(n), sp(n-1) \oplus sp(1)), \quad n > 2; \quad (f_4, so(9)).$$

Let us now use iso-involutive decompositions which we can construct by means of an iso-involutive rotation generated by a vector $\zeta \in \mathfrak{m}_1^+ \subset \mathfrak{g}_1^+$.

In this way we then obtain basis involutive decompositions of involutive pairs of rank 1 for the simple compact Lie algebras pointed out above (see [L.V. Sabinin, 69]).

All such iso-involutive decompositions have been determined in [L.V. Sabinin, 65, 69]. Under the notations there S_3 coincides with σ_1.

Below is the table which we need.

$(\mathfrak{g}, \mathfrak{h})$	$(\mathfrak{g}_1^+, \mathfrak{h}_1^+)$
$(so(n+1), so(n))$	$(so(n-1) \oplus so(2), so(n-1))$
$(su(n+1), u(n))$	$(su(n-1) \oplus su(2) \oplus u(1), su(n-1) \oplus u(1) \oplus u(1))$
$(sp(n+1), sp(n) \oplus sp(1))$	$(sp(n-1) \oplus sp(2), sp(n-1) \oplus sp(1) \oplus sp(1))$
$(f_4, so(9))$	$(so(9), so(8))$

We should select those of the above pairs for which the rotating vector generates a three-dimensional mirror. There is the only possibility $(su(n), u(n-1))$, $n > 4$, with the principal mirror

$$(\mathfrak{g}_1^+, \mathfrak{h}_1^+) \cong (su(n-2) \oplus su(2) \oplus u(1), su(n-2) \oplus u(1) \oplus u(1)).$$

(Note that here $\mathfrak{h} = \mathfrak{g}_2^+$).

Thus we have the theorem:

IV.7.4. Theorem 15. *Let $(\mathfrak{g}, \mathfrak{h})$ be an involutive pair with a simple compact Lie algebra $\mathfrak{g}$ non-isomorphic to $so(n)$ or $su(3)$ with a two-dimensional mobil mirror. Then*

$$(\mathfrak{g}, \mathfrak{h}) \cong (su(n), su(n-1)), \quad n > 4;$$

$$(\mathfrak{g}_1^+, \mathfrak{h}_1^+) \cong (su(n-2) \oplus su(2) \oplus u(1), su(n-2) \oplus u(1) \oplus u(1)),$$

$$(\mathfrak{h}, \mathfrak{h}_1^+) \cong (u(n-1), su(n-2) \oplus u(1) \oplus u(1))$$

$$= (su(n-1), u(n-2)) + (u(1), u(1))$$

with the natural embeddings.

The reformulation in the language of homogeneous spaces is evident.

CHAPTER IV.8

HOMOGENEOUS RIEMANNIAN SPACES WITH
SIMPLE COMPACT LIE GROUPS OF MOTIONS
AND TWO-DIMENSIONAL IMMOBILE MIRRORS

IV.8.1. It remains to determine all pairs $(\mathfrak{g}, \mathfrak{h})$ with a simple compact Lie algebra $\mathfrak{g}$ and two-dimensional immobile mirrors.

In this case $\mathfrak{h}^- = \{0\}$ (see IV.4.2). Taking into account that $\mathfrak{g}^+ = \mathfrak{b} \oplus \mathfrak{c}$, where $\mathfrak{b}$ is a three-dimensional simple ideal, we have $\mathfrak{h} = \mathfrak{z} \oplus \mathfrak{c}$, $\mathfrak{b} = \mathfrak{z} \dotplus \mathfrak{m}^+$, where $[\mathfrak{m}^+\mathfrak{m}^+]$ is one-dimensional.

Consequently the problem is reduced to the classification of simple compact Lie algebras together with their principal involutive automorphisms.

We now use such a classification. See Part I, II or [L.V. Sabinin, 69, 70].

As a result we have:

IV.8.2. Theorem 16. *Let $\mathfrak{g}$ be a simple compact Lie algebra, and let a pair $(\mathfrak{g}, \mathfrak{h})$ have an immobile two-dimensional mirror.*

Then for $(\mathfrak{g}, \mathfrak{h})$ we have the following possibilities:

$$(so(n), so(n-3) \oplus so(2)), \qquad (so(n), so(n-4) \oplus su(2) \oplus so(2)),$$

$$(su(n), su(n-2) \oplus u(1) \oplus u(1)), \qquad (sp(n), sp(n-1) \oplus u(1)),$$

$$(g_2, su(2) \oplus so(2)), \qquad (f_4, sp(3) \oplus u(1)),$$

$$(e_6, so(6) \oplus u(1)), \qquad (e_7, so(12) \oplus u(1)),$$

$$(e_8, e_7 \oplus u(1)), \qquad (su(3), so(2))$$

with the natural embeddings.

The reformulation from the language of Lie algebras into the language of Lie groups and homogeneous spaces is evident.

APPENDIX ONE

Когда волшебница Геката
Варила студень в Час Заката,
Смешав кагор и мандрагор,
Родился Я - Квантасмагор.

Квантасмагор

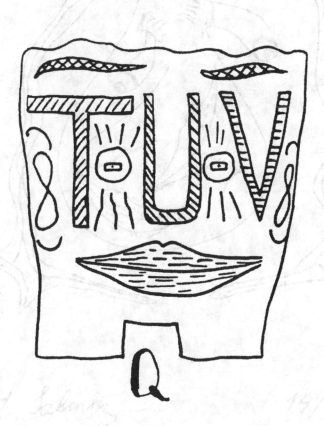

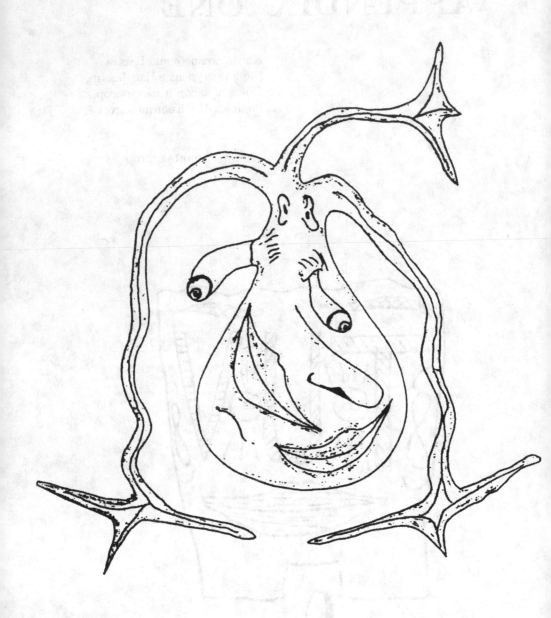

APPENDIX 1

ON THE STRUCTURE OF T, U, V-ISOSPINS
IN THE THEORY OF HIGHER SYMMETRY

A.1.1. Introduction. One of the aims of this brief appendix is to give an invariant description of the subalgebras of T, U, V-isospins of the algebra $su(3)$ connected with the unitary symmetry group $SU(3)$ [Nguen-Van-Hew, 67].

It is surprising, indeed, that in the general theory of the unitary symmetry these subgroups are introduced in a non-invariant manner by means of some generators of a concrete matrix representation, although if T, U, V-isospins structure has a physical meaning it should be described by means of its Lie algebra (or group) only, without any concrete representation, that is, in an invariant way.

An invariant description of T, U, V-isospins can be based on the so called Mirror Geometry (or Involutive Calculus) which has been constructed for the purposes of differential geometry, the geometry of homogeneous spaces, theoretical physics and presented in our works, see the Bibliography of this monograph.

Firstly, the case of $su(3)$ is considered and some special algebraic structures are obtained. Secondly, these structures are applied to the theory of higher symmetry, especially to the problem of T, U, V-isospins subalgebras of some Lie algebras, and solutions of problems appeared are given.

A.1.2. Notations. The following notations are used in this appendix:

$\mathrm{Int}(\mathfrak{g})$ is the group of all inner automorphisms of $\mathfrak{g}$.

$\mathrm{Int}_{\mathfrak{g}}(\mathfrak{a})$ is the restriction of $\mathrm{Int}(\mathfrak{g})$ to $\mathfrak{a} \subset \mathfrak{g}$.

$\mathrm{Int}_{\mathfrak{g}}^{\mathfrak{b}}(\mathfrak{a})$ $= \{\varphi \mid \varphi = f|_{\mathfrak{b}},\ f \in \mathrm{Int}_{\mathfrak{g}}(\mathfrak{a})\}$.

$\mathfrak{g} = \mathfrak{a} \oplus \mathfrak{b}$ is the direct sum of the ideals $\mathfrak{a}$ and $\mathfrak{b}$.

$\mathfrak{p} = \mathfrak{a} + \mathfrak{b}$ is the sum of the subspaces $\mathfrak{a}$ and $\mathfrak{b}$ of the linear space $\mathfrak{p}$.

$\mathfrak{g}/\mathfrak{l}$ is an involutive pair, that is, the pair $(\mathfrak{g}, \mathfrak{l})$ such that $\mathfrak{l}$ is a subalgebra of some involutive automorphism S of $\mathfrak{g}$.

$\mathfrak{g}/\mathfrak{l} = \mathfrak{a}/\mathfrak{b} + \mathfrak{c}/\mathfrak{d}$ $\Leftrightarrow \mathfrak{g} = \mathfrak{a} \oplus \mathfrak{c}, \quad \mathfrak{l} = \mathfrak{b} \oplus \mathfrak{d}, \quad \mathfrak{b} \subset \mathfrak{a}, \quad \mathfrak{d} \subset \mathfrak{c}$.

$\mathbb{R}$ is the field of real numbers.

$\cong$ is the simbol of an isomorphism.

$\mathfrak{g} \dot{-} \mathfrak{l} = \dfrac{\mathrm{Id} - S}{2}\mathfrak{g}$, where S is an isomorphism of $\mathfrak{g}$ $\left(\mathfrak{l} = \dfrac{\mathrm{Id} + S}{2}\mathfrak{g}\right)$.

A.1.3. Let $\mathfrak{g} \cong su(3)$. Then $\mathfrak{g}$ is a linear representation of the adjoint group $\mathrm{Int}(\mathfrak{g}) \cong SU(3)$. The operator of the hypercharge, $Y = Y_T$, commutes with the T-isospin subalgebra ($Y_T X = 0$ for each $X \in T$) as is well known. It is easily seen that $\mathrm{Int}_{\mathfrak{g}}(\mathbb{R}Y)$ is a compact subgroup in $\mathrm{Int}(\mathfrak{g})$ because each proper number of the hypercharge operator Y is either an integer or a half-integer in conformity with physical arguments of the theory, consequently there exists $S_1 \in \mathrm{Int}_{\mathfrak{g}}(\mathbb{R}Y)$ such that $S_1 \neq \mathrm{id}_{\mathfrak{g}}$, $S_1^2 = \mathrm{id}_{\mathfrak{g}}$ (so called an involutive automorphism [S. Helgason, 62, 78], or, more briefly, invomorphism [L.V. Sabinin, 65a,b]). Then $S_1 \eta = \eta$ for $\eta \in T \oplus \mathbb{R}Y$. As is well known, the maximal subalgebra $\mathfrak{l}_1$ of elements immobile under the action of the inner involutive automorphism S_1 (the so called involutive algebra) has the form $\mathfrak{l}_1 \cong su(2) \oplus u(1)$ in the case of $\mathfrak{g} \cong su(3)$ [S. Helgason, 62,78]. Consequently $\mathfrak{l}_1 = T \oplus \mathbb{R}Y$ (such an algebra is said to be principal [L.V. Sabinin, 67a,b], because $T \cong su(2)$, and central [L.V. Sabinin, 67a,b], because $\mathbb{R}Y \cong u(1)$ is a non-trivial centre in $\mathfrak{l}_1$).

Moreover, $\mathfrak{l}_1$ is a principal unitary involutive algebra [L.V. Sabinin, 67a,b, 70a], that is, $S_1 \in \mathrm{Int}_{\mathfrak{g}}(T) \cong SU(2)$ (for $SU(3)$, as is well known, there exist no other inner involutive automorphisms besides those mentioned above). We note also that $S_1 \in \mathrm{Int}_{\mathfrak{g}}(\mathbb{R}Y)$, which is obvious. In this way we have an interior description of the T-isospin subalgebra and the hypercharge operator Y in $\mathfrak{g}$ by means of two methods:

1. $T \oplus \mathbb{R}Y$ is an involutive algebra of a principal unitary involutive automorphism S_1, $S_1 \in \mathrm{Int}_{\mathfrak{g}}(T) \cong SU(2)$.

2. $T \oplus \mathbb{R}Y$ is an involutive algebra of a central involutive automorphism S_1, $S_1 \in \mathrm{Int}_{\mathfrak{g}}(\mathbb{R}Y)$.

It is reasonable here to illustrate the physical meaning of the involutive automorphism S_1, namely, $S_1 = E_T^2$, where E_T is the operator of T-charge symmetry [Nguen-Van-Hew, 67], [K. Nishijima, 65].

$\mathfrak{g}/T \oplus \mathbb{R}Y \cong su(3)/su(2) \oplus u(1)$ is an involutive pair with the natural embedding [S. Helgason, 62, 78], so that the description above is the unique description. The subalgebras of U and V-isospins are introduced in [Nguen-Van-Hew, 67], [U. Rumer, A. Fet, 70] in a formal way by means of giving basis operators in some representation, although their connection with infinitesimal rotations about the coordinate axes of unitary three-dimensional space has been taken into account.

Reasoning analogously to the case of T-isospin shows us that $\mathfrak{l}_2 = U \oplus \mathbb{R}Y_U$ (here Y_U is an operator commuting with U) is the involutive algebra of the involutive automorphism $S_2 \in \mathrm{Int}_{\mathfrak{g}}(U) \cong SU(2)$, $S_2 \in \mathrm{Int}_{\mathfrak{g}}(\mathbb{R}Y_U)$. We note that $S_2 = E_U^2$, where E_U is the operator of U-charge symmetry [Nguen-Van-Hew, 67]. Similarly $\mathfrak{l}_3 = V \oplus \mathbb{R}Y_V$ is the involutive algebra of the involutive automorphism $S_2 \in \mathrm{Int}_{\mathfrak{g}}(V) \cong SU(2)$, $S_3 \in \mathrm{Int}_{\mathfrak{g}}(\mathbb{R}Y_V)$.

It is now easily seen that S_1, S_2, S_3 are pair-wise commutative and the product of arbitrary two of them equals to the third of them. (Indeed, $E_V T = U$ for physical reason and $E_V \mathfrak{l}_1 = \mathfrak{l}_2$, which implies $E_V^{-1} S_2 E_V = S_1$, or $S_2 E_V = E_V S_1$. But $S_1 E_V S_1 = E_V^{-1}$, $E_V^2 = S_3$, consequently $S_2 S_3 = S_2(E_V^2) = E_V S_1 E_V = S_1(S_1 E_V S_1) E_V = S_1 E_V^{-1} E_V = S_1$, etc.). Thus the discrete commutative group (the so called involutive group) $\{\mathrm{id}_{\mathfrak{g}}, S_1, S_2, S_3\} \subset \mathrm{Int}(\mathfrak{g})$ is obtained. In particular, this means that $\mathfrak{l}_\alpha \dot{-} \mathfrak{l}_0 = \{\eta \mid \eta \in \mathfrak{l}_\alpha,\ S_\beta \eta = -\eta,\ \beta \neq \alpha\}$, $\alpha = 1, 2, 3$, (where $\mathfrak{l} = \mathfrak{l}_\nu \cap \mathfrak{l}_\mu,\ \nu \neq \mu$) are pair-wise orthogonal subspaces.

The involutive group $\{\mathrm{id}_\mathfrak{g}, S_1, S_2, S_3\}$ generates the involutive sum decomposition [L.V. Sabinin, 65a,b,68] $\mathfrak{g} = \mathfrak{l}_1 + \mathfrak{l}_2 + \mathfrak{l}_3$, where $\mathfrak{l}_\alpha$ is an involutive algebra of S_α. This sum is hyper-involutive [L.V. Sabinin, 67a,b], that is, there exists $\rho \in \mathrm{Int}(\mathfrak{g})$, $\rho^3 = \mathrm{id}_\mathfrak{g}$, such that $\rho\,\mathfrak{l}_1 = \mathfrak{l}_2$, $\rho\,\mathfrak{l}_2 = \mathfrak{l}_3$, $\rho\,\mathfrak{l}_3 = \mathfrak{l}_1$. Indeed, $\rho = E_U E_V$ may be taken as that one. In this way the algebra $\mathfrak{g} \cong su(3)$ is decomposed into an hyper-involutive sum decomposition of principal unitary involutive algebras $\mathfrak{l}_1, \mathfrak{l}_2, \mathfrak{l}_3$ connected with T, U, V-isospin algebras.

Let us now consider $\mathfrak{l}_0 = \mathfrak{l}_1 \cap \mathfrak{l}_2 = \mathfrak{l}_2 \cap \mathfrak{l}_3 = \mathfrak{l}_3 \cap \mathfrak{l}_1 \cong u(1) \oplus u(1)$. $\mathfrak{l}_\alpha/\mathfrak{l}_0$ is an involutive pair [L.V. Sabinin, 65a,b, 68] because $\mathfrak{g} = \mathfrak{l}_1 + \mathfrak{l}_2 + \mathfrak{l}_3$ is an involutive decomposition. In particular, in this case $\mathfrak{l}_1/\mathfrak{l}_0 \cong T/(\mathbb{R}T_3) + (\mathbb{R}Y_T)/(\mathbb{R}Y_T) \cong su(2)/u(1) + u(1)/u(1)$ (the notational form $\mathfrak{m}/\mathfrak{n} = \mathfrak{a}/\mathfrak{b} + \mathfrak{c}/\mathfrak{d}$ means $\mathfrak{m} = \mathfrak{a} \oplus \mathfrak{c}$, $\mathfrak{n} = \mathfrak{b} \oplus \mathfrak{d}$, $\mathfrak{b} \subset \mathfrak{a}$, $\mathfrak{c} \subset \mathfrak{d}$). $\rho Y_T = Y_U$ because $\rho\,\mathfrak{l}_1 = \mathfrak{l}_2$ (analogously, by the way, $E_V Y_T = Y_U$ and so on).

Thus the two-dimensional plane $\mathfrak{l}_0$ contains Y_T and $Y_U = \rho Y_T$ which are of the same length in the Cartan metric of $\mathfrak{g}$ because $\rho \in \mathrm{Aut}(\mathfrak{g})$. Therefore after a suitable choice of T_3 (it must be, of course, orthogonal to Y_T because of the commutativity of T and Y_T) we have $\rho Y_T = Y_U = -(T_3 + \frac{1}{2}Y_T)$ and if we set $Y_U = -Q$ then we have $Q = T_3 + \frac{1}{2}Y_T$ (which is the well known formula of Gell–Mann and Nishijima that the charge is equal to the component of isospin plus half of the hypercharge [Nguen-Van-Hew, 67], [L.V. Sabinin, 69].

By the previous arguments it follows that the operators of the charge Q and the hypercharge are conjugate under the automorphism $\rho \in \mathrm{Int}(\mathfrak{g}) \cong SU(3)$ (up to sign), that is, $\rho Y = -Q$. This last result might, of course, be taken as a preliminary knowledge, since the symmetry of the charge and the hypercharge is well known physical phenomenon.

It is curious to note that the signs of the charge's and the hypercharge's operators are not satisfactorily chosen in the standard theory. It would need to be called $(-Y)$ the hypercharge operator because of the conjugacy of $(-Y)$ and Q under ρ (or E_V). It is, of course, possible to change the sign of the charge operator Q instead.

Thus if we would like to generalize the concept of T, U, V-isospins for an algebra $\mathfrak{g}$ in the theory of higher symmetry we would have the following construction.

The algebra Lie $\mathfrak{g} = \mathfrak{l}_1 + \mathfrak{l}_2 + \mathfrak{l}_3$ is an hyper-involutive sum [L.V. Sabinin, 67a,b, 70a] of involutive algebras $\mathfrak{l}_\alpha$ generated by inner involutive automorphisms $S_\alpha \in \mathrm{Int}(\mathfrak{g})$. Moreover, $T \subset \mathfrak{l}_1$, $U \subset \mathfrak{l}_2$, $V \subset \mathfrak{l}_3$ and $\rho\Gamma = U$, $\rho U = V, \rho V = T$, $S_1 \in \mathrm{Int}_\mathfrak{g}(T)$, $S_2 \in \mathrm{Int}_\mathfrak{g}(U)$, $S_3 \in \mathrm{Int}_\mathfrak{g}(V)$, where ρ is the conjugating automorphism of the hyper-involutive decomposition; $\rho^3 = \mathrm{id}_\mathfrak{g}$, $\rho\,\mathfrak{l}_1 = \mathfrak{l}_2$, $\rho\,\mathfrak{l}_2 = \mathfrak{l}_3$, $\rho\,\mathfrak{l}_3 = \mathfrak{l}_1$.

Furthermore, it is natural to suppose that the algebra of T-isospin is embedded into $\mathfrak{l}_1$ analogously to the case of $SU(3)$ theory, namely, $\mathfrak{l}_1 = T \oplus \tilde{\mathfrak{l}}_1$. But then we have $\mathfrak{l}_2 = U \oplus \tilde{\mathfrak{l}}_2$, $\mathfrak{l}_3 = V \oplus \tilde{\mathfrak{l}}_3$ by virtue of the conjugacy of $\mathfrak{l}_1, \mathfrak{l}_2, \mathfrak{l}_3$ and T, U, V, respectively, under acting of ρ.

Thus $\mathfrak{l}_\alpha$ are the principal unitary involutive algebras of the hyper-involutive sum decomposition [L.V. Sabinin, 67a,b, 70a].

The above having been said, there has arisen the problem of defining all simple compact algebras $\mathfrak{g}$ admitting hyper-involutive sum decompositions $\mathfrak{g} = \mathfrak{l}_1 + \mathfrak{l}_2 + \mathfrak{l}_3$ with inner involutive automorphisms S_α and to classify all these decompositions (in particular, to clarify all algebras admitting the principal unitary hyper-involutive sum decompositions).

All principal unitary hyper-involutive sum decompositions of simple compact Lie algebras are given below in Table 1 [L.V. Sabinin, 71a,b].

Table 1

$\mathfrak{g}$	Principal unitary involutive pair $\mathfrak{g}/\mathfrak{l}_\alpha$, $\alpha = 1, 2, 3$	$\mathfrak{l}_\alpha/\mathfrak{l}_0$ $(\mathfrak{l}_0 = \mathfrak{l}_\nu \cap \mathfrak{l}_\mu, \ \nu \neq \mu)$ $\alpha = 1, 2, 3$
$so(n), \ n > 5$	$so(n)/so(n-4) \oplus so(4)$	$so(n-4)/so(n-6) \oplus so(2)$ $+so(4)/so(2) \oplus so(2)$
$su(n), \ n > 2$	$su(n)/s(u(n-2) \oplus u(2))$	$su(n-2)/u(n-3)$ $+su(2)/u(1) + u(1)/u(1)$
$sp(3)$	$sp(3)/sp(2) \oplus sp(1)$	$sp(2)/sp(1) \oplus sp(1) + sp(1)/sp(1)$
f_4	$f_4/sp(3) \oplus sp(1)$	$sp(3)/u(3) + su(2)/u(1)$
e_6	$e_6/su(6) \oplus su(2)$	$su(6)/s(u(3) + su(3)) + su(2)/u(1)$
e_7	$e_7/so(12) \oplus su(2)$	$so(12)/u(6) + su(2)/u(1)$
g_2	$g_2/so(4)$	$so(4)/so(2) \oplus so(2)$

(with the natural embedding).

It is seen in Table 1 that the principal unitary hyper-involutive decompositions may be obtained for almost all types of simple compact Lie algebras except $so(3)$, $so(5)$, $su(2)$, $sp(n)$ $(n \neq 3)$.

It is remarkable that in each of the cases $T + U + V = N$ is a sublagebra in $\mathfrak{g}$ and $\mathfrak{n} \cong su(3)$. Moreover, analogously to the case of $su(3)$ symmetry $\mathfrak{n} = \mathfrak{l}_1 + \mathfrak{l}_2 + \mathfrak{l}_3$ is the principal unitary hyper-involutive sum induced on $\mathfrak{n}$ by involutive automorphisms $S_\alpha \in \mathrm{Int}(\mathfrak{g})$, where $\mathfrak{l}_1 = T \oplus \mathbb{R}Y_T$, $\mathfrak{l}_2 = U \oplus \mathbb{R}Y_U$, $\mathfrak{l}_3 = V \oplus \mathbb{R}Y_V$.

In addition, $\mathfrak{l}_0 \cap N = \mathbb{R}Y_T + \mathbb{R}Y_U = \mathbb{R}Y + \mathbb{R}Q$, that is, it constitutes the two-dimensional plane of the charge and the hypercharge.

In such a way the cases considered give us the theory of the higher symmetry by means of an enlargement of $su(3)$ symmetry theory (indeed, $su(3) \cong \mathfrak{n} \subset \mathfrak{g}$).

A.1.4. The requirement on embedding of the T-isospin algebra into $\mathfrak{l}_1 \subset \mathfrak{g}$ could be weakened but in such a way that the same situation as above would be held for $\mathfrak{g} \cong su(3)$. We require the existence of the involutive pair $\mathfrak{l}_1/T \oplus \tilde{\mathfrak{l}}_1$ such that $\mathrm{Int}_{\mathfrak{g}}(T) \cong SU(2)$, $\mathrm{Int}_{\mathfrak{l}_1}(T) \cong SO(3)$. In this case $\mathfrak{l}_1$ is called a special unitary involutive algebra in $\mathfrak{g}$ [L.V. Sabinin, 67a,b] and S_1 is called a special unitary involutive automorphism [L.V. Sabinin, 67a,b].

Every principal unitary involutive automorphism is evidently a particular case of a special unitary involutive automorphism. With this new conditions the problem of T, U, V-isospin structure is the problem of classifying all simple compact algebras together with all their special unitary hyper-involutive sum decompositions [L.V. Sabinin, 71a,b] $\mathfrak{g} = \mathfrak{l}_1 + \mathfrak{l}_2 + \mathfrak{l}_3$, which is one such that $\mathfrak{l}_\alpha$ is a special unitary involutive algebra in $\mathfrak{g}$. There is given below the Table 2 of all special unitary hyper-involutive sum decompositions which are not taken into account in Table 1 [L.V. Sabinin, 71a,b].

Table 2

$\mathfrak{g}$	Principal unitary involutive pair $\mathfrak{g}/\mathfrak{l}_\alpha,\ \alpha = 1, 2, 3$	$\mathfrak{l}_\alpha/\mathfrak{l}_0$ $(\mathfrak{l}_0 = \mathfrak{l}_\nu \cap \mathfrak{l}_\mu,\ \nu \neq \mu)$ $\alpha = 1, 2, 3$
f_4	$f_4/so(9)$	$so(9)/so(8)$
e_6	$e_6/so(10) \oplus so(2)$	$so(10)/so(8) \oplus so(2) + so(2)/so(2)$
e_7	$e_7/so(12) \oplus su(2)$	$so(12)/so(8) \oplus so(4) + su(2)/su(2)$
e_8	$e_8/so(16)$	$so(16)/so(8) \oplus so(8)$
$so(n),\ n \geq 12$	$so(n)/so(8) \oplus so(n-8)$	$so(8)/so(4) \oplus so(4)$ $+so(n-8)/so(n-12) \oplus so(4)$
$su(n),\ n \geq 6$	$su(n)/s(u(4) \oplus u(n-4))$	$su(4)/s(u(2) \oplus u(2))$ $+su(n-4)/s(u(n-6) \oplus u(2))$ $+u(1)/u(1)$
$sp(n),\ n \geq 3$	$sp(n)/sp(2) \oplus sp(n-2)$	$sp(2)/sp(1) \oplus sp(1)$ $+sp(n-2)/sp(n-3) \oplus sp(1)$

(with the natural embedding).

Amongst the cases pointed out above the case $\mathfrak{g} \cong e_6$ seems us the most remarkable. Indeed, despite that one important property of $su(3)$ symmetry, namely $\mathfrak{l}_\alpha = T \oplus \tilde{\mathfrak{l}}_\alpha$, is not holding, the other important property is correct, that is, $\mathfrak{l}_\alpha$ contains a one-dimensional centre $\mathfrak{z}_\alpha \cong so(2)$, $\mathfrak{l}_\alpha = \tilde{\mathfrak{l}}_\alpha \oplus \mathfrak{z}_\alpha \cong so(10) \oplus so(2)$. Such a one-dimensional centre has been connected with the hypercharge Y in $su(3)$ theory. Thus in this case it is natural to suppose $\mathfrak{z}_1 = \mathbb{R}Y$, where Y is the hypercharge operator. Then ρY generates the centre of $\mathfrak{l}_2$ and it is natural to set $Q = -\rho Y$, where Q is the charge operator. $\mathbb{R}Y \oplus \mathbb{R}Q$ constitutes the centre of $\mathfrak{l}_0 \cong so(8) \oplus so(2)$. Y and Q have equal lengths and, because of $\rho^3 = \mathrm{id}_\mathfrak{g}$, the angle between them equals $2\pi/3$. In such a way we have the construction being analogous to the $su(3)$ symmetry case. Furthermore, by means of the Gell-Mann–Nishijima formula $Q = T_3 + \frac{1}{2}Y$ the isospin operator T_3 can be obtained. It is easily verified that $\mathbb{R}T_3$ is $so(2)$ with the natural embedding into $so(10) \cong \tilde{\mathfrak{l}}_1$, and, furthermore, taking $T \cong so(3)$ with the natural embedding into $so(10) \cong \tilde{\mathfrak{l}}_1$, we can uniquely restore the subalgebra of T-isospin. Then ρT and $\rho^2 T$ give us the subalgebras of U and V-isospins in $\tilde{\mathfrak{l}}_2$ and $\tilde{\mathfrak{l}}_3$, respectively. It can be checked that $T + U + V = \mathfrak{m}$ is a subalgebra in $\mathfrak{g}$ and $\mathfrak{m} \cong su(3)$. But the embedding of $\mathfrak{m}$ into e_6 is different from the embedding of $\mathfrak{n} \cong su(3)$ into e_6 in the case of the principal unitary hyper-involutive sum.

A.1.5. Finally, the requirements about the choice of the invoalgebra $\mathfrak{l}_\alpha$ of T, U, V-isospins of the hyper-involutive sum decomposition can be weakened more than above. Namely, it is possible to require only S_α to be inner involutive automorphisms, that is, $S_1 \in \mathrm{Int}_\mathfrak{g}(T)$, $S_2 \in \mathrm{Int}_\mathfrak{g}(U)$, $S_3 \in \mathrm{Int}_\mathfrak{g}(V)$. Then, mathematically speaking, the problem is to obtain all hyper-involutive sum decompositions of the simple compact algebras generated by inner involutive automorphisms. We note that for the algebras g_2, f_4, e_6, e_8 there are no inner involutive automorphisms besides special unitary involutive automorphisms [S. Helgason, 62,78], [L.V. Sabinin, 69] and this is why everything just mentioned is contained in Table 1 and 2. For e_7 there exists no more hyper-involutive decomposition besides those in Table 1 and 2, namely $\mathfrak{g} = \mathfrak{l}_1 = \mathfrak{l}_2 + \mathfrak{l}_3$, where $\mathfrak{g}/\mathfrak{l}_\alpha \cong e_7/e_6 \oplus u(1)/\{0\}$. But that is not suitable for our purposes because then $S_1 \in \mathrm{Int}_\mathfrak{g}(\mathbb{R}Y) \subset \mathrm{Int}_\mathfrak{g}(\mathfrak{l}_0) = \mathrm{Int}_{e_7}(f_4)$, and consequently we have the strict morphism $\mathrm{Int}_{e_7}(f_4) \to \mathrm{Int}_{e_7}^{f_4}(f_4) \cong \mathrm{Int}(f_4)$. But the last relation is impossible, for it is well known that a connected group of type F_4 is isomorphic to $\mathrm{Int}(f_4)$ [S. Helgason, 62,78]. Thus there is only the problem of obtaining all hyper-involutive sum decompositions with inner involutive automorphisms S_α for the classical algebras $so(n)$, $su(n)$, $sp(n)$, which is not difficult to do by means of the well known matrix realization of the classical algebras and the knowledge of all inner involutive automorphisms. We shall not concern this particular problem in this appendix.

A.1.6. We note, finally, that it is possible to approach the problem of isospins with the most minimal conditions, that is, to require the conjugacy of the charge Q and the hypercharge Y under $\varphi \in \mathrm{Int}(\mathfrak{g})$, $\varphi^4 = \mathrm{id}_\mathfrak{g}$ (strictly speaking $\varphi Y = -Q$ is our assumption). From the physical point of view φ is the operator E_V of the charge V-symmetry. Having taken the involutive automorphisms $S_1 \in \mathrm{Int}_\mathfrak{g}(\mathbb{R}Y)$, $S_2 \in \mathrm{Int}_\mathfrak{g}(\mathbb{R}Q)$, $S_3 = \varphi^2$ we see that $S_\alpha S_\beta = S_\gamma$ ($\alpha \neq \beta$, $\beta \neq \gamma$, $\gamma \neq \alpha$). This is a reason why we can construct the iso-involutive sum [L.V. Sabinin, 65a,b] $\mathfrak{g} = \mathfrak{l}_1 + \mathfrak{l}_2 + \mathfrak{l}_3$ such that $\varphi \mathfrak{l}_1 = \mathfrak{l}_2$, $\varphi \in \mathrm{Int}_\mathfrak{g}(\mathfrak{l}_3 \dot{-} \mathfrak{l}_0)$. It is the so called iso-involutive sum [L.V. Sabinin, 65a,b]. If we now require additionally $\mathfrak{l}_1 = T \oplus \tilde{\mathfrak{l}}_1$ then $\mathfrak{l}_2 = U \oplus \tilde{\mathfrak{l}}_2$ (by the way, $S_1 \in \mathrm{Int}_\mathfrak{g}(T)$, $S_2 \in \mathrm{Int}_\mathfrak{g}(U)$ which is easy to obtain by means of the description of all iso-involutive sum decompositions for the simple compact algebras, $\mathfrak{g} = \mathfrak{l}_1 + \mathfrak{l}_2 + \mathfrak{l}_3$ with the principal unitary involutive algebras $\mathfrak{l}_1$ and $\mathfrak{l}_2$). If we do not take into account the cases already obtained in Table 1 then, as is known [L.V. Sabinin, 69], $\mathfrak{l}_3$ is a special unitary involutive algebra. We have classified all such cases in the paper [L.V. Sabinin, 69]. From the physical point of view that could give us, probably, the models of theory of higher symmetry with T and U-isospins but without V-isospin (more precisely speaking, V exists but is not isomorphic to $su(2)$).

I am very much thankful to Professor Smorodinski (Moscow) for some useful notes and advice.

APPENDIX TWO

Лети Пегас,
В межзвездьи весь,
С Планеты сбей
Эпохи Спесь.
С Планеты сдуй
Эпохи Хлам.
Лети, ликуй,
К другим Мирам.

Квантасмагор

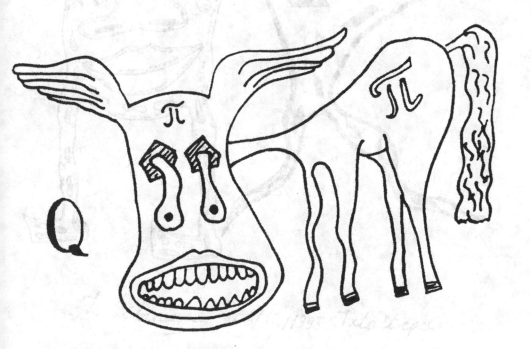

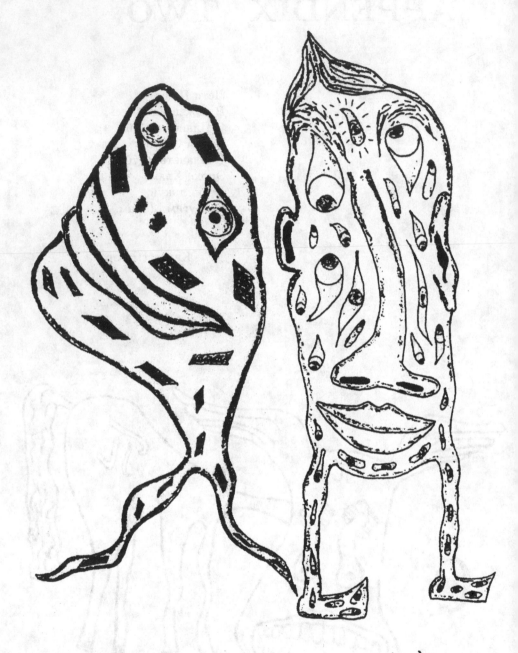

APPENDIX 2

DESCRIPTION OF THE CONTENT

A.2.1. PART I.

The presentation is mainly of a geometric nature with the use of tensor algebra.

I.1. Preliminaries

This chapter contains preliminaries. The basis definitions of principal and special involutive automorphisms, and of involutive, iso-involutive and hyper-involutive sums are introduced. Henceforth only compact Lie algebras are considered.

I.2. Curvature tensor of an involutive pair. Classical involutive pairs of index 1

The curvature tensor of the involutive pair $\mathfrak{g}/\mathfrak{l}$ is introduced (the name is justified by obvious analogy with symmetric spaces). It is shown that the curvature tensor uniquely defines an elementary involutive pair (I.2.4 Theorem 1). The curvature tensors for the classical involutive pairs

$$so(n+1)/so(n), \qquad su(n+1)/su(n), \qquad sp(n+1)/sp(n) \oplus sp(1)$$

are determined. (I.2.5 Theorem 2, I.2.6 Theorem 3, I.2.7 Theorem 4.)

I.3. Iso-involutive sums of Lie algebras

Different notions concerning iso-involutive groups and iso-involutive sums are introduced (for example, iso-involutive sums of type 1 (I.3.6 Definition 19), of index 1 (I.3.14 Definition 21) etc.). It is proved that any involutive automorphism $S_1 \neq \mathrm{id}$ may be included into some iso-involutive group $\chi(S_1, S_2, S_3; \varphi)$ (I.3.13 Theorem 9, I.3.16 Theorem 10).

I.4. Iso-involutive base and structure equations

In a compact Lie algebra, for an involutive sum $\mathfrak{g} = \mathfrak{l}_1 + \mathfrak{l}_2 + \mathfrak{l}_3$, the canonical base invariant with respect to the involutive sum is introduced. In this base the equations of structure and Jacobi identities are written out. Further, these relations are not used in full generality. Iso-involutive sums of type 2 and 3 are introduced (I.4.7 Definition 23).

247

I.5. Iso-involutive sums of types 1 and 2

Involutive sums of types 1 and 2 and of lower index 1 are considered. This leads us to the description of elementary involutive pairs of types $1, 2$ and lower index 1. In particular, then, under some additional conditions, we obtain $\mathfrak{g} \cong so(n)$, $sp(n)$ (I.5.4 Theorem 14, I.5.12 Theorem 17, I.5.14 Theorem 18).

I.6. Iso-involutive sums of lower index 1

It is shown that if the iso-involutive group $\chi(S_1, S_2, S_3; \varphi)$ is of lower index 1 but not of type 1 then its derived iso-involutive group $\chi^{(1)}(S_1, S_2, S_3; \varphi)$ is of lower index 1 and of type 1(I.6.1 Theorem 19).

Furthermore, it is shown that for a simple compact Lie algebra an iso-involutive sum of lower index 1 is of type 1 or 2 (I.6.2 Theorem 20). The proof is rather complicated and it is desirable to simplify it.

On the basis of these results and results of I.5 some characteristics of classical Lie algebras are given by means of involutive decompositions of lower index 1 and by dimension of $\mathfrak{l}_3 - (\mathfrak{l}_2 \cap \mathfrak{l}_3)$ (I.6.3 Theorem 21, I.6.4 Theorem 22, I.6.5 Theorem 23, I.6.6 Theorem 24). Some other results on iso-involutive sums of lower index 1 are proved as well (I.6.9 Theorem 27, I.6.10 Theorem 28).

I.7. Principal central involutive automorphism of type U

It is shown that if $\mathfrak{g}$ is simple compact and $\mathfrak{g}/\mathfrak{l}$ is a principal non-central involutive pair of type U then $\mathfrak{g}/\mathfrak{l} \cong su(n)/s(u(2) \oplus u(n-2))$ with the natural embeddings (I.7.2 Theorem 29). The proof relates to the inclusion of the involutive algebra $\mathfrak{l}$ into an iso-involutive sum of index 1, $\mathfrak{g} = \mathfrak{l}_1 + \mathfrak{l}_2 + \mathfrak{l}$, and to the use of I.6.4 Theorem 22).

I.8. Principal unitary involutive automorphism of index 1

It is proved that if $\mathfrak{g}$ is simple compact and $\mathfrak{g}/\mathfrak{l}$ is a principal non-central involutive pair of type U and of index 1 then $\mathfrak{g}/\mathfrak{l} \cong sp(n+1)/sp(n) \oplus sp(1)$ with the natural embeddings (I.8.4 Theorem 33).

Some other results are proved as well (I.8.1 Theorem 30, I.8.2 Theorem 31, I.8.3 Theorem 32).

A.2.2. PART II.

II.1. Hyper-involutive decomposition of a simple compact Lie algebra

Some sufficient conditions of the existence of hyper-involutive decomposition (II.1.2 Lemma 1) are given.

The notion of hyper-involutive base (II.1.3 Definition 24), the canonical base of a hyper-involutive sum, is introduced and its existence is proved by a direct construction.

The structure equations and Jacobi identities of Lie algebra $\mathfrak{g}$ are written out in a canonical base. All essential relations on structure constants are obtained in the hyper-involutive base, see (II.12).

As a result all information about $\mathfrak{g} = \mathfrak{l}_1 + \mathfrak{l}_2 + \mathfrak{l}_3$ is reduced to the information about $\mathfrak{l}_1/\mathfrak{l}_2 \cap \mathfrak{l}_3$.

The important notion of simple hyper-involutive decomposition is introduced (II.1.7 Definition 25).

II.2. Some auxiliary results

The main result here is that if $\mathfrak{g}/\mathfrak{l}$ is an elementary involutive pair and the automorphism $p \neq \mathrm{id}$, $p^3 = \mathrm{id}$, has the property $p\xi = \xi$ for $\xi \in \mathfrak{l}$ then p is an inner automorphism, $p \in \mathrm{Int}_{\mathfrak{g}}(\mathfrak{l})$.

Some strengthenings of this result are given.

II.3. Principal involutive automorphisms of type O

It is shown that if $\mathfrak{g}$ is a simple compact Lie algebra and $\mathfrak{g}/\mathfrak{l}$ is a principal involutive pair of type O then either $\mathfrak{g}/\mathfrak{l} \cong so(n)/so(3) \oplus so(n-3)$ or $\mathfrak{g}/\mathfrak{l} \cong su(3)/so(3)$.

The construction is related to some (basis for $\mathfrak{l} = \mathfrak{b} \oplus \tilde{\mathfrak{l}}$) hyper-involutive decomposition such that $\chi(S_1, S_2, S_3; p) \subset \mathrm{Int}_{\mathfrak{g}}(\mathfrak{b}) \cong so(3)$ and uses the results on involutive pairs of index 1 (see I.6).

The simple hyper-involutive decomposition gives $\mathfrak{g}/\mathfrak{l} \cong su(n)/so(3) \oplus so(n-3)$ and non-simple hyper-involutive decomposition gives $\mathfrak{g}/\mathfrak{l} \cong su(3)/so(3)$.

II.4. Fundamental theorem

It is proved that any compact simple and semi-simple Lie algebra $\mathfrak{g}$ has a principal unitary involutive automorphism. Along with this the auxiliary notion of associated involutive automorphism is introduced (II.4.1 Definition 27).

In this section the structure of a simple unitary special subalgebra $\mathfrak{k}$ of the Lie algebra $\mathfrak{g}$ (II.1.17 Definition 8) is clarified, $\mathfrak{k} \cong so(m)$. The proof is based on the following two assertions:

1) if an involutive algebra $\mathfrak{l}$ of a Lie algebra $\mathfrak{g}$ has a principal unitary involutive automorphism then $\mathfrak{g}$ has a special unitary involutive automorphism;

2) if $\mathfrak{g}$ has a special unitary involutive automorphism then $\mathfrak{g}$ has a principal unitary involutive automorphism as well.

II.5. Principal di-unitary involutive automorphism

It is shown that if $\mathfrak{g}$ is a simple compact Lie algebra, and if $\mathfrak{g}/\mathfrak{l}$ is a principal di-unitary involutive pair then either $\mathfrak{g}/\mathfrak{l} \cong g_2/so(4)$ or $\mathfrak{g}/\mathfrak{l} \cong so(n)/so(4) \oplus so(n-4)$. The proof relates to the construction of some (basis for $\mathfrak{l} = \mathfrak{p} \oplus \mathfrak{q} \oplus \tilde{\mathfrak{l}}$) hyper-involutive decomposition in such a way that the hyper-involutive group $\chi(S_1, S_2, S_3; p) \subset \mathrm{Int}_{\mathfrak{g}}(\Delta(\mathfrak{p} \oplus \mathfrak{q}) \cong so(3)$ and also makes use of II.3.

The simple hyper-involutive decomposition gives then $\mathfrak{g} \cong so(n)/so(4) \oplus so(n-4)$; otherwise $\mathfrak{g}/\mathfrak{l} \cong g_2/so(4)$ (singular case).

II.6. Singular principal di-unitary involutive automorphism

For this case the numerical values of the structure constants are determined in a hyper-involutive base of the basis decomposition (in particular, this proves the existence and uniqueness of such a Lie algebra).

This construction shows (in the case g_2) how the program of classification of simple compact Lie algebras according to the classical theory of invariants should be realized.

Other results, for example, that $su(3)$ may be included into g_2 as a maximal subalgebra in such a way that the pair $(g_2, su(3))$ is not an involutive pair (II.6.9 Theorem 30), are presented as well.

II.7. Mono-unitary non-central principal involutive automorphism

First, some auxiliary results having a proper significance are proved (II.7.11 Theorem 31, II.7.2 Theorem 32, II.7.3 Theorem 33).

Furthermore, some principle of the iso-involutive duality for principal mono-unitary non-central involutive automorphisms and special (non-principal) subalgebras and involutive automorphisms is presented. The essence of this construction consists in the following: any non-identity principal unitary involutive automorphism of a simple ideal in a principal unitary involutive algebra $\mathfrak{l} \subset \mathfrak{g}$ generates a special unitary involutive automorphism of $\mathfrak{g}$ and any non-identity principal unitary involutive automorphism of a special simple subalgebra in $\mathfrak{g}$ generates a principal unitary automorphism of $\mathfrak{g}$.

Pairs of dual involutive automorphisms thus obtained are commuting and generate the so called basis (II.7.4 Definition 29) iso-involutive decompositions, $\mathfrak{g} = \mathfrak{l}_1 + \mathfrak{l}_2 + \mathfrak{l}_3$, where $\mathfrak{l}_1 \cong \mathfrak{l}_2$ are principal unitary iso-involutive algebras, and $\mathfrak{l}_3$ is a special unitary involutive algebra (II.7.5 Theorem 34).

Finally, a simple algebraic classification of exceptional (that is, mono-unitary non-central, not of index 1) involutive automorphisms is introduced (II.7.46 Definition 30).

II.8. Exceptional principal involutive automorphisms of types f and e

For exceptional principal involutive automorphisms of types f and e the basis iso-involutive sums are constructed. They uniquely define these involutive automorphisms (II.8.1 Theorem 35, II.8.2 Theorem 36, II.8.3 Theorem 37, II.8.4 Theorem 38).

II.9. Classification of simple special unitary subalgebras

First, the construction of the basis iso-involutive decomposition from II.7 is extended up to some non-unitary principal involutive automorphisms (II.9.1 Theorem 40 generalizing II.7.5 Theorem 34). This results in the uniquely determined basis iso-involutive sums for $so(n)$, $su(n)$, $sp(n)$ (II.9.2 Theorem 41, II.9.3 Theorem 42, II.9.4 Theorem 43).

Finally, II.9.5 (Theorem 44) completes the justification of the principle of iso-involutive duality: given a special subalgebra its dual principal involutive automorphisms are uniquely restored.

This principle of iso-involutive duality allows us, furthermore, to determine all special unitary subalgebras and all special unitary involutive automorphisms, having given all principal unitary involutive automorphisms (II.9.10 Theorem 45).

II.10. Hyper-involutive reconstruction of basis decompositions

In this section it is shown that almost all simple compact Lie algebras (except $sp(n)$) have hyper-involutive decompositions with principal unitary involutive automorphisms. These decompositions, so called basis hyper-involutive sums, are described in II.10.2 Theorem 46, II.10.5 Theorem 47, II.10.9 Theorem 48, II.10.11 Theorem 49.

II.11. Special hyper-involutive sums

It is shown that for exceptional Lie algebras of types f and e one may construct hyper-involutive sums using only special unitary involutive algebras (II.11.2 Theorem 50, II.11.3 Theorem 51, II.11.4 Theorem 52, II.11.5 Theorem 53).

These decompositions are of value, showing how using Lie algebras of type $so(m) \cong \mathfrak{l}_\alpha$ one may construct Lie algebras of types f and e by means of hyper-involutive sums.

Furthermore, it is shown that for classical Lie algebras $so(n), su(n), sp(n)$, under some restrictions on dimensions, there exist hyper-involutive sums with special unitary involutive algebras (II.11.10 Theorem 54, II.11.12 Theorem 55, II.11.14 Theorem 56).

In these constructions $\mathfrak{l}_0 = \mathfrak{l}_\alpha \underset{\alpha \neq \beta}{\cap} \mathfrak{l}_\beta$ appears to be connected with $so(8)$.

A.2.3. PART III.

This Part contains some geometric applications of Mirror Geometry.

III.1. Notations, definitions and some preliminaries

Some definitions are given. The majority of them are reformulations of I.1 from the language of Lie algebras into the language of Lie groups and homogeneous spaces. The notion of involutive sum is transformed into the notion of involutive product.

III.2. Symmetric spaces of rank 1

There are given some characteristics of symmetric spaces of rank 1 related to geodesic mirrors (which are of the constant curvature). Such characteristics uniquely define the type of symmetric space of rank 1.

III.3. Principal symmetric spaces

Principal involutive automorphisms of Lie algebras generate one new remarkable class of symmetric spaces. The classification of principal symmetric spaces according to groups of motions is given. As well some of their properties connected with mirrors are considered (III.3.1 Theorem 10, III.3.2 Theorem 11, III.3.3 Theorem

12, III.3.4 Theorem 13, III.3.5 Theorem 14, III.3.6 Theorem 15, III.3.7 Theorem 16, III.3.8 Theorem 17, III.3.9 Theorem 18, III.3.10 Theorem 19, III.3.11 Theorem 20, III.3.12 Theorem 21, III.3.13 Theorem 22, III.3.14 Theorem 23).

III.4. Essentially special symmetric spaces

The classification of essentially special symmetric spaces is given and some of their properties connected with mirrors are considered (III.4.1 Theorem 24, III.4.2 Theorem 25, III.4.3 Theorem 26, III.4.4 Theorem 27, III.4.5 Theorem 28, III.4.6 Theorem 29).

III.5. Some theorems on simple compact Lie groups

In this section we demonstrate how the mirror geometry can be applied to proving some results in the theory of compact Lie groups, in particular, to the problem of universal covering for simple compact Lie groups of types G_2, F_4 (see III.5.2 Theorem 31, III.5.3 Theorem 32) and to the problem of determination of inner involutive automorphisms (III.5.6 Theorem 33, III.5.7 Theorem 34).

III.6. Tri-symmetric and hyper-tri-symmetric spaces

The concept of involutive product of homogeneous spaces is introduced. From the differential geometric point of view this relates to constructing of spaces by means of system of commuting mirrors. See the Bibliography. Here we restrict ourselves by prerequisites needed for the classification of tri-symmetric spaces with compact simple groups of motions only.

The definitions of tri-symmetric and hyper-tri-symmetric spaces are given in a convenient form (III.6.3 Definition 42, III.6.5 Definition 43). Auxiliary theorems III.6.8 Theorem 36, III.6.9 Theorem 37, III.6.10 Theorem 38 are considered.

III. 7. Tri-symmetric spaces with exceptional compact groups of motions

All non-trivial tri-symmetric spaces with exceptional simple compact Lie groups of motions are determined (III.7.3 Theorem 39, III.7.5 Theorem 40, III.7.7 Theorem 41, III.7.9 Theorem 42, III.7.11 Theorem 43). The result is obtained by means of hyper-involutive decompositions of II.10.

All spaces here are hyper-symmetric. For any type of a simple group, except E_7, there exists one such space. For E_7 there exist two tri-symmetric spaces, one with principal unitary and the other with the central mirrors. All spaces here are with irreducible groups of rotations.

III.8. Tri-symmetric spaces with groups of motions $SO(n)$, $SU(n)$, $Sp(n)$

All non-trivial tri-symmetric spaces with the classical groups $SO(n)$, $SU(n)$, $Sp(n)$ are determined. For any type of simple classical group there exists only one such space, which is a space with the central mirror. All spaces here are hyper-tri-symmetric with irreducible groups of rotations (III.8.7 Theorem 44, III.8.13 Theorem 45, III.8.15 Theorem 46).

A.2.4. PART IV.

This Part is devoted to the homogeneous Riemannian spaces with mirrors. Along with the general theory of subsymmetric spaces the classification of all such spaces with simple compact groups of motions and two-dimensional mirrors is given here.

IV.1. Subsymmetric Riemannian homogeneous spaces

This Chapter contains some preliminaries and the basis concepts of mirror subsymmetric spaces.

IV.2. Subsymmetric homogeneous spaces and Lie algebras

Here the structure of homogeneous subsymmetric spaces is studied in the language of Lie algebras.

IV.3. Mirror subsymmetric Lie triplets of Riemannian type.

Lie–Riemannian triplets and quadruplets are considered (IV.3.1, IV.3.2) and studied in terms of Lie algebras. In this case a canonical base is introduced and the structure equations of the corresponding Lie algebras with respect to this canonical base are obtained.

This leads to some preliminary classification of Lie–Riemannian triplets of order two, see IV.3.13 Theorem 3 and Remarks IV.3.14–IV.3.16.

IV.4. Mobile mirrors. Iso-involutive decompositions

The important concept of mobile mirror is introduced and studied. This leads to the triplet of mirrors produced by an initially given mirror by means of rotations. In terms of the correponding Lie algebra this is reduced to the involutive groups and corresponding iso-involutive decompositions (see Part I, II).

The concept of tri-symmetric space (Part III) and related matters are discussed here once more, see IV.4.9 Definition 71, IV.4.10–IV.4.13.

IV.5. Homogeneous Riemannian spaces with two-dimensional mirrors

Making use of the classification (IV.3.13 Theorem 1) we obtain that a two-dimensional mirror of a Riemannian homogeneous space G/H with a simple compact Lie group G is of elliptic type (see IV.5.2 Theorem 6).

Furthermore, it is obtained (see IV.5.6) that in this case a two-dimensional mirror is of orthogonal or unitary type since the involutive automorphism of a two-dimensional mirror is principal orthogonal or principal unitary (see Part I, II, III).

In IV.5.11 Theorem 9 there is established for a simple compact Lie group G that if a two-dimensional mirror of G/H is of unitary type then G/H is a tri-symmetric space (generated by the two-dimensional mirror described above).

IV.6. Homogeneous Riemannian spaces with groups SO(n), SU(3) and two-dimensional mirrors

Here all such spaces are classified (IV.6.5 Theorem 11, IV.6.8 Theorem 12, IV.6.10 Theorem 13). The classification is based upon Part I, II, III of this monograph.

IV.7. Homogeneous Riemannian spaces with simple compact Lie groups G ≠ SO(n), SU(3) and two-dimensional mirrors

All such spaces are classified here (IV.7.2 Theorem 14, IV.7.4 Theorem 15). The classification is based upon Part I, II, III of this monograph, mostly upon Part III.

IV.8. Homogeneous Riemannian spaces with simple compact Lie groups of motions and two-dimensional mirrors

In this Chapter all spaces of such type are classified (IV.8.2 Theorem 16).

A.2.5. Appendix 1. On the structure of T, U, V-isospin in the theory of higher symmetry.

Here it is shown how the notion of involutive sum appears in a natural way from the analysis of T, U, V-isospin construction in the theory of unitary symmetry of elementary particles.

The invariant description (not depending on the choice of representation) of subalgebras of T, U, V-isospin by means of the construction of hyper-involutive sum of principal unitary subalgebras is given. At the same time the invariant description of the operators Q (charge) and Y (hypercharge) is presented.

Furthermore, these constructions are generalized to Lie algebras of possible theories of higher symmetry. It is shown that mathematical problems appeared here relate to principal unitary and special unitary hyper-involutive sums in compact Lie algebras (see Part II).

Finally, the problem of the theory of higher symmetry is considered under the minimal assumption of the conjugacy of the operators Q (charge) and Y (hypercharge) and it is shown that appearing here constructions of T and V-isospins relate to iso-involutive decompositions of simple compact Lie algebras.

A.2.6. Appendix 2. Description of the content. Appendix 2 contains a brief review of main results of this treatise.

A.2.7. Appendix 3. Definitions. Appendix 3 contains a list of definitions of this treatise.

A.2.8. Appendix 4. Theorems. Appendix 4 contains a list of main theorems of this treatise.

A.2.9. Bibliography. It contains cited works, the articles of author, and the treatises of general nature.

A.2.10. Index. Index contains a list of new concepts indicating the number of page and definition, where a concept appears for the first time.

APPENDIX THREE

Между межей ментальной меты
Я хвост расчесывал кометы.

Квантасмагор

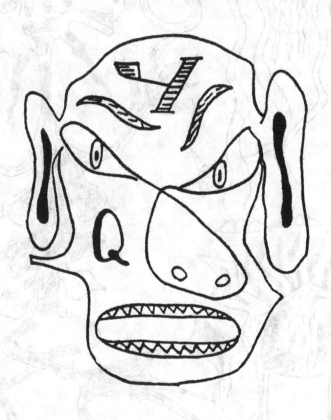

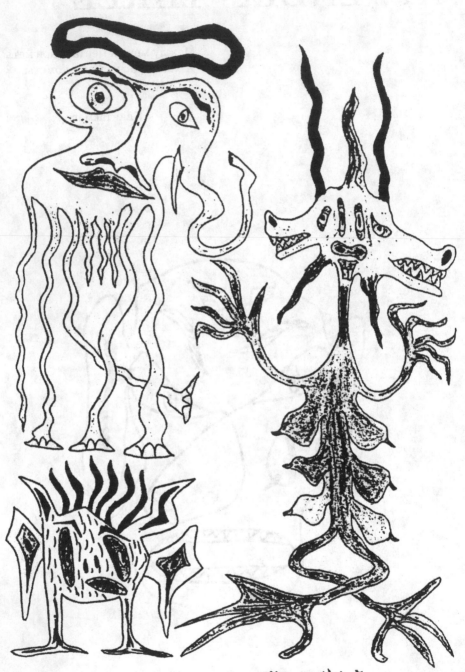

I. Adirondack — 1988

APPENDIX 3

DEFINITIONS

A.3.1. PART I.

I.1.10. Definition 1. We say that an involutive automorphism A of a Lie algebra $\mathfrak{g}$ is principal if its involutive algebra $\mathfrak{l}$ has a simple three-dimensional ideal $\mathfrak{b}$.

Respectively, we say in this case that $\mathfrak{l}$ is a principal involutive algebra and $\mathfrak{g}/\mathfrak{l}$ is a principal involutive pair.

I.1.11. Definition 2. Let A be a principal involutive automorphism of a Lie algebra $\mathfrak{g}$ with an involutive algebra $\mathfrak{l}$, and let $\mathfrak{b}$ be a simple three-dimensional ideal of $\mathfrak{l}$.

We say that A is principal orthogonal (or of type O) if $\mathrm{Int}_{\mathfrak{g}}(\mathfrak{b}) \cong SO(3)$, and that A is principal unitary (or of type U) if $\mathrm{Int}_{\mathfrak{g}}(\mathfrak{b}) \cong SU(2)$.

Respectively, we distinguish between orthogonal and unitary principal involutive algebras $\mathfrak{l}$ and involutive pairs $\mathfrak{g}/\mathfrak{l}$.

I.1.12. Definition 3. We say that an involutive automorphism A of a Lie algebra $\mathfrak{g}$ is central if its involutive algebra $\mathfrak{l}$ has a non-trivial centre.

In this case we say that $\mathfrak{l}$ is a central involutive algebra and $\mathfrak{g}/\mathfrak{l}$ is a central involutive pair.

I.1.13. Definition 4. A principal involutive automorphism A of a Lie algebra $\mathfrak{g}$ is called principal di-unitary (or of type U^2) if its involutive algebra $\mathfrak{l} = \mathfrak{p} \oplus \mathfrak{q} \oplus \tilde{\mathfrak{l}}$ (direct sum decomposition of ideals), where $\mathrm{Int}_{\mathfrak{g}}(\mathfrak{p}) \cong SU(2)$, $\mathrm{Int}_{\mathfrak{g}}(\mathfrak{q}) \cong SU(2)$.

In this case we also say that $\mathfrak{l}$ is a principal di-unitary involutive algebra and $\mathfrak{g}/\mathfrak{l}$ is a principal di-unitary involutive pair.

I.1.14. Definition 5. A unitary but not di-unitary principal involutive automorphism A of a Lie algebra $\mathfrak{g}$ is called mono-unitary (or of type U^1).

In this case we also say that $\mathfrak{l}$ is a principal mono-unitary involutive algebra and $\mathfrak{g}/\mathfrak{l}$ is a principal mono-unitary involutive pair.

I.1.15. Definition 6. An involutive automorphism A of a Lie algebra $\mathfrak{g}$ is said to be special if its involutive algebra $\mathfrak{l}$ has a principal involutive automorphism of type O.

In this case we also say that $\mathfrak{l}$ is a special involutive algebra and $\mathfrak{g}/\mathfrak{l}$ is a special involutive pair.

I.1.16. Definition 7. Let $\mathfrak{l}$ be a special involutive algebra of a Lie algebra $\mathfrak{g}$, and $\mathfrak{b}$ be a three-dimensional simple ideal of its principal involutive algebra such that $\mathrm{Int}_{\mathfrak{l}}(\mathfrak{b}) \cong SO(3)$.

We say that an involutive algebra $\mathfrak{l}$ is orthogonal special (or of type O) if $\mathrm{Int}_{\mathfrak{g}}(\mathfrak{b}) \cong SO(3)$ and is unitary special (or of type U) if $\mathrm{Int}_{\mathfrak{g}}(\mathfrak{b}) \cong SU(2)$.

Respectively, we distinguish between orthogonal and unitary special involutive automorphisms and involutive pairs.

I.1.17. Definition 8. Let $\mathfrak{g}$ be a compact simple Lie algebra and $\mathfrak{l} \subset \mathfrak{g}$ its involutive algebra of an involutive automorphism A.

An ideal $\mathfrak{k}$ of $\mathfrak{l}$ is called a special unitary (or of type U) subalgebra of an involutive automorphism A in $\mathfrak{g}$ if there exists a principal orthogonal involutive pair $\mathfrak{k}/\mathfrak{b} \oplus \tilde{\mathfrak{k}}$ such that $A \in \mathrm{Int}_{\mathfrak{g}}(\mathfrak{b}) \cong SU(2)$.

I.1.18. Definition 9. Let $\mathfrak{g}/\mathfrak{l}$ be an involutive pair of an involutive automorphism A, and $\mathfrak{t} \subset \mathfrak{m} = \mathfrak{g} \dot{-} \mathfrak{l}$ be a maximal subalgebra in $\mathfrak{m}$. Then $\min\limits_{\mathfrak{t} \subset \mathfrak{m}}(\dim \mathfrak{t})$ and $\max\limits_{\mathfrak{t} \subset \mathfrak{m}}(\dim \mathfrak{t})$, respectively, are called the lower and the upper indices of an involutive automorphism A, an involutive algebra $\mathfrak{l}$, and involutive pair $\mathfrak{g}/\mathfrak{l}$.

If

$$\min_{\mathfrak{t} \subset \mathfrak{m}}(\dim \mathfrak{t}) = \max_{\mathfrak{t} \subset \mathfrak{m}}(\dim \mathfrak{t}) = r$$

then we say that r is the index (rank) of an involutive automorphism A, an involutive algebra $\mathfrak{l}$, and involutive pair $\mathfrak{g}/\mathfrak{l}$.

An involutive pair $\mathfrak{g}/\mathfrak{l}$ is called *irreducible* if $\mathrm{ad}_{\mathfrak{g}}^{\mathfrak{g}-\mathfrak{l}}(\mathfrak{l})$ is irreducible.

I.1.19. Definition 10. We say that an involutive pair $\mathfrak{g}/\mathfrak{l}$ is elementary if either $\mathfrak{g}$ is simple and semi-simple or $\mathfrak{g} = \mathfrak{p} \times \mathfrak{p}$ (direct product of ideals), where $\mathfrak{p}$ is simple and semi-simple, and $\mathfrak{l} = \triangle(\mathfrak{p} \times \mathfrak{p}) = \{(\xi, \xi) \mid \xi \in \mathfrak{p}\}$ is the diagonal algebra of the canonical involutive automorphism in $\mathfrak{p} \times \mathfrak{p}$.

Evidently an elementary involutive pair is irreducible.

I.1.20. Definition 11. We say that an involutive pair $\mathfrak{g}/\mathfrak{l}$ is a sum of involutive pairs $\mathfrak{g}_1/\mathfrak{l}_1, \ldots, \mathfrak{g}_m/\mathfrak{l}_m$ and write

$$\mathfrak{g}/\mathfrak{l} = \mathfrak{g}_1/\mathfrak{l}_1 + \cdots + \mathfrak{g}_m/\mathfrak{l}_m = \sum_{\alpha=1}^{m} (\mathfrak{g}_\alpha/\mathfrak{l}_\alpha)$$

if there are direct sum decompositions of ideals:

$$\mathfrak{g} = \mathfrak{g}_1 \oplus \cdots \oplus \mathfrak{g}_m, \qquad \mathfrak{l} = \mathfrak{l}_1 \oplus \cdots \oplus \mathfrak{l}_m \quad (\mathfrak{l}_\alpha \subset \mathfrak{g}_\alpha).$$

If $\mathfrak{g}$ is compact then an involutive pair $\mathfrak{g}/\mathfrak{l}$ has a unique decomposition:

$$\mathfrak{g}/\mathfrak{l} = \mathfrak{g}_0/\mathfrak{l}_0 + \sum_{\alpha=1}^{m} (\mathfrak{g}_\alpha/\mathfrak{l}_\alpha),$$

where $\mathfrak{g}_0$ is a centre of $\mathfrak{g}$ and $\mathfrak{g}_1/\mathfrak{l}_1, \ldots, \mathfrak{g}_m/\mathfrak{l}_m$ are elementary involutive pairs. See, for example, [S. Helgason 62,78].

I.1.21. Definition 12. Let $S_1, S_2, S_3 \in \mathrm{Aut}(\mathfrak{g})$ be pair-wise different commuting involutive automorphisms of a Lie algebra $\mathfrak{g}$ such that the product of any two of them is equal to the third. Then Id, S_1, S_2, S_3 constitute a discrete subgroup $\chi(S_1, S_2, S_3) \subset \mathrm{Aut}(\mathfrak{g})$ which is called an involutive group of the algebra $\mathfrak{g}$.

I.1.22. Definition 13. An involutive group $\chi(S_1, S_2, S_3)$ of a Lie algebra $\mathfrak{g}$ is called an iso-involutive group and is denoted $\chi(S_1, S_2, S_3\,;\,\varphi)$ if

$$S_3 = \varphi^2, \qquad \varphi \in \mathrm{Int}_{\mathfrak{g}}(\{t\zeta\}_{t\in\mathbb{R}}), \qquad S_1\zeta = -\zeta, \qquad S_2\zeta = -\zeta, \quad \zeta \neq 0.$$

In this case obviously $\varphi^{-1}S_1\varphi = S_2$, $\varphi^{-1}S_2\varphi = S_1$.

I.1.24. Definition 14. An involutive group $\chi(S_1, S_2, S_3)$ of a Lie algebra $\mathfrak{g}$ is called a hyper-involutive group and is denoted $\chi(S_1, S_2, S_3\,;\,p)$ if there exists $p \in \mathrm{Aut}(\mathfrak{g})$ such that

$$p^{-1}S_1\,p = S_2, \qquad p^{-1}S_2\,p = S_3, \qquad p^{-1}S_3\,p = S_1, \qquad p^3 = \mathrm{Id}.$$

I.1.25. Definition 15. We say that a Lie algebra $\mathfrak{g}$ is an involutive sum (invosum) of subalgebras $\mathfrak{l}_1, \mathfrak{l}_2, \mathfrak{l}_3$ if $\mathfrak{g} = \mathfrak{l}_1 + \mathfrak{l}_2 + \mathfrak{l}_3$ and $\mathfrak{l}_1, \mathfrak{l}_2, \mathfrak{l}_3$ are involutive algebras of involutive automorphisms S_1, S_2, S_3, respectively, of an involutive group $\chi(S_1, S_2, S_3)$ of $\mathfrak{g}$.

I.1.26. Definition 16. An involutive sum $\mathfrak{g} = \mathfrak{l}_1 + \mathfrak{l}_2 + \mathfrak{l}_3$ of a Lie algebra $\mathfrak{g}$ is called an iso-involutive sum (iso-invosum), or iso-involutive decomposition, if the corresponding involutive group $\chi(S_1, S_2, S_3)$ is an iso-involutive group $\chi(S_1, S_2, S_3\,;\,\varphi)$.

I.1.27. Definition 17. An involutive sum of a Lie algebra $\mathfrak{g}$, $\mathfrak{g} = \mathfrak{l}_1 + \mathfrak{l}_2 + \mathfrak{l}_3$, is called hyper-involutive sum (hyper-invosum), or hyper-involutive decomposition, if the corresponding involutive group $\chi(S_1, S_2, S_3)$ is a hyper-involutive group $\chi(S_1, S_2, S_3\,;\,p)$.

I.2.2. Definition 18. Let $\mathfrak{g}$ be a Lie algebra, and $\mathfrak{l}$ be its involutive algebra of an involutive automorphism S. A multilinear operator

$$R\colon (\xi, \eta, \zeta) \mapsto R(\xi; \eta, \zeta) = -[\xi\,[\eta\,\zeta]],$$

$$R\colon \mathfrak{m} \times \mathfrak{m} \times \mathfrak{m} \to \mathfrak{m}, \qquad \mathfrak{m} = \mathfrak{g} \dot{-} \mathfrak{l},$$

(I.10)

is called the curvature tensor of the involutive automorphism S, involutive algebra $\mathfrak{l}$, involutive pair $\mathfrak{g}/\mathfrak{l}$.

I.3.6. Definition 19. An iso-involutive group $\chi(S_1, S_2, S_3\,;\,\varphi)$ is said to be of type 1 if $\varphi|_{\mathfrak{l}_3}$ (the restriction of φ to the involutive algebra $\mathfrak{l}_3$ of the involutive automorphism S_3) is the identity automorphism.

In this case we also say that $\mathfrak{g} = \mathfrak{l}_1 + \mathfrak{l}_2 + \mathfrak{l}_3$ is an iso-involutive sum of type 1.

I.3.7. Definition 20. Let $\chi(S_1, S_2, S_3; \varphi)$ be an iso-involutive group of a Lie algebra $\mathfrak{g}$, where $\varphi \in \mathrm{Int}_\mathfrak{g}(\{t\zeta\})$, and $\mathfrak{l}_3$ be the involutive algebra of the involutive automorphism S_3. And let $\chi(\sigma_1, \sigma_2, \sigma_3; \psi)$ be an iso-involutive group of the Lie algebra $\mathfrak{l}_3$ such that $\sigma_\alpha = S_\alpha|_{\mathfrak{l}_3}$ ($\alpha \neq 3$), that is, the restrictions of S_1 and S_2 to $\mathfrak{l}_3$, $\sigma_3 = \varphi|_{\mathfrak{l}_3}$, where $\varphi = \theta^2$, $\theta \in \mathrm{Int}_\mathfrak{g}(\{t\zeta\}_{t\in\mathbb{R}})$, $\psi = \theta|_{\mathfrak{l}_3}$.

Then we say that $\chi(\sigma_1, \sigma_2, \sigma_3; \psi)$ is a derived iso-involutive group for $\chi(S_1, S_2, S_3; \varphi)$ and denote it $\chi^{(1)}(S_1, S_2, S_3; \varphi)$.

One may consider the derived iso-involutive group $\chi^{(2)}(S_1, S_2, S_3; \varphi)$ of $\chi^{(1)}(S_1, S_2, S_3; \varphi)$ and so on.

I.3.14. Definition 21. We say that an iso-involutive group $\chi(S_1, S_2, S_3; \varphi)$ and its involutive sum $\mathfrak{g} = \mathfrak{l}_1 + \mathfrak{l}_2 + \mathfrak{l}_3$ is of lower index 1 if $\varphi \in \mathrm{Int}_\mathfrak{g}(\{t\zeta\})$, where $\{t\zeta\}_{t\in\mathbb{R}}$ is a maximal one-dimensional subalgebra in $\mathfrak{m} = \mathfrak{g} \dot{-} \mathfrak{l}$.

I.4.1. Definition 22. We say that a base of a Lie algebra $\mathfrak{g}$ is invariant with respect to an iso-involutive group $\chi(S_1, S_2, S_3; \varphi)$ if in this base S_1, S_2, S_3, φ have canonical forms.

We also say that a base of a Lie algebra $\mathfrak{g}$ is an iso-involutive base (iso-invobase) if it is invariant (up to the multiplication by (± 1)) with respect to $\chi(S_1, S_2, S_3; \varphi)$ and all its derived iso-involutive groups.

I.4.7. Definition 23. An iso-involutive group $\chi(S_1, S_2, S_3; \varphi)$ and the corresponding iso-involutive sum not of type 1 is said to be of type 2 if

$$\theta|_{\mathfrak{g} \dot{-} \mathfrak{l}_3} = \pm \frac{1}{\sqrt{2}}(\mathrm{Id} + \varphi|_{(\mathfrak{g} \dot{-} \mathfrak{l}_3)}),$$

and of type 3 otherwise.

A.3.2. PART II.

II.1.3. Definition 24. A base in a Lie algebra $\mathfrak{g}$ is called a hyper-involutive base (hyper-invobase) of a hyper-involutive group $\chi(S_1, S_2, S_3; p)$, and of the corresponding hyper-involutive sum $\mathfrak{g} = \mathfrak{l}_1 + \mathfrak{l}_2 + \mathfrak{l}_3$, if its restriction to $\mathfrak{m}_1 \dot{+} \mathfrak{m}_2 \dot{+} \mathfrak{m}_3$ is invariant under the action of p, and S_1, S_2, S_3 have diagonal forms in this base.

In the case of a semi-simple Lie algebra $\mathfrak{g}$, by a hyper-involutive base we mean an orthonormal hyper-involutive base only.

II.1.7. Definition 25. We say that the hyper-involutive decomposition (II.4) is prime if the restriction of the automorphism p to $\mathfrak{l}_0$ is the identity automorphism, and is non-prime otherwise.

II.3.3. Definition 26. A hyper-involutive decomposition of a Lie algebra $\mathfrak{g}$ is said to be basis for a principal involutive automorphism S of type O with the corresponding involutive algebra $\mathfrak{l} = \mathfrak{b} \oplus \tilde{\mathfrak{l}}$ if the involutive automorphisms S_1, S_2, S_3 of the hyper-involutive decomposition belong to $\mathrm{Int}_\mathfrak{g}(\mathfrak{b}) \cong SO(3)$, $pS = Sp$, and $p\eta = \eta$ for $\eta \in \tilde{\mathfrak{l}}$.

II.4.1. Definition 27. Let $\mathfrak{g}$ be a compact Lie algebra, $\mathfrak{l} = \mathfrak{b} \oplus \tilde{\mathfrak{l}}$, $\mathfrak{b} \cong so(3)$, be an involutive algebra of a principal involutive automorphism A of type O, $\mathfrak{r} = \mathfrak{p} \oplus \mathfrak{q} \oplus \tilde{\tilde{\mathfrak{l}}}$, $\mathfrak{p} \cong \mathfrak{q} \cong so(3)$, be an involutive algebra of a di-unitary involutive automorphism J, $JA = AJ$, and let $\mathfrak{b}$ be the diagonal in $\mathfrak{p} \oplus \mathfrak{q}$ of the canonical involutive automorphism $\mathfrak{p} \leftrightarrows \mathfrak{q}$. Then we say that J is an associated involutive automorphism of A.

Respectively, we say that $\mathfrak{r}$ is an associated involutive algebra of $\mathfrak{l}$, and $\mathfrak{g}/\mathfrak{r}$ is an associated involutive pair of $\mathfrak{g}/\mathfrak{l}$.

II.5.2. Definition 28. A hyper-involutive decomposition of a Lie algebra $\mathfrak{g}$ is said to be basis for a principal involutive automorphism S of type $U^{(2)}$ with the involutive algebra $\mathfrak{l} = \mathfrak{p} \oplus \mathfrak{q} \oplus \tilde{\mathfrak{l}}$ ($\mathrm{Int}_{\mathfrak{g}}(\mathfrak{p}) \cong SU(2)$, $\mathrm{Int}_{\mathfrak{g}}(\mathfrak{q}) \cong SU(2)$) if the involutive automorphisms S_1, S_2, S_3 of this hyper-involutive decomposition belong to $\mathrm{Int}_{\mathfrak{g}}(\mathfrak{b})$ ($\mathfrak{b}$ being a diagonal in $\mathfrak{p} \oplus \mathfrak{q}$), $\mathfrak{p}S = S\mathfrak{p}$, and $\mathfrak{p}\eta = \eta$ for $\eta \in \tilde{\mathfrak{l}}$.

II.7.4. Definition 29. Let $\mathfrak{g}$ be a simple compact Lie algebra, and $\mathfrak{l}_1 = \mathfrak{p} \oplus \tilde{\mathfrak{l}}$ be its principal involutive algebra of an involutive automorphism $S_1 \in \mathrm{Int}_{\mathfrak{g}}(\mathfrak{p})$, $S_1 \neq \mathrm{Id}$. Let, furthermore, $\mathfrak{q} \oplus \tilde{\tilde{\mathfrak{l}}}$ be a principal unitary involutive algebra in $\tilde{\mathfrak{l}}$ of an involutive automorphism $S_2 \in \mathrm{Int}_{\mathfrak{g}}(\mathfrak{q})$; $\mathfrak{l}_2$ being the involutive algebra of the involutive automorphism $S_2 \neq S_1$, $\mathfrak{l}_3$ being the involutive algebra of the involutive automorphism $S_3 = S_1 S_2 = S_2 S_1$. Then the involutive sum $\mathfrak{g} = \mathfrak{l}_1 + \mathfrak{l}_2 + \mathfrak{l}_3$ is called the basis involutive sum for a principal unitary involutive algebra $\mathfrak{l}_1$.

II.7.6. Definition 30. Let $\mathfrak{g}$ be a simple compact Lie algebra, $\mathfrak{l} = \mathfrak{p} \oplus \tilde{\mathfrak{l}}$ being its involutive algebra of a principal non-central mono-unitary automorphism $S \in \mathrm{Int}_{\mathfrak{g}}(\mathfrak{p})$, where S is not of index 1. We call such an involutive automorphism *exceptional principal* (respectively, we speak of an exceptional principal involutive algebra and exceptional involutive pair)

Moreover, we say in this case that $\mathfrak{g}$ is of type:

(1) f_4 if $\tilde{\mathfrak{l}}$ has a principal mono-unitary non-central involutive automorphism of index 1;

(2) e_6 if $\tilde{\mathfrak{l}}$ has a principal central unitary involutive automorphism;

(3) e_7 if $\tilde{\mathfrak{l}}$ has a principal non-central di-unitary involutive automorphism;

(4) e_8 if $\tilde{\mathfrak{l}}$ has only an exceptional principal involutive automorphism.

II.10.6. Definition 31. The hyper-involutive decompositions described by II.10.2 (Theorem 46), II.10.5 (Theorem 47) are called the basis hyper-involutive sums for the principal exceptional involutive automorphisms of types f_4, e_6, e_7, e_8.

II.10.12. Definition 32. The hyper-involutive decompositions described by II.10.9 (Theorem 48), II.10.11 (Theorem 49) are called basis hyper-involutive for a principal central involutive automorphism of type U (or for the Lie algebra $su(n)$) and basis hyper-involutive for a principal involutive automorphism of type $U^{(2)}$ (or for the Lie algebra $so(n)$), respectively.

II.11.6. Definition 33. The hyper-involutive decompositions described in II.11.2 (Theorem 50), II.11.3 (Theorem 51), II.11.4 (Theorem 52), II.11.5 (Theorem 53) are called special hyper-involutive sums for the Lie algebras of types f_4, e_6, e_7, e_8, respectively.

II.11.15. Definition 34. The hyper-involutive decompositions described in II.11.10 (Theorem 54), II.11.12 (Theorem 55), II.11.14 (Theorem 56) are called the special hyper-involutive sums for the Lie algebras of type $so(n)$, $su(n)$, $sp(n)$, respectively.

A.3.3. PART III.

III.1.1. Definition 35. Let G be a Lie group, S be its automorphism such that $S^2 = \mathrm{Id}$, and let H be a maximal connected subgroup of G immobile under the action of S. Then we say that S is an involutive automorphism, H is an involutive (or characteristic) group of S, and G/H is an involutive pair of S.

We note that an involutive automorphism S of a Lie group G uniquely generates an involutive automorphism $\ln S$ of the Lie algebra $\ln G$; the converse is true (at least locally).

III.1.2. Definition 36. An involutive automorphism S of a Lie group G is said to be principal if $\ln S$ is a principal involutive automorphism of $\ln G$.

In this case the characteristic group H of the involutive automorphism S, the involutive pair G/H, and its corresponding symmetric space are said to be principal.

(Compare with I.1.10 Definition 1.)

III.1.3. Definition 37. A principal involutive automorphism S of a Lie group G is said to be principal orthogonal (of type O) if $\ln S$ is principal orthogonal (of type O), and principal unitary (of type U) if $\ln S$ is principal unitary (of type U).

Correspondingly we distinguish between orthogonal and unitary principal involutive groups, involutive pairs, and symmetric spaces.

(Compare with I.1.11 Definition 2.)

III.1.4. Definition 38. An involutive automorphism of a Lie group G is said to be central if $\ln S$ is central.

Correspondingly we define a central involutive group, involutive pair, symmetric space.

(Compare with I.1.12 Definition 3.)

III.1.5. Definition 39. A principal involutive automorphism S of a Lie group G is said to be principal di-unitary (of type $U^{(2)}$) if $\ln S$ is principal di-unitary (of type $U^{(2)}$).

Correspondingly we define a principal di-unitary group, involutive pair, symmetric space.

(Compare with I.1.13 Definition 4.)

III.1.6. Definition 40. An involutive automorphism S of a Lie group G is said to be mono-unitary (or of type $U^{(1)}$) if $\ln S$ is mono-unitary.

Correspondingly we define a mono-unitary involutive group, involutive pair, symmetric space.

(Compare with I.1.14 Definition 5.)

III.1.7. Definition 41. An involutive automorphism S of a Lie group G is said to be special if $\ln S$ is special.

Correspondingly we define a special involutive group, involutive pair, symmetric space.

(Compare with I.1.15 Definition 6.)

III.1.8. Definition 42. An involutive automorphism S of a Lie group G is said to be special orthogonal (of type O) if $\ln S$ is special orthogonal (of type O) and special unitary (of type U) if $\ln S$ is special unitary (of type U).

Respectively, we distinguish between a special orthogonal and special unitary involutive group, involutive pair, symmetric spaces.

(Compare with I.1.16 Definition 7.)

III.1.9. Definition 43. Let G be a simple compact Lie group, Q being its subgroup. We say that Q is a special unitary subgroup of G if $\ln_G Q$ is a special unitary subalgebra of $\ln G$.

(Compare with I.1.17 Definition 8.)

III.1.10. Definition 44. Let G/H be an involutive pair of an involutive automorphism S of a Lie group G. By the lower (upper) index of an involutive pair G/H, involutive automorphism S, and involutive group H we mean the lower (upper) index of the involutive automorphism $\ln S$.

If the lower and upper indices of an involutive pair coincide then we call it simply the index of the involutive automorphism, involutive group, involutive pair, or symmetric space, respectively.

(Compare with I.1.18 Definition 9.)

III.1.11. Definition 45. We say that a Lie group G is an involutive product of involutive groups $H_1, H_2, H_3 \subset G$, writing in that case

$$G = H_1 \boxtimes H_2 \boxtimes H_3,$$

if there exists the involutive decomposition

$$\ln G = \ln_G H_1 + \ln_G H_2 + \ln_G H_3$$

of the Lie algebras $\ln_G H_1, \ln_G H_2, \ln_G H_3 \subset \ln G$.

In this case we also say that there is the involutive decomposition of a Lie group G,

$$G = H_1 \boxtimes H_2 \boxtimes H_3,$$

into the involutive product of involutive groups $H_1, H_2, H_3 \subset G$.

(Compare with I.1.25 Definition 15.)

III.1.12. Definition 46. An involutive product (decomposition) of a Lie group G,

$$G = H_1 \boxtimes H_2 \boxtimes H_3,$$

is said to be iso-involutive if

$$\ln G = \ln_G H_1 + \ln_G H_2 + \ln_G H_3$$

is an iso-involutive sum.

(Compare with I.1.26 Definition 16.)

III.1.13. Definition 47. An involutive product $G = H_1 \boxtimes H_2 \boxtimes H_3$ is called hyper-involutive if the involutive decomposition

$$\ln G = \ln_G H_1 + \ln_G H_2 + \ln_G H_3$$

is hyper-involutive.

(Compare with I.1.27 Definition 17.)

III.1.14. Definition 48. Let G be a Lie group. By the curvature tensor of an involutive pair G/H (or symmetric space G/H) we mean the curvature tensor of an involutive pair $\ln G / \ln_G H$.

Correspondingly we speak of the curvature tensor of an iso-involutive group and of an involutive automorphism of a Lie group G.

(Compare with I.2.2 Definition 18.)

III.1.15. Definition 49. An iso-involutive decomposition of a Lie group G,

$$G = H_1 \boxtimes H_2 \boxtimes H_3,$$

is of the type 1 if

$$\ln G = \ln_G H_1 + \ln_G H_2 + \ln_G H_3$$

is an iso-involutive sum of type 1.

(Compare with I.3.6 Definition 19.)

III.1.16. Definition 50. A hyper-involutive decomposition of a Lie group G,

$$G = H_1 \boxtimes H_2 \boxtimes H_3,$$

is said to be simple if the hyper-involutive sum

$$\ln G = \ln_G H_1 + \ln_G H_2 + \ln_G H_3$$

is simple.

Otherwise we say that this hyper-involutive decomposition is general.

(Compare with II.1.7 Definition 25.)

III.1.17. Definition 51. Let G be a simple compact Lie group, H be its involutive group of a principal involutive automorphism S. We say that the involutive automorphism S, involutive group H, involutive pair G/H (or symmetric space G/H) are exceptional principal if $\ln S$ is an exceptional principal involutive automorphism.

Moreover, we say that S is of type:

1) F_4 if $\ln S$ is of type f_4;

2) E_6 if $\ln S$ is of type e_6;

3) E_7 if $\ln S$ is of type e_7;

4) E_8 if $\ln S$ is of type e_8.

(Compare with II.7.6 Definition 30.)

III.1.18. Definition 52. Let $M = G/H$ be a homogeneous reductive space [S. Kobayashi, K. Nomizu 63,69], and H be its stabilizer of a point $o \in M$. We say that a subsymmetry S_o is a geodesic subsymmetry if it is generated by an involutive automorphism S of a Lie group G in such a way that $\ln S \in \mathrm{Int}_{\ln G}\{t\zeta\}$, where $\zeta \in \mathfrak{m} = \ln G \dot{-} \ln_G H$ (meaning that $\mathfrak{m}$ is a reductive complement to $\ln_G H$, that is, $\ln G = \mathfrak{m} \dot{+} \ln_G H$, $[\mathfrak{m} \, \ln_G H] \subset \mathfrak{m}$).

Correspondingly we speak of a geodesic mirror [L.V. Sabinin, 58a,59a,59b].

III.1.21. Definition 53. We say that a mirror W of a homogeneous space G/H is special (unitary, orthogonal) if the corresponding subsymmetry generates in G a special (unitary, orthogonal) involutive automorphism.

Respectively, we speak of a special (unitary, orthogonal) subsymmetry.

III.1.22. Definition 54. We say that a mirror of a homogeneous space G/H is principal (unitary, orthogonal) if the corresponding subsymmetry generates in G a principal (unitary, orthogonal) involutive automorphism.

Respectively, we speak of a principal (unitary, orthogonal) subsymmetry.

III.1.23. Definition 55. A unitary special symmetric space G/H is said to be essentially unitary special if it is not unitary principal.

III.1.24. Definition 56. An essentially special but not principal symmetric space is called strictly special.

III.6.2. Definition 57. We say that a homogeneous space G/H is an involutive product of homogeneous spaces $G_1 \overset{*}{/} H_1$, $G_2 \overset{*}{/} H_2$, and $G_3 \overset{*}{/} H_3$ and write

$$G/H = G_1 \overset{*}{/} H_1 \boxtimes G_2 \overset{*}{/} H_2 \boxtimes G_3 \overset{*}{/} H_3$$

if

$$G = G_1 \boxtimes G_2 \boxtimes G_3, \qquad H = H_1 \boxtimes H_2 \boxtimes H_3.$$

We say in this case that $G_\alpha \overset{*}{/} H_\alpha$ ($\alpha = 1, 2, 3$) are mirrors in G/H.

III.6.3. Definition 58. An involutive product

$$G/H = G_1 \overset{*}{/} H_1 \boxtimes G_2 \overset{*}{/} H_2 \boxtimes G_3 \overset{*}{/} H_3$$

is called a tri-symmetric space [L.V. Sabinin, 61] with the mirrors $G_\alpha \overset{*}{/} H_\alpha$ ($\alpha = 1, 2, 3$) if

$$\mathfrak{k} = (\ln_G G_1) \cap (\ln_G G_2) \cap (\ln_G G_3) \subset \ln_G H;$$

moreover, it is said to be trivial if $\mathfrak{k} = \ln_G H$, semi-trivial if $\mathfrak{k} = \ln_G H_\alpha$ for some $\alpha = 1, 2, 3$, and non-trivial otherwise.

III.6.5. Definition 59. A tri-symmetric space

$$G/H = G_1 \overset{*}{/} H_1 \boxtimes G_2 \overset{*}{/} H_2 \boxtimes G_3 \overset{*}{/} H_3$$

is called hyper-tri-symmetric if

$$G = G_1 \boxtimes G_2 \boxtimes G_3, \qquad H = H_1 \boxtimes H_2 \boxtimes H_3$$

are hyper-involutive products with the common conjugating automorphism p ($p^3 = \mathrm{id}$),

$$pG_1 = G_2, \qquad pG_2 = G_3, \qquad pG_3 = G_1,$$
$$pH_1 = H_2, \qquad pH_2 = H_3, \qquad pH_3 = H_1.$$

A.3.4. PART IV.

IV.1.2. Definition 60. Let M be a C^∞-smooth manifold, and G be a Lie group. We say that there is defined a representation, or action T, of a Lie group G on a manifold M if a map

$$G \ni g \overset{T}{\mapsto} T_g \in \mathrm{Diff}\, M$$

with the properties

$$T_{g_1} \circ T_{g_2} = T_{g_1 g_2}, \qquad (T_g)^{-1} = T_{g^{-1}}, \qquad T_e = \mathrm{id}_M$$

is given.

Here e means the identity element of G.

IV.1.3. Definition 61. An action (representation) is called faithful (or effective) if

$$T_g = \mathrm{id}_M \implies g = e.$$

An action is called transitive if for any $x, y \in M$ there is $g \in G$ such that $T_g x = y$.

IV.1.4. Definition 62. A manifold M is called a homogeneous space if a faithful transitive smooth action T of a Lie group G is defined on M.

In this case we write (M, T, G).

IV.1.5. Definition 63. Let (M, T, G) be a smooth homogeneous space. A closed subgroup

$$G_x = \{g \in G \mid T_g x = x, \ x \in M\}$$

of a group G is called a stationary subgroup (stabilizer, or isotropy group) of a point $x \in M$.

IV.1.7. Definition 64. A homogeneous space (M, G, T) is called a Riemannian homogeneous space if there is given a Riemannian metric g on M (that is, a tensor field which is twice covariant, symmetric, and positive-definite at any point) and T_a ($a \in G$) are isometries of this metric (that is, for any smooth curve $\gamma \colon [0, 1] \to M$ its length $\ell(\gamma)$ coincides with the length $\ell(T_a \circ \gamma)$ of the transformed curve).

We use in this case the notation (M, G, T, g).

IV.1.9. Definition 65. A diffeomorphism $s \colon M \to M$ of a homogeneous space (M, G, T) is called a mirror subsymmetry (reflection) if:

1. there exists $x \in M$ such that $sx = x$;
2. $s \circ s = \mathrm{id}_M$;
3. s is an automorphism of the action T, that is, $s \circ T_g \circ s^{-1} = T_{\sigma(g)}$, where $\sigma \colon G \to G$ is a smooth map.

We often write s_x instead of s, explicitly indicating a point $x \in M$ which is immobile under the action of s_x.

IV.1.11. Definition 66. The set $\mathfrak{k}$ of all immobile points of a mirror subsymmetry s_x is called a mirror. This set is a submanifold of M, and may be not connected.

Considering $G^+ = \{g \in G \mid \sigma g = g\}$ we have a subgroup of G which is called a mirror subgroup of a mirror subsymmetry s_x. (See IV.1.9 Definition 65.)

IV.1.15. Definition 67. A homogeneous space (M, G, T) with a mirror subsymmetry s_x is called a mirror subsymmetric homogeneous space (or a homogeneous space with a mirror).

The dimension of the component of connectedness of a point x of the mirror $\mathfrak{k}$ generated by a mirror subsymmetry s_x is called the order of the mirror $\mathfrak{k}$ and of the mirror subsymmetry s_x.

IV.1.17.   Definition 68. A homogeneous mirror subsymmetric space (M, G, T) with a Riemannian metric g such that s_x is its isometry is called a homogeneous Riemannian mirror subsymmetric space and is denoted (M, G, T, s_x, g).

IV.2.11. Definition 69. A mirror subsymmetric triplet $(\mathfrak{g}, \mathfrak{h}, \sigma)$ is said to be reductive if

$$\mathfrak{g} = \mathfrak{m} + \mathfrak{h}, \qquad (\operatorname{ad}\mathfrak{h})\mathfrak{m} \subset \mathfrak{m}, \qquad \sigma\mathfrak{m} \subset \mathfrak{m}.$$

Respectively, a homogeneous mirror subsymmetric space (G, H, s) is said to be reductive if its corresponding triplet of Lie algebras, $(\mathfrak{g}, \mathfrak{h}, \sigma)$ (here $\sigma = s_{*,e}$), is reductive.

IV.4.8. Definition 70. If three mirror subsymmetries $\underset{x}{s_1}$, $\underset{x}{s_2}$, $\underset{x}{s_3}$ (with the immobile point $x \in M$) of a homogeneous Riemannian space $M = G/H$ (H being the stabilizer of $x \in M$) with the properties

$$(\underset{x}{s_i})^2 = \operatorname{id}, \qquad \underset{x}{s_i} \circ \underset{x}{s_j} = \underset{x}{s_k} \quad (i \neq j,\ j \neq k,\ k \neq i) \tag{IV.53}$$

are given then the group generated by these subsymmetries is called an involutive group and is denoted $\mathrm{I.Gr}(\underset{x}{s_1}, \underset{x}{s_2}, \underset{x}{s_3})$.

For the homogeneous mirror subsymmetric space $M = G/H$ with an involutive group of automorphisms $\mathrm{I.Gr}(\underset{x}{s_1}, \underset{x}{s_2}, \underset{x}{s_3})$ we use the notation $(G, H, \mathrm{I.Gr}(\underset{x}{s_1}, \underset{x}{s_2}, \underset{x}{s_3}))$, or $(G, H, \underset{x}{s_1}, \underset{x}{s_2}, \underset{x}{s_3})$, for short.

Correspondingly, in the iso-involutive case we write $(G, H, \mathrm{I.I.Gr}(\underset{x}{s_1}, \underset{x}{s_2}, \underset{x}{s_3}, \varphi))$, or more briefly $(G, H, \underset{x}{s_1}, \underset{x}{s_2}, \underset{x}{s_3}, \varphi)$.

IV.4.9. Definition 71. A homogeneous space M/G is called tri-symmetric if it has an involutive group $\mathrm{I.Gr}(\underset{x}{s_1}, \underset{x}{s_2}, \underset{x}{s_3})$ such that in some neighbourhood of $x \in M$ the intersection of the corresponding mirrors consists of a one point,

$$K_x^{(0)} = K_x^{(1)} \cap K_x^{(2)} \cap K_x^{(3)} = \{x\}.$$

IV.4.11. Definition 72. A mirror triplet $(\mathfrak{g}, \mathfrak{h}, \mathrm{I.Gr}(\sigma_1, \sigma_2, \sigma_3))$ is said to be tri-symmetric if

$$\mathfrak{g}_1^+ \cap \mathfrak{g}_2^+ = \mathfrak{g}_2^+ \cap \mathfrak{g}_3^+ = \mathfrak{g}_3^+ \cap \mathfrak{g}_1^+ = \mathfrak{g}_1^+ \cap \mathfrak{g}_2^+ \cap \mathfrak{g}_3^+ \subset \mathfrak{h}$$

(here $\mathfrak{g}_i^+ = \{x \in \mathfrak{g} \mid \sigma_i x = x\}$).

APPENDIX FOUR

Без раздумий брыкнув безделушки,
Расцветая Венками Сонетов,
Ты остатками черствой краюшки
Накорми Хвостоглазье Кометы.

Квантасмагор

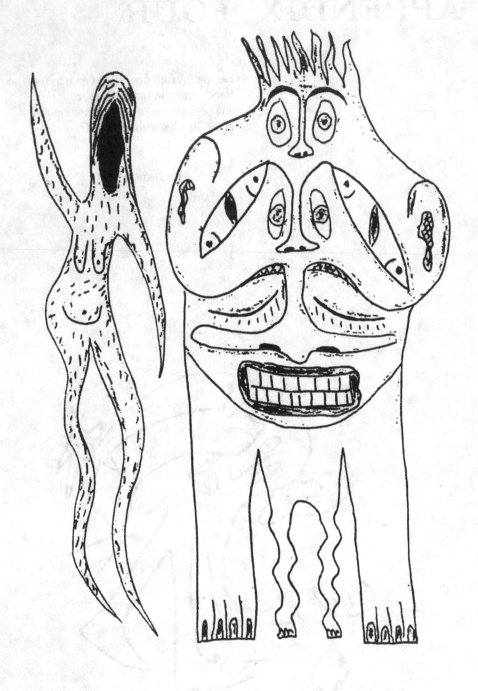

Fabruary - 30/4/83

APPENDIX 4

THEOREMS

A.4.1. PART I.

I.2.4. Theorem 1. *An exact (in particular, elementary) involutive pair* $\mathfrak{g}/\mathfrak{l}$
is uniquely defined by its curvature tensor.

Taking a base $X_1, \ldots, X_n$ *on* $\mathfrak{m} = \mathfrak{g} - \mathfrak{l}$ *we have*

$$R(X_i; X_j, X_k) = -[X_i \, [X_j \, X_k]]$$
$$= R^q_{i,jk} X_q. \tag{I.11}$$

Thus we also say that $R^q_{i,jk}$ *is a curvature tensor for the involutive pair* $\mathfrak{g}/\mathfrak{l}$.

I.2.5. Theorem 2. *There exists a unique (up to isomorphism) elementary
involutive pair* $\mathfrak{g}/\mathfrak{l}$ *such that*

$$R^q_{i,jk} = \delta^q_k g_{ij} - \delta^q_j g_{ik},$$

where $g_{ij} = g_{ji}$ *is positive-definite* $(i, j, k, q = 1, \ldots, n; \; n > 1)$.
In this case

$$\mathfrak{g}/\mathfrak{l} \cong so(n+1)/so(n)$$

(with the natural embedding).

I.2.6. Theorem 3. *There exists a unique (up to isomorphism) elementary
involutive pair* $\mathfrak{g}/\mathfrak{l}$ *such that*

$$R^q_{i,jk} = 2b^q_i b_{jk} + b^q_k b_{ji} - b^q_j b_{ki} + \delta^q_k g_{ij} - \delta^q_j g_{ik},$$

where $g_{ij} = g_{ji}$ *is positive-definite,*

$$b_{ij} = -b_{ji} \qquad b_{ij} = b^s_i g_{sj}, \qquad b^i_q b^q_j = -\delta^i_j$$
$$(i, j, q, s = 1, \ldots, 2n).$$

In this case

$$\mathfrak{g}/\mathfrak{l} \cong su(n+1)/u(n)$$

(with the natural embedding).

I.2.7. Theorem 4. *There exists a unique (up to isomorphism) elementary involutive pair* $\mathfrak{g}/\mathfrak{l}$ *such that*

$$R^q_{i\,,\,jk} = 2b^q_{i\lambda}b^\lambda_{jk} + b^q_{k\lambda}b^\lambda_{ji} - b^q_{j\lambda}b^\lambda_{ki} + \delta^q_k g_{ij} - \delta^q_j g_{ik},$$

where $g_{ij} = g_{ji}$ *is positive-definite,*

$$b^\lambda_{ij} = -b^\lambda_{ji}, \qquad b^\lambda_{ij} = b^s_{i\lambda}g_{sj}, \qquad b^i_{q\lambda}b^q_{j\mu} + b^i_{q\mu}b^q_{j\lambda} = -2\delta^i_j\delta_{\lambda\mu}$$

$$(\,i, j, k, q, s = 1, \ldots, 4m; \quad \lambda, \mu = 1, 2, 3\,).$$

In this case
$$\mathfrak{g}/\mathfrak{l} \cong sp(n+1)/sp(n) \oplus sp(1)$$

(with the natural embedding).

I.3.2. Theorem 5. *Any involutive group is defined by two non-trivial commuting involutive automorphisms.*

I.3.3. Theorem 6. *Any involutive group* $\chi(S_1, S_2, S_3) \subset \mathrm{Aut}(\mathfrak{g})$, $\mathfrak{g}$ *being a Lie algebra, generates an involutive sum* $\mathfrak{g} = \mathfrak{l}_1 + \mathfrak{l}_2 + \mathfrak{l}_3$, *where* $\mathfrak{l}_\alpha$ *is the involutive algebra of the involutive automorphism* S_α $(\alpha = 1, 2, 3)$.

Moreover,
$$\mathfrak{l}_1 \cap \mathfrak{l}_2 = \mathfrak{l}_2 \cap \mathfrak{l}_3 = \mathfrak{l}_3 \cap \mathfrak{l}_1 = \mathfrak{l}_1 \cap \mathfrak{l}_2 \cap \mathfrak{l}_3 = \mathfrak{l}_0,$$

and $\mathfrak{l}_\alpha/\mathfrak{l}_0$ $(\alpha = 1, 2, 3)$ *are involutive pairs of the involutive automorphisms* $\sigma_\alpha = S_\mu|_{\mathfrak{l}_\alpha}$ $(\alpha \neq \mu, \ \alpha = 1, 2, 3)$, *that is, the restrictions of the involutive automorphisms* S_μ *on* $\mathfrak{l}_\alpha$.

In addition, $\mathfrak{l}_\alpha = \mathfrak{l}_0 \dot{+} \mathfrak{m}_\alpha$ $(\alpha = 1, 2, 3)$, *where* $\mathfrak{m}_\alpha = \{\zeta \in \mathfrak{l}_\alpha \mid \sigma_\alpha \zeta = -\zeta\}$, *and* $\mathfrak{l}_0, \mathfrak{m}_1, \mathfrak{m}_2, \mathfrak{m}_3$ *are pair-wise orthogonal with respect to the Cartan metric of the Lie algebra* $\mathfrak{g}$.

I.3.5. Theorem 7. *Let* $\chi(S_1, S_2, S_3; \varphi)$ *be an iso-involutive group of a Lie algebra* $\mathfrak{g}$, *and* $\mathfrak{g} = \mathfrak{l}_1 + \mathfrak{l}_2 + \mathfrak{l}_3$ *be the corresponding involutive sum.*

Then $\varphi\mathfrak{l}_1 = \mathfrak{l}_2$, $\varphi\mathfrak{l}_2 = \mathfrak{l}_1$, *and the automorphisms* $\varphi|_{\mathfrak{l}_3}$, $\varphi|_{\mathfrak{l}_0}$, *the restrictions of* φ *on* $\mathfrak{l}_3$ *and on* $\mathfrak{l}_0 = \mathfrak{l}_1 \cap \mathfrak{l}_2 \cap \mathfrak{l}_3$, *respectively, are involutive automorphisms.*

I.3.8. Theorem 8. *If* $\chi(S_1, S_2, S_2; \varphi)$ *is not of type 1 then* $\chi^{(1)}(S_1, S_2, S_3; \varphi)$ *exists, otherwise it does not exist.*

I.3.13. Theorem 9. *Let* $\mathfrak{g}$ *be a compact semi-simple Lie algebra. If* $\mathfrak{l}_1$ *is its involutive algebra of an involutive automorphism* $S_1 \neq \mathrm{Id}$ *then there exists an iso-involutive group* $\chi(S_1, S_2, S_3; \varphi)$ *of* $\mathfrak{g}$ *and the corresponding involutive decomposition* $\mathfrak{g} = \mathfrak{l}_1 + \mathfrak{l}_2 + \mathfrak{l}_3$.

I.3.16. Theorem 10. *Let* $\mathfrak{g}$ *be a compact semi-simple Lie algebra,* $\mathfrak{g}/\mathfrak{l}_1$ *be an involutive pair of an involutive automorphism* S_1 *of lower index 1, and let* $\{t\zeta\}_{t \in \mathbb{R}}$ *be a one-dimensional maximal subalgebra in* $\mathfrak{m} = \mathfrak{g} \dot{-} \mathfrak{l}_1$.

Then there exists an iso-involutive group $\chi(S_1, S_2, S_3; \varphi)$ *and the corresponding involutive sum of lower index 1 such that* $\varphi \in \mathrm{Int}_\mathfrak{g}(\{t\zeta\})$.

I.4.2. Theorem 11. Let $\chi(S_1, S_2, S_3; \varphi)$ be an iso-involutive group of a Lie algebra $\mathfrak{g}$, then there exists an involutive base of $\mathfrak{g}$ which is invariant with respect to $\chi(S_1, S_2, S_3; \varphi)$.

If, in addition, $\mathfrak{g}$ is compact semi-simple then there exists an involutive base which is orthogonal with respect to the Cartan metric of $\mathfrak{g}$.

I.5.2. Theorem 12. Let $\mathfrak{g} = \mathfrak{l}_1 + \mathfrak{l}_2 + \mathfrak{l}_3$ be an iso-involutive decomposition of type 1 for a compact semi-simple Lie algebra $\mathfrak{g}$. Then the involutive algebra $\mathfrak{l}_3$ possesses a non-trivial centre $\mathfrak{z} \subset \mathfrak{m}_3 = \mathfrak{l}_3 \dot{-} \mathfrak{l}_0$.

I.5.3. Theorem 13. Let $\mathfrak{g} = \mathfrak{l}_1 + \mathfrak{l}_2 + \mathfrak{l}_3$ be an iso-involutive sum of type 1. If $\mathfrak{g}$ is compact and $\mathfrak{g}/\mathfrak{l}_1$ is an elementary involutive pair, then $\mathfrak{l}_3$ has the unique one-dimensional centre $\mathfrak{z} \subset \mathfrak{l}_3 \dot{-} \mathfrak{l}_0$. Moreover

$$\mathrm{ad}_\mathfrak{g}^{(\mathfrak{g} \dot{-} \mathfrak{l}_3)}(\mathfrak{z}) = \{t\,(\varphi|_{(\mathfrak{g} \dot{-} \mathfrak{l}_3)})\}_{t \in \mathbb{R}}.$$

I.5.4. Theorem 14. Let $\mathfrak{g}$ be a compact Lie algebra, $\mathfrak{g}/\mathfrak{l}_1$ be an elementary involutive pair, and let $\mathfrak{g} = \mathfrak{l}_1 + \mathfrak{l}_2 + \mathfrak{l}_3$ be an iso-involutive sum of lower index 1 and of type 1, then

$$\mathfrak{g}/\mathfrak{l}_1 \cong so(n+1)/so(n), \qquad \mathfrak{g}/\mathfrak{l}_2 \cong so(n+1)/so(n),$$

$$\mathfrak{g}/\mathfrak{l}_3 \cong so(n+1)/so(n-1) \oplus so(2),$$

$$\mathfrak{l}_1/\mathfrak{l}_0 \cong so(n)/so(n-1), \qquad \mathfrak{l}_2/\mathfrak{l}_0 \cong so(n)/so(n-1),$$

$$\mathfrak{l}_3/\mathfrak{l}_0 \cong so(2)/\{0\} + so(n-1)/so(n-1)$$

with the natural embeddings.

I.5.7. Theorem 15. If a compact Lie algebra $\mathfrak{g}$ has an iso-involutive decomposition $\mathfrak{g} = \mathfrak{l}_1 + \mathfrak{l}_2 + \mathfrak{l}_3$ of type 2 and $\mathfrak{g}/\mathfrak{l}_1$ is elementary then the maximal subalgebra of elements immobile under the action of a conjugate automorphism φ has a non-trivial central ideal

$$\mathfrak{z} \subset \mathfrak{q}_3 = \mathfrak{n}_3 \dot{-} \mathfrak{n}_0 \quad (\mathfrak{n}_0 = \mathfrak{n}_3 \cap \mathfrak{l}_1).$$

I.5.9. Theorem 16. Let $\mathfrak{g}$ be compact, $\mathfrak{g}/\mathfrak{l}_1$ be an elementary involutive pair, and let $\mathfrak{g} = \mathfrak{l}_1 + \mathfrak{l}_2 + \mathfrak{l}_3$ be the iso-involutive sum of an iso-involutive group $\chi(S_1, S_2, S_3; \varphi)$ of lower index 1 and of type 2. Then

$$\mathrm{ad}_\mathfrak{g}^{(\mathfrak{g} \dot{-} \mathfrak{l}_3)}(\{t\zeta\}) = \{t\,(\varphi|_{(\mathfrak{g} \dot{-} \mathfrak{l}_3)})\}.$$

I.5.12. Theorem 17. Let $\mathfrak{g}$ be a compact Lie algebra, $\mathfrak{g}/\mathfrak{l}_1$ be an elementary involutive pair, and let $\mathfrak{g} = \mathfrak{l}_1 + \mathfrak{l}_2 + \mathfrak{l}_3$ be an iso-involutive sum of type 2 and of *lower index 1, and* $\dim(\mathfrak{l}_3 \dot{-} \mathfrak{l}_0) = 2$. *Then*

$$\mathfrak{g}/\mathfrak{l}_1 \cong su(n+1)/su(n) \oplus u(1),$$

$$\mathfrak{g}/\mathfrak{l}_2 \cong su(n+1)/su(n) \oplus u(1),$$

$$\mathfrak{g}/\mathfrak{l}_3 \cong su(n+1)/su(n-1) \oplus su(2) \oplus u(1),$$

$$\mathfrak{l}_1/\mathfrak{l}_0 \cong su(n)/su(n-1) \oplus u(1) + u(1)/u(1),$$

$$\mathfrak{l}_2/\mathfrak{l}_0 \cong su(n)/su(n-1) \oplus u(1) + u(1)/u(1),$$

$$\mathfrak{l}_3/\mathfrak{l}_0 \cong su(2)/u(1) + su(n-1) \oplus u(1)/su(n-1) \oplus u(1)$$

with the natural embeddings.

I.5.14. Theorem 18. *Let* $\mathfrak{g}$ *be a compact Lie algebra,* $\mathfrak{g}/\mathfrak{l}_1$ *be an elementary involutive pair, and let* $\mathfrak{g} = \mathfrak{l}_1 + \mathfrak{l}_2 + \mathfrak{l}_3$ *be an involutive sum of type 2 and of lower index 1, and* $\dim(\mathfrak{l}_3 \dot{-} \mathfrak{l}_0) = 4$. *Then*

$$\mathfrak{g}/\mathfrak{l}_1 \cong sp(n+1)/sp(n) \oplus sp(1),$$

$$\mathfrak{g}/\mathfrak{l}_2 \cong sp(n+1)/sp(n) \oplus sp(1),$$

$$\mathfrak{g}/\mathfrak{l}_3 \cong sp(n+1)/sp(n-1) \oplus sp(2),$$

$$\mathfrak{l}_1/\mathfrak{l}_0 \cong sp(n)/sp(n-1) \oplus sp(1) + sp(1)/sp(1),$$

$$\mathfrak{l}_2/\mathfrak{l}_0 \cong sp(n)/sp(n-1) \oplus sp(1) + sp(1)/sp(1),$$

$$\mathfrak{l}_3/\mathfrak{l}_0 \cong sp(2)/sp(1) \oplus sp(1) + sp(n-1)/sp(n-1)$$

with the natural embeddings.

I.6.1. Theorem 19. *Let* $\mathfrak{g}$ *be a simple compact Lie algebra, and* $\mathfrak{g} = \mathfrak{l}_1 + \mathfrak{l}_2 + \mathfrak{l}_3$ *be an iso-involutive sum of lower index 1 and not of type 1 generated by an iso-involutive group* $\chi(S_1, S_2, S_3; \varphi)$. *Then* $\chi^{(1)}(S_1, S_2, S_3; \varphi)$ *and its corresponding iso-involutive sum* $\mathfrak{l}_3 = \mathfrak{l}_0 + \mathfrak{n}_2 + \mathfrak{n}_3$ *is of lower index 1 and of type 1 (with the conjugating isomorphism* $\theta \in \mathrm{Int}_\mathfrak{g}(\{t\zeta\})$, $\theta^2 = \varphi$).

I.6.2. Theorem 20. *Let* $\mathfrak{g}$ *be a simple compact Lie algebra,* $\mathfrak{g} = \mathfrak{l}_1 + \mathfrak{l}_2 + \mathfrak{l}_3$ *be an iso-involutive sum of lower index 1 and not of type 1. Then it is of type 2.*

I.6.3. Theorem 21. *Let* $\mathfrak{g}$ *be a compact Lie algebra,* $\mathfrak{g} = \mathfrak{l}_1 + \mathfrak{l}_2 + \mathfrak{l}_3$ *be the iso-involutive sum of an iso-involutive group* $\chi(S_1, S_2, S_3; \varphi)$, $\mathfrak{g}/\mathfrak{l}_1$ *being an elementary involutive pair,* $\dim(\mathfrak{l}_3 \dot{-} \mathfrak{l}_0) = 1$. *Then*

$$\mathfrak{g}/\mathfrak{l}_1 \cong so(n+1)/so(n), \qquad \mathfrak{g}/\mathfrak{l}_2 \cong so(n+1)/so(n),$$

$$\mathfrak{g}/\mathfrak{l}_3 \cong so(n+1)/so(n-1) \oplus so(2),$$

$$\mathfrak{l}_1/\mathfrak{l}_0 \cong so(n)/so(n-1), \qquad \mathfrak{l}_2/\mathfrak{l}_0 \cong so(n)/so(n-1),$$

$$\mathfrak{l}_3/\mathfrak{l}_0 \cong so(2)/\{0\} + so(n-1)/so(n-1)$$

with the natural embeddings.

I.6.4. **Theorem 22.** *Let $\mathfrak{g}$ be a simple compact Lie algebra, $\mathfrak{g} = \mathfrak{l}_1 + \mathfrak{l}_2 + \mathfrak{l}_3$ being the iso-involutive sum of the iso-involutive group $\chi(S_1, S_2, S_3 ; \varphi)$, $\dim(\mathfrak{l}_3 \dot{-} \mathfrak{l}_0) = 2$. Then*

$$\mathfrak{g}/\mathfrak{l}_1 \cong su(n+1)/su(n) \oplus u(1), \qquad \mathfrak{g}/\mathfrak{l}_2 \cong su(n+1)/su(n) \oplus u(1),$$

$$\mathfrak{g}/\mathfrak{l}_3 \cong su(n+1)/su(n-1) \oplus su(2) \oplus u(1),$$

$$\mathfrak{l}_1/\mathfrak{l}_0 \cong su(n)/su(n-1) \oplus u(1) + u(1)/u(1),$$

$$\mathfrak{l}_2/\mathfrak{l}_0 \cong su(n)/su(n-1) \oplus u(1) + u(1)/u(1),$$

$$\mathfrak{l}_3/\mathfrak{l}_0 \cong su(2)/u(1) + su(n-1) \oplus u(1)/su(n-1) \oplus u(1)$$

with the natural embeddings.

I.6.5. **Theorem 23.** *Let $\mathfrak{g}$ be a simple compact Lie algebra, $\mathfrak{g} = \mathfrak{l}_1 + \mathfrak{l}_2 + \mathfrak{l}_3$ be an iso-involutive sum of lower index 1, and $\dim(\mathfrak{l}_3 \dot{-} \mathfrak{l}_0) = 4$. Then*

$$\mathfrak{g}/\mathfrak{l}_1 \cong sp(n+1)/sp(n) \oplus sp(1),$$

$$\mathfrak{g}/\mathfrak{l}_2 \cong sp(n+1)/sp(n) \oplus sp(1),$$

$$\mathfrak{g}/\mathfrak{l}_3 \cong sp(n+1)/sp(n-1) \oplus sp(2),$$

$$\mathfrak{l}_1/\mathfrak{l}_0 \cong sp(n)/sp(n-1) \oplus sp(1) + sp(1)/sp(1),$$

$$\mathfrak{l}_2/\mathfrak{l}_0 \cong sp(n)/sp(n-1) \oplus sp(1) + sp(1)/sp(1),$$

$$\mathfrak{l}_3/\mathfrak{l}_0 \cong sp(2)/sp(1) \oplus sp(1) + sp(n-1)/sp(n-1)$$

with the natural embeddings.

I.6.6. **Theorem 24.** *Let $\mathfrak{g}$ be a simple compact Lie algebra, $\mathfrak{g} = \mathfrak{l}_1 + \mathfrak{l}_2 + \mathfrak{l}_3$ be an involutive sum of lower index 1 and not of type 1. Then*

$$\mathfrak{l}_3/\mathfrak{l}_0 = \tilde{\mathfrak{l}}_3/\tilde{\mathfrak{l}}_0 + \tilde{\tilde{\mathfrak{l}}}_0/\tilde{\mathfrak{l}}_0,$$

$$\tilde{\mathfrak{l}}_3/\tilde{\mathfrak{l}}_0 \cong so(m+1)/so(m)$$

with the natural embedding, and $\tilde{\mathfrak{l}}_3$ is the special unitary involutive subalgebra of the involutive automorphism S_3.

I.6.7. **Theorem 25.** *Let $\mathfrak{g}$ be a simple compact Lie algebra, and $\mathfrak{g} = \mathfrak{l}_1 + \mathfrak{l}_2 + \mathfrak{l}_3$ be the iso-involutive sum of lower index 1 generated by an iso-involutive group $\chi(S_1, S_2, S_3; \varphi)$ ($\varphi \in \mathrm{Int}_{\mathfrak{g}}(\{t\zeta\})$). If $\mathfrak{k}$ is the maximal subalgebra of elements immobile under the action of φ then $\mathfrak{k} \cap (\mathfrak{l}_3 \dot{-} \mathfrak{l}_0) = \{t\zeta\}$ and*

$$\mathrm{ad}_{\mathfrak{g}}^{(\mathfrak{g} \dot{-} \mathfrak{l}_3)}(\{t\zeta\}) = \{t\,(\varphi|_{(\mathfrak{g} \dot{-} \mathfrak{l}_3)})\}.$$

I.6.8. Theorem 26. Let $\mathfrak{g}$ be a simple compact Lie algebra, and $\mathfrak{g} = \mathfrak{l}_1 + \mathfrak{l}_2 + \mathfrak{l}_3$ be the iso-involutive sum of lower index 1 with the conjugating automorphism $\varphi \in \mathrm{Int}_{\mathfrak{g}}(\{t\zeta\})$. Then, for any $x \in \mathfrak{l}_1 \dot{-} \mathfrak{l}_0$,

$$\mathfrak{b} = \{\alpha x + \beta\varphi x + \gamma\zeta\}$$

is a three-dimensional simple compact algebra and $\mathrm{Int}_{\mathfrak{g}}(\mathfrak{b}) \cong SO(3)$.

I.6.9. Theorem 27. Let $\mathfrak{g}$ be a compact Lie algebra, and $\mathfrak{g}/\mathfrak{l}_1$ be an elementary involutive pair of lower index 1, then all one-dimensional subalgebras $\{tp\} \subset \mathfrak{g} \dot{-} \mathfrak{l}_1$ are conjugated in $\mathrm{Int}_{\mathfrak{g}}(\mathfrak{l}_1)$, and consequently $\mathfrak{g}/\mathfrak{l}_1$ is of index 1.

I.6.10. Theorem 28. Let $\mathfrak{g}$ be a simple compact Lie algebra, $\mathfrak{g} = \mathfrak{l}_1 + \mathfrak{l}_2 + \mathfrak{l}_3$ be an iso-involutive sum of lower index 1. Then $\dim(\mathfrak{l}_3 \dot{-} \mathfrak{l}_0) \neq 3$.

I.7.2. Theorem 29. Let $\mathfrak{g}$ be a simple compact Lie algebra, and let $\mathfrak{l} = \mathfrak{b} \oplus_{\mathfrak{z}} \oplus \tilde{\mathfrak{l}}$ ($\mathrm{Int}_{\mathfrak{g}}(\mathfrak{b}) \cong SU(2)$, $\{0\} \neq \mathfrak{z}$ being the center in $\mathfrak{l}$) be a principal unitary central involutive algebra of a principal unitary central involutive automorphism S. Then

$$\mathfrak{g}/\mathfrak{l} \cong su(n)/su(2) \oplus u(1) \oplus su(n-2)$$

with the natural embedding.

I.8.1. Theorem 30. Let $\mathfrak{g}$ be a simple compact Lie algebra, and $\mathfrak{g}/\mathfrak{l}$ be an involutive pair of index 1,

$$\mathfrak{l} = \mathfrak{m} \oplus \mathfrak{n},$$

where $\mathfrak{m} \neq \{0\}$, $\mathfrak{n} \neq \{0\}$.

Then either

$$\mathfrak{g}/\mathfrak{l} \cong su(n+1)/su(n) \oplus u(1)$$

or

$$\mathfrak{g}/\mathfrak{l} \cong sp(n+1)/sp(n) \oplus sp(1)$$

with the natural embeddings.

I.8.2. Theorem 31. Let $\mathfrak{g}$ be a compact Lie algebra, and $\mathfrak{g}/\mathfrak{l}$ be an elementary principal involutive pair, $\dim \mathfrak{l} = 3$. Then $\mathfrak{g}/\mathfrak{l}$ is a principal orthogonal involutive pair.

I.8.3. Theorem 32. Let $\mathfrak{g}$ be a compact simple Lie algebra, and $\mathfrak{g}/\mathfrak{l}$ be a principal central involutive pair of index 1 for an involutive automorphism S. Then S is a principal involutive automorphism of type U and

$$\mathfrak{g}/\mathfrak{l} \cong su(3)/su(2) \oplus u(1)$$

with the natural embedding.

I.8.4. Theorem 33. Let $\mathfrak{g}$ be a simple compact Lie algebra, and $\mathfrak{g}/\mathfrak{l}$ be a principal non-central involutive pair of type U and of index 1. Then

$$\mathfrak{g}/\mathfrak{l} \cong sp(n+1)/sp(n) \oplus sp(1)$$

with the natural embedding.

A.4.2. PART II.

II.2.1. Theorem 1. Let $\mathfrak{g}/\mathfrak{l}$ be an involutive pair of an involutive automorphism S, and q be an automorphism of $\mathfrak{g}$ such that $qS = Sq$ and $q\eta = \eta$ for any $\eta \in \mathfrak{l}$. Then

$$q\mathfrak{m} = \mathfrak{m} \quad (\mathfrak{m} = \mathfrak{g} \dot{-} \mathfrak{l}), \qquad q^{-1}\mathfrak{m} = \mathfrak{m},$$
$$\left(q \pm q^{-1} \right) \operatorname{ad}_{\mathfrak{g}}^{\mathfrak{g}-\mathfrak{l}}(y) = \operatorname{ad}_{\mathfrak{g}}^{\mathfrak{g}-\mathfrak{l}}(y) \left(q \pm q^{-1} \right)$$

for any $y \in \mathfrak{l}$, and $\left(q - q^{-1} \right)$ is a differentiation of the Lie algebra $\mathfrak{g}$.

II.2.2. Theorem 2. If $\mathfrak{g}$ is a semi-simple compact Lie algebra, $\mathfrak{g}/\mathfrak{l}$ an irreducible (in particular, elementary) involutive pair of an involutive automorphism S, and q an automorphism of $\mathfrak{g}$ such that $q \neq \operatorname{Id}$, $q \neq S$, $q\eta = \eta$ for $\eta \in \mathfrak{l}$, then $q \in \operatorname{Int}_{\mathfrak{g}}(\mathfrak{z})$, where $\mathfrak{z}$ is a non-trivial centre in $\mathfrak{l}$.

II.2.3. Theorem 3. Let $\mathfrak{g}$ be a semi-simple compact Lie algebra, and $\mathfrak{l}$ be its involutive algebra of an involutive automorhism S. If $q \in \operatorname{Aut} \mathfrak{g}$ $(q \neq \operatorname{Id})$, $q\eta = \eta$ for any $\eta \in \mathfrak{l}$, and $q + \operatorname{Id}$ is invertible (in particular, $q^{2n+1} = \operatorname{Id}$) then $q \in \operatorname{Int}_{\mathfrak{g}}(\mathfrak{z})$, where $\mathfrak{z}$ is a non-trivial centre of $\mathfrak{l}$.

II.2.4. Theorem 4. Let $\mathfrak{g}$ be a compact Lie algebra, $\mathfrak{l} \subset \mathfrak{g}$ be its involutive algebra (of an involutive automorphism S) such that the centre $\mathfrak{g}_0$ of $\mathfrak{g}$ belongs to $\mathfrak{l}$, $\mathfrak{g}_0 \in \mathfrak{l}$. If $q \in \operatorname{Aut} \mathfrak{g}$ $(q \neq \operatorname{Id})$, $q\eta = \eta$ for $\eta \in \mathfrak{l}$, and $q + \operatorname{Id}$ is invertible (in particular, $q^{2n+1} = \operatorname{Id}$) then $q \in \operatorname{Int}_{\mathfrak{g}}(\mathfrak{z})$, where $\mathfrak{z}$ is a (non-trivial) centre of $\mathfrak{l}$.

II.3.4. Theorem 5. Let $\mathfrak{g}$ be a simple compact Lie algebra, and $\mathfrak{l} = \mathfrak{b} \oplus \tilde{\mathfrak{l}}$ be its principal involutive algebra of an involutive automorphism S of type O.
If $\tilde{\mathfrak{l}} \neq \{0\}$ then $\mathfrak{g}$ possesses a prime hyper-involutive decomposition basis for S.

II.3.6. Theorem 6. If $\mathfrak{g}$ is a simple compact Lie algebra, $\mathfrak{l} = \mathfrak{b} \oplus \tilde{\mathfrak{l}}$ ($\mathfrak{b} \cong so(3)$) being a principal involutive algebra of an involutive automorphism S of type O, then $\mathfrak{g}$ has:

a) either a prime hyper-involutive decomposition basis for S;

b) or a non-prime hyper-involutive decomposition basis for S. In this case

$$\mathfrak{l} = \mathfrak{b}, \qquad \mathfrak{l}_\alpha = \mathfrak{p}_\alpha \oplus \mathfrak{z}_\alpha \quad (\alpha = 1, 2, 3),$$

where $\mathfrak{p}_\alpha$ are simple and three-dimensional, $\mathfrak{z}_\alpha$ are one-dimensional, and $\mathfrak{l}_0 = \mathfrak{z}_\alpha \oplus \mathfrak{w}_\alpha$, $\mathfrak{w}_\alpha$ are one-dimensional, moreover

$$\mathfrak{w}_\alpha \subset \mathfrak{p}_\alpha, \qquad \mathfrak{l}_\alpha \cap \mathfrak{l} = \mathfrak{p}_\alpha \cap \mathfrak{l} = \mathfrak{q}_\alpha \subset \mathfrak{p}_\alpha \dot{-} \mathfrak{w}_\alpha$$

(where $\mathfrak{q}_\alpha$ are one-dimensional), $\dim \mathfrak{g} = 8$, and S_α are principal central involutive automorphisms of type U.

II.3.8. Theorem 7. *Let $\mathfrak{g}$ be a simple compact Lie algebra, and $\mathfrak{l} = \mathfrak{b} \oplus \tilde{\mathfrak{l}}$ be its principal involutive algebra of an involutive automorphism S of type O.*
If $\mathfrak{g}$ has a prime hyper-involutive decomposition $\mathfrak{g} = \mathfrak{l}_1 + \mathfrak{l}_2 + \mathfrak{l}_3$ basis for S then

$$\tilde{\mathfrak{l}} = \mathfrak{l}_0 = \mathfrak{l}_1 \cap \mathfrak{l}_2 = \mathfrak{l}_2 \cap \mathfrak{l}_3 = \mathfrak{l}_3 \cap \mathfrak{l}_1,$$

and $\mathfrak{b} \cap \mathfrak{l}_\alpha$ is a one-dimensional centre of $\mathfrak{l}_\alpha$ ($\alpha = 1,2,3$).

II.3.10. Theorem 8. *Let $\mathfrak{g}$ be a simple compact Lie algebra, and $\mathfrak{l} = \mathfrak{b} \oplus \tilde{\mathfrak{l}}$ be its principal involutive algebra of an involutive automorphism S of type O.*
If $\mathfrak{g}$ has a prime hyper-involutive decomposition $\mathfrak{g} = \mathfrak{l}_1 + \mathfrak{l}_2 + \mathfrak{l}_3$ basis for S then

$$\mathfrak{g}/\mathfrak{l} \cong so(m)/(so(m-3) \oplus so(3))$$

$$\mathfrak{g}/\mathfrak{l}_\alpha \cong so(m)/(so(m-2) \oplus so(2))$$

$$(\alpha = 1,2,3; \ m > 4)$$

with the natural embeddings.

II.3.11. Theorem 9. *Let $\mathfrak{g}$ be a simple compact Lie algebra, $\mathfrak{l} = \mathfrak{b} \oplus \tilde{\mathfrak{l}}$ be its principal involutive algebra of an involutive automorphism S of type O, and let $\tilde{\mathfrak{l}} \neq \{0\}$. Then*

$$\mathfrak{g}/\mathfrak{l} \cong so(m)/(so(m-3) \oplus so(3)), \quad m > 4,$$

with the natural embedding.

II.3.14. Theorem 10. *Let $\mathfrak{g}$ be a simple compact Lie algebra, $\mathfrak{l} = \mathfrak{b} \oplus \tilde{\mathfrak{l}}$ be the principal involutive algebra of an involutive automorphism S of type O, and let $\mathfrak{g}$ have no prime hyper-involutive decomposition basis for S.*
Then $\mathfrak{b} = \mathfrak{l}$, $\dim \mathfrak{g} = 8$, and

$$\mathfrak{g}/\mathfrak{l} \cong su(3)/so(3)$$

with the natural embedding.

II.4.2. Theorem 11. *If $\mathfrak{g}$ is a simple compact Lie algebra, $\mathfrak{g}/\mathfrak{b} \oplus \tilde{\mathfrak{l}}$ ($\mathfrak{b} \cong so(3)$) a principal involutive pair of type O of an involutive automorphism A, $\tilde{\mathfrak{l}} \neq \{0\}$, then there exists in $\mathfrak{g}$ an involutive automorphism J associated with A.*

II.4.3. Theorem 12. *If a simple compact Lie algebra* $\mathfrak{g}$ *has an orthogonal principal involutive automorphism* $A \neq \mathrm{Id}$ *then* $\mathfrak{g}$ *has a unitary principal involutive automorphism* J *such that* $JA = AJ$.

II.4.4. Theorem 13. *Let* $\mathfrak{g}$ *be a simple compact Lie algebra, and let its involutive algebra* $\mathfrak{l}$ *of an involutive automorphism* A $(A \neq \mathrm{Id})$ *have a principal involutive automorphism* ψ, *then* $\mathfrak{g}$ *has a unitary special involutive automorphism.*

II.4.6. Theorem 14. *Let* $\mathfrak{g}$ *be a simple compact Lie algebra,* $\mathfrak{l}$ *be its unitary special non-principal involutive algebra, and* $\mathfrak{b} \oplus \mathfrak{v}$ *be a principal orthogonal involutive algebra of* $\mathfrak{l}$, $\mathrm{Int}_{\mathfrak{g}}(\mathfrak{b}) \cong SU(2)$, $\mathrm{Int}_{\mathfrak{l}}(\mathfrak{b}) \cong SO(3)$. *Then* $\mathfrak{l} = \mathfrak{k} \oplus \tilde{\mathfrak{l}}$, $\mathfrak{k}$ *has a principal involutive algebra* $\mathfrak{b} \oplus \mathfrak{p}$, $\mathfrak{p} \subset \mathfrak{v}$, *and*

$$\mathfrak{k}/\mathfrak{b} \oplus \mathfrak{p} \cong so(m)/(so(3) \oplus so(m-3)), \quad m > 4,$$

with the natural embeddings.

II.4.7. Theorem 15. *If* $\mathfrak{g}$ *is a simple compact Lie algebra,* $\mathfrak{k}$ *being its special unitary involutive subalgebra of an involutive automorphism* S, *then* $\mathfrak{k}$ *contains a simple ideal* $\mathfrak{q} \cong so(m)$ $(m > 2)$ *and* $\mathfrak{q}$ *is a special unitary involutive subalgebra of the involutive automorphism* S.

II.4.8. Theorem 16. *If a simple compact Lie algebra* $\mathfrak{g}$ *has a unitary special involutive algebra* $\mathfrak{l} \not\cong so(7)$ *then it also has a principal involutive algebra.*

II.4.9. Theorem 17. *If* $\mathfrak{g}$ *is a simple compact Lie algebra which has a unitary special involutive algebra* $\mathfrak{l}$ *of an involutive automorphism* S, *but has no principal involutive algebras, then* $\mathfrak{g} = \mathfrak{l}_1 + \mathfrak{l}_2 + \mathfrak{l}$, *where* $\mathfrak{l}_1$ *and* $\mathfrak{l}_2$ *are unitary special involutive algebras of the involutive automorphisms* S_1 *and* S_2, *respectively,* $S_1 S_2 = S_2 S_1 = S$, $\mathfrak{l} \cong so(7)$, $\mathfrak{l}_1 \cong so(7)$, $\mathfrak{l}_2 \cong so(7)$, $\mathfrak{l}_0 = \mathfrak{l}_1 \cap \mathfrak{l}_2 = \mathfrak{l}_1 \cap \mathfrak{l} = \mathfrak{l}_2 \cap \mathfrak{l} = \mathfrak{p} \oplus \mathfrak{q} \oplus \mathfrak{f}$, *where* $\mathfrak{p}$, $\mathfrak{q}$, $\mathfrak{f}$ *are simple three-dimensional Lie algebras,*

$$\mathfrak{l}/\mathfrak{l}_0 \cong so(7)/(so(3) \oplus so(4)),$$

$$\mathfrak{l}_1/\mathfrak{l}_0 \cong so(7)/(so(3) \oplus so(4)),$$

$$\mathfrak{l}_2/\mathfrak{l}_0 \cong so(7)/(so(3) \oplus so(4))$$

with the natural embeddings; $S_1 \in \mathrm{Int}_{\mathfrak{g}}(\mathfrak{p})$, $S_2 \in \mathrm{Int}_{\mathfrak{g}}(\mathfrak{q})$, $S \in \mathrm{Int}_{\mathfrak{g}}(\mathfrak{b})$, *where* $\mathfrak{b}$ *is a diagonal of a canonical involutive automorphism of* $\mathfrak{p} \oplus \mathfrak{q}$.

II.4.11. Theorem 18. *There are no Lie algebras satisfying the conditions of II.4.9 (Theorem 17).*

II.4.12. Theorem 19. *There are no simple compact Lie algebras* $\mathfrak{g}$ *which have a special unitary involutive algebra* $\mathfrak{l} \cong so(7)$, *but have no principal involutive algebras.*

II.4.13. Theorem 20. *If a simple compact Lie algebra* $\mathfrak{g}$ *has a unitary special involutive automorphism (algebra) then it has a principal involutive automorphism (algebra).*

II.4.14. Theorem 21. *A simple semi-simple compact Lie algebra has a principal involutive automorphism.*

II.4.15. Theorem 22. *If $\mathfrak{g}$ is a simple compact non-commutative Lie algebra and $\dim \mathfrak{g} \neq 3$ then $\mathfrak{g}$ has a unitary principal involutive automorphism.*

II.5.4. Theorem 23. *Let $\mathfrak{g}$ be a simple compact Lie algebra, and $\mathfrak{l} = \mathfrak{p} \oplus \mathfrak{q} \oplus \tilde{\mathfrak{l}}$ be its principal involutive algebra of an involutive automorphism S of type $U^{(2)}$ $(\mathrm{Int}_{\mathfrak{g}}(\mathfrak{p}) \cong SU(2), \mathrm{Int}_{\mathfrak{g}}(\mathfrak{q}) \cong SU(2))$. If $\tilde{\mathfrak{l}} \neq \{0\}$ then $\mathfrak{g}$ has a prime hyper-involutive decomposition basis for S.*

II.5.6. Theorem 24. *If $\mathfrak{g}$ is a simple compact Lie algebra and $\mathfrak{l}$ is its principal involutive algebra of an involutive automorphism S of type $U^{(2)}$ then either $\mathfrak{g}$ has a prime hyper-involutive decomposition basis for S or $\mathfrak{g}$ has a non-prime hyper-involutive decomposition basis for S such that $\mathfrak{l} = \mathfrak{p} \oplus \mathfrak{q}$, $\mathfrak{l}_\alpha = \mathfrak{r}_\alpha \oplus \mathfrak{w}_\alpha$, $\mathrm{Int}_{\mathfrak{g}}(\mathfrak{r}_\alpha) \cong SU(2)$, $\mathrm{Int}_{\mathfrak{g}}(\mathfrak{w}_\alpha) \cong SU(2)$, $\mathfrak{l}_0 = \mathfrak{t}_\alpha \oplus \mathfrak{n}_\alpha$, where $\mathfrak{t}_\alpha \subset \mathfrak{r}_\alpha$, and $\mathfrak{n}_\alpha \subset \mathfrak{w}_\alpha$, $\mathfrak{t}_\alpha$ and $\mathfrak{n}_\alpha$ are one-dimensional, $\mathfrak{l}_\alpha \cap \mathfrak{l} = \mathfrak{a}_\alpha \oplus \mathfrak{c}_\alpha$, where $\mathfrak{a}_\alpha$ and $\mathfrak{c}_\alpha$ are one-dimensional and $\mathfrak{a}_\alpha \subset \mathfrak{p}_\alpha$, $\mathfrak{c}_\alpha \subset \mathfrak{q}$, $\dim \mathfrak{g} = 14$. The involutive automorphisms S and S_1, S_2, S_3 are conjugated in $\mathrm{Int}(\mathfrak{g})$.*

II.5.8. Theorem 25. *Let $\mathfrak{g}$ be a simple compact Lie algebra, $\mathfrak{l} = \mathfrak{p} \oplus \mathfrak{q} \oplus \tilde{\mathfrak{l}}$ be its principal di-unitary involutive algebra of an involutive automorphism S (that is, $\mathrm{Int}_{\mathfrak{g}}(\mathfrak{p}) \cong SU(2)$, $\mathrm{Int}_{\mathfrak{g}}(\mathfrak{q}) \cong SU(2)$), and let $\mathfrak{g}$ have a prime hyper-involutive decomposition basis for S. Then there exists in $\mathfrak{g}$ a principal involutive automorphism A of type O such that S is an associated involutive automorphism for A. Moreover, the prime hyper-involutive decomposition basis for S is also a prime hyper-involutive decomposition basis for A.*

II.5.9. Theorem 26. *Let $\mathfrak{l} = \mathfrak{p} \oplus \mathfrak{q} \oplus \tilde{\mathfrak{l}}$ be the principal involutive algebra of an involutive automorphism S of type $U^{(2)}$ (that is, $\mathrm{Int}_{\mathfrak{g}}(\mathfrak{p}) \cong SU(2)$, $\mathrm{Int}_{\mathfrak{g}}(\mathfrak{q}) \cong SU(2)$) of a simple compact Lie algebra $\mathfrak{g}$ having a prime hyper-involutive decomposition basis for S. Then*

$$\mathfrak{g}/\mathfrak{p} \oplus \mathfrak{q} \oplus \tilde{\mathfrak{l}} \cong so(m)/(so(4) \oplus so(m-4)), \quad m > 4,$$

with the natural embedding.

II.5.10. Theorem 27. *If $\mathfrak{g}$ is a simple compact Lie algebra, $\mathfrak{l} = \mathfrak{p} \oplus \mathfrak{q} \oplus \tilde{\mathfrak{l}}$ is its principal involutive algebra of type $U^{(2)}$ (that is, $\mathrm{Int}_{\mathfrak{g}}(\mathfrak{p}) \cong SU(2)$, $\mathrm{Int}_{\mathfrak{g}}(\mathfrak{q}) \cong SU(2)$), and $\tilde{\mathfrak{l}} \neq \{0\}$ then*

$$\mathfrak{g}/\mathfrak{l} \cong so(m)/(so(4) \oplus so(m-4)), \quad m > 5,$$

with the natural embedding.

II.6.4. Theorem 28. *Under the conditions and notations of II.5.6 (Theorem 24), if $\mathfrak{g}$ has only non-prime hyper-involutive decomposition basis for S then one may regard $\mathfrak{b} \cap \mathfrak{l}_\alpha = \mathfrak{b} \cap \mathfrak{r}_\alpha$, where $\mathfrak{b}$ is the diagonal of $\mathfrak{l} = \mathfrak{p} \oplus \mathfrak{q}$ generating the hyper-involutive decomposition basis for S.*

II.6.6. Theorem 29. *There exists a unique simple compact Lie algebra* $\mathfrak{g}$ *with a principal involutive automorphism* S *of type* $U^{(2)}$ *having only non-prime hyper-involutive decomposition basis for* S.

The embedding of the involutive algebra $\mathfrak{l}$ *of the involutive automorphism* S *into* $\mathfrak{g}$ *is unique, up to transformations from* $\mathrm{Int}(\mathfrak{g})$.

II.6.9. Theorem 30. *The Lie algebra* $su(3)$ *can be embedded into* g_2 *in such a way that* $su(3)$ *is a maximal subalgebra of* g_2, *but the pair* $(\,g_2\,,\,su(3)\,)$ *is not an involutive pair.*

In the notations of II.5.6 (Theorem 24)

$$su(3) \cong (\mathfrak{r}_1 + \mathfrak{r}_2 + \mathfrak{r}_3) \subset \mathfrak{g} \cong g_2 .$$

II.7.1. Theorem 31. *Let* $\mathfrak{g}$ *be a simple compact Lie algebra,* $\mathfrak{l} = \mathfrak{p} \oplus \mathfrak{q} \oplus \tilde{\mathfrak{l}}$ *be its principal di-unitary involutive algebra of an involutive automorphism* S $(S \in \mathrm{Int}_\mathfrak{g}(\mathfrak{p}),\ S \in \mathrm{Int}_\mathfrak{g}(\mathfrak{q}))$. *If* $\tilde{\mathfrak{l}}$ *is a special unitary subalgebra of the involutive automorphism* S *then either* $\mathfrak{g} \cong so(8)$ *or* $\mathfrak{g} \cong so(12)$.

II.7.2. Theorem 32. *If* $\mathfrak{g}$ *is a simple compact Lie algebra and* $\mathfrak{k} \cong so(6)$ *is its special unitary subalgebra of an involutive automorphism* S *then* S *is a central involutive automorphism.*

II.7.3. Theorem 33. *Let* $\mathfrak{g}$ *be a simple compact Lie algebra, and* $\mathfrak{l}_1 = \mathfrak{p} \oplus \tilde{\mathfrak{l}}$ *be its principal non-central mono-unitary involutive algebra of an involutive automorphism* $S_1 \in \mathrm{Int}_\mathfrak{g}(\mathfrak{p}) \cong SU(2)$. *Then* $\tilde{\mathfrak{l}}$ *is simple and* $\dim \tilde{\mathfrak{l}} > 3$.

II.7.5. Theorem 34. *Let* $\mathfrak{g}$ *be a simple compact Lie algebra, and* $\mathfrak{l}_1 = \mathfrak{p} \oplus \tilde{\mathfrak{l}}$ *be its principal mono-unitary non-central involutive algebra of an involutive automorphism* $S_1 \in \mathrm{Int}_\mathfrak{g}(\mathfrak{p})$. *Then the basis for* $\mathfrak{l}_1$ *involutive sum* $\mathfrak{g} = \mathfrak{l}_1 + \mathfrak{l}_2 + \mathfrak{l}_3$ *of the involutive automorphisms* $S_1, S_2, S_3 = S_1 S_2 = S_2 S_1$ *exists and is iso-involutive with a conjugating automorphism* $\varphi \in \mathrm{Int}_\mathfrak{g}(\{\,t\zeta\,\}),\ \zeta \in \mathfrak{l}_3 \dot{-} \mathfrak{l}_0$ ($\mathfrak{l}_0 = \mathfrak{l}_\alpha \cap \mathfrak{l}_\beta,\ \alpha \neq \beta$).

Moreover,

$$\mathfrak{l}_3/\mathfrak{l}_0 = \mathfrak{k}/\mathfrak{l}_0^{(1)} + \mathfrak{l}_0^{(2)}/\mathfrak{l}_0^{(2)} ,$$

where $\mathfrak{k}$ *is a simple special unitary subalgebra of the involutive automorphism* $S_3 = \varphi^2$ *and* $\mathfrak{k}/\mathfrak{l}_0^{(1)}$ *is a principal di-unitary involutive pair.*

II.8.1. Theorem 35. *Let* $\mathfrak{g}$ *be a Lie algebra of type* f_4, *then there exists an iso-involutive decomposition*

$$\mathfrak{g} = \mathfrak{l}_1 + \mathfrak{l}_2 + \mathfrak{l}_3 \qquad (\mathfrak{l}_\alpha \cap \mathfrak{l}_\beta = \mathfrak{l}_0,\quad \alpha \neq \beta)$$

basis for an exceptional involutive algebra $\mathfrak{l}_1$, *where*

$$\mathfrak{l}_1/\mathfrak{l}_0 \cong sp(3)/(\,sp(2) \oplus sp(1)\,) + sp(1)/sp(1),$$

$$\mathfrak{l}_2/\mathfrak{l}_0 \cong sp(3)/(\,sp(2) \oplus sp(1)\,) + sp(1)/sp(1),$$

$$\mathfrak{l}_3/\mathfrak{l}_0 \cong so(9)/(\,so(4) \oplus so(5)\,)$$

with the natural embeddings, and $\dim \mathfrak{g} = 52$.

II.8.2. Theorem 36. *Let* $\mathfrak{g}$ *be a Lie algebra of type* e_6, *then there exists an iso-involutive decomposition* $\mathfrak{g} = \mathfrak{l}_1 + \mathfrak{l}_2 + \mathfrak{l}_3$ $(\mathfrak{l}_\alpha \cap \mathfrak{l}_\beta = \mathfrak{l}_0,\ \alpha \neq \beta)$ *basis for an exceptional involutive algebra* $\mathfrak{l}_1$, *where*

$$\mathfrak{l}_1/\mathfrak{l}_0 \cong su(6)/(\,su(4) \oplus su(2) \oplus u(1)\,) + su(2)/su(2),$$

$$\mathfrak{l}_2/\mathfrak{l}_0 \cong su(6)/(\,su(4) \oplus su(2) \oplus u(1)\,) + su(2)/su(2),$$

$$\mathfrak{l}_3/\mathfrak{l}_0 \cong so(10)/(\,so(4) \oplus so(6)\,) + u(1)/u(1)$$

with the natural embedding, and $\dim \mathfrak{g} = 78$.

II.8.3. Theorem 37. *Let* $\mathfrak{g}$ *be a Lie algebra of type* e_7. *Then there exists an iso-involutive decomposition*

$$\mathfrak{g} = \mathfrak{l}_1 + \mathfrak{l}_2 + \mathfrak{l}_3 \qquad (\mathfrak{l}_\alpha \cap \mathfrak{l}_\beta = \mathfrak{l}_0,\quad \alpha \neq \beta)$$

basis for an exceptional involutive algebra $\mathfrak{l}_1$, *where*

$$\mathfrak{l}_\alpha/\mathfrak{l}_0 \cong so(12)/(\,so(4) \oplus so(8)\,) + su(2)/su(2) \quad (\alpha = 1,2,3)$$

with the natural embedding, and $\dim \mathfrak{g} = 133$.

II.8.4. Theorem 38. *Let* $\mathfrak{g}$ *be a Lie algebra of type* e_8. *Then there exists an iso-involutive decomposition*

$$\mathfrak{g} = \mathfrak{l}_1 + \mathfrak{l}_2 + \mathfrak{l}_3 \quad (\mathfrak{l}_\alpha \cap \mathfrak{l}_\beta = \mathfrak{l}_0,\quad \alpha \neq \beta)$$

basis for an exceptional involutive algebra $\mathfrak{l}_1$, *where*

$$\mathfrak{l}_1/\mathfrak{l}_0 \cong e_7/(\,so(12) \oplus so(3)\,) + su(2)/su(2),$$

$$\mathfrak{l}_2/\mathfrak{l}_0 \cong e_7/(\,so(12) \oplus so(3)\,) + su(2)/su(2),$$

$$\mathfrak{l}_3/\mathfrak{l}_0 \cong so(16)/(\,so(4) \oplus so(12)\,)$$

with the natural embeddings, and $\dim \mathfrak{g} = 248$.

II.8.5. Theorem 39. *The simple compact non-one-dimensional Lie algebras are isomorphic to*

$$so(n) \quad (n \neq 4,2,1), \qquad g_2, \qquad su(n) \quad (n > 1),$$

$$sp(n) \quad (n > 0), \qquad f_4, \quad e_6, \quad e_7, \quad e_8.$$

All these algebras are pair-wise non-isomorphic and uniquely determined by the type of a principal unitary involutive automorphism, except for the cases

$$so(3) \cong su(2) \cong sp(1), \qquad so(5) \cong sp(2), \qquad so(6) \cong su(4).$$

II.9.1. Theorem 40. *Let $\mathfrak{g}$ be a simple compact Lie algebra, $\mathfrak{l}_1 = \mathfrak{p} \oplus \tilde{\mathfrak{l}}$ be its principal unitary involutive algebra of an involutive automorphism $S_1 \in \mathrm{Int}_\mathfrak{g}(\mathfrak{p})$, and let $\tilde{\mathfrak{l}}$ have a simple ideal $\mathfrak{a}$ such that $\dim \mathfrak{a} > 3$.*

Then the basis for $\mathfrak{l}_1$ involutive sum $\mathfrak{g} = \mathfrak{l}_1 + \mathfrak{l}_2 + \mathfrak{l}_3$ of the involutive automorphisms $S_1, S_2, S_3 = S_1 S_2 = S_2 S_1$ exists, being iso-involutive with the conjugating automorphism $\varphi \in \mathrm{Int}_\mathfrak{g}(\{t\zeta\})$, $\zeta \in \mathfrak{l}_3 \dot{-} \mathfrak{l}_0$ ($\mathfrak{l}_0 = \mathfrak{l}_\alpha \cap \mathfrak{l}_\beta$, $\alpha \neq \beta$).

Furthermore,

$$\mathfrak{l}_3/\mathfrak{l}_0 = \mathfrak{k}/\mathfrak{l}_0^{(1)} + \mathfrak{l}_0^{(2)}/\mathfrak{l}_0^{(2)},$$

where $\mathfrak{k}$ is the simple special unitary subalgebra of the involutive automorphism $S_3 = \varphi^2$, and $\mathfrak{k}/\mathfrak{l}_0^{(1)}$ is a principal unitary involutive pair.

II.9.2. Theorem 41. *Let $\mathfrak{g}$ be a simple compact Lie algebra, $\mathfrak{l}_1 = \mathfrak{p} \oplus \mathfrak{m} \oplus \tilde{\mathfrak{l}}$ be its principal di-unitary involutive algebra of an involutive automorphism S_1 ($S_1 \in \mathrm{Int}_\mathfrak{g}(\mathfrak{p})$, $S_1 \in \mathrm{Int}_\mathfrak{g}(\mathfrak{m})$), and let $\tilde{\mathfrak{l}}$ have a simple ideal $\mathfrak{a}$, $\dim \mathfrak{a} > 3$.*

Then $\mathfrak{g} \cong so(n)$ ($n > 8$), and there exists the iso-involutive decomposition $\mathfrak{g} = \mathfrak{l}_1 + \mathfrak{l}_2 + \mathfrak{l}_3$ basis for $\mathfrak{l}_1$ ($\mathfrak{l}_0 = \mathfrak{l}_\alpha \cap \mathfrak{l}_\beta$, $\alpha \neq \beta$).

Moreover,

$$\mathfrak{l}_1/\mathfrak{l}_0 \cong so(n-4)/(so(n-8) \oplus so(4)) + so(4)/so(4),$$

$$\mathfrak{l}_2/\mathfrak{l}_0 \cong so(n-4)/(so(n-8) \oplus so(4)) + so(4)/so(4),$$

$$\mathfrak{l}_3/\mathfrak{l}_0 \cong so(8)/(so(4) \oplus so(4)) + so(n-8)/so(n-8)$$

with the natural embeddings into $\mathfrak{g} \cong so(n)$.

II.9.3. Theorem 42. *Let $\mathfrak{g}$ be a simple compact Lie algebra, $\mathfrak{l}_1 = \mathfrak{p} \oplus \tilde{\mathfrak{l}} \oplus \mathfrak{z}$ be its central unitary involutive algebra ($\dim \mathfrak{z} = 1$) of an involutive automorphism $S_1 \in \mathrm{Int}_\mathfrak{g}(\mathfrak{p})$, and let $\tilde{\mathfrak{l}}$ have a simple ideal $\mathfrak{a}$, $\dim \mathfrak{a} > 3$.*

Then $\mathfrak{g} \cong su(n)$ ($n > 4$) and there exists the iso-involutive decomposition basis for $\mathfrak{l}_1$, $\mathfrak{g} = \mathfrak{l}_1 + \mathfrak{l}_2 + \mathfrak{l}_3$ ($\mathfrak{l}_0 = \mathfrak{l}_\alpha \cap \mathfrak{l}_\beta$, $\alpha \neq \beta$).

Moreover,

$$\mathfrak{l}_1/\mathfrak{l}_0 \cong su(n-2)/(su(2) \oplus su(n-4) \oplus u(1)) + su(2)/su(2) + u(1)/u(1),$$

$$\mathfrak{l}_2/\mathfrak{l}_0 \cong su(n-2)/(su(2) \oplus su(n-4) \oplus u(1)) + su(2)/su(2) + u(1)/u(1),$$

$$\mathfrak{l}_3/\mathfrak{l}_0 \cong so(6)/(so(4) \oplus so(2)) + su(n-4)/su(n-4) + u(1)/u(1)$$

with the natural embeddings into $\mathfrak{g} \cong su(n)$.

II.9.4. Theorem 43. *Let $\mathfrak{g}$ be a simple compact Lie algebra, and $\mathfrak{l}_1 = \mathfrak{p} \oplus \tilde{\mathfrak{l}}$ be its principal mono-unitary non-central involutive algebra of index 1 of an involutive automorphism $S_1 \in \mathrm{Int}_\mathfrak{g}(\mathfrak{p})$.*

Then $\mathfrak{g} \cong sp(n)$ and there exists the iso-involutive decomposition

$$\mathfrak{g} = \mathfrak{l}_1 + \mathfrak{l}_2 + \mathfrak{l}_3 \quad (\mathfrak{l}_0 = \mathfrak{l}_\alpha \cap \mathfrak{l}_\beta, \quad \alpha \neq \beta)$$

basis for $\mathfrak{l}_1$. Moreover,

$$\mathfrak{l}_1/\mathfrak{l}_0 \cong sp(n-1)/(sp(n-2) \oplus sp(1)) + sp(1)/sp(1),$$

$$\mathfrak{l}_2/\mathfrak{l}_0 \cong sp(n-1)/(sp(n-2) \oplus sp(1)) + sp(1)/sp(1),$$

$$\mathfrak{l}_3/\mathfrak{l}_0 \cong so(5)/so(4) + sp(n-2)/sp(n-2)$$

with the natural embedding into $\mathfrak{g} \cong sp(n)$.

II.9.5. Theorem 44. *Let* $\mathfrak{g}$ *be a simple compact Lie algebra,* $\mathfrak{k}$ *be its simple special subalgebra of an involutive automorphism* S_3, $\dim \mathfrak{k} > 3$.

Then there exist the principal unitary automorphisms S_1 *and* S_2 *such that* $S_1 S_2 = S_2 S_1 = S_3$, *and there exists the iso-involutive decomposition*

$$\mathfrak{g} = \mathfrak{l}_1 + \mathfrak{l}_2 + \mathfrak{l}_3 \quad (\mathfrak{l}_0 = \mathfrak{l}_\alpha \cap \mathfrak{l}_\beta, \quad \alpha \neq \beta)$$

basis for $\mathfrak{l}_1$, *where* $\mathfrak{l}_1, \mathfrak{l}_2, \mathfrak{l}_3$ *are the involutive algebras of the involutive automorphisms* S_1, S_2, S_3, *respectively, and*

$$\mathfrak{l}_3/\mathfrak{l}_0 = \mathfrak{k}/\mathfrak{l}_0^{(1)} + \mathfrak{l}_0^{(2)}/\mathfrak{l}_0^{(2)},$$

where $\mathfrak{k}/\mathfrak{l}_0^{(1)}$ *is a principal di-unitary invoutive pair.*

II.9.10. Theorem 45. *Let* $\mathfrak{g}$ *be a simple compact Lie algebra,* $\mathfrak{k}$ *be its special unitary simple subalgebra,* $\dim \mathfrak{k} > 3$, *and let* $\mathfrak{l} = \mathfrak{k} \oplus \mathfrak{m}$ *be its special unitary involutive algebra.*

Then $\mathfrak{k} \cong so(5), so(6), so(8), so(9), so(10), so(12), so(16)$ *and, respectively,*

$$\mathfrak{g}/\mathfrak{l} \cong sp(n)/(sp(2) \oplus sp(n-2)) \quad (n > 2),$$

$$\cong su(n)/(su(4) \oplus su(n-4) \oplus u(1)) \quad (n > 4),$$

$$\cong so(n)/(so(8) \oplus so(n-8)) \quad (n > 8),$$

$$\cong f_4/so(9),$$

$$\cong e_6/(so(10) \oplus so(2)),$$

$$\cong e_7/(so(12) \oplus so(3)),$$

$$\cong e_8/so(16)$$

with the natural embeddings.

II.10.2. Theorem 46. *Let* $\mathfrak{g}$ *be a simple compact Lie algebra, and let* $\tilde{\mathfrak{l}}_1 = \mathfrak{p}^{(1)} \oplus \tilde{\tilde{\mathfrak{l}}}_1$ *be its exceptional principal involutive algebra of an involutive automorphism* $\tilde{S}_1 \in \mathrm{Int}_\mathfrak{g}(\mathfrak{p}^{(1)}) \cong SU(2)$.

Then there exists the hyper-involutive decomposition $\mathfrak{g} = \tilde{\mathfrak{l}}_1 + \tilde{\mathfrak{l}}_2 + \tilde{\mathfrak{l}}_3$ *of the exceptional principal involutive automorphisms* $\tilde{S}_1, \tilde{S}_2, \tilde{S}_3$ *such that*

$$\tilde{\mathfrak{l}}_\alpha/\tilde{\mathfrak{l}}_0 = \mathfrak{l}^{(\alpha)}/{}'\mathfrak{l}_0 \oplus \mathfrak{w}^{(\alpha)} + \mathfrak{p}^{(\alpha)}/\mathfrak{z}^{(\alpha)}, \qquad \dim \mathfrak{w}^{(\alpha)} = \dim \mathfrak{z}^{(\alpha)} = 1,$$

the conjugating automorphism $p = \text{Id}$ on $'\mathfrak{l}_0$ and is non-trivial on $\mathfrak{w}^{(\alpha)} \oplus \tilde{\mathfrak{z}}^{(\alpha)} = \mathfrak{t}$, $p\mathfrak{t} = \mathfrak{t}$.

Moreover,

$$p\mathfrak{p}^{(1)} = \mathfrak{p}^{(2)}, \qquad p\mathfrak{p}^{(2)} = \mathfrak{p}^{(3)}, \qquad p\mathfrak{p}^{(3)} = \mathfrak{p}^{(1)},$$

and

$$\tilde{S}_\alpha \in \text{Int}_\mathfrak{g}(\mathfrak{b} \cap \mathfrak{p}^{(\alpha)}) \subset \text{Int}_\mathfrak{g}(\mathfrak{b}) \cong SO(3), \qquad p \in \text{Int}_\mathfrak{g}(\mathfrak{b}),$$

where

$$\mathfrak{p}^{(1)} + \mathfrak{p}^{(2)} + \mathfrak{p}^{(3)} = \mathfrak{n} \cong su(3), \qquad \mathfrak{n}/\mathfrak{b} \cong su(3)/so(3)$$

with the natural embedding.

II.10.5. Theorem 47. *Under the assumptions and notations of II.10.2 (Theorem 46) for the Lie algebras of types f_4, e_6, e_7, e_8, respectively, $\mathfrak{l}^{(\alpha)}/'\mathfrak{l}_0 \oplus \mathfrak{w}^{(\alpha)}$ is*

$$sp(3)/u(3), \qquad su(6)/su(3) \oplus su(3) \oplus u(1),$$

$$so(12)/u(6), \qquad e_7/e_6 \oplus so(2)$$

with the natural embeddings.

II.10.9. Theorem 48. *Let $\mathfrak{g}$ be a Lie algebra, $\mathfrak{g} \cong su(n)$, $n > 2$, and $\tilde{\mathfrak{l}}_1 = \mathfrak{p}^{(1)} \oplus \tilde{\tilde{\mathfrak{l}}}_1$ be its principal unitary unvolutive algebra of an involutive automorphism*

$$\tilde{S}_1 \in \text{Int}_\mathfrak{g}(\mathfrak{p}^{(1)}) \cong SU(2),$$

that is,

$$\tilde{\mathfrak{l}}_1 \cong su(2) \oplus su(n-2) \oplus u(1).$$

Then there exists the hyper-involutive decomposition $\mathfrak{g} = \tilde{\mathfrak{l}}_1 + \tilde{\mathfrak{l}}_2 + \tilde{\mathfrak{l}}_3$ of the principal unitary involutive automorphisms $\tilde{S}_1$, $\tilde{S}_2$, $\tilde{S}_3$ such that

$$\tilde{\mathfrak{l}}_\alpha/\tilde{\mathfrak{l}}_0 \cong \tilde{\tilde{\mathfrak{l}}}_\alpha/\tilde{\tilde{\mathfrak{l}}}_0 + \mathfrak{p}^\alpha/\mathfrak{z}^{(\alpha)} + \mathfrak{w}^{(\alpha)}/\mathfrak{w}^{(\alpha)}$$

$$\cong su(n-2)/u(n-3) + su(2)/u(1) + u(1)/u(1)$$

with the natural embeddings.

Moreover, the conjugating automorphism p is the identity automorphism on $\tilde{\tilde{\mathfrak{l}}}_0$ and non-trivial on $\mathfrak{t} = \mathfrak{w}^{(\alpha)} \oplus \mathfrak{z}^{(\alpha)}$, $p\mathfrak{t} = \mathfrak{t}$, and

$$p\mathfrak{p}^{(1)} = \mathfrak{p}^{(2)}, \qquad p\mathfrak{p}^{(2)} = \mathfrak{p}^{(3)}, \qquad p\mathfrak{p}^{(3)} = \mathfrak{p}^{(1)}.$$

In addition, $p \in \text{Int}_\mathfrak{g}(\mathfrak{b})$, where

$$\mathfrak{b} \subset \mathfrak{p}^{(1)} + \mathfrak{p}^{(2)} + \mathfrak{p}^{(3)} = \mathfrak{n} \cong su(3), \qquad \mathfrak{n}/\mathfrak{b} \cong su(3)/so(3),$$

with the natural embeddings.

II.10.11.   Theorem 49. *Let $\mathfrak{g}$ be a Lie algebra, $\mathfrak{g} \cong so(n)$, $n > 5$, and let $\tilde{\mathfrak{l}}_1 = \mathfrak{p} \oplus \mathfrak{q} \oplus \tilde{\tilde{\mathfrak{l}}}_1$ be its principal di-unitary involutive algebra of an involutive automorphism $\tilde{S}_1$,*

$$\tilde{S}_1 \in \mathrm{Int}_{\mathfrak{g}}(\mathfrak{p}) \cong SU(2), \qquad \tilde{S}_1 \in \mathrm{Int}_{\mathfrak{g}}(\mathfrak{q}) \cong SU(2),$$

that is, $\tilde{\mathfrak{l}}_1 \cong su(2) \oplus su(2) \oplus so(n-4)$.

Then there exists a hyper-involutive decomposition $\mathfrak{g} = \tilde{\mathfrak{l}}_1 + \tilde{\mathfrak{l}}_2 + \tilde{\mathfrak{l}}_3$ of principal di-unitary involutive automorphisms $\tilde{S}_1$, $\tilde{S}_2$, $\tilde{S}_3$ such that

$$\tilde{\mathfrak{l}}_\alpha/\tilde{\mathfrak{l}}_0 = \tilde{\tilde{\mathfrak{l}}}_\alpha/\tilde{\tilde{\mathfrak{l}}}_0 + \mathfrak{p}^{(\alpha)}/\mathfrak{z}^\alpha + \mathfrak{q}^{(\alpha)}/\mathfrak{w}^{(\alpha)}$$

$$\cong so(n-4)/(\,so(n-6) \oplus so(2)\,) + su(2)/so(2) + su(2)/so(2)$$

with the natural embeddings.

The conjugating automorphism p is the identity automorphism on $\tilde{\tilde{\mathfrak{l}}}_0$, non-trivial on $\mathfrak{t} = \mathfrak{w}^{(\alpha)} + \mathfrak{z}^{(\alpha)}$, and $p\,\mathfrak{t} = \mathfrak{t}$.

Moreover,

$$p\,\mathfrak{p}^{(1)} = \mathfrak{p}^{(2)}, \qquad p\,\mathfrak{p}^{(2)} = \mathfrak{p}^{(3)}, \qquad p\,\mathfrak{p}^{(3)} = \mathfrak{p}^{(1)},$$

$$p\,\mathfrak{q}^{(1)} = \mathfrak{q}^{(2)}, \qquad p\,\mathfrak{q}^{(2)} = \mathfrak{q}^{(3)}, \qquad p\,\mathfrak{q}^{(3)} = \mathfrak{q}^{(1)}.$$

In addition, $p \in \mathrm{Int}_{\mathfrak{g}}(\mathfrak{b}) \cong SO(3)$, where

$$\mathfrak{b} \subset \mathfrak{p}^{(1)} + \mathfrak{p}^{(2)} + \mathfrak{p}^{(3)} = \mathfrak{n} \cong su(3), \qquad \mathfrak{n}/\mathfrak{b} \cong su(3)/so(3),$$

with the natural embeddings.

II.11.2.   Theorem 50. *Let $\mathfrak{g} \cong f_4$, and $\mathfrak{l}_1$ be its special non-principal unitary involutive algebra of an involutive automorphism S_1.*

Then there exists a hyper-involutive decomposition $\mathfrak{g} = \mathfrak{l}_1 + \mathfrak{l}_2 + \mathfrak{l}_3$ such that

$$\mathfrak{g}/\mathfrak{l}_\alpha \cong f_4/so(9), \qquad \mathfrak{l}_\alpha/\mathfrak{l}_0 \cong so(9)/so(8)$$

with the natural embeddings. (Here $\mathfrak{l}_0 = \mathfrak{l}_\alpha \cap \mathfrak{l}_\beta$, $\alpha \neq \beta$).

Moreover, the conjugating automorphism p belongs to $\mathrm{Int}_{\mathfrak{g}}(\mathfrak{b}) \subset \mathrm{Int}_{\mathfrak{g}}(\mathfrak{f})$, where the involutive pair $\mathfrak{f}/\mathfrak{b}$ is isomorphic to $su(3)/so(3)$ with the natural embedding, and the subalgebra $\mathfrak{f}$ is the maximal subalgebra of elements of $\mathfrak{g}$ commuting with $\Delta \subset \mathfrak{l}_0$, where Δ is a diagonal in

$$\mathfrak{r} \cong su(2) \oplus su(2) \oplus su(2) \oplus su(2)$$

$$\cong so(4) \oplus so(4) \subset so(8)$$

$$\cong \mathfrak{l}_0 \subset \mathfrak{l}_\alpha \cong so(9)$$

with the natural embeddings.

II.11.3. Theorem 51. *Let* $\mathfrak{g} \cong e_6$, *and* $\mathfrak{l}_1$ *be its special non-principal unitary involutive algebra of an involutive automorphism* S_1.
Then there exists a hyper-involutive decomposition $\mathfrak{g} = \mathfrak{l}_1 + \mathfrak{l}_2 + \mathfrak{l}_3$ *such that*

$$\mathfrak{g}/\mathfrak{l}_\alpha \cong e_6/so(10) \oplus u(1),$$

$$\mathfrak{l}_\alpha/\mathfrak{l}_0 \cong so(10)/so(8) \oplus so(2) + u(1)/u(1)$$

with the natural embeddings ($\mathfrak{l}_0 = \mathfrak{l}_\alpha \cap L_\beta$, $\alpha \neq \beta$ *).*

Moreover, the conjugating automorphism p *belongs to* $\mathrm{Int}_\mathfrak{g}(\mathfrak{b}) \subset \mathrm{Int}_\mathfrak{g}(\mathfrak{f})$, *where the involutive pair is* $\mathfrak{f}/\mathfrak{b} \cong su(3)/so(3)$ *with the natural embedding, and the subalgebra* $\mathfrak{f}$ *is a diagonal of the canonical involutive automorphism* $\mathfrak{f}' \rightleftarrows \mathfrak{f}''$, *where* $\mathfrak{f}' \oplus \mathfrak{f}''$ *is the maximal subalgebra of elements from* $\mathfrak{g}$ *commuting with the subalgebra* $\Delta \subset \mathfrak{l}_0$, Δ *being a diagonal in*

$$\mathfrak{r} \cong su(2) \oplus su(2) \oplus su(2) \oplus su(2) \cong so(4) \oplus so(4) \subset so(8)$$

$$\cong \mathfrak{l}_0' \subset \mathfrak{l}_0 \cong so(8) \oplus so(2) \subset so(10) \cong \mathfrak{l}_\alpha' \subset \mathfrak{l}_\alpha \cong so(10) \oplus u(1)$$

with the natural embedding (that is, Δ *is isomorphic to a diagonal in* $so(3) \oplus so(3)$ *with the natural embedding into* $so(8) \cong \mathfrak{l}'_0$ *).*

II.11.4. Theorem 52. *Let* $\mathfrak{g} \cong e_7$, *and* $\mathfrak{l}_1$ *be its special unitary involutive algebra of an involutive automorphism* S_1. *Then there exists a hyper-involutive decomposition* $\mathfrak{g} = \mathfrak{l}_1 + \mathfrak{l}_2 + \mathfrak{l}_3$ *of involutive automorphisms* S_1, S_2, S_3 *such that*

$$\mathfrak{l}_\alpha = \tilde{\mathfrak{l}}_\alpha \oplus \mathfrak{p}^{(\alpha)} \cong so(12) \oplus su(2), \qquad \mathfrak{g}/\mathfrak{l}_\alpha \cong e_7/so(12) \oplus su(2),$$

$$\mathfrak{l}_\alpha/\mathfrak{l}_0 \cong so(12)/so(8) \oplus so(4) + su(2)/su(2)$$

with the natural embedding.

The subalgebra of all elements of $\mathfrak{g} \cong e_7$ *which commute with a diagonal in* $\mathfrak{p}^{(1)} + \mathfrak{p}^{(2)} + \mathfrak{p}^{(3)}$ *is* $\tilde{\mathfrak{g}} \cong f_4$, *and the involutive automorphisms* S_1, S_2, S_3 *induce in* $\tilde{\mathfrak{g}}$ *the hyper-involutive decomposition described by II.11.2 (Theorem 50).*

The hyper-involutive decompositions for $\mathfrak{g}$ *and* $\tilde{\mathfrak{g}}$ *have the same conjugating automorphism* p.

II.11.5. Theorem 53. *Let* $\mathfrak{g} \cong e_8$, *and* $\mathfrak{l}_1$ *be its special non-principal unitary involutive algebra of an involutive automorphism* S_1. *Then there exists the hyper-involutive decomposition* $\mathfrak{g} = \mathfrak{l}_1 + \mathfrak{l}_2 + \mathfrak{l}_3$ *of the involutive automorphisms* S_1, S_2, S_3 *such that*

$$\mathfrak{g}/\mathfrak{l}_\alpha \cong e_8/so(16), \qquad \mathfrak{l}_\alpha/\mathfrak{l}_0 = \mathfrak{l}_\alpha/\mathfrak{l}'_0 \oplus \mathfrak{l}''_0 \cong so(16)/so(8) \oplus so(8)$$

with the natural embeddings.
If

$$\mathfrak{p} \cong su(2) \subset so(4) \subset so(8) \cong \mathfrak{l}'_0$$

with the natural embeddings then the subalgebra of all elements from $\mathfrak{g} \cong e_8$ *commuting with* $\mathfrak{p}$ *is* $\tilde{\mathfrak{g}} \cong e_7$, *and the involutive automorphisms* S_1, S_2, S_3 *induce in* $\tilde{\mathfrak{g}}$ *the hyper-involutive decomposition described in II.11.4 (Theorem 52).*

The hyper-involutive decompositions for $\mathfrak{g}$ *and* $\tilde{\mathfrak{g}}$ *have the same conjugating automorphism* p.

II.11.10. Theorem 54. *Let* $\mathfrak{g} \cong so(n)$ $(n \geq 12)$, *and* $\mathfrak{l}_1 \cong so(8) \oplus so(n-8)$ *be its special unitary involutive algebra of an involutive automorphism* S_1.

Then there exists the hyper-involutive decomposition $\mathfrak{g} = \mathfrak{l}_1 + \mathfrak{l}_2 + \mathfrak{l}_3$ *of the involutive automorphisms* S_1, S_2, S_3 *such that*

$$\mathfrak{g}/\mathfrak{l}_\alpha \cong so(n)/so(8) \oplus so(n-8),$$

$$\mathfrak{l}_\alpha/\mathfrak{l}_0 = \mathfrak{k}_\alpha/\mathfrak{m}_\alpha \oplus \mathfrak{n}_\alpha = \mathfrak{l}'_\alpha/\mathfrak{l}'_0 \oplus \mathfrak{t}_\alpha$$

$$\cong so(8)/so(4) \oplus so(4) + so(n-8)/so(n-12) \oplus so(4)$$

with the natural embeddings.

Moreover,

$$S_\alpha \in \mathrm{Int}_{\mathfrak{g}}(\mathfrak{b}^{(\alpha)}) \subset \mathrm{Int}_{\mathfrak{g}}(\mathfrak{b}) \cong SO(3),$$

where $\mathfrak{b}^{(\alpha)}$ *is the maximal subalgebra of elements in* $\mathfrak{k}_\alpha \cong so(8)$ *commuting with a diagonal* $\mathfrak{d}$ *of*

$$\mathfrak{m}_\alpha \oplus \mathfrak{n}_\alpha \cong so(4) \oplus so(4),$$

and $\dim \mathfrak{b}^{(\alpha)} = 1$.

II.11.12. Theorem 55. *Let* $\mathfrak{g} \cong su(n)$ $(n \geq 6)$, *and*

$$\mathfrak{l}_1 \cong su(4) \oplus su(n-4) \oplus u(1)$$

be its special unitary involutive algebra of an involutive automorphism S_1.

Then there exists the hyper-involutive decomposition $\mathfrak{g} = \mathfrak{l}_1 + \mathfrak{l}_2 + \mathfrak{l}_3$ *of the involutive automorphisms* S_1, S_2, S_3 *such that*

$$\mathfrak{g}/\mathfrak{l}_\alpha \cong su(n)/su(4) \oplus su(n-4) \oplus u(1),$$

$$\mathfrak{l}_\alpha/\mathfrak{l}_0 = \mathfrak{k}_\alpha/\mathfrak{m}_\alpha \oplus \mathfrak{n}_\alpha \oplus \mathfrak{z}_\alpha + \mathfrak{l}'_\alpha/\mathfrak{l}'_0 \oplus \mathfrak{t}_\alpha \oplus \mathfrak{w}^{(\alpha)} + \mathfrak{v}^{(\alpha)}/\mathfrak{v}^{(\alpha)}$$

$$\cong su(4)/su(2) \oplus su(2) \oplus u(1)$$

$$+ su(n-4)/su(n-6) \oplus su(2) \oplus u(1) + u(1)/u(1)$$

$$(\mathfrak{l}_0 = \mathfrak{l}_\alpha \cap \mathfrak{l}_\beta, \ \alpha \neq \beta)$$

with the natural embeddings.

Moreover,

$$S_\alpha \in \mathrm{Int}_{\mathfrak{g}}(\mathfrak{h}^{(\alpha)}) \subset \mathrm{Int}_{\mathfrak{g}}(\mathfrak{h}) \cong SO(3), \qquad \dim \mathfrak{h}^{(\alpha)} = 1,$$

$$\mathfrak{h} \cong \Delta(so(3) \oplus so(3)) \subset \Delta(su(3) \oplus su(3))$$

$$\cong \tilde{\mathfrak{b}} \subset \mathfrak{n}' \oplus \mathfrak{n}'' \cong su(3) \oplus su(3) \subset su(6) \cong \tilde{\mathfrak{k}} \subset \mathfrak{g} \cong su(n)$$

with the natural embeddings, where $\tilde{\mathfrak{b}}$ *is the maximal subalgebra of elements from* $\tilde{\mathfrak{k}}$ *commuting with the diagonal*

$$\Delta(\mathfrak{m}_\alpha \oplus \mathfrak{n}_\alpha \oplus \mathfrak{t}_\alpha) \cong \Delta(su(2) \oplus su(2) \oplus su(2)) \subset su(6) \cong \tilde{\mathfrak{k}}$$

with the natural embeddings.

II.11.14. Theorem 56. Let $\mathfrak{g} \cong sp(n)$ $(n \geq 3)$, and $\mathfrak{l}_1 \cong sp(2) \oplus sp(n-2)$ be its special unitary involutive algebra of an involutive automorphism S_1.

Then there exists the hyper-involutive decomposition $\mathfrak{g} = \mathfrak{l}_1 + \mathfrak{l}_2 + \mathfrak{l}_3$ of the involutive automorphisms S_1, S_2, S_3 such that

$$\mathfrak{g}/\mathfrak{l}_\alpha \cong sp(n)/sp(2) \oplus sp(n-2),$$

$$\mathfrak{l}_\alpha/\mathfrak{l}_0 = \mathfrak{k}_\alpha/\mathfrak{m}_\alpha \oplus \mathfrak{n}_\alpha + \mathfrak{l}'_\alpha/\mathfrak{l}'_0 \oplus \mathfrak{t}_\alpha$$

$$\cong sp(2)/sp(1) \oplus sp(1) + sp(n-2)/sp(n-3) \oplus sp(1),$$

$$\mathfrak{l}_0 = \mathfrak{l}_\alpha \cap \mathfrak{l}_\beta, \quad \alpha \neq \beta,$$

with the natural embeddings.

A.4.3. PART III.

III.1.20. Theorem 1. Let G/H be a symmetric space of an involutive automorphism S $(H = \{h \in G \mid Sh = h\})$ with a simple compact Lie group G.

Then in G/H there exist non-trivial (i.e., containing more than one point) geodesic mirrors.

III. 2.1. Theorem 2. A symmetric space G/H with a simple compact Lie group G is of rank 1 if and only if it has a geodesic mirror of rank 1.

III. 2.2. Theorem 3. Let $V = G/H$ be a symmetric space with a simple compact Lie group G, and W be its geodesic mirror of rank 1. Then

$$W \cong SO(m+1)/SO(m)$$

with the natural embedding.

That is, W is of the constant curvature.

III.2.3. Theorem 4. A symmetric space G/H with a simple compact Lie group G is of rank 1 if and only if G/H has a geodesic mirror of the constant curvature.

III.2.4. Theorem 5. Let G/H be a symmetric space with a simple compact Lie group G, and let its geodesic mirror W of index 1 be one-dimensional. Then

$$G/H \cong SO(n+1)/SO(n)$$

with the natural embedding.

III.2.5. Theorem 6. Let G/H be a symmetric space with a simple compact Lie group G, and let its geodesic mirror W of rank 1 be two-dimensional. Then

$$G/H \cong SU(n+1)/U(n)$$

with the natural embedding.

III.2.6. Theorem 7. *Let G/H be a symmetric space with a simple compact Lie group G, and let its geodesic mirror W of rank 1 be four-dimensional. Then*

$$W \cong Sp(n+1)/Sp(n) \times Sp(1)$$

with the natural embedding.

III.2.7. Theorem 8. *If G/H is a symmetric space of rank 1 with a simple compact Lie group G then G/H has no three-dimensional mirrors.*

III.2.8. Theorem 9. *Let G/H be a symmetric space of rank 1, and let G be a simple compact Lie group. If $H = H' \times H''$, $H' \neq \{e\}$, $H'' \neq \{e\}$, then either*

$$G/H \cong SU(n+1)/U(n)$$

or

$$G/H \cong Sp(n+1)/Sp(n) \times Sp(1)$$

with the natural embeddings.

III.3.1. Theorem 10. *Let G/H be an irreducible symmetric space, G being a compact Lie group, $\dim H = 3$. Then $H \cong SO(3)$.*

III.3.2. Theorem 11. *Let G be a simple compact Lie group, $\dim G > 3$. Then there exists a principal unitary symmetric space G/H.*

III.3.3. Theorem 12. *If G/H is a principal orthogonal symmetric space with a simple compact Lie group G then either*

$$G/H \cong SO(n)/SO(n-3) \times SO(3)$$

or

$$G/H \cong SU(3)/SO(3)$$

with the natural embeddings.

III.3.4. Theorem 13. *Let $V = G/H$ be a principal orthogonal symmetric space with a compact simple Lie group G. Then for any point x of $V = G/H$ there exist three mirrors $V^{(1)}$, $V^{(2)}$, $V^{(3)}$ of the same dimension, passing through x, being generated by a discrete commutative group of subsymmetries $\{ \mathrm{id}, S_x^{(1)}, S_x^{(2)}, S_x^{(3)} \}$, $S_x^{(i)} S_x^{(j)} = S_x^{(k)}$ $(i \neq j, j \neq k, k \neq i)$.*

Moreover, there exists an inner automorphism $p \in H$ generated by an element of H such that $p^3 = \mathrm{id}$, $p V^{(1)} = V^{(2)}$, $p V^{(2)} = V^{(3)}$, $p V^{(3)} = V^{(1)}$.

III.3.5. Theorem 14. *If G/H is a principal orthogonal symmetric space with a simple compact Lie group G then G/H has a mirror W isomorphic either to a space of the constant curvature or to a direct product of spaces of the constant curvature and a one-dimensional space.*

III.3.6. Theorem 15. *Let G/H be a principal di-unitary symmetric space with a compact simple Lie group G.*

Then either

$$G/H \cong SO(n)/SO(n-4) \times SO(4)$$

or

$$G/H \cong G_2/SO(4)$$

with the natural embeddings.

III.3.7. Theorem 16. *Let G/H be a principal mono-unitary non-central symmetric space with a compact simple Lie group G. If H is semi-simple then $H = H' \times H''$, where $H' \cong SU(2)$, H'' is simple, and $\dim H'' > 3$.*

III.3.8. Theorem 17. *Let G be a simple compact Lie group, G/H be a principal mono-unitary symmetric space of type F_4, then*

$$G/H \cong F_4/Sp(3) \times Sp(1)$$

with the natural embedding.

III.3.9. Theorem 18. *Let G be a simple compact Lie group, G/H be a principal mono-unitary symmetric space of type E_6, then*

$$G/H \cong E_6/SU(2) \times SU(6)$$

with the natural embedding.

III.3.10. Theorem 19. *Let G be a simple compact Lie group. If G/H is a principal mono-unitary symmetric space of type E_7 then*

$$G/H \cong E_7/SU(2) \times SO(12)$$

with the natural embedding.

III.3.11. Theorem 20. *Let G be a simple compact Lie group. If G/H is a principal mono-unitary symmetric space of type E_8 then*

$$G/H \cong E_8/SU(2) \times E_7$$

with the natural embedding.

III.3.12. Theorem 21. *Let G/H be a principal unitary symmetric space with a simple compact Lie group G, and let $G \ncong SO(k)$ $(k = 5,6,7,8)$.*

Then G/H has a special geodesic mirror

$$W = \widetilde{G}/\widetilde{H} \cong SO(m)/SO(4) \times SO(m-4)$$

(with the natural embedding) such that $m = 5, 6, 8, 9, 10, 12, 16$.

III.3.13.   Theorem 22. *Under the assumptions and notations of III.3.12 (Theorem 21), if*

$$W \cong SO(5)/SO(4),$$
$$\cong SO(6)/SO(4) \times SO(2),$$
$$\cong SO(8)/SO(4) \times SO(4),$$
$$\cong SO(9)/SO(4) \times SO(5),$$
$$\cong SO(10)/SO(4) \times SO(6),$$
$$\cong SO(12)/SO(4) \times SO(8),$$
$$\cong SO(16)/SO(4) \times SO(12)$$

with the natural embeddings then correspondingly

$$G/H \cong Sp(n)/Sp(1) \times Sp(n-1), \quad n > 2;$$
$$\cong SU(n)/SU(2) \times U(n-2), \quad n > 4;$$
$$\cong SO(n)/SO(4) \times SO(n-4), \quad n > 8;$$
$$\cong F_4/SU(2) \times Sp(3);$$
$$\cong E_6/SU(2) \times SU(6);$$
$$\cong E_7/SU(2) \times SO(12);$$
$$\cong E_8/SU(2) \times E_7$$

with the natural embeddings.

(Here the groups are given up to local isomorphisms).

III.3.15.   Theorem 23. *Let G/H be a symmetric space with a simple compact Lie group G, $H = H' \times H''$, and let $W = \tilde{G}^W/(H' \times H'')^W$ be its mirror with a simple Lie group $\tilde{G} \subset G$, $H' \subset H$ being isomorphic to $SU(2) \times SU(2)$ (locally). Then W is a special mirror and G/H is a principal unitary symmetric space.*

III.4.1.   Theorem 24. *If $V = G/H$ is an essentially special symmetric space with a simple compact Lie group G then for any point $x \in V = G/H$ there exist two principal mirrors W and W', and two related commutative principal subsymmetries, S and S', respectively, such that $W \cap W' = \{x\}$ (locally), W and W' are conjugated by an element from H, and*

$$\dim W + \dim W' = \dim G/H.$$

III.4.2. Theorem 25. *Let* $V = G/H$ *be a symmetric space generated by an involutive automorphism* S *with a simple compact Lie group* G, *and let* S' *and* S'' *be its principal subsymmetries at a point* $x \in V$ *with the mirrors* W', W'', *respectively, such that* $S'S'' = S$ *is a symmetry at the point* x.

If $\tilde{G}^{W'}/\tilde{H}^{W'}$ *and* $\tilde{G}^{W''}/\tilde{H}^{W''}$ *are principal symmetric spaces then* G/H *is an essentially special symmetric space.*

III.4.3. Theorem 26. *If* G/H *is an essentially special symmetric space with a simple compact Lie group* G *and its principal unitary non-exceptional mirror* $W = \tilde{G}^W/\tilde{H}^W$ *is a principal unitary symmetric space then*

$$W \cong SO(n-4)/SO(4) \times SO(n-8), \quad n \geq 8;$$

$$\cong SU(n-2)/SU(2) \times U(n-4), \quad n > 4;$$

$$\cong Sp(n-1)/Sp(n-2) \times Sp(1), \quad n \geq 2,$$

and, respectively,

$$G/H \cong SO(n)/SO(8) \times SO(n-8),$$

$$\cong SU(n)/SU(4) \times U(n-4),$$

$$\cong Sp(n)/Sp(2) \times Sp(n-2).$$

(The groups are indicated up to local isomorphisms).

III.4.4. Theorem 27. *If* G/H *is a special symmetric space with a simple compact Lie group* G *and its principal exceptional mirror* $W = \tilde{G}^W/\tilde{H}^W$ *is a principal unitary symmetric space then*

$$W \cong Sp(3)/Sp(2) \times Sp(1),$$

$$\cong SU(6)/S(U(4) \times U(2)),$$

$$\cong SO(12)/SO(8) \times SO(4),$$

$$\cong E_7/SO(12) \times SU(2)$$

and, respectively,

$$G/H \cong F_4/SO(9),$$

$$\cong E_6/SO(10) \times SO(2),$$

$$\cong E_7/SO(12) \times SU(2),$$

$$\cong E_8/SO(16).$$

(The groups are indicated up to local isomorphisms.)

III.4.5. Theorem 28. *If G/H is an essentially special symmetric space with a simple compact Lie group G then*

$$G/H \cong SO(n)/SO(8) \times SO(n-8),$$
$$\cong SU(n)/S(U(4) \times U(n-4)),$$
$$\cong Sp(n)/Sp(2) \times Sp(n-2),$$
$$\cong F_4/SO(9),$$
$$\cong E_6/SO(10) \times SO(2),$$
$$\cong E_7/SO(12) \times SU(2),$$
$$\cong E_8/SO(16).$$

(The groups are indicated up to local isomorphisms.)

III.4.6. Theorem 29. *If G/H is a principal essentially special symmetric space with a simple compact Lie group G then*

$$G/H \cong SO(12)/SO(8) \times SO(4),$$
$$\cong SU(6)/S(U(4) \times U(2)),$$
$$\cong Sp(3)/Sp(2) \times Sp(1),$$
$$\cong E_7/SO(12) \times SU(2).$$

III.5.1. Theorem 30. *If G is a simple compact Lie group, G/H is an involutive pair, $H \neq G$, and $\dim H = 3$ then $H \cong SO(3)$.*

III.5.2. Theorem 31. *If G is a simple compact connected Lie group and $\ln G \cong g_2$ then $G \cong \mathrm{Int}(g_2)$.*

III.5.3. Theorem 32. *If G is a simple compact connected Lie group and $\ln G = f_4$ then $G \cong \mathrm{Int}(f_4)$.*

III.5.6. Theorem 33. *A non-trivial inner automorphism of a compact simple Lie algebra $\mathfrak{g}$ of type g_2 is unique, up to the conjugacy by inner automorphisms. That automorphism is a principal unitary involutive automorphism of $\mathfrak{g}$.*

III.5.7. Theorem 34. *Let $\mathfrak{g}$ be a simple compact Lie algebra of type f_4, and S be its non-trivial inner involutive automorphism. Then S is either principal unitary or special unitary non-principal.*

III.5.9. Theorem 35. *Let $\mathfrak{g}$ be a simple compact Lie algebra of type e_6 or e_8, S being a non-trivial inner involutive automorphism. Then S is either principal unitary or special unitary non-principal.*

III.6.8. Theorem 36. *Let G be a simple Lie group, and*

$$G/H = G_1 \overset{*}{/} H_1 \boxtimes G_2 \overset{*}{/} H_2 \boxtimes G_3 \overset{*}{/} H_3$$

be a tri-symmetric non-trivial space with a compact Lie group H.
   Then locally

$$G_\alpha = \tilde{H}_\alpha \times \tilde{G}_\alpha \quad (\alpha = 1, 2, 3),$$

where $\tilde{G}_\alpha \neq \{e\}$, $\tilde{H}_\alpha \subset H_\alpha$, and

$$G_\alpha \overset{*}{/} H_\alpha = \tilde{G}_\alpha \overset{*}{/} (H_\alpha \cap \tilde{G}_\alpha), \qquad H_\alpha \cap \tilde{G}_\alpha \subset G_1 \cap G_2 \cap G_3.$$

III.6.9. Theorem 37. *If a tri-symmetric space*

$$G/H = G_1 \overset{*}{/} H_1 \boxtimes G_2 \overset{*}{/} H_2 \boxtimes G_3 \overset{*}{/} H_3$$

with a simple Lie group G and a compact Lie group H:
   a) *is non-trivial then G_1, G_2, G_3 can not be simple non-commutative Lie groups;*
   b) *has at least one simple and non-commutative Lie group G_α $(\alpha = 1, 2, 3)$ then G/H is trivial or semi-trivial.*

III.6.10. Theorem 38. *Let*

$$G/H = G_1 \overset{*}{/} H_1 \boxtimes G_2 \overset{*}{/} H_2 \boxtimes G_3 \overset{*}{/} H_3$$

be a tri-symmetric space with a compact Lie group H, and let a symmetric space $G_\alpha \overset{*}{/} G_0$, where $G_0 = G_1 \cap G_2 \cap G_3$, be irreducible for some $\alpha = 1, 2, 3$.
   Then G/H is semi-trivial or trivial.

III.7.3. Theorem 39. *If G is a compact simple Lie group isomorphic to G_2 then there exists a unique non-trivial tri-symmetric space*

$$G/H = G_1 \overset{*}{/} H_1 \boxtimes G_2 \overset{*}{/} H_2 \boxtimes G_3 \overset{*}{/} H_3 \cong G_2/SU(3)$$

with the principal mirrors

$$G_\alpha \overset{*}{/} H_\alpha \cong SO(4) \overset{*}{/} SU(2) \times SO(2) = SU(2)/SO(2) \quad (\alpha = 1, 2, 3)$$

(with the natural embeddings).
   This space is hyper-tri-symmetric and has an irreducible Lie group H.

III.7.5. Theorem 40. *If G is a compact simple Lie group isomorphic to F_4 then there exists the unique non-trivial tri-symmetric space*

$$G/H = G_1 \overset{*}{/} H_1 \boxtimes G_2 \overset{*}{/} H_2 \boxtimes G_3 \overset{*}{/} H_3 \cong F_4/SU(3) \times SU(3)$$

with the principal mirrors

$$G_\alpha \overset{*}{/} H_\alpha \cong SU(2) \times Sp(3) \overset{*}{/} SU(2) \times U(3) \cong Sp(3)/U(3)$$

(with the natural embeddings).

This space is hyper-tri-symmetric and has an irreducible group H.

III.7.7. Theorem 41. *If a compact simple Lie group G is isomorphic to E_6 then there exists the unique non-trivial symmetric space*

$$G/H = G_1 \overset{*}{/} H_1 \boxtimes G_2 \overset{*}{/} H_2 \boxtimes G_3 \overset{*}{/} H_3 \cong E_6/SU(3) \times SU(3) \times SU(3)$$

with the principle central mirrors

$$G_\alpha \overset{*}{/} H_\alpha \cong SU(2) \times SU(6) \overset{*}{/} SU(2) \times SU(3) \times SU(3) \times U(1)$$
$$\cong SU(6)/SU(3) \times SU(3) \times U(1)$$

(with the natural embeddings).

This space is hyper-tri-symmetric and has an irreducible group H.

III.7.9. Theorem 42. *If a compact simple Lie group G is isomorphic to E_7 then there exist only two non-trivial tri-symmetric spaces (up to isomorphism)*

$$G/H = G_1 \overset{*}{/} H_1 \boxtimes G_2 \overset{*}{/} H_2 \boxtimes G_3 \overset{*}{/} H_3,$$

namely:

1. $G/H \cong E_7/SU(3) \times SU(6)$ *with the principal mirrors*

$$G_\alpha \overset{*}{/} H_\alpha \cong SO(12) \times SU(2) \overset{*}{/} U(6) \times SU(2) \cong SO(12)/U(6);$$

2. $G/H \cong E_7/F_4 \times SO(3)$ *with the central mirrors*

$$G_\alpha \overset{*}{/} H_\alpha \cong E_6 \times U(1) \overset{*}{/} F_4 \times U(1) \cong E_6/F_4$$

(with the natural embeddings).

These spaces are hyper-tri-symmetric and have an irreducible Lie group H.

III.7.11. Theorem 43. *If a compact Lie group G is isomorphic to E_8 then there exists the unique non-trivial tri-symmetric space*

$$G/H = G_1 \overset{*}{/} H_1 \boxtimes G_2 \overset{*}{/} H_2 \boxtimes G_3 \overset{*}{/} H_3 \cong E_8/E_6 \times SU(3)$$

with the principal mirrors

$$G_\alpha \overset{*}{/} H_\alpha \cong E_7 \times SU(2) \overset{*}{/} E_6 \times U(1) \times SU(2 \cong E_7/E_6 \times U(1)$$

(with the natural embeddings).

This space is hyper-tri-symmetric and has an irreducible group H.

III.8.7. Theorem 44. *If a compact simple Lie group G is isomorphic to $SO(n)$ then all non-trivial non-symmetric tri-symmetric spaces with the group of motions G and the maximal Lie subgroup H have the form*

$$G/H = G_1 \overset{*}{/} H_1 \boxtimes G_2 \overset{*}{/} H_2 \boxtimes G_3 \overset{*}{/} H_3 \cong SO(4m)/SO(3) \times Sp(m) \quad (m > 2)$$

with the central mirrors

$$G_\alpha \overset{*}{/} H_\alpha \cong U(2m) \overset{*}{/} Sp(m) \times U(1) \cong SU(2m)/Sp(m)$$

(with the natural embeddings).

These spaces are hyper-tri-symmetric and have an irreducible group H.

III.8.13. Theorem 45. *If a compact simple Lie group G is isomorphic to $Sp(n)$ ($n > 2$) then any non-trivial tri-symmetric non-symmetric space with the group of motions G has the form:*

$$G/H = G_1 \overset{*}{/} H_1 \boxtimes G_2 \overset{*}{/} H_2 \boxtimes G_3 \overset{*}{/} H_3 \cong Sp(n)/SO(3) \times SO(n)$$

with the central mirrors

$$G_\alpha \overset{*}{/} H_\alpha \cong U(n) \overset{*}{/} SO(n) \times U(1) \cong SU(n)/SO(n)$$

(with the natural embeddings).

This space is tri-symmetric and has an irreducible isotropy group H.

III.8.15. Theorem 46. *If G is a simple compact Lie group isomorphic to $SU(n)$ then any non-trivial tri-symmetric non-symmetric space with the group of motions G has the form:*

$$G/H = G_1 \overset{*}{/} H_1 \boxtimes G_2 \overset{*}{/} H_2 \boxtimes G_3 \overset{*}{/} H_3$$

$$\cong SU(2m)/SO(3) \times SU(m) \quad (m > 2)$$

with the central mirrors

$$G_\alpha \overset{*}{/} H_\alpha \cong S(U(m) \times U(m)) \overset{*}{/} \Delta(SU(m) \times SU(m)) \times U(1)$$

$$\cong SU(m) \times SU(m)/\Delta(SU(m) \times SU(m))$$

(with the natural embeddings).

This space is hyper-tri-symmetric and has an irreducible isotropy group H.

A.4.4. PART IV.

IV.1.19. Theorem 1. *Any homogeneous Riemannian space with a non-trivial stationary group G_x has an isometric mirror subsymmetry s_x (i.e., it is a mirror subsymmetric space).*

IV.1.20. Theorem 2. *A mirror of a mirror subsymmetry of a Riemannian space is a totally geodesic submanifold.*

IV.3.13.   Theorem 3. *Let $(\mathfrak{g}, \mathfrak{h}, \mathfrak{m}, \sigma)$ be an exact mirror subsymmetric quadruplet of Riemannian type and of order two. Then for the triplet $(\mathfrak{g}^+, \mathfrak{h}^+, \mathfrak{m}^+)$ we have $\dim \mathfrak{m}^+ = 2$ and there are the following possibilities:*

I. $\mathfrak{g}^+ = \mathfrak{h}^+ \oplus \mathfrak{m}^+$ (direct sum of ideals), moreover

a) *either $[\mathfrak{m}^+\mathfrak{m}^+] = \mathfrak{n} \neq \{0\}$ is a one-dimensional ideal in $\mathfrak{m}^+$ and then $B|_{\mathfrak{n}} = 0$ (solvable case)*

b) *or $[\mathfrak{m}^+\mathfrak{m}^+] = \{0\}$, that is, $\mathfrak{m}^+$ is an abelian ideal of $\mathfrak{g}^+$ (abelian case).*

II. $\mathfrak{g}^+ = \tilde{\mathfrak{h}}^+ \oplus (\mathfrak{z} \dotplus \mathfrak{m}^+)$ (direct sum of ideals of $\mathfrak{g}^+$), $\mathfrak{h}^+ = \tilde{\mathfrak{h}}^+ \oplus \mathfrak{z}$ (direct sum of ideals of $\mathfrak{h}^+$), $\dim \mathfrak{z} = 1$, moreover

a) *either $[\mathfrak{m}^+\mathfrak{m}^+] = \mathfrak{z}$, $B|_{\mathfrak{m}^+} = c\,g|_{\mathfrak{m}^+}$ ($c > 0$, elliptic case),*

b) *or $[\mathfrak{m}^+\mathfrak{m}^+] = \mathfrak{z}$, $B|_{\mathfrak{m}^+} = c\,g|_{\mathfrak{m}^+}$ ($c < 0$, hyperbolic case),*

c) *or $[\mathfrak{m}^+\mathfrak{m}^+] = \{0\}$, $B|_{\mathfrak{m}^+} = c\,g|_{\mathfrak{m}^+}$ ($c = 0$, parabolic case).*

IV.4.4. Theorem 4. *If the mirror of a mirror subsymmetric triplet of Riemannian type (G, H, s) is mobile then there exists a one-dimensional subgroup $T = \{h(t)\}_{t\in\mathbb{R}} \cong SO(2)$ (that is, a closed one-dimensional subgroup), $T \subset H$, such that*

$$s \circ T_{h(t)} \circ s^{-1} = T_{[h(t)]^{-1}}.$$

IV.4.6.   Theorem 5. *Any mirror subsymmetry s_1 of a point $z \in M$ of a mirror subsymmetric homogeneous space $M = G/H$ (H being compact) can be included into the iso-involutive discrete group I.I.$\mathrm{Gr}(s_1, s_2, s_3;\, \varphi)$ with the mirror subsymmetries s_1, s_2, s_3 of the point $z \in M$.*

IV.5.2. Theorem 6. *A two-dimensional mirror of a Riemannian homogeneous space G/H with a simple compact Lie group G is of elliptic type.*

IV.5.4.   Theorem 7. *The centre of the group of rotations of a symmetric Riemannian space G/H with a simple Lie group G is at most one-dimensional.*

IV.5.5. Theorem 8. *Let $(\mathfrak{g}, \mathfrak{h})$ be an involutive pair of Lie algebras, where $\mathfrak{g}$ is simple and $\mathfrak{h}$ is its compact subalgebra. Then the centre of $\mathfrak{h}$ is at most one-dimensional.*

IV.5.12. Theorem 9. *Let $\mathfrak{g}$ be a simple compact Lie algebra, $(\mathfrak{g}, \mathfrak{h})$ be a doublet of Lie algebras with a mobile mirror $\mathfrak{m}^+$ $(\mathfrak{g}^+ = \mathfrak{m}^+ \dotplus \mathfrak{h}^+)$, $\dim \mathfrak{m}^+ = 2$, generated by a principal unitary involutive automorphism σ.*

Then there exists the iso-involutive group,

$$\sigma_1 = \sigma, \quad \sigma_2 = \varphi \circ \sigma_1 \circ \varphi^{-1}, \quad \sigma_3 = \varphi^2, \quad \varphi, \quad \varphi^{-1}, \quad \text{id},$$

such that $\mathfrak{g}$ and $\mathfrak{h}$ are invariant under its action, and there exists the corresponding iso-involutive decomposition

$$\mathfrak{g} = \mathfrak{g}_1^+ + \mathfrak{g}_2^+ + \mathfrak{g}_3^+ \quad (\mathfrak{g}_i^+ = \{x \in \mathfrak{g} \mid \sigma_i x = x\})$$

which is tri-symmetric, that is,

$$\mathfrak{g}_1^+ \cap \mathfrak{g}_2^+ = \mathfrak{g}_2^+ \cap \mathfrak{g}_3^+ = \mathfrak{g}_3^+ \cap \mathfrak{g}_1^+ = \mathfrak{g}^0 \subset \mathfrak{h}.$$

IV.5.13. Theorem 10. *Let $\mathfrak{g}$ be a simple compact Lie algebra non-isomorphic to $so(n)$ or $su(3)$, and $(\mathfrak{g}, \mathfrak{h})$ be a doublet having a two-dimensional mobile mirror of an involutive automorphism σ.*

Then there exists the iso-involutive group,

$$\sigma_1 = \sigma, \quad \sigma_2 = \varphi \circ \sigma_1 \circ \varphi^{-1}, \quad \sigma_3 = \varphi^2, \quad \varphi, \quad \varphi^{-1}, \quad \text{id},$$

and the corresponding iso-involutive decomposition $\mathfrak{g} = \mathfrak{g}_1^+ + \mathfrak{g}_2^+ + \mathfrak{g}_3^+$ such that the pair $(\mathfrak{g}, \mathfrak{h})$ is invariant under the action of this group and is tri-symmetric.

IV.6.5. Theorem 11. *Let $(\mathfrak{g}, \mathfrak{h})$ be a pair with a simple compact Lie algebra $\mathfrak{g} \cong so(n)$ and with a two-dimensional mobile mirror $(\mathfrak{g}^+, \mathfrak{h}^+)$ of orthogonal type $(\mathfrak{g}^+ = \mathfrak{m}^+ \dotplus \mathfrak{h}^+$, $\dim \mathfrak{m}^+ = 2$, $\mathfrak{h} \supsetneq \mathfrak{h}^+)$.*

Then the following cases are possible:

1). $(\mathfrak{g}, \mathfrak{h}) \cong (so(n), so(n-1))$ *is an involutive pair,*

   $(\mathfrak{g}^+, \mathfrak{h}^+) \cong (so(3) \oplus so(n-3), so(2) \oplus so(n-3))$,

   $(\mathfrak{h}, \mathfrak{h}^+) \cong (so(n-1), so(2) \oplus so(n-3))$

   (with the natural embeddings).

In this case the mirror $(\mathfrak{g}^+, \mathfrak{h}^+)$ can be moved either in a tri-symmetric or in a non-tri-symmetric way.

2). $(\mathfrak{g}, \mathfrak{h}) \cong (so(n), so(2) \oplus so(n-2))$ *is an involutive pair,*

   $(\mathfrak{g}^+, \mathfrak{h}^+) \cong (so(3) \oplus so(n-3), so(2) \oplus so(n-3))$,

   $(\mathfrak{h}, \mathfrak{h}^+) \cong (so(2) \oplus so(n-2), so(2) \oplus so(n-3))$

   (with the natural embeddings).

In this case the mirror $(\mathfrak{g}^+, \mathfrak{h}^+)$ can be moved in a non-tri-symmetric way only.

3). $(\mathfrak{g}, \mathfrak{h}) \cong (so(5), u(1) \oplus su(2)) \cong (sp(2), u(1) \oplus su(2))$,

   $(\mathfrak{g}^+, \mathfrak{h}^+) \cong (so(3) \oplus so(2), so(2) \oplus so(2))$,

   $(\mathfrak{h}, \mathfrak{h}^+) \cong (u(1) \oplus su(2), u(1) \oplus u(1))$

   (with the natural embeddings).

Here $(\mathfrak{g}, \mathfrak{h})$ *is not an involutive pair with the non-maximal subalgebra* $\mathfrak{h} \subset \mathfrak{g}$. *In this case the mirror* $(\mathfrak{g}^+, \mathfrak{h}^+)$ *can be moved only in a tri-symmetric way.*

IV.6.8. Theorem 12. *Let* $(\mathfrak{g}, \mathfrak{h})$, $\mathfrak{g} \cong su(3)$, *be a pair with a two-dimensional mobile mirror* $(\mathfrak{g}^+, \mathfrak{h}^+)$ $(\mathfrak{g}^+ = \mathfrak{m}^+ + \mathfrak{h}^+$, $\dim \mathfrak{m}^+ = 2$, $\mathfrak{h} \neq \mathfrak{h}^+)$ *of orthogonal type. Then there exist only the following possibilities:*

1. $(\mathfrak{g}, \mathfrak{h}) \cong (su(3), su(2) \oplus u(1))$ *is an involutive pair,*
$(\mathfrak{g}^+, \mathfrak{h}^+) \cong (so(3), so(2))$,
$(\mathfrak{h}, \mathfrak{h}^+) \cong (su(2) \oplus u(1), so(2) \oplus u(1))$,
with the natural embeddings.

2. $(\mathfrak{g}, \mathfrak{h}) \cong (su(3), so(3))$ *is an involutive pair,*
$(\mathfrak{g}^+, \mathfrak{h}^+) \cong (so(3), so(2))$,
$(\mathfrak{h}, \mathfrak{h}^+) \cong (so(3), so(2))$,
with the natural embeddings.

3. $(\mathfrak{g}, \mathfrak{h}) \cong (su(3), su(2))$ *is a non-involutive pair,*
$(\mathfrak{g}^+, \mathfrak{h}^+) \cong (so(3), so(2))$,
$(\mathfrak{h}, \mathfrak{h}^+) \cong (su(2), so(2))$,
with the natural embeddings.
This is a tri-symmetric case.

4. $(\mathfrak{g}, \mathfrak{h}) \cong (su(3), so(2) \oplus u(1))$ *is a non-involutive pair,*
$(\mathfrak{g}^+, \mathfrak{h}^+) \cong (so(3), so(2))$,
$(\mathfrak{h}, \mathfrak{h}^+) \cong (su(2) \oplus u(1), so(2))$,
with the natural embeddings.
This is a tri-symmetric case.

IV.6.10. Theorem 13. *Let* $\mathfrak{g} \cong so(n)$, $n > 4$, *and a pair* $(\mathfrak{g}, \mathfrak{h})$ *has a two-dimensional mobile mirror of unitary type. Then this pair is involutive and*

$$(\mathfrak{g}, \mathfrak{h}) \cong (so(6), u(3)) \cong (su(4), u(3)).$$

Furthermore,

$$(\mathfrak{g}^+, \mathfrak{h}^+) \cong (so(4) \oplus so(2), u(1) \oplus su(2) \oplus su(2)),$$

$$(\mathfrak{h}, \mathfrak{h}^+) \cong (u(3), u(2) \oplus u(1)) \cong (su(3), u(2)) + (u(1), u(1))$$

with the natural embeddings.
This case is tri-symmetric.

IV.7.2. Theorem 14. *Let* $\mathfrak{g}$ *be a simple compact Lie algebra non-isomorphic to* $so(n)$ *or* $su(3)$, *and let* $(\mathfrak{g}, \mathfrak{h})$ *be a non-symmetric pair with a mobile two-dimensional mirror. Then*

$$(\mathfrak{g}, \mathfrak{h}) \cong g_2/su(3),$$

$$(\mathfrak{g}^+, \mathfrak{h}^+) \cong (su(2) \oplus su(2), so(2) \oplus su(2))$$

$$\cong (su(2), so(2)) + (su(2), su(2)),$$

$$(\mathfrak{h}, \mathfrak{h}^+) \cong (su(3), u(2))$$

with the natural embeddings.

In this case the mirror is of unitary type and the case is tri-symmetric.

IV.7.4. Theorem 15. *Let $(\mathfrak{g}, \mathfrak{h})$ be an involutive pair with a simple compact Lie algebra $\mathfrak{g}$ non-isomorphic to $so(n)$ or $su(3)$ with a two-dimensional mobil mirror. Then*

$$(\mathfrak{g}, \mathfrak{h}) \cong (su(n), su(n-1)), \quad n > 4;$$

$$(\mathfrak{g}_1^+, \mathfrak{h}_1^+) \cong (su(n-2) \oplus su(2) \oplus u(1), \, su(n-2) \oplus u(1) \oplus u(1)),$$

$$(\mathfrak{h}, \mathfrak{h}_1^+) \cong (u(n-1), \, su(n-2) \oplus u(1) \oplus u(1))$$

$$= (su(n-1), u(n-2)) + (u(1), u(1))$$

with the natural embeddings.

IV.8.2. Theorem 16. *Let $\mathfrak{g}$ be a simple compact Lie algebra, and let a pair $(\mathfrak{g}, \mathfrak{h})$ have an immobile two-dimensional mirror.*

Then for $(\mathfrak{g}, \mathfrak{h})$ we have the following possibilities:

$$(so(n), so(n-3) \oplus so(2)), \qquad (so(n), so(n-4) \oplus su(2) \oplus so(2)),$$

$$(su(n), su(n-2) \oplus u(1) \oplus u(1)), \qquad (sp(n), sp(n-1) \oplus u(1)),$$

$$(g_2, su(2) \oplus so(2)), \qquad (f_4, sp(3) \oplus u(1)),$$

$$(e_6, so(6) \oplus u(1)), \qquad (e_7, so(12) \oplus u(1)),$$

$$(e_8, e_7 \oplus u(1)), \qquad (su(3), so(2))$$

with the natural embeddings.

BIBLIOGRAPHY

Я - смехач, и охульник,
Я - звонкий Предтеча,
Провозвестник Свершений,
Низвергатель Миров.
Разгорайся Светильник
Диковинной Речи,
Огнезарных Дыханий
И неведомых слов.

Квантасмагор

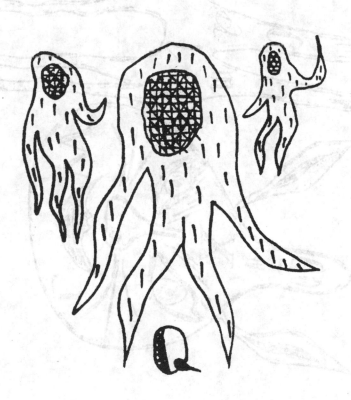

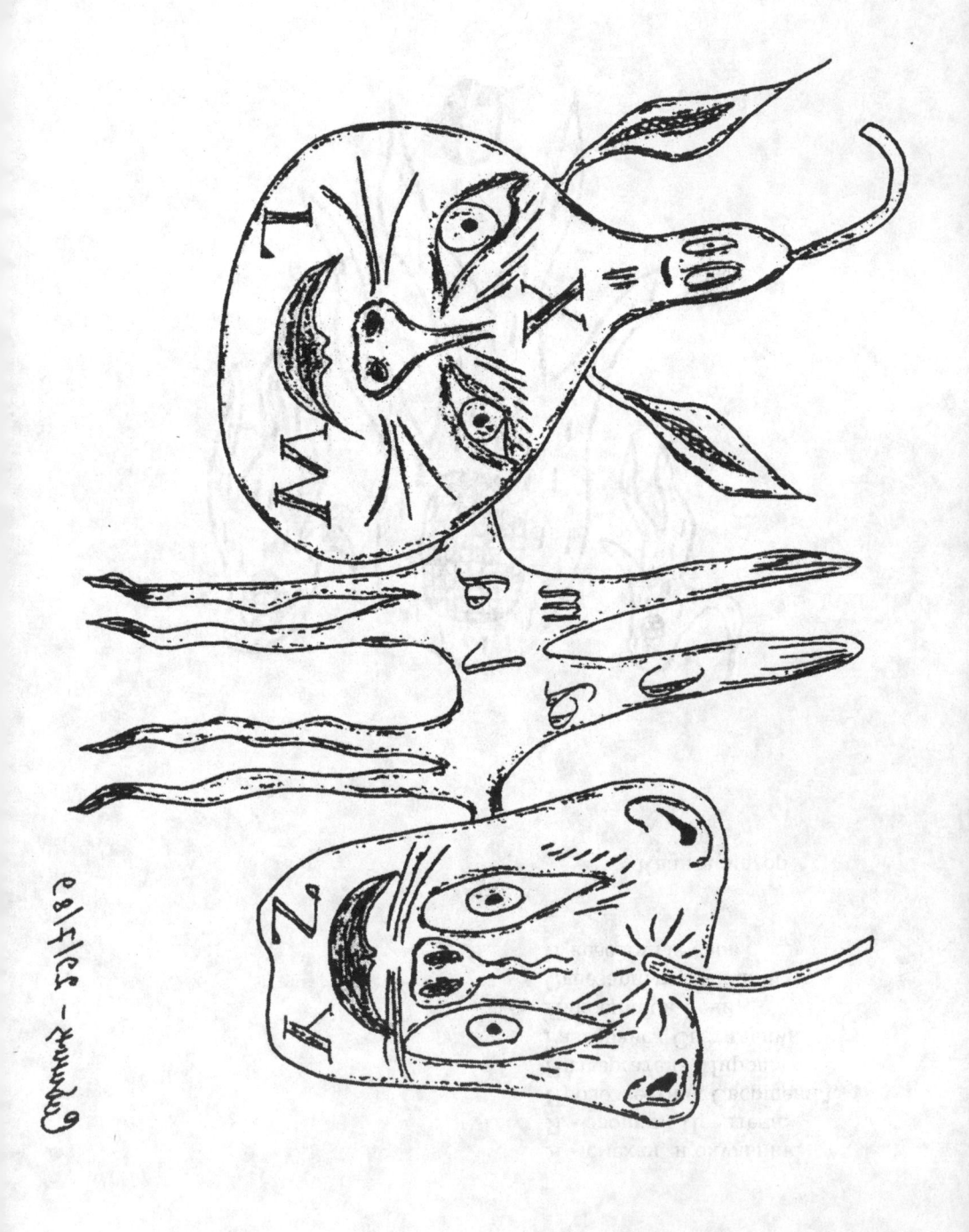

BIBLIOGRAPHY

È. Cartan
[49]   *Geometry of Lie groups and Symmetric spaces*, IL (Foreign Literature). Moscow,
            1949. (Russian)
[52]   *La géométrie des groupes de transformations*, (French), Oeuvres complétes, pt. 1 2
            (1952), Paris: Gauthiers-Villars, 673–692, **MR** 14 (1953), p. 353.

I.G. Chebotarev
[40]   *Theory of Lie groups*, GITTL. Moscow, 1940. (Russian)

C. Chevalley
[48]   *Theory of Lie groups I*, IL (Foreign Literature). Moscow, 1948. (Russian)
[58a]   *Theory of Lie groups II*, IL (Foreign Literature). Moscow, 1958. (Russian)
[58b]   *Theory of Lie groups III*, IL (Foreign Literature). Moscow, 1958. (Russian)

Collection of the papers
[64]   *The elementary particles and gage fields*, Mir. Moscow, 1964. (Russian)

Collection of the papers
[67]   *The group theory and the elementary particles*, Mir. Moscow, 1967. (Russian)

E.V. Dynkin
[47]   *Structure of semisimple Lie algebras*, (Russian), Advances in Math. Sci. (Uspehi
            Mat. Nauk) 2 (1947), 59–127.

L.P. Eisenhart
[26]   *Continuous Groups of Transformations*, Princeton University Press, 1926.
[27]   *Non-Riemannian Geometry*, Colloquium Publications of the Amer. Math. Soc. 8,
            New York, 1927.

F. Engel, S. Lie
   See S. Lie, F. Engel

F.R. Gantmacher
[39a]   *Canonical representation of automorphisms of a complex semisimple Lie group*,
            (Russian), Mat. Sb. Ac. Sci. USSR 5 (1939), no. 47, 101–104.
[39b]   *On the classification of real simple Lie groups*, (Russian), Mat. Sb. Ac. Sci. USSR
            5 (1939), no. 47, 217–249.

N. Jacobson
[62]   *Lie Algebras*, Wiley Interscience Publishers, New York, 1962, **MR** 26 (1963) #1345.

S. Helgason
[62]   *Differential Geometry and Symmetric Spaces*, Academic Press, New York, 1962,
            MR 26 (1963) #2986.
[78]   *Differential Geometry, Lie Groups and Symmetric Spaces*, Academic Press, New
            York, 1978.

W. Killing

[1888]   *Die Zusammensetzang der stetigen endlichen Transfomationsgruppen I*, Math. Ann. **31** (1888), 252–290. (German)

[1889a]   *Die Zusammensetzang der stetigen endlichen Transfomationsgruppen II*, Math. Ann. **33** (1889), 1–48. (German)

[1889b]   *Die Zusammensetzang der stetigen endlichen Transfomationsgruppen III*, Math. Ann. **34** (1889), 57–122. (German)

[1890]   *Die Zusammensetzang der stetigen endlichen Transfomationsgruppen IV*, Math. Ann. **36** (1890), 161–189. (German)

S. Kobayashi, K. Nomizu

[63]   *Foundations of differential geometry*, vol. 1, Interscience-Wiley, New York-London, 1963, xi + 329 p.p., **MR** 27 (1964) #2945.

[69]   *Foundations of differential geometry*, vol. 2, Interscience-Willey, New York-London, 1969, xiii + 470 p.p., **MR** 38 (1969) #6501.

S. Lie, F. Engel

[1888]   *Theorie de Transformationsgruppen I*, (German), Teubingen, Leipzig, 1888, Reprinted in 1930.

[1890]   *Theorie de Transformationsgruppen II*, (German), Teubingen, Leipzig, 1890, Reprinted in 1930.

[1893]   *Theorie de Transformationsgruppen III*, (German), Teubingen, Leipzig, 1893, Reprinted in 1930.

A.I. Malcev

[45]   *On the theory of Lie groups in the large*, (Russian), Mat. Sb. Ac. Sci. USSR **16** (1945), no. 58, 163–190.

O.V. Manturov

[72]   *Homogeneous Riemannian spaces with irreducible groups of rotations*, Transactions of Sem. on Vector and Tensor Analysis **13** (1966), 68-145, Moscow University. (Russian).

A.A. Matevosian

[81]   *Nonsymmetric compact pairs of Lie algebras of class 2*, VINITI, 1981, pp. 2–7, Dep. No.238-81 (Russian).

[82]   *On compact pairs of Lie algebras with involutive automorphisms of class 2*, (Russian), Problems of homological algebra. Jaroslavl' University (1982), 62–66.

[85]   *Homogeneous spaces of simple compact Lie groups with two-dimensional mirrors*, Principle of Inclusion and Invariant Tensors. Moscow District Pedagogical Institute. VINITI, 1985, pp. 74–77, Dep. No.426-B86 (Russian).

[86]   *On the theory of Homogeneous spaces with two-dimensional mirrors*, (Russian), Webs and Quasigroups. Kalinin University Press (1986), 50–53.

[87a]   *Homogeneous Riemannian spaces with two-dimensional mirrors*, (Russian) Dr.Ph. dissertation. Friendship of Nations University, 1987.

[87b]   *Homogeneous Riemannian spaces with two-dimensional mirrors*, (Russian) Sinopsis of Dr.Ph. dissertation. Friendship of Nations University, 1987.

Nguen-Van-Hew

[67]   *Lectures on the unitary simmetry theory of the elementary particles*, Atomizdat. Moscow, 1967. (Russian)

K. Nishijima

[65]   *The fundamental particles*, Mir. Moscow, 1965. (Russian)

K. Nomizu

[58]   *Lie groups and Differential geometry*, Japan Math. Soc., 1958.

M.N. Nutsubidze

[85]   *Homogeneous Riemannian spaces with mirrors of co-dimension 2*, Principle of Inclusion and Invariant Tensors. Moscow District Pedagogical Institute., 1985, pp. 85–91, VINITI Dep. No.426-B86 (Russian).

[86]   *On mirrors of co-dimension 2 in homogeneous Riemannian spaces*, (Russian), Webs and Quasigroups. Kalinin University Press (1986), 78–82.

L.S. Pontryagin

[73]   *Continuous Groups*, 3^d ed., (Russian), Nauka, Moscow, 1973, 519 p.p., **MR** 50 (1975) #10141; English translation of the 2^d Russian edition: Gordon and Breach Science Publishers, Inc., New York-London-Paris, 1966, xv + 543 p.p., **MR** 34 (1967) #1439.

P.K. Rashevski

[50]   *Symmetric spaces of affine connection with torsion*, (Russian), Proceedings of Sem. on Vector and Tensor Analysis **8** (1950), Moscow University, 82–92, **MR** 12 (1951), p. 534.

[51]   *On the geometry of homogeneous spaces*, (Russian), Dokl. Akad. Nauk SSSR **80** (1951), 169–171, **MR** 13 (1952), p. 383.

[52]   *On the geometry of homogeneous spaces*, (Russian), Proceedings of Sem. on Vector and Tensor Analysis **9** (1952), Moscow University, 49–74, **MR** 14 (1953) p.795.

[53]   *Riemanian Geometry and Tensor Analysis*, GITTL. Moscow, 1953, (Russian).

B.A. Rosenfeld

[55]   *Non-Euclidean Geometries*, GITTL. Moscow, 1955, (Russian).

[57]   *On the theory of symmetric spaces of rank 1*, (Russian), Mat. Sb. Ac. Sci. USSR **41** (1957), no. 83, 373–380.

U. Rumer, A. Fet

[70]   *Theory of the unitary symmetry*, Nauka. Moscow, 1970, (Russian).

L.V. Sabinin

[58a]   *On the geometry of subsymmetric spaces*, Scientific reports of Higher School. ser. Phys.–Math. Sci. **3** (1958), 46–49. (Russian)

[58b]   *On the structure of the groups of motions of homogeneous Riemannian spaces with axial symmetry*, Scientific reports of Higher School. ser. Phys.–Math. Sci. **6** (1958), 127–138. (Russian)

[59a]   *Mirror symmetries of Riemannian spaces*, Synopsis of Dr. Ph. in Math. dissertation, Moscow University Press, 1959, 7 pp.. (Russian)

[59b]   *Mirror symmetries of Riemannian spaces*, Dr. Ph. in Math. dissertation, Moscow University, 1959. (Russian)

[59c]   *The geometry of homogeneous Riemannian spaces and intrinsic geometry of symmetric spaces*, Reports of Ac. of Sci. of USSR (Math.) **129** (1959), no. 6, 1238–1241, (Russian) **MR** 28#1556.

[60]   *On the explicit expression of forms of connection for quasisymmetric spaces*, Reports of Ac. of Sci. of USSR (Math.) **132** (1960), no. 6, 1273–1276, (Russian) **MR** 22#9941.

[61a]   *Some algebraic identities in the theory of homogeneous spaces*, Siberian Math. Journal **2** (1961), no. 2, 279–281, (Russian) **MR** 23#A2482.

[61b]   *On the geometry of tri-symmetric Riemannian spaces*, Siberian Math. Journal **2** (1961), no. 2, 266–278, (Russian) **MR** 24#A2350.

[65a]   *On isoinvolutive decompositions of Lie algebras*, Reports of Ac. of Sci. of USSR (Math.) **165** (1965), no. 5, 1003–1006, (Russian) **MR** 33#5788.

[65b]   *Isoinvolutive decompositions of Lie algebras*, Soviet Math. Dokl. **165** (1965), no. 5, Amer. Math. Soc., 1554–1557, (English) **MR** 33#5788.

[67a]   *On principal invomorphisms of Lie algebras*, Reports of Ac. of Sci. of USSR **175** (1967), no. 1, 31–33, (Russian) **MR** 36#224.

[67b]   *Principal invomorphisms of Lie algebras*, Soviet Math. Dokl. **8** (1967), 807–809, Amer. Math. Soc. (English) **MR** 36#224.

[68]   *On involutive sums of Lie algebras*, Transactions of Sem. on Vector and Tensor Analysis **14** (1968), 94–113, Moscow University. (Russian) **MR** 40#7392.

[69]   *Involutive duality in simple compact Lie algebras*, Proceedings of Geometric seminar. Inst. of Sci. Information of Ac. of Sci. of the USSR **2** (1969), 277–298, (Russian) **MR** 41#8596.

[70a]   *Principal invomorphisms of compact Lie algebras*, Transactions of Sem. on Vector and Tensor Analysis **15** (1970), 188–226, Moscow University. (Russian) **MR** 45#5279.

[70b]   *On the classification of tri-symmetric spaces*, Reports of Ac. of Sci of the USSR (Math.) **194** (1970), no. 3, 518–520. (Russian)

[70c]   *On the classification of tri-symmetric spaces*, Soviet Math. Dokl. **11** (1970), no. 5, 1245–1247, Amer. Math. Soc.. (English)

[70d]   *Homogeneous Riemannian spaces with* $(n-1)$-*dimensional mirrors*, Collection of Math. papers. Friendship of Nations University, Moscow (1970), 116–126, (Russian) **MR** 51#4111.

[71a]   *Involutive geometry of Lie algebras*, Synopsis of Dr. Sci. in Math. dissertation, Kazan, 1971, 17 pp.. (Russian)

[71b]   *Involutive geometry of Lie algebras*, Dr. Sci. in Math. dissertation, Kazan University, 1971. (Russian)

[72]   *Tri-symmetric spaces with simple compact groups of motions*, Transactions of Sem. on Vector and Tensor Analysis **16** (1972), 202-226, Moscow University. (Russian) **MR** 48#1110.

[95]   *On the structure of* T, U, V *isospins in the theory of Higher Symmetry*, Herald of Friendship of Nations University (Math.) **2** (1995), no. 1, 130–134. (English)

[97]   *Involutive Geometry of Lie algebras*, Monograph, Friendship of Nations University, Moscow, 1997. (Russian)

Seminar Sophus Lie
[62]   *Theory of Lie algebras. Topology of Lie group*, IL (Foreign Literature). Moscow, 1962. (Russian)

A.A. Sagle, R.E. Walde
[73]   *Introduction to Lie Groups and Lie Algebras*, Academic Press, New York and London, 1973, **MR** 50 (1975) #13374.

P.A. Shirokov
[57]   *On one type of symmetric spaces*, Mat. Sbornik **41** (1957), no. 33, 361–372. (Russian)

J.P. Serre
[69]   *Lie algebras and Lie groups*, Mir. Moscow, 1969.

B.L. Van-der-Warden
[33]   *Die Klassifizierung der einfachen Lie'schen Gruppen*, (German), Math. Z. **37** (1933), 446–462.

V. Varadarajan
[73]   *Lie Groups, Lie Algebras and their Representations*, Prentice-Hall, 1974.

H. Weyl
[25]   *Theorie der Darstellung kontinuierlicher halbeinfacher Gruppen durch lineare Transformationen I, II, III und Nachtrag*, (German), Math. Z. **23** (1925), 271–309.

[26]   *Theorie der Darstellungkontinuierlicher halbeinfacher Gruppen, durch lineare Transformationen I, II, III und Nachtrag*, (German), Math. Z. **24** (1926), 328–395, 789–791.

[47]   *Classical groups, their invariants and representations*, IL (Foreign Literature). Moscow, 1947.

M.P. Zamakhovski
[97a]   *Periodic and trisymmetric spaces of internal type with simple groups of motions*, Izv. Of Russian Ac. Sci. Ser. Mat. **61** (1997), no. 6, 103–118. (Russian)

[97b]   *Periodic and trisymmetric spaces of internal type with simple groups of motions*, Izv. Math. USA **61** (1997), no. 6, 1215–1229.

INDEX

Мир Восторг источал,
В фиолетово-синем Экстазе
Я пригоршнями черпал
Вселенские россыпи звёзд.
И меня обнимал
Свет оранжево-желтых Фантазий,
Элегантной дугой,
Изгибался изысканной Радуги хвост.

Квантасмагор

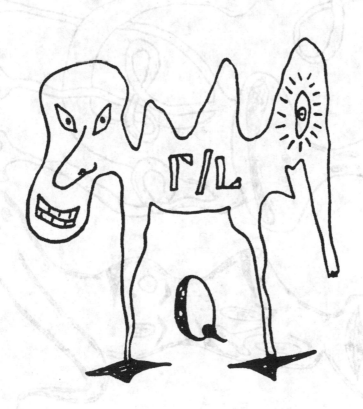

INDEX

Когда раздвоился Хаос
До корня вздыбленных волос,
Взорвался Сингулярий–Конус
И начал бег весёлый Хронос,
Ехидна злая, Энтропия,
Сплела покровы роковые.
Родились Смерть, Распад и Тлен.
Застой и Мрак без перемен.

Но День Брамана уж на склоне,
Грядёт, бредёт Брамана Ночь,
Корова Дхармы ноги клонит,
Не в силах дрёму превозмочь.
Опустошённый, мир унылый
Устало завершает круг,
И Хронос замирает вдруг,
Хаос играет новой силой.
И Жизнь и Смерть,
И Мрак и Тлен,
В круговороте перемен.

Квантасмагор

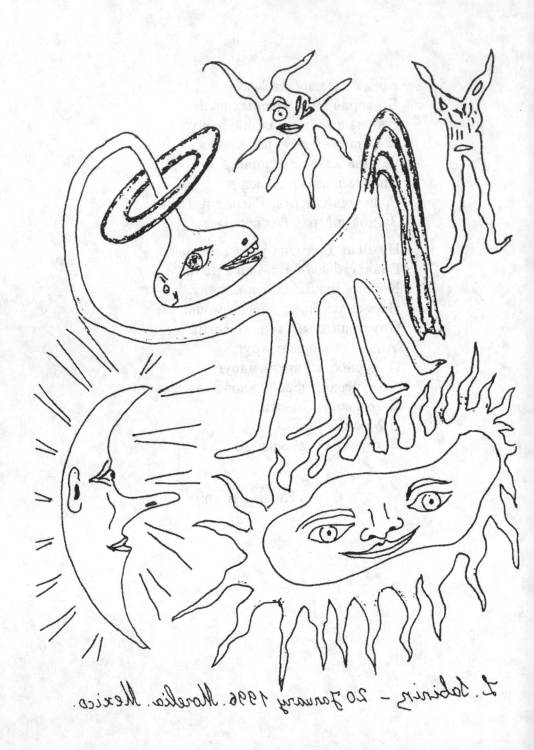

T. Tabiria – 20 January 1996. Morelia. Mexico.